Detection, Identification, and Monitoring of Food Contaminants

Matthew Chidozie Ogwu
Appalachian State University, USA

Sylvester Chibueze Izah
Bayelsa Medical University, Nigeria

IGI Global
Scientific Publishing
Publishing Tomorrow's Research Today

Vice President of Editorial Melissa Wagner
Director of Acquisitions Mikaela Felty
Director of Book Development Jocelynn Hessler
Production Manager Mike Brehm
Cover Design Jose Rosado

Published in the United States of America by
IGI Global Scientific Publishing
701 East Chocolate Avenue
Hershey, PA, 17033, USA
Tel: 717-533-8845 | Fax: 717-533-7115
Website: https://www.igi-global.com E-mail: cust@igi-global.com

Library of Congress Cataloging-in-Publication Data

Names: Ogwu, Matthew Chidozie editor | Izah, Sylvester Chibueze editor
Title: Detection, identification, and mitigation of food contaminants / edited by Matthew Chidozie Ogwu and Sylvester Chibueze Izah.
Description: Hershey, PA : IGI Global Scientific Publishing, [2026] | Includes bibliographical references and index. | Summary: "This book aims to provide a comprehensive overview of contemporary methods, tools, and practices in detecting, identifying, and mitigating food contaminants"-- Provided by publisher.
Identifiers: LCCN 2025027228 (print) | LCCN 2025027229 (ebook) | ISBN 9798337339825 h/c | ISBN 9798337339832 s/c | ISBN 9798337339849 ebook
Subjects: LCSH: Food--Analysis | Food contamination--Prevention | Food adulteration and inspection
Classification: LCC TP372.5 .D48 2026 (print) | LCC TP372.5 (ebook)
LC record available at https://lccn.loc.gov/2025027228
LC ebook record available at https://lccn.loc.gov/2025027229

British Cataloguing in Publication Data
A Cataloguing in Publication record for this book is available from the British Library.

Table of Contents

Detailed Table of Contents

Vivek Pazhamalai, Department of Bio-Engineering, School of Engineering, Vels Institute of Science, Technology and Advanced Studies (VISTAS), Chennai, India
Amjath Ahamed W., Department of Bio-Engineering, School of Engineering, Vels Institute of Science, Technology and Advanced Studies (VISTAS), Chennai, India
Manideepa Govindarajan, Department of Bio-Engineering, School of Engineering, Vels Institute of Science, Technology and Advanced Studies (VISTAS), Chennai, India
Ivo Romauld S., Department of Bio-Engineering, School of Engineering, Vels Institute of Science, Technology and Advanced Studies (VISTAS), Chennai, India
Meenambiga S. S., Department of Bio-Engineering, School of Engineering, Vels Institute of Science, Technology and Advanced Studies (VISTAS), Chennai, India
G. Abirami, Department of Biotechnology, School of Life Sciences, Vels Institute of Science, Technology and Advanced Studies (VISTAS), Chennai, India

Food safety has traditionally focused on visible contaminants like bacteria, pesticides, and heavy metals. However, as food systems evolve and the use of industrial chemicals increases, new, often invisible risks have emerged. These risks, which can be characterized as ghost molecules or emerging contaminants, that may be present in trace amounts, often undetectable by traditional food safety protocols, yet potentially hazardous. Microplastics, PFAS and pharmaceutical residues are leading examples of such contaminants that have infiltrated the modern food supply. The implications of their presence in food ranging from endocrine disruption to carcinogenic effects are still being explored. With increasing consumer awareness and regulatory concerns, the need to address these invisible contaminants is more urgent than ever. This chapter focuses on source, occurrence, detection methods and toxicity of these emerging contaminants in the food chain.

Emerging food contaminants have gained significant attention due to their potential adverse effects on human health and the environment. This paper focuses on three critical categories of emerging contaminants: microplastics, per- and polyfluoroalkyl substances (PFAS), and pharmaceutical residues. Microplastics are ubiquitous in the environment and can enter the food chain, posing risks to aquatic organisms and potentially human consumers. PFAS, often referred to as "forever chemicals," are persistent in the environment and have been linked to various health issues, leading to their classification as emerging pollutants. Additionally, pharmaceutical residues, stemming from agricultural runoff and wastewater discharge, are increasingly found in food matrices, raising concerns about their impact on food safety. This review aims to synthesize current knowledge on the occurrence, pathways, impacts, and detection techniques for these contaminants, highlighting the urgent need for effective risk management strategies to mitigate their effects on food systems and public health.

Food safety is one of the crucial global alarming spreading issue at very fast pace. Therefore, it requires continuous monitoring of contaminants by robust and integrative forensic approaches for the accurate and precise identification and quantification of contaminants in various food matrices. In addition, different types of approaches are employed for the investigation of contaminants in food i.e. conventional analytical and integrative forensic approaches. Conventional analytical approaches are often not capable of identifying and detecting complex emerging contaminants such as pharmaceutical residues specifically antibiotics, pesticides, microplastics, nanomaterials etc. Thus, integrative forensic approaches syndicate different advanced analytical techniques—such as hyphenated techniques (i.e. LC-MS/MS and GC-MS),

PCR, loop-mediated isothermal amplification, and next-generation sequencing. Emerging technologies such as biosensors, Clustered Regularly Interspaced Short Palindromic Repeats-based diagnostics, and metagenomics are also discussed for their potential to enhance sensitivity, specificity, and rapidity in pathogen detection. Each technique is assessed in terms of sensitivity, specificity, operational feasibility, and its integration into food safety risk management frameworks. Special attention is given to validation standards, harmonization efforts, and the challenges of deploying these technologies in low-resource settings. The chapter concludes by identifying emerging trends, such as artificial intelligence-assisted detection and portable diagnostics, which hold promise for revolutionizing microbial monitoring in food systems. By bridging microbiological principles with practical applications and regulatory contexts, offering critical insights for researchers, food safety practitioners, and policymakers.

Heavy metal contamination in food is a growing global concern due to its significant implications for public health and food security. Exposure poses substantial health risks through acute and chronic toxicity, organ-specific effects, and associations with severe health conditions, particularly affecting vulnerable populations. This chapter examines the sources, pathways, and prevalence of toxic metals, including arsenic, lead, mercury, and cadmium, in food systems, with a focus on both natural and anthropogenic contributions. It examines how these metals accumulate in the food chain and outlines their acute and chronic health impacts, particularly among vulnerable populations. The chapter provides an in-depth review of international regulatory standards, dietary risk assessments, and current analytical methodologies, including atomic absorption spectroscopy, inductively coupled plasma mass spectrometry, and portable detection systems. It also presents global monitoring strategies, case studies, and real-world surveillance data that highlight regional disparities and emerging challenges. Innovative approaches such as green chemistry, nanotechnology, and AI-driven tools are discussed as part of future directions. By integrating insights from environmental science, analytical chemistry, toxicology, and public health policy, this chapter provides a comprehensive perspective on the detection, measurement, and mitigation of heavy metals in food, thereby contributing to the development of safer and more resilient food systems worldwide.

This chapter provides a comprehensive overview of mycotoxin and pesticide residue analysis, aiming to elucidate current practices, address challenges, and explore innovative technologies for ensuring food safety. It details analytical methods like LC-MS/MS, GC-MS, HPLC, TLC, and ELISA. Seven key challenges- analytical sensitivity, matrix complexity, cost, regulatory harmonization, environmental impact, emerging contaminants, and scalability are examined, emphasizing barriers to effective detection. Emerging trends, including HRMS, DART-MS, portable biosensors, nanotechnology, AI, and green analytical chemistry techniques like SPME and DLLME, are discussed for their potential to enhance sensitivity, speed, and sustainability. Future directions involve multi-omics, cost-effective platforms, and international collaborations led by Codex Alimentarius, FAO, and WHO to standardize methods and MRLs. By integrating recent studies and regulatory perspectives, this chapter bridges foundational knowledge with forward-looking solutions, fostering interdisciplinary innovation in food safety.

Chromatography plays a pivotal role in food contaminant analysis, enabling the detection, separation, and quantification of harmful substances to ensure food safety and regulatory compliance. Recent advancements in chromatographic techniques—such as gas chromatography (GC), liquid chromatography (LC), and mass spectrometry (MS)—have enhanced the identification of low-abundance contaminants, including chiral compounds, organophosphorus pesticides, and mycotoxins. High-performance liquid chromatography and ultra-high-performance liquid chromatography, coupled with MS, have improved sensitivity and accuracy, particularly in complex food matrices. Additionally, innovations such as microfluidic technologies, immunochromatographic assays, and magnetic covalent organic frameworks are revolutionizing rapid, on-site contaminant detection. The integration of artificial intelligence and green chromatography further optimizes analytical efficiency while reducing environmental impact. Regulatory frameworks and multi-residue methods support global harmonization of food safety standards, ensuring traceability and consumer protection. This chapter highlights the interdisciplinary applications of chromatography in detecting pesticide residues, mycotoxins, veterinary drugs, and food adulterants, emphasizing its critical role in risk assessment and public health. Emerging trends, including portable platforms and non-targeted screening, are also explored, demonstrating how chromatography continues to evolve in response to new food safety challenges.

Water contamination poses a huge risk in both food production and consumption. To monitor water quality, environmental microbes (EMs) can be used as they act as bioindicators. Automating the detection of EMs would be a highly time-saving and vital procedure, particularly since early monitoring of environments can prevent large-scale agricultural contamination. However, such automation remains underdeveloped for industrial use. Hence, a series of comparison experiments were conducted on various deep learning models using the open-source dataset EMDS-6. The pipeline includes segmentation followed by classification using different Convolutional Neural Networks (CNNs) to achieve a fully automated system. Given the limited dataset size, transfer learning, data augmentation, and k-fold cross-validation to mitigate overfitting and bias are employed. It was observed that deeper models like DenseNet121 achieve the highest accuracy (71.67%), whereas lightweight architectures such as MobileNetV2 strike a balance between accuracy and computational efficiency, making them well-suited for resource constrained environments. Finally, feature fusion is applied on the classifier models to assess potential improvements in accuracy which concatenates the feature representations from individual models. While fusion did not consistently outperform standalone models and often showed minimal or negative improvement, combining MobileNetV2 with other models improved performance in most cases. The fusion of Xception and MobileNetV2 yielded an accuracy improvement of 1.90%, while DenseNet121 and MobileNetV2 achieved the highest overall accuracy (71.19%) among all combinations.

also addressed, focusing on the development of cost-efficient, user-friendly, and multiplexed detection platforms.

Preface

Food safety remains a cornerstone of public health, sustainable agriculture, and global food security. With the growing complexity of the global food system, the emergence of new contaminants, climate change, and advances in scientific technologies, the detection, identification, and mitigation of food contaminants demand an integrated, multidisciplinary approach. This book, *Detection, Identification, and Mitigation of Food Contaminants*, offers a comprehensive overview of current and emerging threats in the food system while highlighting scientific, technological, and policy-driven responses. Bringing together contributions from scientists, technologists, environmentalists, and policy experts across the globe, this volume spans microbial, chemical, physical, and emerging contaminants in food. It explores cutting-edge diagnostic tools, innovative mitigation strategies, traditional and indigenous knowledge systems, and risk assessment frameworks that shape food safety interventions.

The book comprises 10 chapters that collectively explore the multifaceted dimensions of food contamination, detection, and mitigation. These chapters cover a wide spectrum of topics, including emerging contaminants such as microplastics, PFAS, pharmaceutical residues, and the influence of climate change on contaminant patterns. The volume presents innovative technologies for identifying and monitoring contaminants, including biosensors, molecular diagnostics, deep learning models, and chromatographic methods. Several chapters address practical mitigation strategies such as the use of nanotechnology, biodegradable packaging materials, natural compounds, and novel decontamination approaches. In addition, the book features contributions that examine predictive risk assessment frameworks, genetically modified crop biosafety, integrative forensic methods, and the role of indigenous and traditional knowledge systems in safeguarding food safety. Collectively, the chapters offer a comprehensive and interdisciplinary perspective on the current and future challenges in food contaminant management.

CHAPTER OVERVIEW

Chapter 1: Emerging Food Contaminants: An Overview of the Growing Threat to Human Health

This opening chapter lays the conceptual groundwork for the volume, introducing the key categories of emerging contaminants—including microplastics, PFAS, and pharmaceutical residues—and highlighting their sources, detection challenges, and health implications. It sets the stage for why a shift in food safety paradigms is urgently needed.

Chapter 2: Invisible Threats on the Plate: Detecting Emerging Contaminants in Food

This chapter delves deeper into the detection of the same key contaminants introduced in Chapter 1. It provides a detailed analysis of their occurrence in food systems and outlines current methodologies for monitoring and risk management.

Chapter 3: Integrative Forensic Methods in Food Contaminant Detection and Identification

Integrating legal, analytical, and scientific perspectives, this chapter explores forensic approaches such as isotopic fingerprinting, GIS tools, and hyphenated techniques (LC-MS/MS, GC-MS). It emphasizes the value of forensic science in food fraud detection, traceability, and consumer protection.

Chapter 4: Detection of Microbial Contaminants in Food and Food Products

This chapter examines microbial contamination, discussing pathogen sources, sampling strategies, and detection methodologies ranging from culture-based assays to molecular diagnostics, with a strong emphasis on global foodborne illness burdens.

Chapter 5: Heavy Metal Contamination in Food: Monitoring, Detection, and Public Health Risks

Focusing on toxic metals such as lead, arsenic, cadmium, and mercury, this chapter evaluates global monitoring strategies and analytical tools (e.g., ICP-MS, AAS). It also examines regulatory frameworks and public health risks associated with chronic exposure.

Chapter 6: Analysis of Mycotoxin and Pesticide Residue in Food: Current Practices and Potential Future Trends

This chapter presents analytical techniques used in the detection of mycotoxins and pesticide residues, discusses challenges like matrix complexity and cost, and surveys emerging innovations like portable biosensors, nanotech, and green chemistry.

Chapter 7: Chromatography in Food Contaminant Analysis: Principles, Advances, and Applications for Food Safety

Chromatographic techniques remain a cornerstone of food contaminant analysis. This chapter reviews foundational principles and recent advances, exploring their utility in detecting both chemical and biological hazards in complex food matrices.

Chapter 8: Classification of Environmental Microorganisms for Monitoring Water Quality for Food Safety Using Deep Learning

Water is an essential input in food production. This chapter explores how environmental microorganisms can serve as bioindicators of water quality and how deep learning and computer vision can automate their classification for improved monitoring.

Chapter 9: Molecular Diagnostics and Genomic Tools for Food Contaminant Detection

Highlighting advanced nucleic acid-based diagnostics like qPCR, NGS, and CRISPR-based assays, this chapter discusses their role in identifying hard-to-detect and unculturable contaminants. It also examines genomic tools in quality control and surveillance.

Chapter 10: Biosensors for Rapid and On-Site Detection of Food Contaminants

The final chapter emphasizes the emerging role of biosensors in rapid, sensitive, and on-site contaminant detection. It reviews technological innovations, practical applications, and future directions for commercial deployment in the food sector.

By bringing together diverse methodologies, global case studies, and interdisciplinary insights, this volume contributes significantly to the scientific and applied literature on food contaminants. It not only synthesizes current research but also identifies knowledge gaps and future directions, thereby advancing the discourse in food safety, environmental health, and sustainable food systems. The book serves as both a foundational text for academic study and a reference guide for applied decision-making, ultimately supporting more informed policies and practices in food contaminant management worldwide. We extend our deepest gratitude to all contributors whose interdisciplinary insights have enriched this book.

Matthew Ogwu
Appalachian State University, USA

Sylvester Izah
Bayelsa Medical University, Nigeria

Section 1
Introduction and Emerging Contaminant Landscape

Chapter 1
Emerging Food Contaminants:
An Overview of the Growing Threat to Human Health

Vivek Pazhamalai
https://orcid.org/0000-0003-2088-7545
Department of Bio-Engineering, School of Engineering, Vels Institute of Science, Technology and Advanced Studies (VISTAS), Chennai, India

Amjath Ahamed W.
Department of Bio-Engineering, School of Engineering, Vels Institute of Science, Technology and Advanced Studies (VISTAS), Chennai, India

Manideepa Govindarajan
https://orcid.org/0009-0007-7312-9875
Department of Bio-Engineering, School of Engineering, Vels Institute of Science, Technology and Advanced Studies (VISTAS), Chennai, India

Ivo Romauld S.
https://orcid.org/0000-0003-0610-0646
Department of Bio-Engineering, School of Engineering, Vels Institute of Science, Technology and Advanced Studies (VISTAS), Chennai, India

Meenambiga S. S.
https://orcid.org/0000-0002-5445-525X
Department of Bio-Engineering, School of Engineering, Vels Institute of Science, Technology and Advanced Studies (VISTAS), Chennai, India

G. Abirami
Department of Biotechnology, School of Life Sciences, Vels Institute of Science, Technology and Advanced Studies (VISTAS), Chennai, India

ABSTRACT

Food safety has traditionally focused on visible contaminants like bacteria, pesticides, and heavy metals. However, as food systems evolve and the use of industrial chemicals increases, new, often invisible risks have emerged. These risks, which can be characterized as ghost molecules or emerging contaminants, that may be present in trace amounts, often undetectable by traditional food safety protocols,

DOI: 10.4018/979-8-3373-3982-5.ch001

yet potentially hazardous. Microplastics, PFAS and pharmaceutical residues are leading examples of such contaminants that have infiltrated the modern food supply. The implications of their presence in food ranging from endocrine disruption to carcinogenic effects are still being explored. With increasing consumer awareness and regulatory concerns, the need to address these invisible contaminants is more urgent than ever. This chapter focuses on source, occurrence, detection methods and toxicity of these emerging contaminants in the food chain.

1. INTRODUCTION

Emerging contaminants (ECs) represent a wide range of chemical substances that have recently become a focus of environmental concern. These pollutants, although increasingly detected in various ecosystems, are not yet comprehensively included in standard regulatory frameworks or environmental monitoring programs. ECs typically consist of compounds such as pharmaceuticals, components of personal care products, synthetic industrial chemicals, nanomaterials, and other newly developed substances associated with modern human activity. What distinguishes these contaminants from conventional pollutants is their resistance to degradation, ability to accumulate in living organisms, and potential to pose unknown or long-term health and ecological risks. Since many ECs are not efficiently removed by conventional wastewater treatment facilities, they often find their way into natural water bodies, sediments, and even the food web, leading to ongoing environmental exposure. These substances enter the environment through multiple pathways, including runoff from agricultural fields, effluents from industrial processes, household wastewater, leachate from solid waste disposal sites, and urban drainage systems. In many cases, these chemicals were not originally considered hazardous, which is why they were neither closely studied nor regulated until recent technological advances made their detection possible.

Urbanization, industrial development, and the widespread use of chemicals in agriculture and manufacturing are accelerating the release of ECs into the environment. For example, heavy metals such as arsenic are naturally present in the Earth's crust but have become more prevalent due to mining activities, use in agrochemicals, and industrial applications. Arsenic, a metalloid with significant industrial value, is used in the production of semiconductors, pesticides, paints, glass products, and even cosmetics. Despite its utility, arsenic is a known human carcinogen and toxicant. In its inorganic form, it is classified as a Group 1 carcinogen by the International Agency for Research on Cancer. Chronic exposure to arsenic can lead to serious health issues including cancer, neurotoxicity, and developmental problems (Singh, Yadav, Sharma, & Singh, 2023).

The mobility of arsenic in the environment is facilitated by various geochemical and biological processes. In agroecosystems, arsenic can bind to soil particles and be taken up by crop roots through mineral uptake pathways. This transfer to edible plant parts makes it a critical food safety issue, especially in regions dependent on contaminated groundwater for irrigation. Additionally, arsenic can bioaccumulate in the tissues of animals consuming contaminated feed or water. Products like milk, eggs, meat, and fish from these animals can serve as secondary sources of arsenic exposure in the human diet (Zakhar, Derco, & Cacho, 2018).

Another major category of emerging contaminants is microplastics (MPs). These are plastic particles smaller than 5 millimeters that result from the breakdown of larger plastic waste or are manufactured at a micro scale for use in various products. Although microplastics are primarily associated with marine pollution, their presence in terrestrial ecosystems and agricultural soils has become an increasing concern. Plastic particles can enter soil through the application of sewage sludge, plastic mulching in agriculture, or littering. Once in the environment, microplastics can alter soil structure, affect microbial communities, and potentially carry toxic chemicals that adsorb onto their surfaces. (Singh, Yadav, Sharma, & Singh, 2023; Zakhar, Derco, & Cacho, 2018)

The COVID-19 pandemic has contributed to a surge in plastic waste generation due to the extensive use of disposable personal protective equipment (PPE), such as face masks, gloves, and plastic containers. Much of this waste has not been properly managed and has made its way into natural ecosystems, exacerbating the problem of microplastic pollution.

2. MICROPLASTICS

Plastics are now a significant component of everyday life and activities. Because of their mobility, elasticity, compact size, and simplicity of delivery, these artificial polymers are widely used in a variety of consumer care items. Due to these special qualities, plastics are especially appealing to the commercial packaging sector. Plastics' non-biodegradability and waste management may be issues, despite the fact that their uses in everyday products have increased (Sridhar, Kannan, Kapoor, & Prabhakar, 2022). Nearly 370 million tons of plastic were produced worldwide in 2019, with more than one-third of that plastic utilized in the US and Europe in throwaway items like garbage bags, eating utensils, and packaging that are intended to be thrown away within three years of manufacture. Despite recycling efforts, a significant amount of waste has collected in the environment and is gradually deteriorated by weathering and aging to micro and nanoscale sizes (Udovicki, Andjelkovic, Cirkovic-Velickovic, & Rajkovic, 2022).

The modern world's increased usage of plastic items has contaminated almost every environmental medium with microplastics, or synthetic plastic particles smaller than 5 mm. Beaches, seas, soils, sediments, and freshwater systems have all been found to accumulate microplastics. Therefore, it is possible that the intake of different foods will eventually bring the global contamination of microplastics back to our dinner table (Kwon et al., 2020).

The general definition of microplastics is synthetic polymers with a maximum size of 5 mm and no lesser limit (Li, Liu, & Chen, 2018). Plastic particles can enter the environment directly or indirectly through the movement of MPs throughout the ecosystem, contaminating human food. Microplastics have a significant impact on the environment as a whole by contaminating food and posing health risks to consumers. Because they are consumed by both marine and terrestrial creatures, microplastics are readily making their way into the human food chain (Al Mamun, Prasetya, Dewi, & Ahmad, 2023).

Drinking water may contain MPs. Additionally, they are present in fish, shellfish, poultry, milk, honey, eggs, salt, sugar, fruits, vegetables, wine, beer, tea, energy drinks, and soft drinks. Plastics in food pose health problems because of their makeup. Furthermore, MPs' toxicity varies according to their sizes and forms (Kadac-Czapska, Trzebiatowska, Knez, Zaleska-Medynska, & Grembecka, 2023).

2.1 Classification of Microplastics

It is classified into two types. Primary and secondary microplastics.

Primary Microplastics: The bigger plastic granules used as starter materials for goods fabrication, sandblasting media, and scrubs in personal care products are all examples of primary microplastics (MPs), which are produced industrially as microbeads of various sizes. When these pellets are manufactured, transported, or used, they "leak" into the environment (Andrady, 2017).

Secondary Microplastics: The destruction of larger objects that have somehow found their way into the marine ecosystem is the source of secondary MPs. It is widely believed that the majority of MPs in the environment are secondary MPs due to the significant volume of macroplastics entering the environment (Efimova, Bagaeva, Bagaev, Kileso, & Chubarenko, 2018).

2.2. Occurrence of Microplastics in Food

Human beings are primarily contaminated with MPs through the consumption of food, such as commercially processed fish, seafood tainted with microplastics, salt from the sea, nectar, alcohol, and culinary ingredients. Additionally, pollutants found in containers and impurities from processing materials might occasionally infect the majority of these food products. Inhaling MP-containing dust and air is the second way to become exposed. Seafood is an essential part of the human diet due to its high nutritional value, so eating these tainted items increases the risk of illness, particularly for little fish consumed whole (Ziani et al., 2023).

2.2. Salt

In addition to being utilized in food preservation, salts are vital components of the diet of humans. As a result, everyday salt intake is unavoidable, just like air and water. According to reports, around 11.6% of the 300 million tons of salt consumed worldwide in 2018 (including table salts and food processing) was consumed for human consumption (Taghipour, Ghayebzadeh, Mousavi, Sharifi, & Payandeh, 2023). The main sources of these are the sea, salty lakes, salty rocks, and salty wells. Usually, the combined effects of sunlight and evaporation cause sea salts to crystallize in salinas, also known as solar work ponds. Numerous saltworks can be found in waters that have been touched by human activity in a number of places, including Europe. As a result, they frequently come into contact with various pollutants. Many saltworks are probably affected by MPs since these regions are also thought to be hotspots for MP pollution of the environment (Peixoto et al., 2019).

2.3. Drinking Water

Numerous investigations worldwide have documented the pollution of water resources with various MP kinds in recent decades. Primary and secondary MPs entering the environment and moving to the aquatic environment are the causes of these diseases. The amount of of MPs in various water supplies vary, nevertheless, and there is a noticeable variance in the concentrations that have been observed (Babaei, Reshadatian, & Feizi, 2024). The United States had the most concentration of MPs in tap water, while Italy and Denmark had the lowest.

2.4. Air

Traffic dust, rubber tire erosion, and synthetic fabrics (such as plastic fibers or pieces of clothing) are thought to be the main sources of primary atmospheric MPs, which winds can carry to other environmental areas. Since 2015, the occurrence of MPs in the atmosphere has drawn more attention. Wet and dry deposition, atmospheric sampling, and dust collection are the three sample techniques that have been utilized to gather atmospheric MPs. As a result, it is not possible to directly compare studies using various sampling techniques (Zhang et al., 2020).

2.5. Detection of Microplastics

Common methods employed for the detection of microplastics are Nile red staining, scanning electron microscopy-energy dispersive spectroscopy (SEM-EDS), visual identification (with the aid of a microscope), Fourier Transform Infrared (FT-IR) spectroscopy. But none of the aforementioned methods can simultaneously provide the chemical (polymer type) and physical (size, color, and form) characteristics. Time constraints, size restrictions, and interference concerns can all restrict the use of a single technique (Ding et al., 2019).

2.6. SEM-EDS

Scanning electron microscopy plus energy-dispersive X-ray spectroscopy (SEM/EDS) and optical microscopy work together to screen a large number of microplastic particles rapidly and effectively with a lower chance of misidentification mistakes. Because it can easily separate plastic material from other substances including lipids, mineral particles, and cellulose fibers—all of which can be separated from plastics under optical microscopy—optical microscopy is the most effective method for analyzing microplastics. However, because certain particles can be mistakenly classified as plastics or as other materials, optical microscope imaging must be used carefully (Wang, Wagner, Ghosal, Bedi, & Wall, 2017).

2.7. FT-IR

By revealing details about the particular interactions found in plastics, FTIR is a potent method for the diagnostic investigation of synthetic polymers. Traditional FTIR is typically used to identify plastic debris and visible MPs, whereas micro-FTIR (μ-FTIR), which is fitted with a microscope to localize MPs on filters, is needed for smaller particles. Theoretically, μ-FTIR can detect MPs as small as 10 μm in diameter (Chen et al., 2020). It measures the amount of infrared (IR) light that the MP sample absorbs, enabling the analysis of its molecular makeup. The unique identification of a sample (MP) is represented by an infrared spectrum, where the absorption peaks match the frequencies of vibration between the atoms' bonds. Since every polymer material is a distinct atom-by-atom arrangement, no two compounds have precisely the same infrared spectra (Veerasingam et al., 2021).

2.8. SERS

The most commonly method used for the detection of chemical composition of microplastic is raman spectroscopy. This reliable technique is frequently used to detect trace amounts of a wide range of chemicals, including pharmaceuticals, contaminants, biological molecules dynamite, and microscopic plastic particles. SERS's 1974 revelation garnered a lot of interest due to its powerful augmentation of the inadequate Raman signal (Mikac et al., 2023).

Because of its exceptional chemical specificity and high sensitivity, surface-enhanced Raman spectroscopy (SERS) finds extensive application in ecological, food safety, and health surveillance. When the specimen's molecules are adsorbed on the perforated metal surface (often gold or silver nanoparticles), the Raman signal of the sample itself—which is typically quite weak—can be significantly increased. Based on this idea, it appears that SERS technology may detect microplastics and nanoplastics with smaller sizes and low concentrations in aqueous solutions (Lv et al., 2020).

2.9. Nail red Staining

Environmental specimens can be quickly screened for the occurrence of MPs using fluorescence microscopy when MPs are fluorescently stained with the hydrophobic dye Nile red (NR), which is widely used in histopathology. MPs can be categorized as "polar" or "hydrophobic" based on their polymer properties and the solvatochromic nature of NR, which causes its emission spectra to change based on the polarity of its surroundings. Because NR's interactions with various polymers varied depending on the plastic's chemical makeup, NR staining of particles has proven to be an effective method for polymer identification based on fluorescence (Meyers et al., 2022). Even though NR by itself is currently unable to identify particles (i.e., distinguish between different types of polymers), it is recognized as the most effective straightforward strategy for reducing the number of particles that require confirmation by more reliable methods like FTIR or Raman spectroscopies. Despite its widespread use in studies pertaining to the detection and measurement of MPs in various sample types (mostly environmental), there is no accepted technique for applying NR (Ribeiro, Duarte, & da Costa, 2024).

2.10. Toxicity of Microplastics on Organisms

A wide variety of aquatic taxa are vulnerable to these new contaminants due to the abundance of microplastics in aquatic settings. There have been reports of over 690 aquatic wildlife species consuming macro- or microplastics. Following ingestion, microplastics may have unknown effects on aquatic organisms' health (Wang, Gao, Jin, Li, & Na, 2019). The consumption of microplastic particles by marine animals may result in both deadly and sublethal consequences, including toxicosis in freshwater and marine fish species, according to recent ecotoxicological studies. Numerous researchers have documented a number of detrimental impacts on both freshwater and marine animals, such as altered fertility, abnormal reproduction, inflammatory reactions, gut inflammation, metabolic disorders, immune response impairments, hormonal interruption, and decreased mortality rates (Benson et al., 2022). It is possible for people to absorb MP particles by food or breathing. Eating seafood represents a single way that people can be exposed to MP. The majority of people eat bivalves, oysters, and tiny fish as entire creatures (such as sardines, anchovies, and sprat), and microplastic particles are frequently discovered concentrated in the digestive track (Jinadasa, Uddin, & Fowler, 2023).

3. PFAS

The heterogeneous group of fluorinated synthetic compounds known as per-and polyfluoroalkyl substances (PFAS) are distinguished by the presence of a perfluorinated methylene group (–CF2–) or at least one perfluorinated methyl group (–CF3), as well as a variable number of carbon atoms, fluorination degree, and other chemical groups. Since PFAS can be found in a wide range of consumer goods and industrial industries, they are practically always present in the environment (Panieri, Baralic, Djukic-Cosic, Buha Djordjevic, & Saso, 2022). The two most prevalent PFAS, per-fluorooctanoic acid (PFOA) and per-fluorooctane sulfonic acid (PFOS), are found in a variety of environmental matrices across the globe (Podder, Sadmani, Reinhart, Chang, & Goel, 2021).

3.1. Characteristics

The excellent chemical and thermal stability of PFAS is one of their defining characteristics. They are interesting for many technological applications since they do not break down at temperatures of several hundred degrees or when exposed to harsh chemicals. The remarkable stability of the carbon–fluorine bond, for instance, is the cause of this feature. It is much higher than the bond energy of the carbon–oxygen single bond. In addition to making the trifluoromethyl groups (-CF3) useful building blocks, the fluorine atom's strong electronegative characteristic also explains why many PFAS are water, oil, and dirt resistant (Brunn et al., 2023).

3.2. Sources and Occurrence

Both commercial and residential sources emit PFAS into the atmosphere. PFAS are found in everyday items such as industrial domestic goods and food processing. Chemicals used in food packaging are the main source of PFAS. These include noodle cups, popcorn bags, grease-resistant paper, candy wrappers, and containers and wrappers for fast food (O'Connor et al., 2022). One significant way that humans come into contact to PFAS is through the consumption of polluted water, agricultural products, and poultry. According to numerous studies, PFAS can build up in animals, soil and water used for food production, and other food packaging materials and/or processing machinery that contain PFAS can all contaminate food (Eze et al., 2024).

According to recent research, PFAS species can bioaccumulate in agricultural crops, which helps spread the chemicals to people and animals through food consumption. Hydrophobic interactions with organic materials dominate the sorption behavior of PFAS. Numerous investigations have demonstrated that whereas less hydrophobic PFASs are more common in aqueous solutions, highly hydrophobic PFASs have a tendency to sorb to suspended solids (TSS).

Because various PFASs have been found in PFAS-impregnated clothing, shoes, and upholstery, humans are susceptible to PFAS exposure. Furthermore, it has been demonstrated that humans are exposed to PFAS through the ambient intake of fish, shellfish, fruits, and their products (Nahar, Zulkarnain, & Niven, 2023; Johnson, 2022; Shen, Lee, Ra, Suk, & Moon, 2018).

3.3. Detection and Analysis

GC-MS, LC-MS, and HPLC-MS/MS are examples of chromatographic techniques combined with mass spectrometry that are the foundation of the current gold standard detection procedures for PFASs. In 2009, the EPA established "Method 537," an official analytical technique that uses liquid chromatography-tandem mass spectrometry (LC-MS/MS) enabled by solid phase extraction (SPE) to analyze 14 distinct PFASs, including PFOA and PFOS, in drinking water. The EPA announced "Method 533," a revised analytical technique, in November 2019 to focus more on the detection of "short chain" PFASs (Ryu, Li, De Guise, McCutcheon, & Lei, 2021).

The gold-standard method for determining targeted PFAS has been liquid chromatography (LC) in conjunction with mass spectrometric (MS) or tandem mass spectrometric (MS/MS) detection. Novel PFASs have been consistently found in a range of natural matrices thanks to recent advancements in high resolution mass spectrometry (HRMS). The structure of unknown PFAS molecules from over twenty established classes16 has recently been proposed using time-of-flight mass spectrometry (TOF-MS), Fourier transform ion cyclotron resonance mass spectrometry (FTICR-MS), and orbitrap instruments. These methods also contribute to the growing field of arising PFAS discovery through suspect and non-target screening (Androulakakis, Alygizakis, Bizani, & Thomaidis, 2022).

One strategy to manage PFAS collectively is combustion ion chromatography (CIC). More precise techniques are required to quantify all forms of PFAS entirely because CIC is able to determine the overall quantity of thermally degraded fluorine and cannot differentiate between inorganic and organic sources. There aren't many options besides LC-MS and CIC. Notably, by generating volatile derivatives, certain non-volatile substances can be examined using traditional GC (Yanagisawa, Obayashi, Furuta, & Fujimaki, 2025).

3.4. Toxicity

One significant way that humans are exposed to PFAS is through the consumption of tainted water, crops, and livestock. Many studies demonstrate that PFAS buildup in animals, soil and water used to grow food, and other food packaging materials and/or processing equipment/machines that contain PFAS can all contaminate food (Eze et al., 2024). Harmful consequences of newly discovered per- and polyfluoroalkyl compounds. The renal glands, hepatocytes lymph nodes, and brain of many living things are among the organs where PFAS tend to build up. PFAS have a number of negative health effects on humans, including immune-mediated (asthma, reduced vaccination response), heart disease (hypertension), cancer (breast, liver, testicular, kidney), endocrine (thyroid disease), fertile (decreased fertility), and developmental (Manojkumar, Pilli, Rao, & Tyagi, 2023). Kids are affected by the effects of PFASs. Children who are exposed to PFOA have an increased chance of acquiring asthma. The same is true for PFOS exposure, which has been linked to documented pulmonary function impairment. Young children are more likely than adults to be exposed to PFASs because of their smaller stature, greater respiratory rates, and hand-to-mouth and crawling habits that necessitate contact with contaminated surfaces, like floors. Children are exposed to PFASs primarily through inhalation, ingestion of dust, dirt, food, water, and breast milk, and cutaneous exposure to contaminated air or materials. There have also been reports of children's serum having higher levels of PFASs than adults (Lee, Smaoui, Duffill, Marandi, & Varzakas, 2025; Starnes, Rock, Jackson, & Belcher, 2022).

Because PFAS interfere with testosterone signaling, they are frequently linked to reproductive disorders such male infertility. The substantial burden of PFAS on the fetus throughout crucial phases of development is demonstrated by the increased risk of adipokine hormone dysregulation and lower levels of growth factors and sex hormones during fetal and postnatal development that result from prenatal exposure to PFAS. Pregnancy-related placental transfer of PFAS is also visible, and the offspring may have metabolic syndromes, chronic renal disorders, and abnormal thyroid signaling. When PFAS in breastmilk are passed from mother to kid through infant feeding, it is anticipated that an infant's exposure to PFAS will increase even more (Baluyot, Reyes, & Velarde, 2021).

Because of their extensive production and strong tendency to migrate from their source locations, PFAS are found in soil and water bodies all over the world. The apoplastic, symplastic, and transmembrane pathways are the three routes via which nutrients can reach plant roots. Additionally, foreign chemicals like PFAS can enter plant roots through similar mechanisms. Apart from the buildup of PFAS in plants, their possible phytotoxicity might have detrimental effects on the health of the ecosystem and food security, so it is important to pay attention to this. Values like IC50 or EC50, which represent the concentration of PFAS that results in a 50% reduction or inhibitory impact on plant growth, can be used to demonstrate the phytotoxicity of PFAS (Adu, Ma, & Sharma, 2023).

In the United States, water samples from military installations were found to contain PFAS pollution in 2337 locations across 49 states as of January 2021. Because PFOA and PFOS molecules are found in blood, serum, and urine, the majority of people in the US are exposed to PFAS. With measurable amounts of PFAS in soils, hay, milk, and human blood samples, a nearby ICLS was shut down in 2017 after PFAS were found in water samples from a well water district in Maine. Levels of PFAS in milk samples from one of the New Mexico dairies have been identified by the US Food and Drug Administration to pose a risk to human health (Jha et al., 2021).

3.5. Mitigation

Reducing the amount of PFAS in food products is beneficial for public health since most human cultures have a strong tradition of consuming animal products, and this is unlikely to change very soon. The effects of thermal processing (such as boiling, steaming, and frying) on PFAS loads in a range of blue food species (freshwater fish and shellfish) have been the subject of numerous research. Given PFAS's chemical stability and heat tolerance, it seems unlikely that they will break down under typical cooking circumstances. But according to the theories of Del Gobbo et al. (2008), Kim et al. (2020), Luo et al. (2019), and Sungur et al. (2018), cooking heat breaks the protein-PFAS connection in animal tissue, causing PFAS to be released (Vendl et al., 2022)

PFAS pollution prevention measures are highly sought after and should be implemented at the sources of pollution rather than after PFAS have entered marine and estuarine environments. Water treatment facilities, as well as factories and enterprises that utilize and generate PFAS, are responsible for handling PFAS waste. Although there are a number of PFAS remediation strategies available today, adsorption on natural and renewable material-based adsorbents, agricultural wastes, biochar, and several combinations of materials are some of the most widely utilized and acknowledged approaches for PFAS contamination mitigation (Bilela et al., 2023).

Lowering the amount of PFAS in drinking water is necessary to lower the risk of female PFAS intake. Because MPs adsorb to PFAS, improving the removal of MPs from drinking water has proven to be an effective way to lower PFAS. Eliminating PFAS from drinking water is a difficult task because it is a newly discovered water contaminant. PFAS are hard to get rid of with conventional drinking water treatment (Du et al., 2023). One of the most significant and useful of them for treating PFAS in drinking water is the adsorption of PFAS by activated carbon (AC) (Pranić, Carlucci, van der Wal, & Dykstra, 2025).

In a mixture of phosphate buffer (pH 7.2), the photo-transformation rates in aqueous solutions and the mineral content of clay of PFOSA, a sulfonamide precursor to the more stable carboxylated PFAS, were assessed and contrasted. By interacting with superoxide anions and hydroxide radicals, the natural clay mineral montmorillonite promoted the breakdown of PFOA to the shorter chain carboxylates (Dickman & Aga, 2022).

The granular byproduct of the thermochemical conversion of biomass in a low-oxygen environment (pyrolysis) is called biochar. In addition to its potential for carbon sequestration and, particularly in agriculture, its capacity to rebuild and support soil health, it has been acknowledged for its capacity to eliminate new pollutants like heavy metals and trace organic compounds from environmental media (Ramos & Ashworth, 2024).

3.6. Future Challenges

The restriction of historical PFAS, including PFOA and PFOS, has led to the development of new per and polyfluorinated species. Furthermore, because long-chain PFAS can persist in the environment, build up in lifeforms, and cause adverse effects, their manufacture has been reduced or stopped in the United States (Ehsan et al., 2023).

4. PHARMACEUTICALS AND RESIDUE CONTAMINATION IN FOOD PRODUCTS

4.1. Antibiotic Residues in Food-Producing Animals

The persistence of antibiotic residues in edible animal products has emerged as a significant food safety concern within the domain of food technology. These residues, when not adequately managed through withdrawal periods and proper veterinary oversight, can pose serious health risks to consumers (Singh, Shukla, Tandia, Kumar, & Paliwal, 2014). Among the most well-documented effects are allergic and anaphylactic responses, particularly in individuals with hypersensitivity to specific antibiotics such as penicillin. Studies estimate that approximately 7% of the general population may be susceptible to drug-induced hypersensitivity, with even trace concentrations (e.g., 0.6 μg of penicillin) being sufficient to trigger reactions in sensitive individuals (Kempe et al., 1999). The presence of such residues in milk (as low as 0.01 IU/mL) is especially concerning from a food processing and public health standpoint.

Chronic dietary intake of residues from antibiotics like oxytetracycline (OTC) can lead to hematological and hepatic alterations, including leucocytosis, atypical lymphocyte development, toxic granulations in granulocytes, and impaired coagulation. Additionally, OTC has a strong affinity for calcium-rich tissues, contributing to dental discoloration in children and prenatal exposure effects in neonates. These factors raise critical challenges for food technologists responsible for quality assurance and consumer safety (Lee SueJan, Lin ChinShiung, & Chen MaoChiang, 1996).

From a food toxicology and risk assessment perspective, β-lactam antibiotics are known to accumulate at significantly higher concentrations in liver and kidney tissues compared to muscle, emphasizing the need for rigorous residue monitoring in organ meats. Furthermore, fluoroquinolone antibiotics such as enrofloxacin—used in poultry production—act through bacterial DNA gyrase inhibition and have demonstrated teratogenic and embryotoxic effects in laboratory models (Kempe et al., 1999). This has prompted regulatory actions by agencies such as the U.S. FDA, which proposed banning enrofloxacin use in poultry not due to direct toxicity but due to its role in fostering antimicrobial resistance in zoonotic pathogens.

Environmental contamination through improperly disposed carcasses has also been observed to impact wildlife. Field studies have shown accumulation of enrofloxacin and ciprofloxacin in avian scavengers like cinereous and Egyptian vultures, correlating with significant mortality. These findings underscore the broader ecological and food chain risks associated with residual antibiotics in animal-derived foods (Lemus, Blanco, Grande, Arroyo, & Garcia-Montijano, 2008).

For food technologists, the mitigation of such risks requires integrated residue control strategies, robust screening techniques, and compliance with international residue limit standards to ensure the production of microbiologically and chemically safe food products for consumers (Singh, Shukla, Tandia, Kumar, & Paliwal, 2014).

4.1.1 Antibiotic Resistance: A Food Technological Perspective on Human Health Risks

Antibiotic resistance has become a significant public health crisis with direct implications for food safety and food processing. Regulatory bodies such as the World Health Organization (WHO) and the U.S. Centers for Disease Control and Prevention (CDC) have classified antibiotic resistance as one of the most urgent global health threats. From a food technology standpoint, the issue is particularly concerning due to the integration of antibiotics into food production systems, especially within animal husbandry and intensive livestock farming.

Bacterial resistance develops as a direct result of antimicrobial misuse or overuse in both clinical and agricultural contexts. The introduction of antibiotics in animal feed as growth promoters and for prophylactic use has accelerated the selection pressure on microbial populations. Consequently, many bacterial strains—once responsive to conventional treatments—have developed multidrug resistance. Infections caused by resistant pathogens such as Salmonella, Escherichia coli, Staphylococcus aureus, and Campylobacter jejuni are increasingly linked to contaminated food products, creating severe challenges for food technologists, quality control specialists, and public health authorities.

Antibiotics used in food-producing animals are often the same or structurally similar to those used in human medicine. Therefore, cross-resistance between human and animal bacterial pathogens is inevitable. Resistant strains may be transferred to humans through the food chain, especially when meat, milk, or eggs are not processed, handled, or cooked correctly. For instance, foodborne outbreaks of Salmonella typhimurium DT104, resistant to multiple antibiotics like tetracyclines, chloramphenicol, and ampicillin, have been traced back to livestock treated with these drugs. A similar trend has been seen in Campylobacter species resistant to fluoroquinolones, correlating with their use in poultry production.

The emergence of vancomycin-resistant enterococci (VRE) is a direct result of the use of avoparcin, a glycopeptide antibiotic, as a feed additive in animal production. These resistant bacteria colonize animal intestines and can transfer resistance genes to human pathogens via the food chain, particularly affecting immunocompromised individuals in clinical settings. From a food microbiology standpoint, such developments increase the complexity of ensuring microbial safety in food products.

Regulatory frameworks in food technology emphasize adherence to Maximum Residue Limits (MRLs) and enforcement of appropriate withdrawal intervals. For example, antibiotics like lincomycin and spectinomycin—used to treat mycoplasma infections in poultry—must be discontinued at least 48 hours before slaughter to avoid contamination in final products. Inadequate adherence to such guidelines contributes to the accumulation of antimicrobial-resistant bacteria in the food supply.

WHO and other international food safety agencies have recommended restricting the use of antibiotics critical for human health in food animals, especially as growth promoters. Food technologists and meat processors must integrate robust residue monitoring programs, enforce withdrawal periods, and adopt alternative biosecurity and husbandry practices to minimize antimicrobial usage.

The ongoing challenge of antimicrobial resistance underscores the need for multidisciplinary action across food processing, veterinary science, public health, and regulatory sectors. Ensuring microbiological integrity in animal-derived food products is essential to safeguard both consumer health and the efficacy of therapeutic agents. (Lemus, Blanco, Grande, Arroyo, & Garcia-Montijano, 2008)

4.2. Veterinary Pharmaceuticals and Residue Contamination in Animal-Derived Food Products

Veterinary pharmaceuticals constitute a broad class of bioactive compounds specifically formulated for therapeutic, prophylactic, and nutritional applications in animal health management. These compounds encompass antimicrobials, antiparasitics, hormones, vaccines, and immunomodulators, and are integral to maintaining homeostasis, optimizing productivity, and enhancing the physiological performance of domesticated and farmed animals. Their utilization is particularly critical in high-density livestock production systems where pathogenic exposure is elevated due to intensified animal husbandry practices.

A significant biotechnological concern associated with veterinary drug administration is the persistence of pharmacological residues in edible animal tissues and by-products. These xenobiotic residues may be traced in meat, milk, eggs, and aquatic products, often resulting from suboptimal compliance with established pharmacokinetic withdrawal intervals. Veterinary drug residues are defined as any pharmacologically active substance, including parent compounds, metabolites, excipients, or degradation intermediates, that persist post-therapy within biological matrices of treated animals. The bioaccumulation and biomagnification of such residues have raised public health concerns due to their potential to induce cytotoxic, genotoxic, immunotoxic, and endocrinological perturbations in human consumers.

In intensive animal agriculture, antimicrobial agents are routinely administered not only for curative interventions but also as metaphylactic and growth-promoting agents. According to global predictive models, the annual consumption of veterinary antibiotics is projected to escalate significantly—from approximately 63,151 metric tons in 2010 to over 105,596 metric tons by the year 2030—reflecting the intensification of biotechnological inputs in animal agriculture. This surge underscores the necessity for robust pharmacovigilance frameworks and residue surveillance systems to mitigate the risk of chemical contaminants in the food supply chain.

Regulatory authorities have instituted rigorous residue monitoring protocols and established maximum residue limits (MRLs) based on toxicological thresholds and acceptable daily intake (ADI) values. In the United States, the Food and Drug Administration (FDA) enforces veterinary drug tolerances under Title 21 CFR Part 556, while the European Medicines Agency (EMA) regulates residue levels in accordance with Commission Regulation (EU) No. 37/2010. These regulations rely on biotechnological methodologies including enzyme-linked immunosorbent assay (ELISA), liquid chromatography-mass spectrometry (LC-MS/MS), and radioimmunoassay for residue quantification.

Moreover, environmental ecotoxicology studies have shown that veterinary drug residues, excreted via urine and feces, may infiltrate aquatic and terrestrial ecosystems. Such contamination can alter microbial diversity, interfere with biogeochemical cycles, and facilitate the development of resistant environmental microbiota, further complicating the biosphere-human interface.

To address these challenges, integrative strategies involving pharmacogenomics, molecular diagnostics, residue biosensors, and bioremediation are being explored. A comprehensive biotechnological review encompassing residue sources, detection methodologies, risk assessments, and regulatory interventions is essential to inform policy formulation and ensure biosafety in the agro-food sector. Bridging the current knowledge gaps through advanced research will fortify our ability to manage veterinary pharmaceutical residues as a pressing public health concern and preserve ecological integrity in the face of intensifying agricultural biotechnology.

4.3. Pharmaceuticals and Personal Care Products (PPCPS): Environmental Presence and Impact

Pharmaceuticals and personal care products (PPCPs) refer to a diverse group of chemical substances, including human and veterinary medicines such as pain relievers, antibiotics, contraceptives, and cholesterol-lowering drugs. The constant introduction of these substances into the environment—through household use, industrial discharge, and improper disposal—poses a growing concern due to their bioactive nature and potential ecological and human health risks (Li et al., 2024; Singh, Yadav, Sharma, & Singh, 2023; Zakhar, Derco, & Cacho, 2018; Rubio-Armendáriz et al., 2022; Briffa, Sinagra, & Blundell, 2020; Silva et al., 2021). Ebele et al. (Hawthorne,

Grabanski, Martin, & Miller, 2000) describe pharmaceuticals as therapeutic agents used to treat or prevent diseases in humans and animals, which may sometimes be misused or overprescribed.

Collectively, PPCPs consist of compounds with diverse biological effects, many of which can interfere with hormonal systems or exert toxic effects even at low concentrations (Selvaraj, Shanmugam, Sampath, Larsson, & Ramaswamy, 2014). The pharmaceutical category includes substances such as antihyperlipidemic agents, stimulants (e.g., caffeine), painkillers (e.g., ibuprofen, paracetamol, ketoprofen, naproxen), psychiatric medications (e.g., carbamazepine, primidone), and antibiotics (e.g., sulfamethoxazole, chloramphenicol, trimethoprim, ciprofloxacin). Cardiovascular drugs such as propranolol, atenolol, and metoprolol, as well as lipid regulators like gemfibrozil and bezafibrate, are also common. On the personal care side, compounds include synthetic fragrances (e.g., polycyclic musks), antimicrobial agents (e.g., triclosan), sunscreen ingredients (e.g., methylbenzylidene camphor), preservatives (e.g., parabens), insect repellents (e.g., DEET), and hormones such as estrone (E1), estradiol (E2), and ethinylestradiol (EE2) (Omar, Ahmad, Aris, & Yusoff, 2016). These substances are frequently detected in various environmental compartments, primarily due to widespread daily usage. Research conducted by Yang et al. At a wastewater treatment plant in Gwinnett County identified common PPCPs like sulfamethoxazole, primidone, caffeine, and DEET in treated effluent, with concentrations between 10 and 100 nanograms per liter. Similarly, findings by Vlachagrammi and Athanasios (Petrovic & Barceló, 2006) highlight the frequent presence of compounds such as DEET, caffeine, triclosan, coprostanol, cholesterol, tri(2-chloroethyl) phosphate, and 4-nonylphenol in environmental samples, with non-prescription medications and steroids among the most commonly detected

4.4. Detection Techniques for PPCPS in Environmental Analysis

The detection of trace levels of endocrine-disrupting compounds (EDCs), which include pharmaceuticals and personal care products (PPCPs), has long relied on two primary analytical approaches: gas chromatography (GC) and liquid chromatography (LC). GC is typically paired with detectors such as flame ionization (FID), electron capture (ECD), or mass spectrometry (MS), whereas LC is often used alongside detectors like diode array (DAD), fluorescence, or MS . Each technique offers specific advantages depending on the analyte's chemical behavior. However, traditional methods sometimes lack the sensitivity or selectivity needed for accurate detection in complex environmental samples. To overcome these limitations, researchers have increasingly turned to advanced MS-based techniques—such as ion trap MS, triple quadrupole MS, Orbitrap, and time-of-flight MS—for improved precision and lower detection thresholds. Of these, GC-MS and LC-MS/MS remain the most widely used tools for monitoring pharmaceutical residues, estrogenic compounds, and alkylphenols in environmental contexts . Identifying these compounds in clean water is a complex task due to the wide array of potential contaminants and the low concentrations at which they appear . Mass spectrometry has emerged as the leading technique in this field, largely because of its capability to both identify and quantify trace pollutants in matrices like wastewater sludge. While earlier studies employed single-quadrupole LC-MS systems with some success, modern environmental monitoring often depends on more advanced setups like triple quadrupole mass spectrometers (QqQ) for accurate and specific detection of target substances [189,190]. Tandem mass spectrometry (MS/MS), including QqQ and quadrupole time-of-flight (QTOF), is widely implemented in PPCP analysis, frequently in combination with methods such as pressurized liquid extraction (PLE), LC, and either single or double-stage MS. This combination allows for the detection of compounds at very low concentrations, often in the microgram-per-kilogram (μg/kg) range when analyzing dry weight samples . As noted by Diaz-Cruz et al, LC-MS^2 systems operating in selected reaction monitoring (SRM) mode typically offer better detection limits than GC-MS systems, although LC-MS techniques may require more adaptable sample preparation protocols.

4.5. ELISA Method for Detecting SEM Contaminants in Food

The analytical process used in this ELISA-based screening method for semicarbazide (SEM) detection parallels key steps in conventional LC–MS/MS analysis. Traditional methods involve breaking down animal tissue, releasing bound SEM through acid hydrolysis, chemically tagging SEM with o-nitrobenzaldehyde (o-NBA), followed by neutralization and extraction using ethyl acetate. The ELISA protocol streamlines this process by integrating two important modifications that significantly reduce sample preparation time. Key Improvements in Sample Processing (Cooper, Samsonova, Plumpton, Elliott, & Kennedy, 2007)

4.5.1. Elimination of Full Tissue Homogenization

Unlike LC–MS/MS, which typically requires complete tissue homogenization, the ELISA method allows for only coarse mincing of samples. This is made possible through the use of Protease Type XIII, derived from Aspergillus saitoi, which remains active even in highly acidic environments (pH ~1.4). During the acid hydrolysis and derivatization step, this enzyme breaks down the tissue, allowing complete solubilization of both free and bound SEM. This approach not only shortens preparation time but also reduces the risk of false negatives, as SEM from all cellular compartments is released effectively. Although derivatization is commonly carried out overnight to ease workflow management, this duration could be shortened to 2–3 hours with occasional mixing, which helps speed up the tissue degradation. Further improvements, such as reducing the ELISA incubation period, could allow the entire assay to be completed in a single working day.

4.5.2. Efficient Extraction via Solid-Phase Columns:

Instead of using conventional liquid–liquid extraction with ethyl acetate, this method employs solid-phase extraction (SPE) using mixed-mode cation exchange columns. These columns are suitable for use under acidic conditions, allowing the sample supernatant to be applied directly without the need to adjust the pH

4.5.3. Relevance to Composite Food Testing

In 2003, it was discovered that SEM in processed meat products often originated not from the meat itself but from external ingredients such as coatings (e.g., breadcrumbs). In response, the AFSSA (French Agency for Food Safety), along with the RIKILT Institute of Food Safety in the Netherlands, issued a recommendation that when analyzing processed foods, only the animal-derived portion should be tested. For example, in breaded meat products, the focus should be solely on the meat component. They also advised that when a sample exceeds SEM limits in initial screening, further analysis should target only bound SEM. This involves removing any free SEM before performing the confirmatory test (Cooper, Samsonova, Plumpton, Elliott, & Kennedy, 2007). The ELISA method described here is ideal for this type of screening, as it accurately measures total SEM (free and bound) without requiring the time-consuming pre-washing steps needed for confirmation of tissue-bound SEM.

4.5.4. Detection of SEM From Different Sources

Research indicates that in poultry meat exposed to nitrofurazone, about 75% of the SEM present is in the tissue-bound form (Cooper, Samsonova, Plumpton, Elliott, & Kennedy, 2007). However, SEM can also originate from other sources, such as food additives like carrageenan or azodicarbonamide, where it generally appears in a free (extractable) form. In some cases, SEM has been found in unexpected places—for instance, in crayfish meat where known contaminants weren't suspected. In one such study, the total SEM concentration averaged 4.2 μg/kg, while the bound SEM accounted for just 0.5 μg/kg (Saari & Peltonen, 2004). This highlights the versatility of the ELISA method, making it suitable not only for detecting

SEM resulting from nitrofuran drug use but also for screening contamination from alternative or unknown sources.

4.6. Chromatography Technique

Gas chromatography (GC) is the most widely applied technique for isolating compounds with moderate to low polarity and semi-volatile properties, making it suitable for many POPs. GC separation efficiency depends on the compound's volatility and its interaction with the column's stationary phase. However, complex food samples often contain structurally similar compounds, making complete separation a challenge. In such cases, comprehensive two-dimensional gas chromatography (GC×GC) significantly enhances separation by incorporating two distinct columns based on different physicochemical principles, improving peak capacity and resolution.

Although GC is effective for many POPs, per- and polyfluoroalkyl substances (PFAS), due to their unique chemical characteristics, require analysis using liquid chromatography–tandem mass spectrometry (LC–MS/MS) (Alqarni, 2024). LC techniques are often paired with atmospheric pressure ionization methods such as electrospray ionization (ESI) and atmospheric pressure chemical ionization (APCI), which are suitable for non-volatile, thermally unstable compounds often found in food matrices.

In POP detection, mass spectrometry can be operated in various ionization modes. The electron capture detector (ECD) is cost-effective and still employed for detecting halogenated contaminants such as polychlorinated biphenyls (PCBs) and organochlorine pesticides (OCPs). However, MS, particularly in selected ion monitoring (SIM) mode, has largely replaced conventional detectors due to its superior specificity. While electron ionization (EI) provides reproducible spectra, it may lead to excessive fragmentation in sensitive food samples (Alqarni, 2024). In such cases, negative chemical ionization (NCI) offers a gentler alternative with enhanced selectivity.

Recent advances such as triple quadrupole tandem MS (MS/MS) and atmospheric pressure gas chromatography (APGC) have further enhanced detection sensitivity and selectivity, particularly for dioxins, PCBs, and pharmaceutical residues in high-fat food matrices like vegetable oils and fish (Alqarni, 2024). These innovations are crucial in ensuring compliance with food safety regulations and maintaining public health standards

4.7. Persistent Organic Pollutants (POPS) and Their Impact on Food Safety

In recent years, growing concern has emerged over chemical contamination in the food supply, particularly from pesticide residues and environmental pollutants. Industrial expansion and the intensification of agriculture have contributed significantly to the release of various contaminants into the environment, many of which eventually enter the food chain. Because of their widespread presence and potential to cause serious health issues, managing chemical contaminants in food has become a global public health priority (Hojjati-Najafabadi, Mansoorianfar, Liang, Shahin, & Karimi-Maleh, 2022). A significant focus has been directed toward a particular class of harmful synthetic chemicals known as Persistent Organic Pollutants (POPs). These substances, primarily generated through human activities, are organic in nature, resistant to degradation, capable of long-distance transport, and tend to accumulate in living organisms over time. Types and Sources of POPs POPs are generally categorized into three main groups: Pesticides: This includes organochlorine pesticides (OCPs) like DDT and its derivatives. (Hojjati-Najafabadi, Mansoorianfar, Liang, Shahin, & Karimi-Maleh, 2022)

By-products: Substances unintentionally generated from manufacturing or combustion processes, such as dioxins (PCDDs), furans (PCDFs), and polycyclic aromatic hydrocarbons (PAHs). Although PAHs are not classified as POPs under the Stockholm Convention, they are included under the Aarhus Protocol and often considered POPs in various scientific assessments due to their environmental persistence and fat-solubility. Environmental Stability and Bioaccumulation Many POPs are halogenated compounds—containing chlorine, bromine, or fluorine—which strengthens their resistance to chemical, biological, and photolytic breakdown. Even non-halogenated POPs are extremely stable due to their molecular structure. As a result, once introduced into the environment, these compounds persist for years or even decades. These pollutants are capable of long-range transport via atmospheric and water pathways, often reaching remote locations like the Arctic where they were never directly used. This widespread distribution means that both humans and wildlife across the globe are at risk of chronic exposure to POPs, sometimes for an entire lifetime.

4.7.1 Health Risks Associated With Pops

POPs are linked to numerous adverse health outcomes, particularly with long-term or high-level exposure. These include hormonal imbalances, cardiovascular disease, reproductive disorders, immune dysfunction, developmental abnormalities, diabetes, and various forms of cancer. Although the use of many POPs peaked during the 1970s, stringent regulations have since led to substantial declines in their levels, especially in industrialized countries. Nevertheless, ongoing exposure—especially in developing nations—remains a major concern due to the persistent nature of these compounds and their presence in commonly consumed foods. Fish and seafood are among the most contaminated sources (Hojjati-Najafabadi, Mansoorianfar, Liang, Shahin, & Karimi-Maleh, 2022). POPs enter the food chain through environmental contamination caused by agricultural practices, industrial discharge, and pesticide use. For instance, legacy pesticides like DDT and dieldrin, once widely used to boost agricultural productivity, have left persistent residues in soil, water, and crops. Similarly, chemicals like PCBs, hexachlorobenzene (HCB), dioxins, and furans are often detected in food products due to their environmental ubiquity and resistance to degradation. Regulatory and International Efforts To reduce public health risks, several national and global regulatory bodies—including the WHO, EPA, EFSA, and FDA—have established strict guidelines and safety limits for POPs in food. The Stockholm Convention on POPs, initiated in 2001, provides a legally binding global framework for eliminating or restricting the production and release of these toxic substances. Initially addressing 12 POPs known as the "dirty dozen," the list has expanded to include additional compounds over the years. Programs like the Global Monitoring Plan by the United Nations Environment Programme (UNEP) have shown measurable reductions in contaminants such as DDT, PCBs, and dioxins, especially in breast milk samples from monitored populations. (Hojjati-Najafabadi, Mansoorianfar, Liang, Shahin, & Karimi-Maleh, 2022)

4.7.2. Challenges in Risk Assessment

One major challenge in food safety today is evaluating the health risks posed by dietary exposure to POPs. This requires comprehensive data on their toxicity and the levels at which they are present in food. However, for many POPs, toxicological data is limited, often derived from predictive models, lab animal research, or observational human studies. While acute toxicity levels can help determine safety thresholds for some compounds, chronic low-level exposure makes it difficult to establish clear cause-effect relationships for others. For genotoxic substances like some PAHs, no safe exposure level can be confidently established, and a margin of exposure model is typically used instead. (Hojjati-Najafabadi, Mansoorianfar, Liang, Shahin, & Karimi-Maleh, 2022) In response, the European Commission has set legal maximum levels for several PAHs across different food categories to mitigate potential health impacts. Analytical Techniques for POP Detection To estimate dietary intake of POPs, many nations conduct Total Diet Studies (TDS) using advanced analytical techniques. These methods must be sensitive enough to detect contaminants at trace levels while also being reliable and reproducible. Sample preparation is a critical step and varies depending on the type of food being tested.

5. CONCLUSION

Numerous emerging contaminants are introduced to people and the natural environment at large through a variety of identified and unidentified means. These pollutants consistently create new and pressing problems for ecosystems, including soil, water, and air, as well as for human health. Furthermore, novel chemical outputs proliferate and typically surpass the capabilities of risk assessment, inspection, hazard remediation, and existing prevention strategies. Additionally, as a result of innovative farming methods, polluted environments, technological progress, and global warming, the range of harmful residues in food is always growing which has potential risk in human health. To reduce potential risks and ensure proper tracking of all kinds of chemicals in the surroundings, which will aid in improving laws on this issue, it is therefore becoming more and more crucial to learn about developing contaminants.

REFERENCES

Adu, O., Ma, X., & Sharma, V. K. (2023). Bioavailability, phytotoxicity and plant uptake of per-and polyfluoroalkyl substances (PFAS): A review. *Journal of Hazardous Materials*, *447*, 130805. PMID: 36669401

Al Mamun, A., Prasetya, T. A. E., Dewi, I. R., & Ahmad, M. (2023). Microplastics in human food chains: Food becoming a threat to health safety. *The Science of the Total Environment*, *858*, 159834. PMID: 36461575

Alqarni, A. M. (2024). Analytical methods for the determination of pharmaceuticals and personal care products in solid and liquid environmental matrices: A review. *Molecules (Basel, Switzerland)*, *29*(16), 3900. PMID: 39202981

Andrady, A. L. (2017). The plastic in microplastics: A review. *Marine Pollution Bulletin*, *119*(1), 12–22. PMID: 28449819

Androulakakis, A., Alygizakis, N., Bizani, E., & Thomaidis, N. S. (2022). Current progress in the environmental analysis of poly-and perfluoroalkyl substances (PFAS). *Environmental Science. Advances*, *1*(5), 705–724.

Babaei, A. A., Reshadatian, N., & Feizi, R. (2024). A state of the art-mini review on the sources, contamination, analysis, and consequences of microplastics in water. *Results in Engineering*, •••, 102827.

Baluyot, J. C., Reyes, E. M., & Velarde, M. C. (2021). Per-and polyfluoroalkyl substances (PFAS) as contaminants of emerging concern in Asia's freshwater resources. *Environmental Research*, *197*, 111122. PMID: 33823192

Benson, N. U., Agboola, O. D., Fred-Ahmadu, O. H., De-la-Torre, G. E., Oluwalana, A., & Williams, A. B. (2022). Micro (nano) plastics prevalence, food web interactions, and toxicity assessment in aquatic organisms: A review. *Frontiers in Marine Science*, *9*, 851281.

Bilela, L. L., Matijošytė, I., Krutkevičius, J., Alexandrino, D. A., Safarik, I., Burlakovs, J., & Carvalho, M. F. (2023). Impact of per-and polyfluorinated alkyl substances (PFAS) on the marine environment: Raising awareness, challenges, legislation, and mitigation approaches under the One Health concept. *Marine Pollution Bulletin*, *194*, 115309. PMID: 37591052

Briffa, J., Sinagra, E., & Blundell, R. (2020). Heavy metal pollution in the environment and their toxicological effects on humans. *Heliyon*, *6*(9). PMID: 32964150

Brunn, H., Arnold, G., Körner, W., Rippen, G., Steinhäuser, K. G., & Valentin, I. (2023). PFAS: Forever chemicals—persistent, bioaccumulative and mobile. Reviewing the status and the need for their phase out and remediation of contaminated sites. *Environmental Sciences Europe*, *35*(1), 1–50.

Chen, Y., Wen, D., Pei, J., Fei, Y., Ouyang, D., Zhang, H., & Luo, Y. (2020). Identification and quantification of microplastics using Fourier-transform infrared spectroscopy: Current status and future prospects. *Current Opinion in Environmental Science & Health*, *18*, 14–19.

Cooper, K. M., Samsonova, J. V., Plumpton, L., Elliott, C. T., & Kennedy, D. G. (2007). Enzyme immunoassay for semicarbazide—The nitrofuran metabolite and food contaminant. *Analytica Chimica Acta*, *592*(1), 64–71. PMID: 17499072

Dickman, R. A., & Aga, D. S. (2022). A review of recent studies on toxicity, sequestration, and degradation of per-and polyfluoroalkyl substances (PFAS). *Journal of Hazardous Materials*, *436*, 129120. PMID: 35643010

Ding, J., Li, J., Sun, C., Jiang, F., Ju, P., Qu, L., & He, C. (2019). Detection of microplastics in local marine organisms using a multi-technology system. *Analytical Methods*, *11*(1), 78–87.

Du, M., Pu, Q., Li, X., Yang, H., Hao, N., Li, Q., & Li, Y. (2023). Perfluoroalkyl and polyfluoroalkyl substances (PFAS) adsorbed on microplastics in drinking water: Implications for female exposure, reproductive health risk and its mitigation strategies through in silico methods. *Journal of Cleaner Production*, *391*, 136191.

Efimova, I., Bagaeva, M., Bagaev, A., Kileso, A., & Chubarenko, I. P. (2018). Secondary microplastics generation in the sea swash zone with coarse bottom sediments: Laboratory experiments. *Frontiers in Marine Science*, *5*, 313.

Ehsan, M. N., Riza, M., Pervez, M. N., Khyum, M. M. O., Liang, Y., & Naddeo, V. (2023). Environmental and health impacts of PFAS: Sources, distribution and sustainable management in North Carolina (USA). *The Science of the Total Environment*, *878*, 163123. PMID: 37001657

Eze, C. G., Okeke, E. S., Nwankwo, C. E., Nyaruaba, R., Anand, U., Okoro, O. J., & Bontempi, E. (2024). Emerging contaminants in food matrices: An overview of the occurrence, pathways, impacts and detection techniques of per-and polyfluoroalkyl substances. *Toxicology Reports*. PMID: 38645434

Eze, C. G., Okeke, E. S., Nwankwo, C. E., Nyaruaba, R., Anand, U., Okoro, O. J., & Bontempi, E. (2024). Emerging contaminants in food matrices: An overview of the occurrence, pathways, impacts and detection techniques of per-and polyfluoroalkyl substances. *Toxicology Reports*. PMID: 38645434

Hawthorne, S. B., Grabanski, C. B., Martin, E., & Miller, D. J. (2000). Comparisons of Soxhlet extraction, pressurized liquid extraction, supercritical fluid extraction and subcritical water extraction for environmental solids: Recovery, selectivity and effects on sample matrix. *Journal of Chromatography. A*, *892*(1-2), 421–433. PMID: 11045502

Hojjati-Najafabadi, A., Mansoorianfar, M., Liang, T., Shahin, K., & Karimi-Maleh, H. (2022). A review on magnetic sensors for monitoring of hazardous pollutants in water resources. *The Science of the Total Environment*, *824*, 153844. PMID: 35176366

Jha, G., Kankarla, V., McLennon, E., Pal, S., Sihi, D., Dari, B., & Nocco, M. (2021). Per-and polyfluoroalkyl substances (PFAS) in integrated crop–livestock systems: Environmental exposure and human health risks. *International Journal of Environmental Research and Public Health*, *18*(23), 12550. PMID: 34886275

Jinadasa, B. K. K. K., Uddin, S., & Fowler, S. W. (2023). Microplastics (MPs) in marine food chains: Is it a food safety issue? [). Academic Press.]. *Advances in Food and Nutrition Research*, *103*, 101–140. PMID: 36863833

Johnson, G. R. (2022). PFAS in soil and groundwater following historical land application of biosolids. *Water Research*, *211*, 118035. PMID: 35032876

Kadac-Czapska, K., Trzebiatowska, P. J., Knez, E., Zaleska-Medynska, A., & Grembecka, M. (2023). Microplastics in food-a critical approach to definition, sample preparation, and characterisation. *Food Chemistry*, *418*, 135985. PMID: 36989641

Kempe, M., Cederfur, J., Zihui, M., Pei, Y., Wistrand, U. J., Fiaccabrino, G. C., & van Rhijn, H. (1999). *CREAM—Cartridges with Molecularly Imprinted Recognition Elements for Antibiotic residues Monitoring in Milk. Lunds Universitet*. Centre for Chemistry and Chemical Engineering.

Kwon, J. H., Kim, J. W., Pham, T. D., Tarafdar, A., Hong, S., Chun, S. H., & Jung, J. (2020). Microplastics in food: A review on analytical methods and challenges. *International Journal of Environmental Research and Public Health*, *17*(18), 6710. PMID: 32942613

Lee, J. C., Smaoui, S., Duffill, J., Marandi, B., & Varzakas, T. (2025). Research Progress in Current and Emerging Issues of PFASs' Global Impact: Long-Term Health Effects and Governance of Food Systems. *Foods*, *14*(6), 958. PMID: 40231978

Lee SueJan, L. S., Lin ChinShiung, L. C., & Chen MaoChiang, C. M. (1996). Antibiotic residues in milk following treatment with antibiotics for lactating dairy cows.

Lemus, J. A., Blanco, G., Grande, J., Arroyo, B., & Garcia-Montijano, M. (2008). Antibiotics Threaten Wildlife: Circulating Quinolone Residues and Disease in Avian Scavengers. *PLoS One*, *3*(1), e1444. PMID: 18197254

Li, J., Liu, H., & Chen, J. P. (2018). Microplastics in freshwater systems: A review on occurrence, environmental effects, and methods for microplastics detection. *Water Research*, *137*, 362–374. PMID: 29580559

Li, X., Shen, X., Jiang, W., Xi, Y., & Li, S. (2024). Comprehensive review of emerging contaminants: Detection technologies, environmental impact, and management strategies. *Ecotoxicology and Environmental Safety*, *278*, 116420. PMID: 38701654

Lv, L., He, L., Jiang, S., Chen, J., Zhou, C., Qu, J., & Li, C. (2020). In situ surface-enhanced Raman spectroscopy for detecting microplastics and nanoplastics in aquatic environments. *The Science of the Total Environment*, *728*, 138449. PMID: 32353796

Manojkumar, Y., Pilli, S., Rao, P. V., & Tyagi, R. D. (2023). Sources, occurrence and toxic effects of emerging per-and polyfluoroalkyl substances (PFAS). *Neurotoxicology and Teratology*, *97*, 107174. PMID: 36907230

Meyers, N., Catarino, A. I., Declercq, A. M., Brenan, A., Devriese, L., Vandegehuchte, M., & Everaert, G. (2022). Microplastic detection and identification by Nile red staining: Towards a semi-automated, cost-and time-effective technique. *The Science of the Total Environment*, *823*, 153441. PMID: 35124051

Mikac, L., Rigó, I., Himics, L., Tolić, A., Ivanda, M., & Veres, M. (2023). Surface-enhanced Raman spectroscopy for the detection of microplastics. *Applied Surface Science*, *608*, 155239.

Nahar, K., Zulkarnain, N. A., & Niven, R. K. (2023). A review of analytical methods and Technologies for Monitoring per-and Polyfluoroalkyl Substances (PFAS) in water. *Water (Basel)*, *15*(20), 3577.

O'Connor, J., Bolan, N. S., Kumar, M., Nitai, A. S., Ahmed, M. B., Bolan, S. S., & Kirkham, M. B. (2022). Distribution, transformation and remediation of poly- and per-fluoroalkyl substances (PFAS) in wastewater sources. *Process Safety and Environmental Protection*, *164*, 91–108.

Omar, T. F. T., Ahmad, A., Aris, A. Z., & Yusoff, F. M. (2016). Endocrine disrupting compounds (EDCs) in environmental matrices: Review of analytical strategies for pharmaceuticals, estrogenic hormones, and alkylphenol compounds. *Trends in Analytical Chemistry*, *85*, 241–259.

Panieri, E., Baralic, K., Djukic-Cosic, D., Buha Djordjevic, A., & Saso, L. (2022). PFAS molecules: A major concern for the human health and the environment. *Toxics*, *10*(2), 44. PMID: 35202231

Peixoto, D., Pinheiro, C., Amorim, J., Oliva-Teles, L., Guilhermino, L., & Vieira, M. N. (2019). Microplastic pollution in commercial salt for human consumption: A review. *Estuarine, Coastal and Shelf Science*, *219*, 161–168.

Petrovic, M., & Barceló, D. (2006). Liquid chromatography–mass spectrometry in the analysis of emerging environmental contaminants. *Analytical and Bioanalytical Chemistry*, *385*, 422–424. PMID: 16832892

Podder, A., Sadmani, A. A., Reinhart, D., Chang, N. B., & Goel, R. (2021). Per and poly-fluoroalkyl substances (PFAS) as a contaminant of emerging concern in surface water: A transboundary review of their occurrences and toxicity effects. *Journal of Hazardous Materials*, *419*, 126361. PMID: 34157464

Pranić, M., Carlucci, L., van der Wal, A., & Dykstra, J. E. (2025). Kinetic and isotherm study for the adsorption of per-and polyfluoroalkyl substances (PFAS) on activated carbon in the low ng/L range. *Chemosphere*, *370*, 143889. PMID: 39638126

Ramos, P., & Ashworth, D. J. (2024). Per-and poly-fluoroalkyl substances in agricultural contexts and mitigation of their impacts using biochar: A review. *The Science of the Total Environment*, •••, 172275. PMID: 38583608

Ribeiro, F., Duarte, A. C., & da Costa, J. P. (2024). Staining methodologies for microplastics screening. *Trends in Analytical Chemistry*, *172*, 117555.

Rubio-Armendáriz, C., Alejandro-Vega, S., Paz-Montelongo, S., Gutiérrez-Fernández, Á. J., Carrascosa-Iruzubieta, C. J., & Hardisson-de la Torre, A. (2022). Microplastics as emerging food contaminants: A challenge for food safety. *International Journal of Environmental Research and Public Health*, *19*(3), 1174. PMID: 35162198

Ryu, H., Li, B., De Guise, S., McCutcheon, J., & Lei, Y. (2021). Recent progress in the detection of emerging contaminants PFASs. *Journal of Hazardous Materials*, *408*, 124437. PMID: 33162244

Saari, L., & Peltonen*, K. (2004). Novel source of semicarbazide: Levels of semicarbazide in cooked crayfish samples determined by LC/MS/MS. *Food Additives and Contaminants*, *21*(9), 825–832. PMID: 15666975

Selvaraj, K. K., Shanmugam, G., Sampath, S., Larsson, D. J., & Ramaswamy, B. R. (2014). GC–MS determination of bisphenol A and alkylphenol ethoxylates in river water from India and their ecotoxicological risk assessment. *Ecotoxicology and Environmental Safety*, *99*, 13–20. PMID: 24183982

Shen, A., Lee, S., Ra, K., Suk, D., & Moon, H. B. (2018). Historical trends of perfluoroalkyl substances (PFASs) in dated sediments from semi-enclosed bays of Korea. *Marine Pollution Bulletin, 128*, 287–294. PMID: 29571375

Silva, A. L. P., Prata, J. C., Walker, T. R., Duarte, A. C., Ouyang, W., Barcelò, D., & Rocha-Santos, T. (2021). Increased plastic pollution due to COVID-19 pandemic: Challenges and recommendations. *Chemical Engineering Journal, 405*, 126683. PMID: 32834764

Singh, S., Shukla, S., Tandia, N., Kumar, N., & Paliwal, R. (2014). Antibiotic residues: A global challenge. *Pharma Science Monitor, 5*(3).

Singh, S., Yadav, R., Sharma, S., & Singh, A. N. (2023). Arsenic contamination in the food chain: A threat to food security and human health. *Journal of Applied Biology and Biotechnology, 11*, 24–33.

Sridhar, A., Kannan, D., Kapoor, A., & Prabhakar, S. (2022). Extraction and detection methods of microplastics in food and marine systems: A critical review. *Chemosphere, 286*, 131653. PMID: 34346338

Starnes, H. M., Rock, K. D., Jackson, T. W., & Belcher, S. M. (2022). A critical review and meta-analysis of impacts of per-and polyfluorinated substances on the brain and behavior. *Frontiers in Toxicology, 4*, 881584. PMID: 35480070

Taghipour, H., Ghayebzadeh, M., Mousavi, S. M. S., Sharifi, H., & Payandeh, A. (2023). Incidence and exposure to microplastics in table salt present in the Iran market. *Toxicology Reports, 11*, 129–140. PMID: 37520774

Udovicki, B., Andjelkovic, M., Cirkovic-Velickovic, T., & Rajkovic, A. (2022). Microplastics in food: Scoping review on health effects, occurrence, and human exposure. *International Journal of Food Contamination, 9*(1), 7.

Veerasingam, S., Ranjani, M., Venkatachalapathy, R., Bagaev, A., Mukhanov, V., Litvinyuk, D., & Vethamony, P. (2021). Contributions of Fourier transform infrared spectroscopy in microplastic pollution research: A review. *Critical Reviews in Environmental Science and Technology, 51*(22), 2681–2743.

Vendl, C., Pottier, P., Taylor, M. D., Bräunig, J., Gibson, M. J., Hesselson, D., & Nakagawa, S. (2022). Thermal processing reduces PFAS concentrations in blue food–A systematic review and meta-analysis. *Environmental Pollution, 304*, 119081. PMID: 35367104

Wang, W., Gao, H., Jin, S., Li, R., & Na, G. (2019). The ecotoxicological effects of microplastics on aquatic food web, from primary producer to human: A review. *Ecotoxicology and Environmental Safety, 173*, 110–117. PMID: 30771654

Wang, Z. M., Wagner, J., Ghosal, S., Bedi, G., & Wall, S. (2017). SEM/EDS and optical microscopy analyses of microplastics in ocean trawl and fish guts. *The Science of the Total Environment, 603*, 616–626. PMID: 28646780

Yanagisawa, H., Obayashi, K., Furuta, M., & Fujimaki, S. (2025). Comprehensive screening of per-and polyfluoroalkyl substances in consumer products using pyrolysis GC–MS. *Discover Chemistry*, *2*(1), 1–7.

Zakhar, R., Derco, J., & Cacho, F. (2018). An overview of main arsenic removal technologies. *Acta Chimica Slovenica, 11*(2), 107–113.

Zhang, Q., Xu, E. G., Li, J., Chen, Q., Ma, L., Zeng, E. Y., & Shi, H. (2020). A review of microplastics in table salt, drinking water, and air: Direct human exposure. *Environmental Science & Technology*, *54*(7), 3740–3751. PMID: 32119774

Ziani, K., Ioni ă-Mîndrican, C. B., Mititelu, M., Neac u, S. M., Negrei, C., Moro an, E., & Preda, O. T. (2023). Microplastics: a real global threat for environment and food safety: a state of the art review. *Nutrients*, *15*(3), 617. PMID: 36771324

Chapter 2
Invisible Threats on the Plate:
Detecting Emerging Contaminants in Food

Vikanksha Thakur
https://orcid.org/0009-0009-0257-4319
Lovely Professional University, India

Arun Kumar
Lovely Professional University, India

Jatinder Singh
https://orcid.org/0000-0001-7691-6689
Lovely Professional University, India

ABSTRACT

Emerging food contaminants have gained significant attention due to their potential adverse effects on human health and the environment. This paper focuses on three critical categories of emerging contaminants: microplastics, per- and polyfluoroalkyl substances (PFAS), and pharmaceutical residues. Microplastics are ubiquitous in the environment and can enter the food chain, posing risks to aquatic organisms and potentially human consumers. PFAS, often referred to as "forever chemicals," are persistent in the environment and have been linked to various health issues, leading to their classification as emerging pollutants. Additionally, pharmaceutical residues, stemming from agricultural runoff and wastewater discharge, are increasingly found in food matrices, raising concerns about their impact on food safety. This review aims

DOI: 10.4018/979-8-3373-3982-5.ch002

to synthesize current knowledge on the occurrence, pathways, impacts, and detection techniques for these contaminants, highlighting the urgent need for effective risk management strategies to mitigate their effects on food systems and public health.

1. INTRODUCTION TO FOOD CONTAMINANTS

Food safety is an ever-present concern for our societies. Foodborne diseases are responsible for about 420,000 deaths in the world annually with an economic burden of approximately 110 million diseases each year (Green 2022). Governments implement exhaustive monitoring systems to guarantee foodborne diseases do not pose a health threat to the public. To this aim, actions are enforced to reduce or eliminate foodborne pathogens both by the consumer, through proper food handling, and food producers, by complying with directives regarding hygiene practices at all stages of food production. However, the presence of food contaminants is another important aspect to be taken into consideration when assessing food safety, because it goes beyond the mere presence of pathogens, that if overcome should guarantee food safety. Food safety can be compromised by any chemical agent, such as a food contaminant, that may cause food to undergo a biochemical degradation process or alter its composition in such a way that there is an unwanted change in organoleptic characteristics or risk for human health, in particular, through the development of cancer or other chronic pathological conditions. While the absorption of hazardous substances, from highly polluted land, by plants and animals is the origin of classical food contaminants (Angon et al., 2024), such as metals, such substances can also be released into food through processing from packaging materials, by migration processes, and thus also form important classes of food contaminants, known as packaging migrants. However, the presence of compounds in food, which were not added intentionally by the manufacturer throughout the entire production process (Table 1), does not imply a strict relationship with food safety violations. In other words, the presence of food contaminants does not mean that the food does not satisfy legal safety limits or is of poor quality. Nonetheless, the recent increased media coverage regarding the presence of emerging food contaminants, such as microplastics, PFAS, and pharmaceutical residues, has highlighted their occurrence in food and proposed that governments should implement comprehensive monitoring strategies for these compounds to guarantee food safety since their presence in food should not be neglected (Lee et al., 2025).

Table 1. Overview of Emerging Food Contaminants

Contaminant	Sources	Food Affected	Potential Health Risks
Microplastics	Plastic degradation, packaging materials	Seafood, salt, bottled water	Inflammation, oxidative stress
PFAS (Per- and Polyfluoroalkyl Substances)	Industrial runoff, food packaging coatings	Fish, dairy, fruits, vegetables	Endocrine disruption, cancer risk
Pharmaceutical Residues	Improper disposal, livestock medication	Meat, dairy, crops (irrigated water)	Antibiotic resistance, hormonal effects

2. MICROPLASTICS IN THE FOOD SUPPLY

Microplastics (MPs) are plastic particles of less than five millimetres in size, made from diverse polymers that compete in the environment with natural organic polymers and that can reside there for thousands of years. Their ubiquity in the environment is well documented, and the entry of MPs into the food supply through various vectors of environmental contamination is still under investigation (Pironti et al., 2021). MPs have been found in agricultural soils, drinking water, seafood, fish, honey, sugar, milk, sea salt, and beer, making huge headlines during the last few years, since the concentrations of MPs in the food supply are often comparable to or higher than thresholds of adverse health effects, as established for toxic chemicals. Thus, exposure to MPs carried by the food supply is likely to be an important pathway for human inhalation and chronic ingestion. Yet, in contrast to chemical contaminations in the food supply, which have received huge regulatory attention during the last decades, and which are now subject to strict standards and safety assessments, MPs are still largely unregulated and unassessed (Sorbo et al., 2022). For example, up to now, in Europe and the United States, there are no legal limits for MPs in food. The development of sampling strategies, the validation of detection methods, and the elaboration of regulatory frameworks are urgently needed to answer the question of how dangerous MPs in food really are and to protect consumers. MPs in the ecosystem, including agricultural soils and plants, as well as surface water, are taken up by various animals and are also found within tumors or affect reproductive organs. Many MPs contain toxic chemical compounds, and compatible very small-sized MPs have been shown to translocate in food animals. Assessing potential health effects is as challenging as finding suitable detection methods for particular matrices, as neither MPs in drinking water nor particularly low plastic concentrations in food products or food chain–linked products can be sampled easily (Al Mamun et al., 2023).

2.1. Sources of Microplastics

Microplastics (MPs) are contaminants entering the food supply from diverse environmental pathways. The main pathways are: transfer via atmospheric deposition; direct deposition onto crops via irrigation; root uptake by plants; wash-off onto the landscape surface; erosion and sediment deposition in water bodies; food-web transfer in aquatic ecosystems; and biota harvesting from the wild. These pathways involve several individual steps. This chapter organizes them into three groups: those involving atmospheric, land, and water bodies as emission sources (Kittlaus et al., 2024). Other questions include: What is the extent of MP accumulation in the food supply? What are the factors regulating their accumulation in food, and what are the toxic effects of MPs or of their additives in wildlife? What are the detection methodologies for MPs in food? As pollutants, MPs can have research competences from many science and technology fields, for example, engineering, social, health, food, and analytical sciences. If addressed from a meta-science point of view, it can be said that the global research activity, and its evolution during the last decades, have been governed by a fundamental question: what is the size and the toxicity of the hazard represented by MPs? The answer to such a question can be subdivided into three stages, characterized by i) a preliminary stage of dissemination of scientific information about the existence of large plastic items; ii) a pre-science stage of definition of essential physical chemical characteristics of MPs in certain 'hot' matrices, like marine water and sediments; iii) three ongoing stages of current research activities on major matrices and on the food supply, respectively, that make use of refined methods for MP detection (Figure 1). To stimulate questions and answers about the research activity regarding MPs, this chapter is largely inspired by the research questions previously cited (Depoy 2024).

Figure 1. Evolution of microplastics stages

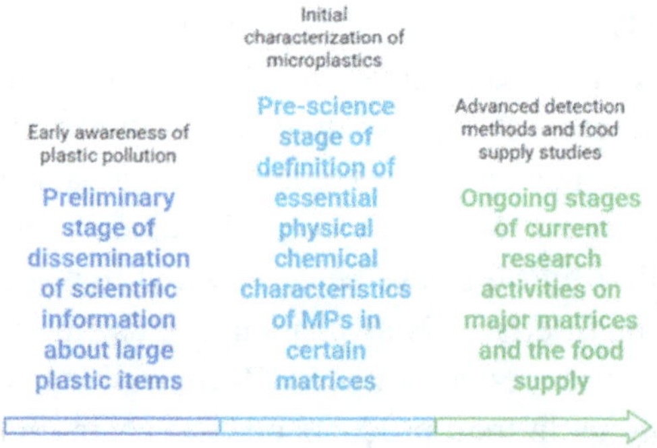

2.2. Health Impacts of Microplastics

The potential adverse effects of MP exposure on human health is an emerging area of research. The potential for MPs to elicit toxicity is primarily driven by polymer type and size, the presence of plastic-associated leachates/toxins, and the presence of adsorbed contaminants on plastic surfaces. Toxic plastics are then capable of causing local inflammatory responses and immune system dysregulation (Ding et al., 2023). However, the influence of these parameters on toxicity has not been extensively studied. Although chronic studies using exposure routes and doses representative of the human MP exposure profile would be necessary to assess the risks posed by MP to human health accurately, very few such studies have been conducted to date. This review mainly discusses in vitro and animal studies using doses higher than estimated human exposure and reports developmental malformations in embryos exposed to PS-NH2 and PS-COOH. In addition, the potential risks of inhalation exposure to MPs due to the COVID-19 pandemic and other sources, such as wildfires and household dust, are also highlighted in perspective. Recent studies indicate that plastic debris elicits noxious effects. Based on the chemistry of MPs and severity of clinical outcomes, different mechanisms inducing cellular stress and injury are proposed: 1) Swellable semi-solid, highly polluted, highly concentrated, and large-sized MPs increase the mechanical burden; 2) plastic-associated leachates like corrosive monomers and plastic-associated leachate toxins like heavy metals and persistent organic pollutants modulate cellular pathways related to apoptosis; 3) very small MPs are engulfed by phagocytosis and elicit localized and systemic inflammation, with resultant injury to tissues and organs and dysregulation of immune responses; and 4) MPs also act as carriers for the transport of pathogenic bacteria and viruses, leading to co-infection (Costa et al., 2023).

2.3. Detection Methods for Microplastics

The detection of microplastics in food and beverages raises concerns regarding the analytical requirements for method development. Scanning Electron Microscopy is the leading technique for analyzing microplastics because it combines imaging with qualitative Elemental Analysis. The characterization of microplastics in food and beverages has shed light on sources of contamination and microplastic transport routes. Historically, Fourier Transform Infrared spectroscopy is the most common technique used for microplastic detection (Guo et al 2022). It has been coupled with thermal extraction and other nozzle devices to render analyses faster. New methods such as X-ray Diffraction with mapping devices have been proposed recently to complement traditional methods. Near Infrared has been shown to be complementary to Fourier Transform Infrared in bulk analyses or Near Infrared hyperspectral imaging combined with machine learning applications. Other methods such as Raman Spectroscopy have gained ground in microplastic detection, especially during the pandemic because it permits analysis in ambient conditions without vacuum. Microscopy and micro-scaled X-ray techniques, while not providing much quantitative information compared with chromatography, remain the only methods available to investigate microplastics in their nanometer size range or for elementary composition information mapping. Thus they remain the best without their high costs and analysis durations. However, microscopic detection remains challenging to automate, and recent attempts such as Near Infrared assisted machine-learning screening remain imperfect with high false-positive rates (Table 2). Though not yet a solitary solution to the detection of microplastics in food samples, the combination of fractal analysis and affinity chromatography permits non-destructive enrichment and quantitative analysis of high-number polymers in those food samples (Fathi-Karkan et al., 2023).

Table 2. Detection and Mitigation Strategies for Emerging Food Contaminants

Contaminant	Detection Methods	Mitigation Approaches
Microplastics	Raman spectroscopy, FTIR analysis	Reducing plastic use, improved water and food filtration
PFAS	LC-MS/MS (Liquid Chromatography-Mass Spectrometry)	Regulation of industrial discharge, banning PFAS in packaging
Pharmaceutical Residues	HPLC (High-Performance Liquid Chromatography), Immunoassays	Advanced wastewater treatment, stricter veterinary control
Microplastics	Nile Red staining, Microscopy analysis	Biodegradable alternatives to plastics
PFAS	SPE (Solid Phase Extraction) combined with LC-MS	Use of safer chemical substitutes in manufacturing
Pharmaceutical Residues	Biosensors, ELISA (Enzyme-Linked Immunosorbent Assay)	Public awareness campaigns on safe drug disposal
Microplastics	Pyrolysis-GC/MS (Gas Chromatography-Mass Spectrometry)	Banning microbeads in cosmetics and food products
PFAS	TOF-MS (Time-of-Flight Mass Spectrometry)	Soil and water remediation technologies (e.g., activated carbon)
Pharmaceutical Residues	LC-HRMS (Liquid Chromatography-High Resolution Mass Spectrometry)	Monitoring pharmaceutical use in aquaculture

2.4. Regulatory Frameworks Addressing Microplastics

The increasing presence of microplastics (MPs), especially in food and the human body, raises consumer concerns regarding food safety and human health. As a result, the public is advocating in favor of regulations constraining the use of plastics and plastic products (Krause 2021). In fact, many countries have implemented bans on specific plastic products such as plates, cutlery, straws, stirrers, and expanded polystyrene food containers. Other countries have also introduced taxes or fees on different plastic bags. Although regulations are currently focusing on the prevention of plastic pollution, they serve to indirectly protect food from harmful particle contamination. More specifically, regulations prohibit the use of poly(vinyl chloride) plasticized with bis(2-ethylhexyl)phthalatc (DEHP), the use of materials and articles containing DEHP in contact with foodstuffs, and the use of foodstuffs containing DEHP. Additional regulations also cover the presence of other phthalate plasticizers in food such as dibutyl phthalate (DBP), benzyl butyl phthalate (BBP), diisobutyl phthalate (DIBP), di-n-pentyl phthalate (DPP), di-n-octyl phthalate (DNOP), isononyl phthalate (INP), and di(2-ethylhexyl) phthalate (DHEP). The above regulations are based on the provisions of the national frameworks, particularly the Federal Food Safety and Veterinary Office Directive for Foodstuffs and Utility Articles and the Swiss Chemicals Risk Reduction Strategy (Vettorazzi et al., 2020). Although existing regulations regarding the use of specific food contact materials do not include MPs, they do prohibit the intentional addition of specific additives. As MPs release plasticizers, these rules also serve as an indirect prohibition of MPs and any other intentionally added substances.

3. PFAS (PER- AND POLYFLUOROALKYL SUBSTANCES)

3.1. Overview of PFAS

Per- and polyfluoroalkyl substances (PFAS) are a class of human-made chemicals that have gained attention for their environment, health, and safety profiles. They are largely characterized by their carbon-fluorine backbone, high stability, and the tendency to bioaccumulate in the environment and living organisms (Han et al., 2021). The term “per-” refers to substances where all fluorine atoms have been replaced by other atoms; these are the most stable substances in the group, such as perfluoro carboxylic acids, perfluoro sulfonic acids, and perfluoroalkanesulfonamides as well as their salts and derivatives. The term “poly-” refers to substances not entirely saturated (PFAS precursors), such as fluorotelomers and polyfluoroalkyl substances; they degrade under environmental conditions to transform ultimately into perfluoro carboxylic acids and perfluoro sulfonic acids.

3.2. Environmental Persistence of PFAS

The vast majority of PFAS do not naturally degrade and can be found in the soil, air, and water. Some PFAS have been found with concentrations in the environmental compartment (e.g., in groundwater, surface water, or soil) at the part-per-trillion level in relative terms and several orders of magnitude above background in absolute terms. Numerous animal and human studies have shown that PFAS can readily bioaccumulate, while the toxicity of PFAS or any effects on population structure are still under dispute (Panieri et al., 2022). In this vein, evidence has been provided, suggesting that PFAS can lead to developmental and reproductive toxicity, carcinogenicity, immune system damage, and endocrine-disrupting activity. Per- and polyfluoroalkyl substances (PFAS) are a family of man-made chemicals that have been produced and used in a variety of industries around the world since the 1940s. Although their manufacturing has been phased out for the most part, consumable products still contain legacy PFAS. PFAS are also being synthesized in other parts of the world, with newer types being used in different industries such as textiles, leather, paper, and metals. PFAS are in the food chain, in food, and in the environment. However, they are not considered as emerging pollutants anymore: they are now known, ubiquitous, recalcitrant, toxic, and bioaccumulative contaminants. Due to their special fluorinated chemical topology and surface activity, PFAS were once considered to have great industrial potential in diverse fields and to have been extensively used to improve life quality (Cao et al., 2022). However, they are recalcitrant to biodegradation and persistent in the environment. Prominent among a long list of globally used, high-volume commodity chemicals are perfluorooctane sulfonate (PFOS) and perfluorooctanoic acid (PFOA), both of which show severe toxicity and bioaccumulation, and for many countries, they represent the most well-studied of all PFAS. The extreme properties of these two PFAS resulted in their demand in a variety of different industrial applications, including surfactants, surfactant-precursors, surfactant-based formulations, and plating agents. PFAS are especially controversial because they resist degradation in the environment and it appears to be any significant detoxification pathway for them in the environment. Scientists have claimed that they do not bioaccumulate and there is a lot of controversy surrounding their potential to biomagnify. They are soluble in both water and lipids, have low vapor pressures, and were detected in a variety of environmental matrices, including water, soil, dust, air, biota and food. More than 4000 different PFAS and related compounds have been identified, and their environmental behavior is very inconsistent mainly due to the vast number of different substituents (Cousins et al., 2020). Although PFAS are soluble in water, studies have shown that a few long-chain PFAS can sorb to particulate matter present in the aquatic environment. In general, PFAS in surface waters are more bioavailable and less affected by sedimentation than those in sediments; therefore, people considering total PFAS in sediments must be indirectly concerned about water quality as well. In soils and ground/surface waters, short-chain PFAS predominate, while PFOS dominates in atmospheric and biological samples. Although it is often claimed that certain long-chain PFAS are toxic to the environment and can cause damage at low concentrations, they do not belong to the group of toxic substances to which strict limits apply in many countries. Based on epidemiological and toxicological studies, human health is currently considered the most affected by PFAS. Studies have demonstrated that the toxicity of PFAS is dominated by the perfluoroalkanesulfonic acids – especially PFOS. To a lesser extent, only the classical perfluoroalkanoic acids with a chain length of from 5 to 7 carbon atoms are toxic to environmental systems. Some of the alkane-carboxylic-acid derivatives with a chain length of 1 to 5 carbon atoms are also toxic to environmental systems (Du Plessis 2025).

3.3. PFAS in Food Packaging

For several years, synthetic PFAS were applied in food packaging materials for their unique water, grease, and oil repellency functions. Currently, PFAS are increasingly being withdrawn from the market at the consumer demand and regulatory levels. Since existing substances are being replaced by alternative chemicals or formulations, new products with good performance must be developed and introduced to the market, which should be free from any concerns. More than 1,500 distinct PFAS have been identified. The unique carbon-fluorine bond gives these substances damage-resistance and waterproofing qualities. However, the C-F bond also makes PFAS resistant to thermal, chemical, and biological breakdown and renders them persistent in the environment. Many PFAS are extremely biopersistent in mammals and humans, accumulating in the liver and other organs. Laboratory animals that were exposed to certain PFAS have shown adverse effects in the growing fetuses and newborns, liver and kidney cancer, increased cholesterol concentrations, adverse effects on the immune response, abnormal thyroid function, birth weight below normal, and effects on male reproductive organs (Spyrakis et al., 2023). Among studies of humans, some reports have found an association between exposure to certain PFAS and increased cholesterol concentrations, fetal developmental effects, effects on the immune response, birth weight below normal, kidney or testicular cancer, and thyroid disease. More generally, exposure to mixtures of PFAS has been associated with increased cholesterol concentration, work-related cholesterol disease, weakened immune response to vaccination, increased risk of pregnancy-induced hypertension, fetal developmental effects, including birth weight below normal, and testicular cancer; study of child populations has linked exposure to PFAS to increased incidence of children with testicular cancer and abnormal bone growth, and testicular function effects in mothers and fathers. PFAS cause some of the same effects in laboratory animals, which were discussed above. These effects in human and animal populations indicate that PFAS may affect the same organ systems after exposure. For this reason, the National Toxicology Program concludes that the available data are sufficient to establish that certain PFAS are presumed to be a hazard to human health. Elucidating why thousands of members make up the PFAS family, one of the most practical applications relies on their oil- and water-repellent characteristics conferred to substrates applied as coating or surface treatments. Because of the increasing awareness regarding consumer demands over the safety of products and their effect over health, regulatory agencies started restricting or banning the use of the longer-chain PFASs (Bălan et al., 2021). However, shorter-chain PFAS are also under scrutiny and increasing regulations regarding their presence are expanding over multiple substrates. Studies have shown the choice of substrate, i.e., paper grade, and origin, recycled or virgin, influence the amount of PFAS present over food contact paper. PFASs can migrate from food-packaging substrates to food and human health implications are considered. Non-stick cookware is also a PFAS application in which coating degradation leads to the leaching of residues to the food where they may end up. It is widely accepted that the presence of hazardous substances in a given consumer item can substantially affect a person's health if consumed over a prolonged period, in particular over the early childhood when rapid growth occurs. Therefore, PFAS contaminations are of the utmost importance when innocent children are involved, and this has boosted interest among scientific researchers. Regulatory agencies have set limits on the allowed concentration of PFAS in food. It has been described that analysis of food-packaging substrates is fundamental in order to assess the potential hazards it entails for humans when packaging foods. The topic has also been considered by relevant authorities which issued a recommendation regarding the risk for consumer safety due to the presence of PFAS in food-contact substances. Specific recommendations are set regarding both the intentional addition of PFAS and undesired levels originating from packaging contamination (Donley et al., 2024). Amongst these, it is recommended to avoid using substrates with significant undesired levels of PFAS that may migrate to food (Figure 2).

Figure 2. Balancing PFA's and benefits

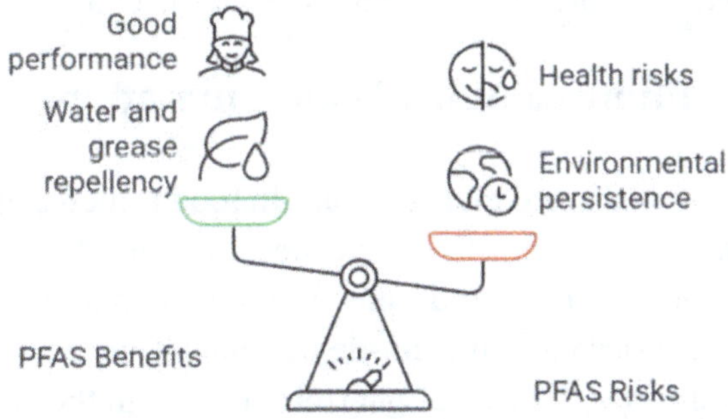

4. PHARMACEUTICAL RESIDUES IN FOOD

Pharmaceuticals are an important class of contaminants that, though added to food for medical reasons, may remain as residues that have consequences on food safety, animal health, and human health. In addition to pharmaceuticals that are deliberately added to food to treat or prevent disease, multiple other pharmaceutical types can contaminate food as a result of complex biogeochemical cycles, which are poorly understood. These pharmaceutical types include human-directed pharmaceuticals that enter the food supply through excretion by consumers, the waste products of wastewater treatment plants, or landfill leachate that leach into agricultural soil or water sources (Webb 2022). Natural and synthetic hormones are steroid-like pharmaceuticals that may enter food through feed, as excreted by food-producing animals. Phyto-pharmaceuticals can contaminate food as residues. While some pharmaceuticals are helpful to consumers, pets, livestock, and wildlife, ultimately impacting human health, the environment, and the agricultural industry when they enter the food supply, other pharmaceutical types are not. The near-universal implementation of antibiotics revolutionized contemporary medicine and veterinary care; however, the use of antibiotics to treat or prevent disease in food-producing animals has contributed to the selective pressure for antibiotic resistant bacteria, making it challenging to treat bacterial infections in both animals and humans. The widespread detection of human-directed pharmaceutical residues in various food products globally has raised concerns regarding potential residue toxicities. Environmental monitoring studies in multiple animal matrices have detected high levels of human-directed pharmaceutical residues. Nontargeted mass spectrometry screening has revealed detectable levels of additional pharmaceutical residues in various food samples. The development of analytical methods that facilitate risk assessment and monitoring efforts may mitigate food safety concerns regarding human-directed pharmaceutical contamination (Garde et al., 2022).

4.1. Sources of Pharmaceutical Contamination

The occurrence of pharmaceutical residues in food is increasingly recognized as an issue of emerging concern. In recent decades, there has been significant growth in the production, prescription, and consumption of marketed drugs, including antibiotics, analgesics, steroids, and anti-depressants. Consequently, this has led to increased detection rates and levels of pharmaceuticals in the environment. Unlike many other chemical contaminants arising from manufacturing processes, the presence of pharmaceutical residues in the environment can exist throughout the entire life cycle of these products, arising from different sources, including the production of raw materials, formulation of drug products, manufacture of active pharmaceutical ingredients, pharmaceutical formulation and packaging, disposal of expired or unused pharmaceutical waste, and excretion of medically unmetabolized pharmaceuticals (Maculewicz et al., 2022). A number of factors, including the therapeutic area of drugs, mode of administration, prescribing behavior, usage frequency, and excretion rate mostly dictate the level and type of pharmaceuticals found in the environment. The volume of global medicinal waste can be around 150 billion parcels, which corresponds to 0.2% of global annual municipal waste generation. Various disposal and treatment methods have been studied to alkaline, neutralization and incineration waste and these methods have been known to be effective. The problem of expired or unshared pharmaceutical waste is compounded because of the infrequent use of local high-temperature incineration plants; it is not uncommon for such waste to be found in nearby landfills. Such waste could, however, be disposed of at special recycling facilities or reduction plants. Compared to other pollutants, pharmaceuticals are more likely to bioaccumulate and cause chronic effects and have generally received less attention. The uneaten portion of animal feed given to farm animals can also be a source of pharmaceuticals found in the environment. Other typical sources of pharmaceuticals found in the environment include a contaminated water supply due to the inefficient removal of pharmaceutical residues in wastewater treatment plants (Ghazal et al., 2022).

4.2. Impact on Human Health

A wide range of pharmaceuticals, including those for human use, veterinary drugs, and nutraceuticals, can contaminate the food chain and the environment. These include antibiotics, nonsteroidal antidepressants, lipid regulators, anxiolytics, antiepileptics, anti-phenomenal drugs, noncardiac prescription drugs, and hormones. Although food analysis has mostly targeted antibiotics, there is rising interest in the examination of endocrine disrupting chemicals such as phytoestrogens and synthetic estrogens in food. Toxicity studies on reproductive development have shown that exposure to xenoestrogens in the early stages of life may result in malformations and reproductive dysfunctions such as cryptorchidism, hypospadias, and genital malformations in boys (Pan et al., 2024). Preclinical and clinical studies have found that early exposure to xenoestrogens has also been associated with neurodevelopmental disorders such as autism and attention deficit hyperactivity disorder. A variety of contaminants, primarily plastics, are present in food packaging. Although their role as a toxic agent has not been adequately studied, due to their omnipresence and harmful nature, their absorption during food manufacturing, storage, transport, and final consumption may constitute a source of risk and form an under-recognized threat to human health. More research is needed to assess with precision the risk impact of contaminants from food packaging. Veterinary dark brown and yellow pigments are typically used in polymers for food and drink contact applications because they do not interrupt the polymer manufacturing process. Furthermore, it is common to see them in food contact products intended to fulfill functionalities, such as opaque thermoplastic elastomers. Postmortem examination reveals the presence of slices of plastic in the stomachs of animals, contributing to questioning the safety of such pigments.

4.3. Detection and Analysis of Pharmaceuticals in Food

Detection and analysis of pharmaceuticals in food is important for the understanding of the degree to which food is exposed to pharmaceutical residues. Pharmaceutical analysis usually occurs on complex samples, such as blood or plasma, where multiple sample preparation steps are required to extract and clean the sample and eliminate potential interferences (Arroyo-Manzanares et al., 2021). Common sample preparation approaches for pharmaceutical analysis in food are liquid–liquid extraction, solid-phase extraction, solid-phase microextraction, the use of disks, and the use of dispersive liquid-phase microextraction. The goal is to increase the sensitivity of the analyze without being exposed to matrix effects and dispersive liquid-phase microextraction, which are easy to apply, inexpensive, and consume less solvent, have gained importance in the last few years. Pharmaceuticals can be determined in food using different analytical methods, such as gas chromatography mass spectrometry, liquid chromatography, high-performance liquid chromatography, and capillary zone electrophoresis. The use of chromatographic separation methods in combination with mass spectrometry has emerged as a powerful analytical tool for the determination of pharmaceuticals in food. Generally speaking, the pharmaceutical concentration levels in food are in the range of ng/g, making pharmaceutical analysis in food particularly challenging. Enzymatic deconjugation, for example, is sometimes required if a specific analysis of free pharmaceuticals is attempted. With the number of pharmaceuticals available in the market increasing rapidly, and pharmaceutical residues in food detected at extremely low concentrations, there is an increasing demand for the development of new analytical tools. The advantages of using enzymatic methods over other approaches are their specificity and sensitivity problem. New advances in solid-phase microextraction allow the incorporation of a large variety of enzymes and are a powerful tool to enhance selectivity and sensitivity prior to liquid chromatography mass spectrometry analysis (Hamidi 2023). Other exciting new innovations are the advances in portable sensing devices employed for different pharmaceutical detections.

4.4. Mitigation Strategies for Pharmaceutical Contamination

Pharmaceutical pollution is currently difficult to remediate for many ecosystems due to the vast diversity of potential pharmaceutical compounds and their metabolites produced by different animals. Further, combinations of compounds can create synergistic effects or antagonistic effects that may lead to unforeseen results when trying to remediate an area containing multiple compounds. Phytoremediation shows promise for some pharmaceuticals in contaminated areas, as hauler plants may uptake, degrade, and even change the chemical structure of pharmaceuticals. However, phytoremediation may not work for all pollutants currently found in treated wastes, including pharmaceuticals. Bioremediation utilizes bacteria to degrade pharmaceutical residues using one or multiple mixtures of specialized degrading bacteria for different groups of chemicals (Chandel et al., 2022). Microbiomes of treated waste and soil surrounding the area may also have the ability to breakdown pharmaceuticals and can be employed in an active or passive nature. Molecular engineering has become an appealing option to help carry out bioremediation by giving regular bacteria specialized competency or creating entirely new bacterial strains. However, these techniques may not be feasible in a practical sense. Bacteriophages are an alternative that can be harvested from the microbes already present in the contaminated soils, as these phages may carry important genes that could help break down these pharmaceutical residues in other microbes. Phage therapy has not only been used to fight infectious bacterial strains that affect both plants and humans but also to enhance plant growth, therefore enhancing the health of the whole ecosystem. Considering that the possible presence of pharmaceutical traces in soil and water is usually not taken into account during planning the location of such facilities at a given site, attention should be paid to the design and operation of incinerators, especially in terms of reducing soil and water contamination via emissions during all incineration phases. On the other hand, it is equally important that in order to implement monitoring measures in the case of discovering pharmaceutical residues in crops, toxicological studies on animals and higher order terrestrial and aquatic consumers are also carried out to demonstrate to what extent the possible pharmaceutical residues accumulated in foodstuffs lead to negative consequences on health and the environment (De Oliveira et al., 2020).

5. COMPARATIVE ANALYSIS OF CONTAMINANTS

The pollution of food with hazardous chemicals has recently caught the public and scientific community's attention. Although there are many chemicals in the food, discussions are growing about the co-occurrence of emerging contaminants. People and animals are exposed to many hazardous chemicals that can antagonize or synergize each other. Due to the emergence of new chemical entities and structural analogues, gaining interest in the potential antagonistic effects or even synergism of different hazard classes is increasing. Work to explore the co-occurrence of different emerging food contaminants is absolutely timely. Three groups of chemicals have emerged in the last decade as the focus of public discussions: microplastics, human, and veterinary pharmaceutical residues, and per- and polyfluoroalkyl substances, resulting in calls for new regulatory frameworks (Zaidi et al., 2025). Although some of these pollutants coexist in food and the environment, studies comparing different pollution classes are rare (Figure 3).

Figure 3. Interactions of emerging food contaminants

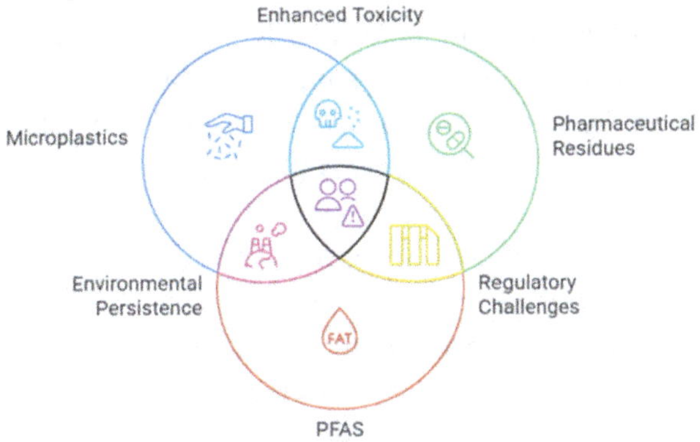

The present chapter will address possible interactions among specific compounds from three classes of emerging contaminants. The first part will provide some considerations on the co-occurrence of different pollutants. The second and third part provides a comparative analysis of microplastics and per- and polyfluoroalkyl substances, microplastics and pharmaceutical residues, and per- and polyfluoroalkyl substances and pharmaceutical residues, respectively. Conclusions will be drawn, and future research challenges suggested in the last part. There is no evidence that microplastics and other mentioned hazard classes, especially per- and polyfluoroalkyl substances, or those from the pharmaceutical class, interact. Therefore, there are no data to assume any effect. Still, it has been previously proposed that controlled release may occur (microplastics controlling the release of per- and polyfluoroalkyl substances) and microplastics transporting other pollutants, similar to the idea of controlled penetration of infiltrating chemicals through microplastics (Parashar et al., 2023).

5.1. Microplastics vs. PFAS

Microplastics and extremely persistent per-/polyfluoroalkyl substances (PFAS) are novel and recognized emerging contaminants. They are difficult-to-remove contaminants of the water source. The presence of PFAS and microplastics in different environmental compartments raises concerns and poses potential risks for human health. PFAS and microplastics can thus gain entry into the human body and then accumulate through the food chain, causing a series of health problems after long-term exposure. PFAS is the common name for a class of over 4700 synthetic, carbon-fluorinated chemicals. Their persistence and mobility are related to the manufacturing process used to produce them, in which fluorinated surfactants are manufactured, applied, discarded as waste, or accidentally released and spread in the environment. Microplastics are defined as solid, synthetic organic polymer particles with a diameter between one nanometer and five millimeters. Microplastics and PFAS have also been detected in arable soil, plants, and other edible plants. They are often cited within the same research articles, as they are both of anthropogenic origin and are environmental contaminants (Table 3). Microplastics have specific risks in addition to the generic risks of toxicity, reactivity, degradation, accumulation, and chronic toxicity, while PFAS have no such specific risks. It should also be noted that their origins differ. Microplastics are emission products and are secondary pollutants as emitted debris fragments into nanoscale particles, but PFAS are primary, unique, synthetic products emitted into the environment (Giroux et al., 2024). Thus, their toxicities, ecotoxicological effects, elimination methods, and degradation methods are also different.

Table 3. Impact of Emerging Food Contaminants on Ecosystem and Human Health

Contaminant	Impact on Ecosystem	Impact on Human Health	Long-term Concerns
Microplastics	Bioaccumulation in aquatic species, soil contamination	Gastrointestinal issues, inflammation	Potential trophic transfer across food chain
PFAS	Persistent soil and water pollution	Hormonal imbalance, immune system suppression	Bioaccumulation and long environmental half-life
Pharmaceutical Residues	Development of drug-resistant pathogens in environment	Antibiotic resistance, endocrine disruption	Threat to public health systems

5.2. Microplastics vs. Pharmaceutical Residues

Microplastics and pharmaceuticals are two emerging groups of contaminants of concern that can coexist in the aquatic environment. Microplastics are ubiquitous in the aquatic environment and were previously reported as a sink for persistent organic pollutants and pharmaceuticals. Besides, the triphenyl phosphates and bisphenols are competing with plasticizers to sorb and desorb from microplastics. The accumulation of pharmaceuticals onto microplastics has a synergistic effect on the toxicity of acute dynamic embryonic response to Ceratophyllum demersum (Mertines et al., 2025). However, how these two contaminants interact with each other in the aquatic environment, especially in the aquatic end members, is still unclear. Comparative studies examine the similarities and differences between microplastics and pharmaceuticals in the aquatic environment, such as source, fate, effect, technique, length scale and the synergistic effect on other organisms. We found that microplastics and pharmaceuticals have different sizes of source, distribution, effects, and analysis techniques. Although both microplastics and pharmaceuticals have been neglected until recently by researchers and policymakers, the first report of microplastics is much later than that of pharmaceuticals. Microplastics and pharmaceuticals have been of great concern in the last decade, however, research on pharmaceuticals is getting less attention recently, as compared to traditional pollutants, microplastics are more novel contaminants. Since the redox potential of the polymer backbone is lower than that of the active groups in organic pollutants, microplastics will not change the active functional groups of pharmaceuticals. But the presence of microplastics will encourage pharmaceutical degradation by microbial communities, since the surface of microplastics is suitable for biofilm growth. The accumulation of pharmaceutical and biofilm on microplastics will inhibit photodegradation of the organic component. Overall, various interactions and the microbial community in the environment largely control the relationship between microplastics and pharmaceuticals (Arienzo et al., 2023).

5.3. PFAS vs. Pharmaceutical Residues

PFAS and pharmaceutical residues differ in many aspects, including their structures and characteristics. The principal chemical structure of PFAS are carbon and fluorine; this is one of the strongest bonds in nature, which means most PFAS are very stable, while pharmaceutical residues are composed of different groups of chemicals, including aliphatic, aromatic, heterocyclic, and chiral compounds. Today, more than 6,000 PFAS and countless pharmaceutical residues have been reported. PFAS can be classified as long-chain and short-chain PFAS. Long-chain and short-chain PFAS are used more steadily in some areas than pharmaceutical residues. Peracetic acid, ozone, chlorination, and hydrolysis can degrade PFAS; advanced oxidation processes, biological degradation, photodegradation, and other methods can degrade pharmaceutical residues (Alalm et al., 2022). UV irradiation can also eliminate both contaminants, but PFAS photodegradation is affected by the presence of pharmaceutical residues. Conventional wastewater treatment plants can treat both, and enzymes can degrade pharmaceutical residues, but not PFAS. In surface water, PFAS occurrences are different from those of pharmaceutical residues. PFAS coexist with pharmaceutical residues in some areas. Groundwater and wastewater treatment plant effluents are the two most contaminated sources of both PFAS and pharmaceutical residues. Recently, several studies have reported short-chain PFAS reminiscence in algae after wastewater treatment plant effluent applications to land because wastewater treatment plants do not completely remove PFAS. Furthermore, some studies point to the risk of land-leachate compost contaminated with pharmaceutical residues. In surface water, short-chain PFAS occurrence is different from that of some pharmaceutical residues. Recently, several studies have reported short-chain PFAS reminiscence in algae after land application of wastewater treatment plant effluents, and some studies point to the risk of leachate compost contaminated with pharmaceutical residues (Khair Biek et al., 2024).

6. CONSUMER AWARENESS AND EDUCATION

The objective of this chapter is to provide a brief overview of consumer perception of food-contaminants and the implementation of communication methods to properly address the associated risks, with a particular emphasis on PFAS, microplastics, and pharmaceutical residues. The demand for food safety has increased dramatically in recent years among consumers. This is because food has evolved, no longer made in small family workshops, but mass-produced in large units. Along with this increase, there is also a growing number of scandals highlighting the contents of some foods sold in stores. Consumers have become concerned about what they are consuming, creating a mistrust of the products sold in supermarkets (Skripnuk et al., 2021). It is expected that consumers will buy safe products, produced under the best conditions and without any chemical or physical contaminants. The strategies of producers in the mass distribution of food have proven to be a double-edged sword. Mass production guarantees to meet the demand of a hungry society, creating an economy of scale, which affects prices negatively. Cheapness has always been the sword of mass distributors, who now use low prices as an advertising tool, creating a reluctance in consumers towards high-quality food. Consumer participation in and knowledge of food safety and quality issues are of fundamental importance for government authorities, industry producers, testing laboratories, and research institutions, as consumers play a significant role in determining whether or not a food safety policy is effective, especially concerning foodborne diseases. Governments, international organizations, and science institutions have invested heavily in educating consumers about the role that consumers can play to protect themselves. Consumers are frequently ignorant of food safety regulations, guidelines, or potential risks in the foods they consume. Limited access to training or supportive information, geographical socialization, or lack of trust from social networks in food safety messages could have affected consumers' food safety knowledge and ultimately their behaviors. Public perception is an important tool in the acceptance of food as safe (Caya et al., 2023).

6.1. Public Perception of Food Contaminants

Across a wide array of health and environmental hazards, consumer perceptions of risk do not always align closely with expert assessments. Indeed, public risk perceptions often reflect a different interplay of factors and personal experiences than those that inform scientific assessments. Food contaminants, including known hazards and emerging pollutants, represent a high-priority area of research that requires careful translation of scientific assessments of risk to the broader population. Contaminants like PFAS, microplastics, and pharmaceutical residues can be present in foods in amounts smaller than those needed to cause direct adverse health effects. Experts, however, are less certain about how long-term exposure to these contaminants, especially in the context of combined exposure to many chemicals in food and other paths, may affect food consumers (Street et al., 2023). Various chemicals have historically been studied and utilized to support product safety, and consumer awareness has drawn media scrutiny. This leads to objections based on lack of feasibility to produce and prepare foods free of various contaminants. Contaminated food may represent a minor pathway for the harmful effects, but this aspect may not always be clear to consumers. These communications must still reconcile demonstrable effects and public perception. For example, consumers are now aware of the negative environment-related effects and ethical stances regarding local sourcing and production proximity. The effect of food safety, however, may sometimes defer consideration when the product description is convincingly useful. For example, a product made using new advanced technology from genetically-modified organisms may be preferred for being more environmentally friendly by reducing greenhouse gases.

6.2. Educational Campaigns on Food Safety

Educational campaigns could manage the public perception of emerging food contaminants. Addressing sensory attributes, trained consumers on the distinction of food products with detectable levels of contaminants are needed. Education campaigns need to be designed for specific contaminants, such as microplastics, PFAS, or pharmaceutical residues, offering knowledge on how to minimize exposure (Al-Horani et al., 2024). For PFAS it is very helpful to distribute information concerning avoiding the intake of affected fish. Although food trade still needs internal and external support, external donators should finance the development of food safety management plans. Internal stakeholders should take responsibility for the facilitation of protocols for the establishment of food quality assurance and control systems. Food contaminant education campaigns are needed to help consumers choose the right food items. Depending on the regions or even countries, such campaigns should be risk-based and depending on regional food habits. Consumers' awareness must also include traditional products for which already some concern has been raised. Seasonal updates, if detection methods prove an increase in a certain contaminant or its exposure route, would improve knowledge transfer to the public. Because parents provide children with most food, additional efforts could involve encouraging personal commitment. Educational cooking workshops are an additional option to increase the knowledge and awareness of parents of preschool children (Sharkey et al., 2023).

7. FUTURE RESEARCH DIRECTIONS

Emerging Food Contaminants, with their potential for significant adverse effects on health and the environment, are an underrepresented area in the food safety and human health disciplines. In support of the concept, interdisciplinary research is essential. From the detection of such contaminants to the determination of their effects, it is critical to invest more resources in applied scientific studies, especially long-term studies, which often present challenges in funding acquisition. Standardized methodologies, as much as possible, and alignment with government guidelines would allow a higher reproducibility and transferability of the knowledge generated, enhancing the research in the area. The advancement of innovations in analytical technique throughput is one area to focus on, such as novel extraction and/or detection methods. There is a growing recognition that there are potential gaps in our understanding of the quantity and composition of contaminants in food contact materials, particularly those used by consumers for everyday purchases (Etxabide et al., 2022). A greater number of studies focused on other types of foods commonly consumed, such as cereals, fruits, and vegetables, searching for a wider variety of derivatives and microplastics, would also be relevant, especially the detection of contaminants in food matrices that have not been analyzed to date. Contaminants that are more rapidly excreted or less bioaccessible may be important to consider in exposure monitoring but may not be a concern for evaluating the potential chronic health effects associated with these chemical exposures. Additional research on mixtures of contaminants, particularly with the additive effects of mixtures, is also important.

7.1. Emerging Technologies for Detection

Emerging food contaminants such as microplastics, PFAS, and pharmaceutical residues have the most stringent regulatory standards. However, conventional detection techniques may not be sensitive enough to meet the increasingly strict regulations due to improvements in analytical detection limits. Therefore, researchers have proposed several innovative detection techniques for contaminants, aiming to improve the detection limit and decrease the cost. The need for rapid quantitative methods to detect low concentrations of pollutants has led research to shift from traditional chromatography techniques to new photonic sensors, integrated devices, paper-based sensors, upconverting fluorescence probes, tube hydrogel and carbon nano-chip-based platforms, and surface plasmon resonance sensors (Caroleo et al., 2022). In particular, paper-based colorimetric sensors and portable detection kits can be easily used for on-site quantitative detection and are popular choices for food safety management. Furthermore, low-cost, portable sensors have also been developed for the detection of PFAS. However, the commercial availability of these compact systems is still limited. The complex structure of foods makes the preparation of food samples a critical and time-consuming step. Microwave-assisted extractions/enrichments, solid-phase extractions, matrix solid-phase dispersions, and accelerated solvent extractions are commonly used for the cleanup of complex matrices, including tissues such as liver or kidney. Compared to conventional techniques, these emerging methods can provide high-throughput and instantaneous sample pretreatment with minimized risks, costs, and residue generation, which is especially beneficial for monitoring the other emerging contaminants. However, the absence of commercial kits or products for those new methodologies makes them less adopted than other conventional pretreatment methods. Nevertheless, developing these emerging detection and cleanup technologies would enhance the analysis accuracy and efficiency and lower sample collection and detection costs, leading to improved risk management for public health (Valskys et al., 2022).

7.2. Long-term Impact Studies

Future research directions are necessary as the existing data on microplastics, PFASs, and pharmaceutical residues in food are not sufficient for toxicological evaluations, exposure assessments, or risk assessments. Currently, there is no harmonized and standardized legislation and sampling protocol for research and monitoring the complex mixture of contaminants in food, which likely differs significantly between food types. Considering long-term food consumption behavior, long-term monitoring for the occurrence of these hazards in food is encouraged. Although the impacts, e.g., toxicity, bioaccumulation, and biomagnification, of microplastics, PFASs, and pharmaceutical residues in target humans, animals, and the environment during short-term exposure have been studied and reviewed, comprehensive long-term studies are lacking. Humans and animals might be continually exposed to these contaminants through affected food, and they can accumulate within the body over time if the absorption is higher than the excretion rate; these may disrupt endocrine functions, immune responses, and neurodevelopment, which result in the incidence of cancers, obesity, diabetes, reproductive dysfunction, thyroid dysfunction, and other diseases (Satarug et al., 2023). In addition, such impacts may occur even after the cessation of exposure due to the long exposure time or the off-target exposure to their metabolites, and effects on the affected individuals' descendants. To better evaluate the impact of chronic exposure to microplastic, PFAS, or pharmaceutical contaminants, more long-term exposure studies should be performed.

8. CONCLUSION

Research on contaminants in food is awakening interest due to their potential health effects but also because of the exposure level and frequency, which are exceeding the tolerable weekly intake by 40-times. Input pathways and sources for some emerging contaminants such as microplastics and pharmaceuticals and their potential adverse health effects in sensitive subpopulations such as children are extensively studied and have already paved the way for industry action or for regulations. New methods are required for fast exploration of wide food groups for detection and quantification of "unknown" contaminants and for risk evaluation, using proteomic and genomic screening technology. Novel full microbiome models are being invented which should include multiple natural barriers in human and animal health, including mucus layer associated with microbiome, microbiome metabolism, and transport processes through epithelial barrier in gut health protection against cumulative exposure. Further research is needed to fill the gaps regarding the knowledge on source, potential adverse health effects, and exposure levels in food of emerging food contaminants for specific commodities and food groups such as fetuses and infants. Micro physiological tissue models using stem cells and organoids could help to detect early subtle effects on cell viability, cell stress, modifying microbiome metabolism, immunotoxicity, cellular differentiation and transport function, alteration of epithelium barrier function with leakage of large molecules, and other adverse effects such as carcinogenic effects.

REFERENCES

Al-Horani, R. A., & Ayyad, A. (2024). Educating the public about toxic chemicals that we unknowingly consume: A potential important role for the practicing pharmacists. *Journal of the American Pharmacists Association, 64*(2), 355–363. PMID: 37940098

Al Mamun, A., Prasetya, T. A. E., Dewi, I. R., & Ahmad, M. (2023). Microplastics in human food chains: Food becoming a threat to health safety. *The Science of the Total Environment, 858*, 159834. PMID: 36461575

Alalm, M. G., & Boffito, D. C. (2022). Mechanisms and pathways of PFAS degradation by advanced oxidation and reduction processes: A critical review. *Chemical Engineering Journal, 450*, 138352.

Angon, P. B., Islam, M. S., Das, A., Anjum, N., Poudel, A., & Suchi, S. A. (2024). Sources, effects and present perspectives of heavy metals contamination: Soil, plants and human food chain. *Heliyon, 10*(7). PMID: 38590838

Arienzo, M., & Donadio, C. (2023). Microplastic–pharmaceuticals interaction in water systems. *Journal of Marine Science and Engineering, 11*(7), 1437.

Arroyo-Manzanares, N., Campillo, N., Lopez-Garcia, I., Hernandez-Cordoba, M., & Viñas, P. (2021). High-resolution mass spectrometry for the determination of mycotoxins in biological samples. A review. *Microchemical Journal, 166*, 106197.

Bălan, S. A., Mathrani, V. C., Guo, D. F., & Algazi, A. M. (2021). Regulating PFAS as a chemical class under the California Safer Consumer Products Program. *Environmental Health Perspectives, 129*(2), 025001. PMID: 33595352

Cao, H., Peng, J., Zhou, Z., Yang, Z., Wang, L., Sun, Y., & Liang, Y. (2022). Investigation of the binding fraction of PFAS in human plasma and underlying mechanisms based on machine learning and molecular dynamics simulation. *Environmental Science & Technology, 57*(46), 17762–17773. PMID: 36282672

Caroleo, F., Magna, G., Naitana, M. L., Di Zazzo, L., Martini, R., Pizzoli, F., & Paolesse, R. (2022). Advances in optical sensors for persistent organic pollutant environmental monitoring. *Sensors (Basel), 22*(7), 2649. PMID: 35408267

Caya, O., & Mosconi, E. (2023). Citizen behaviors, enterprise social media and firm performance. *Information Technology & People, 36*(3), 1298–1325.

Chandel, N., Ahuja, V., Gurav, R., Kumar, V., Tyagi, V. K., Pugazhendhi, A., & Bhatia, S. K. (2022). Progress in microalgal mediated bioremediation systems for the removal of antibiotics and pharmaceuticals from wastewater. *The Science of the Total Environment*, *825*, 153895. PMID: 35182616

Costa, C. H., Chang, K. P., Costa, D. L., & Cunha, F. V. M. (2023). From infection to death: An overview of the pathogenesis of visceral leishmaniasis. *Pathogens (Basel, Switzerland)*, *12*(7), 969. PMID: 37513817

Cousins, I. T., DeWitt, J. C., Glüge, J., Goldenman, G., Herzke, D., Lohmann, R., & Wang, Z. (2020). Strategies for grouping per-and polyfluoroalkyl substances (PFAS) to protect human and environmental health. *Environmental Science. Processes & Impacts*, *22*(7), 1444–1460. PMID: 32495786

De Oliveira, M., Frihling, B. E. F., Velasques, J., Magalhães Filho, F. J. C., Cavalheri, P. S., & Migliolo, L. (2020). Pharmaceuticals residues and xenobiotics contaminants: Occurrence, analytical techniques and sustainable alternatives for wastewater treatment. *The Science of the Total Environment*, *705*, 135568. PMID: 31846817

DePoy, E. (2024). *Introduction to research-e-book: Understanding and applying multiple strategies*. Elsevier Health Sciences.

Ding, R., Ma, Y., Li, T., Sun, M., Sun, Z., & Duan, J. (2023). The detrimental effects of micro-and nano-plastics on digestive system: An overview of oxidative stress-related adverse outcome pathway. *The Science of the Total Environment*, *878*, 163144. PMID: 37003332

Donley, N., Cox, C., Bennett, K., Temkin, A. M., Andrews, D. Q., & Naidenko, O. V. (2024). Forever pesticides: A growing source of PFAS contamination in the environment. *Environmental Health Perspectives*, *132*(7), 075003. PMID: 39046250

Du Plessis, S. H. (2025). *Improving the understanding of the complex phase behaviour of mixtures containing various organics and supercritical carbon dioxide* (Doctoral dissertation, Stellenbosch: Stellenbosch University).

Etxabide, A., Young, B., Bremer, P. J., & Kilmartin, P. A. (2022). Non-permanent primary food packaging materials assessment: Identification, migration, toxicity, and consumption of substances. *Comprehensive Reviews in Food Science and Food Safety*, *21*(5), 4130–4145. PMID: 35791070

Fathi-Karkan, S., Mirinejad, S., Ulucan-Karnak, F., Mukhtar, M., Almanghadim, H. G., Sargazi, S., & Díez-Pascual, A. M. (2023). Biomedical applications of aptamer-modified chitosan nanomaterials: An updated review. *International Journal of Biological Macromolecules*, *238*, 124103. PMID: 36948344

Garde, E., Marín-Vial, P., Pérez, G. E., & Sandvig, E. M. (2022). A review and analysis of the national dog population management program in Chile. *Animals (Basel), 12*(3), 228. PMID: 35158553

Ghazal, H., Koumaki, E., Hoslett, J., Malamis, S., Katsou, E., Barcelo, D., & Jouhara, H. (2022). Insights into current physical, chemical and hybrid technologies used for the treatment of wastewater contaminated with pharmaceuticals. *Journal of Cleaner Production, 361*, 132079.

Giroux, M. S., Siddiqui, S., & Brander, S. M. (2024). Classic contaminants in aquatic ecosystems: POPs, PFAS, heavy metals, and microplastics. In *Aquatic ecotoxicology: Understanding pollutants, aquatic organisms, and their environments* (pp. 43–58). Springer International Publishing.

. Green Jr, C. (2022). *Industry 4.0 Technologies and Biotechnology Biosensors a Game Changer in US Supply Chain Food Safety: An Exploratory Study.* Capella University.

Guo, X., Lin, H., Xu, S., & He, L. (2022). Recent advances in spectroscopic techniques for the analysis of microplastics in food. *Journal of Agricultural and Food Chemistry, 70*(5), 1410–1422. PMID: 35099960

Hamidi, S. (2023). Recent advances in solid-phase extraction as a platform for sample preparation in biomarker assay. *Critical Reviews in Analytical Chemistry, 53*(1), 199–210. PMID: 35192409

Han, J., Kiss, L., Mei, H., Remete, A. M., Ponikvar-Svet, M., Sedgwick, D. M., & Soloshonok, V. A. (2021). Chemical aspects of human and environmental overload with fluorine. *Chemical Reviews, 121*(8), 4678–4742. PMID: 33723999

Khair Biek, S., Khudur, L. S., & Ball, A. S. (2024). Challenges and Remediation Strategies for Per-and Polyfluoroalkyl Substances (PFAS) Contamination in Composting. *Sustainability, 16*(11), 4745.

Kittlaus, S., Kardos, M. K., Dudás, K. M., Weber, N., Clement, A., Petkova, S., & Zoboli, O. (2024). A harmonized Danube basin-wide multi-compartment concentration database to support inventories of micropollutant emissions to surface waters. *Environmental Sciences Europe, 36*(1), 52.

Krause, R. M. (2021). Why are we doing this? Issue framing, problem proximity, and cities' rationale for regulating single-use plastics. *Journal of Environmental Policy and Planning, 23*(4), 482–495.

Lee, J. C., Smaoui, S., Duffill, J., Marandi, B., & Varzakas, T. (2025). Research progress in current and emerging issues of PFASs' global impact: Long-term health effects and governance of food systems. *Foods*, *14*(6), 958. PMID: 40231978

Maculewicz, J., Kowalska, D., Świacka, K., Toński, M., Stepnowski, P., Białk-Bielińska, A., & Dołżonek, J. (2022). Transformation products of pharmaceuticals in the environment: Their fate,(eco) toxicity and bioaccumulation potential. *The Science of the Total Environment*, *802*, 149916. PMID: 34525754

Pan, J., Liu, P., Yu, X., Zhang, Z., & Liu, J. (2024). The adverse role of endocrine disrupting chemicals in the reproductive system. *Frontiers in Endocrinology*, *14*, 1324993. PMID: 38303976

Panieri, E., Baralic, K., Djukic-Cosic, D., Buha Djordjevic, A., & Saso, L. (2022). PFAS molecules: A major concern for the human health and the environment. *Toxics*, *10*(2), 44. PMID: 35202231

. Parashar, N., Mahanty, B., & Hait, S. (2023). Microplastics as carriers of per-and polyfluoroalkyl substances (PFAS) in aquatic environment: Interactions and eco-toxicological effects. *Water Emerging Contaminants & Nanoplastics, 2*(7), N-A.

Pironti, C., Ricciardi, M., Motta, O., Miele, Y., Proto, A., & Montano, L. (2021). Microplastics in the environment: Intake through the food web, human exposure and toxicological effects. *Toxics*, *9*(9), 224. PMID: 34564375

Satarug, S., Vesey, D. A., Gobe, G. C., & Phelps, K. R. (2023). Estimation of health risks associated with dietary cadmium exposure. *Archives of Toxicology*, *97*(2), 329–358. PMID: 36592197

Sharkey, J. R., & Smith, A. (2023). Cooking with the seasons for health (CwS4H): An innovative intervention that links nutrition education, cooking skills, and locally grown produce to increase vegetable intake among limited-resource parent–child dyads in rural Washington. *Nutrients*, *15*(22), 4851. PMID: 38004245

Skripnuk, D. F., Davydenko, V. A., Romashkina, G. F., & Khuziakhmetov, R. R. (2021). Consumer trust in quality and safety of food products in western Siberia. *Agronomy (Basel)*, *11*(2), 257.

Sorbo, A., Pucci, E., Nobili, C., Taglieri, I., Passeri, D., & Zoani, C. (2022). Food safety assessment: Overview of metrological issues and regulatory aspects in the European Union. *Separations*, *9*(2), 53.

Spyrakis, F., & Dragani, T. A. (2023). The EU's per-and Polyfluoroalkyl substances (PFAS) ban: A case of policy over science. *Toxics*, *11*(9), 721. PMID: 37755732

Street, M. E., Shulhai, A. M., Rotondo, R., Giannì, G., & Caffarelli, C. (2023). Current knowledge on the effects of environmental contaminants in early life nutrition. *Frontiers in Nutrition*, *10*, 1120293. PMID: 37324741

Valskys, V., Hassan, H. R., Wołkowicz, S., Satkūnas, J., Kibirkštis, G., & Ignatavičius, G. (2022). A review on detection techniques, health hazards and human health risk assessment of arsenic pollution in soil and groundwater. *Minerals (Basel)*, *12*(10), 1326.

Vettorazzi, A., Lopez de Cerain, A., Sanz-Serrano, J., Gil, A. G., & Azqueta, A. (2020). European regulatory framework and safety assessment of food-related bioactive compounds. *Nutrients*, *12*(3), 613. PMID: 32110982

Webb, W. A. (2022). *An Anthropology with Human Waste Management: Non-Humans, The State, and Matters of Care on the Placencia Peninsula*. University of South Florida.

Zaidi, N., Mir, M. A., Chang, S. K., Abdelli, N., Hasnain, S. M., Ali Khan, M. A., & Andrews, K. (2025). Pharmaceuticals and personal care products as emerging contaminants: Environmental fate, detection, and mitigation strategies. *International Journal of Environmental Analytical Chemistry*, •••, 1–29.

Chapter 3
Integrative Forensic Methods in Food Contaminant Detection and Identification

Baljeet Yadav
https://orcid.org/0000-0003-0624-6784
KR Mangalam University, Gurugram, India

Girraj Sharma
O.P. Jindal Global University, Sonipat, India

Mahipal Singh Sankhla
KR Mangalam University, Gurugram, India

Shivani Sehgal
KR Mangalam University, Gurugram, India

Janmejay Singh Tomar
Institute of Professional Studies, India

Anjali Dubey
https://orcid.org/0009-0002-7703-7680
Medicaps University, India

ABSTRACT

Food safety is one of the crucial global alarming spreading issue at very fast pace. Therefore, it requires continuous monitoring of contaminants by robust and integrative forensic approaches for the accurate and precise identification and quantification of contaminants in various food matrices. In addition, different types of approaches are employed for the investigation of contaminants in food i.e. conventional analytical and integrative forensic approaches. Conventional analytical approaches are often not capable of identifying and detecting complex emerging contaminants such as pharmaceutical residues specifically antibiotics, pesticides, microplastics, nano-materials etc. Thus, integrative forensic approaches syndicate different advanced

DOI: 10.4018/979-8-3373-3982-5.ch003

analytical techniques—such as hyphenated techniques (i.e. LC-MS/MS and GC-MS), molecular techniques (i.e. DNA barcoding), isotopic fingerprinting, biosensor etc. with multidisciplinary tools including chemometrics and geographic information systems (GIS). These advance strategies have several advantages over conventional analytical techniques in terms of accurate and precise identification and quantification of analyte at molecular level in complex food matrices, authentication of food origin, and comprehensive contamination profiling. Furthermore, the introduction of portable lab-on-a-chip technology and real-time data analytics improves on-site detection and rapid decision-making. Moreover, Forensic integrative approaches also significant to tackle some serious problems like tracing of food fraud, labelling of food, unintentional contamination etc. which can help to strengthen regulatory policy and framework, trail of civil cases, delivery of justice, consumer protection. This chapter will explore the collaboration among food scientists, forensic experts, and policymakers globally to develop robust and comprehensive standard operating procedure for handling and analysis of food sample by adopting integrative forensic approach. Moreover, it also covers emerging trends, technological innovations, challenges and ethical considerations in forensic identification and quantification of food contaminants, highlighting the need for interdisciplinary innovations to ensure food integrity, safety, and traceability in an increasingly complex food supply chain. It also aligns with United Nations Sustainable Development Goals (SDGs), particularly SDG 2 (Zero Hunger), SDG 3 (Good Health and Well-being), and SDG 12 (Responsible Consumption and Production).

1. INTRODUCTION

Over the past decade, forensic science has established persistence and instrumental presence within the criminal justice system, emerging as an important pillar of the scientific investigation and delivery of justice of both civil and criminal cases through the different applications of forensic science (G. Sharma et al., 2017). In addition, the expansion of applications of forensic science from management of the crime scene, scientific proceeding of the crime scene, handling of physical evidence to scientific investigation of traditional as well as emerging forms of crime like digital forensic, cybercrime, identity theft, sextortion, forensic accounting, cross border terrorism, clandestine laboratories, environmental forensics, microbial forensic, disputed paternity, food adulteration and contamination etc. (Chango et al., 2024). This expansion emphasizes the crucial role of an integrative forensic science approach, particularly in the qualitative and quantitative analysis of food contaminants and adulterants, thereby ensuring public health and safety (Ziani et al., 2025). Moreover, the detection, identification and quantification of contaminants are done by employing conventional approach as well as integrative forensic methods (Haji et al., 2023). However, the conventional approaches are not well suited for the identification and detection of contaminants at trace level in food (Alahi & Mukhopadhyay, 2017). Therefore, the use of integrative forensic methods is increasing day by day in food forensic for the investigation of contaminants in food. Additionally, these techniques are group of various instrumental techniques and methods such analytical chemistry, molecular biology and forensic data science which are commonly applied for the investigation of contaminants in food at micro level (Ranbir et al., 2022). Additionally, it shows numerous advantages over conventional methods in terms of methods validation parameters, ease of analysis etc. for the analysis of contaminants in food (Adade et al., 2025). Furthermore, Its analytical strength is prominent character of the approach which make it less susceptible to the limitations that is commonly showed by conventional approaches (Adade et al., 2025). For example, DNA profiling and isotopic examination of food product may be instrumental in the identification of geographic location of food and can aid the analysis of contaminants at the molecular level (Consonni et al., 2016). In addition, the introduction of artificial intelligence and machine learning algorithms enhances their predictive thrust and allow the systematic analysis of complex databases for initial indicators of contaminants in food (Ranbir et al., 2022). Furthermore, bioinformatics and digital forensics may also perform a crucial role in the recognition of patterns associated to food contamination issues and may differentiate among intentional and unintentional contamination in food (Soroushianfar et al., 2025).

Thus, the significance of these methods may be increased when emerging global threats including climate change, globalization, and a change in consumer, which collectively insure the uncompromised quality, safety, authenticity, and traceability of food within the modern food supply chain (Michael Alurame Eruaga, 2024; Guanqi & Husnain, 2022); (Bharti et al., 2024)) . Furthermore, it can also help to upgrade policy and regulatory standards, advocating for a transparent and uniform food producer and distributor system (Michael Alurame Eruaga, 2024). The use of forensic science in food safety receptiveness reinforces stakeholders' favour of prudent planning and threat management, hence safeguarding public health and retaining market integrity, as supporting the resilience and adaptability perspectives argued throughout this chapter.

This chapter presents systematic and comprehensive aspect of integrative forensic approaches for the accurate and precise detection, identification and quantification of contaminants in food matrices. Food contamination is one of the significant persistent and escalating problem in the world. Whether it has occurred accidentally or intentionally, it puts One Health approach at risk and poses economic burden on nation. These contaminants may be detected by traditional as well as integrative forensic approaches. Although, traditional analytical methods have their own limitations in terms of effectiveness, sensitivity, accuracy, precision, rigidness, robustness and adaptability to detect emerging contaminants in food. Therefore, there is a dire need to accept integrative forensic approaches that use application of traditional analytical techniques, advanced instrumental analytical techniques such as chromatographic, spectroscopic, molecular biological techniques, biosensors, and artificial intelligence-based data analysis.

The primary aim of this chapter is to recognize the importance of multidisciplinary analytical techniques that bring together applications from forensic science, food science, analytical chemistry, microbiology, and toxicology to detect chemical, biological, and physical contaminants with greater accuracy and reliability. Integrative forensic approaches emphasize the traceability and authenticity of food products, supporting not only consumer safety but also strengthening regulatory frameworks and supply chain transparency. Moreover, it also highlights the key role of emerging technologies—such as lab-on-a-chip systems, next-generation sequencing, portable detection devices, and blockchain traceability—in enhancing the speed, precision, and field applicability of food contaminant detection. By bridging laboratory-based forensic investigations with real-time monitoring and rapid response tools, these approaches enable timely interventions to prevent foodborne outbreaks and mitigate economic losses. Additionally, this chapter also brings into line with the SDGs, particularly SDG:2, SDG:3 and SDG:12 and make sure to provide safe and uncontaminated food supplies for sustainable food systems and public health resilience.

Finally, it contributes to policy framework and recommendations, capacity building, and the development of integrative frameworks that can be practically implemented by regulatory agencies, food industries, and forensic laboratories to safeguard food quality and consumer trust in an increasingly globalized food market.

2. FOOD CONTAMINANTS

Food and consumer safety has become a major issue for any country because of globalization of food trade, the setup of large food industries and use of cheap and adulterated material in the food (Liguori et al., 2022). Therefore, food safety not only harms One Health approach but also creates potential impact on the economy of the country and social instability (Angelos et al., 2017; King et al., 2017). In general, these food safety problems may arise because of various types of food contaminants, which can be mainly grouped into three categories according to their sources such as physical, chemical and biological contaminants. Physical contaminants encompass heavy metal ions, stone, soil, bone chips, micro plastics, hair, jewelry, nails etc. In addition, chemical contaminants such as pharmaceutical compounds i.e. veterinary drugs, antibiotics etc., pesticides, personal care products, food additives, foodborne pathogens, mycotoxins, harmful colorants and dyes (Lee & Yoon, 2021) . Biological contaminants contain bacteria, viruses, parasites, fungi etc. These contaminants enter the food chain via different ways and reach to the living organism and responsible foodborne illness or poisoning (Roos et al., 2018). In addition, according to studies approximately 600 million cases of foodborne illness occur worldwide each year, resulting in an estimated 420,000 deaths, which is about 7.5% of the annual deaths worldwide (56 million deaths) (Lee & Yoon, 2021). These food contaminants have continuously posed a major threat to human health, particularly when exposure occurs over prolonged durations, leading to chronic health conditions and long-term biological effects (Zahir et al., 2025). Thus, these issues are the most important concern for one health approach. Additionally, some of the food contaminants are briefly explained below.

Figure 1. shows different types of food contaminants.

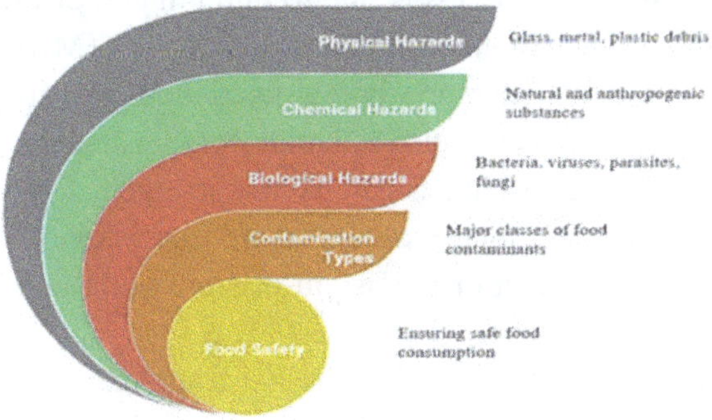

On the one hand, pesticides are commonly used to prevent pest damage to crops (Ahmad et al., 2024); however, residues may remain on ingested products, resulting in consumer exposure (Ahmad et al., 2024). On the other hand, industries may significant producers of heavy metals, pharmaceutical compounds, personal care products etc., which can accumulate in soil and aquatic bodies, potentially impacting the food supply (Osuoha et al., 2023); (Angon et al., 2024); (Alengebawy et al., 2021). Moreover, Food additives and processing aids, regardless of regulation, may occasionally induce undetected responses or interactions that pose dangers to consumer health (Paramasivam et al., 2024). Foodborne illness may result from microorganisms such as bacteria, viruses, fungus, and parasites present in biological pollution (Elbehiry et al., 2023). Foodborne pathogens, including *Salmonella*, *Escherichia coli*, and *Listeria monocytogenes* can contaminate food products due to handling, manufacturing, or distribution mistakes (Elbehiry et al., 2023 ; Bintsis & Department of International Trade, TEI of West Macedonia, Kastoria, Greece,

2017) . Although, it is also important to mention that some incidents of physical contamination can occur during food processing, such as the accidental introduction of glass, metal, or plastic particles into products (Onyeaka et al., 2023).

Food contaminants show great variations from region to region, mainly between low- and middle-income countries (LMICs) and high-income countries (HICs), because of cultural, religious, agricultural practices, industrial regulation, and food safety infrastructure. In LMICs, biological contaminants like *Salmonella*, *E. coli*, and *Vibrio cholerae* are widespread, primarily due to inadequate sanitation, unsafe water, and poor hygiene during food handling. For example, street-vended foods in many Asian and African cities are common sources of microbial contamination. Additionally, mycotoxins such as aflatoxins are a significant concern in tropical regions where high humidity and improper storage of grains and nuts encourage mold growth.

Chemical contaminants also show regional disparities. In LMICs, pesticide residues often exceed permissible levels due to unregulated or excessive pesticide use and limited enforcement of maximum residue limits (MRLs). Heavy metal contamination is also prevalent near poorly regulated industrial areas—for instance, rice grown near mining regions in South Asia may contain arsenic above safe levels. In contrast, HICs face different challenges. While stringent food safety standards minimize biological risks, chemical contaminants like bisphenol A (BPA) from packaging, food additive misuse, or persistent organic pollutants (POPs) can still pose concerns.

Physical contaminants such as stones or metal fragments are more frequent in regions where manual processing or rudimentary equipment is used. Moreover, allergenic contamination through cross-contact is a growing challenge globally but is better managed in HICs due to stricter labeling laws and traceability systems.

Figure 2. Sources of Pollutants as food contaminants

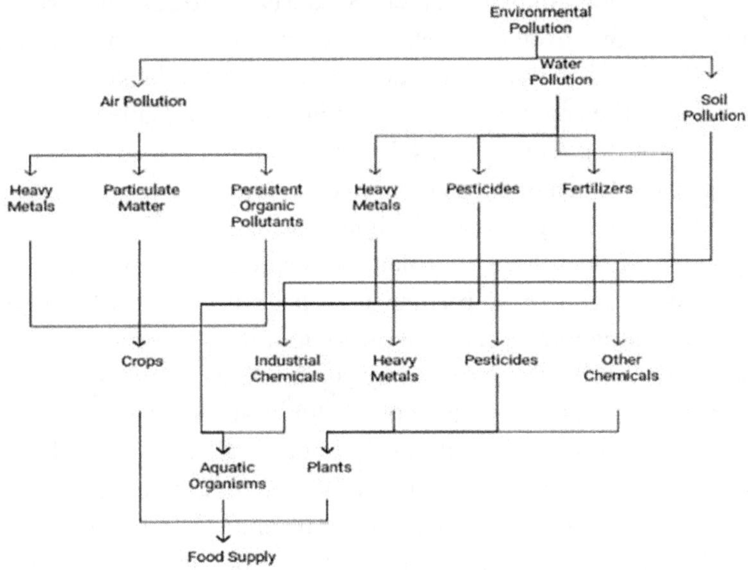

Detection capacity also differs greatly: HICs benefit from advanced surveillance systems and rapid testing methods, while many LMICs struggle with limited laboratory infrastructure, delayed responses, and insufficient monitoring. Understanding these regional variations is vital to designing targeted interventions, strengthening food safety governance, and ensuring safer food for diverse populations worldwide.

An integrated strategy to evaluate food contaminants is necessary for comprehensive study, focusing on varied properties, pathways, and impacts of these contaminants to establish robust detection and control methods (Pakdel et al., 2023). Furthermore, these forensic based approaches can aid in the reduction of contamination in food by using state- of the-art facility, regulatory frameworks and forensic science to ensure an inclusive and operative food safety system (Ahuja et al., 2023; Gbashi & Njobeh, 2024) .

4. FORENSIC SCIENCE AND FOOD SAFETY

Forensic science may play a key role in fixing food safety issues like adulteration, mislabeling, etc. (Saadat et al., 2022) which is diminishing the food quality day by day because of the globalization of food systems (Thorsen et al., 2025). In addition, to fix these issues, forensic science may apply robust strategies to overcome and manage these issues such as identification and detection of food contaminants at trace level (Ling et al., 2024). Therefore, these methods may efficiently identify, monitor, and mitigate contaminants in food (Saadat et al., 2022). These sophisticated methods allow specialists to detect sources of contamination—whether in manufacturing, packaging, or distribution—and determine responsible parties(Bharti et al., 2024) . This enables the enforcement of stringent safety regulations and the implementation of measures to rectify issues that may result in health risks. In addition, forensic science can play an important role in elucidating the dissemination of food contamination and providing valuable insights for implementing measures (Roy & Chakravorty, 2025). It may help to explain both reactive and proactive efforts, as it not only detects toxins but also predicts their pattern, that may be helpful in the identification of predictable cycles empowers industry experts to anticipate and mitigate future risks within the food supply chain (Dhal & Kar, 2025).

On the one hand, forensic science is notable for its ability to generate compelling evidence suitable for judicial proceedings (Bakhtiar et al., 2025). Food forensic can settle different dispute and fix accountability in issues of food including fraud, food substitution, mislabeling, or contamination that results in legal problems (Visciano & Schirone, 2021). In addition, it can also resolve one of the common problem adulteration of food products (Cavallini et al., 2023). Furthermore, it ensures authenticity and market integrity of food by enabling precise identification and verification of food products (Cavallini et al., 2023); (Pacholczyk-Sienicka, 2024). Consequently, to achieve this aim forensic scientists must have sound knowledge of technical capabilities and a comprehensive understanding of the legal angle which regulate food safety (Aimutis et al., 2024).

Thus, the effective implementation of forensic science and regulatory policy may explain how novel advance scientific methods may ensure safety of public health from adulterated food (Chango et al., 2024). Therefore, It will enhance the use and diversity of the available method in the investigation to reduce the risk of contamination in the food products (Dwivedi et al., 2017).

Figure 3. Ensuring food safety through forensic techniques and scientific investigations

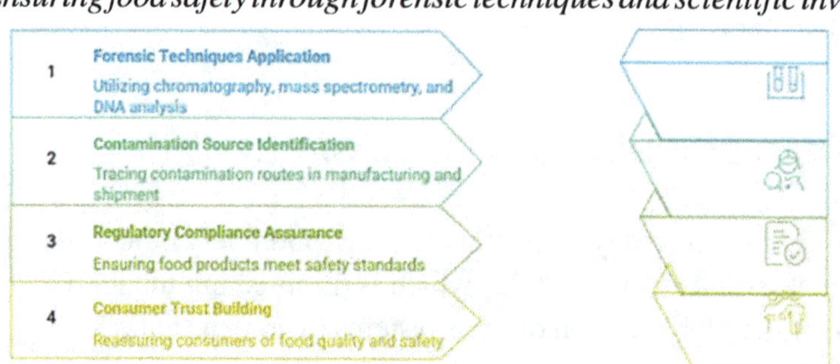

4.1 Role of forensic science in legal proceedings-

Forensic science plays a pivotal role in upholding food safety standards, especially when contamination incidents escalate into legal disputes or criminal cases. When food contamination leads to outbreaks of illness, product recalls, or allegations of fraud and adulteration, forensic investigation provides the scientific basis for identifying the source, proving liability, and ensuring accountability. A crucial aspect of this process is the meticulous maintenance of the chain-of-custody, which documents the collection, handling, transfer, and storage of samples from the moment of seizure to their presentation in court.

A robust chain of custody ensures that the evidence remains untampered, credible, and admissible in legal proceedings. Forensic experts follow strict protocols for labeling, sealing, and transporting samples under controlled conditions to prevent cross-contamination or degradation. Any break in this chain can compromise the integrity of the evidence and weaken the legal case, potentially allowing responsible parties to evade accountability.

Moreover, forensic science integrates analytical techniques such as DNA fingerprinting, chemical residue analysis, and trace element profiling to establish the origin and nature of the contaminant. In cases of intentional adulteration—such as the addition of harmful substances to enhance appearance or profit, these methods provide irrefutable scientific proof of malpractice. In many jurisdictions, forensic evidence has been instrumental in prosecuting food fraud cases, imposing penalties, and enforcing food safety laws.

The collaboration between food safety inspectors, forensic laboratories, and legal authorities strengthens regulatory enforcement and public trust. By ensuring evidence integrity and adherence to legal standards, forensic science acts as a bridge between scientific investigation and judicial outcomes. In an increasingly globalized food supply chain, its role is indispensable for protecting consumers, deterring fraudulent practices, and promoting justice when food safety is compromised.Top of FormBottom of Form

5. INTEGRATIVE FORENSIC APPROACHES / METHODS

The effective method should be used for the detection, identification and quantification of contaminants in food (Bharti et al., 2024). These methods are capable to investigate the complex issues of global food systems. However, as food production and distribution exponentially grow, complex emerging threats are emerged which demand robust, rapid, and flexible detection systems (Abiri et al., 2023). Therefore, integrative forensic approaches may be useful to fulfill aim, and it combines various analytical methods, advanced technology, and interdisciplinary expertise to detect and identify contaminants in food micro level compared to conventional approaches. Moreover, the conventional approaches has limitation to address the emerging threats associated with modern food production and global supply chains (Duan et al., 2025; Manning & Soon, 2014). Thus, integrative approaches is able to create a comprehensive diagnostic framework which is capable of identifying both predicted and unforeseen contaminants in food (Van Der Fels-Klerx et al., 2018). Therefore, the integration of chromatography, spectroscopy with molecular biological techniques escalate the efficiency of developed protocol for the detection of contaminants in food (Chen et al., 2025; Jayamohan et al., 2017). These hyphenated techniques can detect and isolate traces of contaminants, and identify molecular markers associated with novel or distinct hazards, In addition, the use of artificial intelligence and machine learning algorithms in the detection of food contaminants is also useful for various purposes (Kuppusamy et al., 2024) like identification of abnormalities in real time, detect subtle trends, and forecast potential contamination events before they escalate by analyzing the vast amounts of data produced by high-throughput screening (Dhal & Kar, 2025). These beauty of the system capabilities may identify early warning systems significantly and facilitate prompt response by individuals (Dhal & Kar, 2025).

6. ANALYTICAL TECHNIQUES FOR INTEGRATIVE FORENSIC APPROACHES

The use of analytical methods in the food forensic is important and may play crucial role to fulfill the following aim in the investigation of case such as detection, identification, individualization and quantification of contaminants (Gupta et al., 2003). This section of the chapter discusses important techniques, including chromatography, spectroscopy, and molecular biology, and their respective contributions to contemporary forensic food analysis.

Chromatographic methods have important space in the analysis of forensic food because these methods can separate complex mixtures into individual molecular identity and identify contaminants in food (Socas-Rodríguez, 2022). Among all chromatographic methods, gas chromatography (GC) and liquid chromatography (LC) are most common one for the same (Lehotay & Hajšlová, 2002; Bajkacz & Kycia-Słocka, 2020). GC is highly useful for the analysis of volatile molecules, whereas LC is most suitable for analysis of non-volatile materials (Coskun, 2016; Cordero et al., 2012; Poole, 2003) . the separation of analytes is based on the differential behaviour of mobile and stationary phases target analytes (Cheremisinoff, 1996). GC and LC are coupled with any spectroscopic technique known as hyphenated technique i.e.

GC-MS or LC-MS. Hyphenated techniques have high accuracy, precision, efficiency for the analysis of contaminants in food (Hernández-Mesa & Moreno-González, 2022; Sun et al., 2023). Spectroscopic techniques are mainly used for the identification of analyte in the matrix which is based on the interaction of electromagnetic radiation with analyte in the sample (Modupalli et al., 2021). In addition, advanced spectroscopic techniques like fourier transform infrared spectroscopy (FTIR) can be highly beneficial for identification of functional groups in molecules and provide unique chemical spectra like "fingerprints" of compounds (Liberty et al., 2025) . Moreover, nuclear magnetic resonance (NMR) is also has significant place in the lab to study unknown compounds in food products (Kamal et al., 2021).

Forensic food analysis used different type of molecular biological techniques for the detection and identification of foodborne pathogens or genetically modified organisms (GMOs) (Ahmed, 2002; Rafael Audino & Isabella Montenegro, 2022). In addition, the focal point of these identification and detection is based on polymerase chain reaction (PCR) and next-generation sequencing (NGS) which facilitate the rapid and precise analysis of genetic material of pathogen found in food (Salihah et al., 2016); (Nafea et al., 2024).

The integration of chromatography, spectroscopy, and molecular biological techniques in food forensic set up a comprehensive analytical framework that give strong value in terms of the reliability and accuracy of investigation and report of forensic food case (Debnath et al., 2025).

7. CASE STUDIES

Food safety is critical issue in the world which may be addressed by employing scientific approaches to identify and quantify contaminants in food (Saadat et al., 2022). The given case studies explain how advanced forensic methods are crucial to delivery of justice and punishing culprit. These methods also fix accountability and enhance food safety regulations. These cases highlight the necessity of forensic science to collaborate with other disciplines to mitigate and manage risks in food contamination. Furthermore, historical cases of food contamination frequently identify drawbacks in global food supply chains, resulting in regulatory and policy modifications and the adoption of new technologies.

In 2008, melamine case was happened in China and become notable issue of food poisoning in which melamine was intentionally used into baby formula to falsely increase its protein content (Gossner et al., 2009). This case of food poisoning resulted in over 30,000 instances of kidney illness in infants, some of which were lethal (Gossner et al., 2009). To detect melamine accurately various analytical techniques were used (Pei et al., 2011) . These technologies were crucial for identifying the source of contamination, revealing food fraud, and enhancing quality control and governmental oversight. This instance demonstrated the necessity of developing analytical tools and regulatory frameworks to prevent commercial alterations of food goods (Pei et al., 2011).

In 2011, E. *coli* outbreak happened in Europe, associated to contaminated fenugreek seeds from Egypt, ("Shiga Toxin-Producing E. Coli (STEC) O104," n.d.). The forensic investigation was significantly delayed due to initial difficulties in identifying the contamination source, underscoring critical gaps in the robustness of existing traceability systems (Buchholz et al., 2011). Investigators were able to associate the outbreak with a specific seed batch only through the application of advanced whole-genome sequencing technologies and international collaboration, which enabled precise pathogen tracking across borders (European Food Safety Authority (EFSA), 2011). This case point out the undoubtable role of molecular biological techniques and integrative forensic approaches in identifying, detecting and combating microbial threats within complex and globalized food supply chain (Mellmann et al., 2011)

8. REGULATORY FRAMEWORK

Regulation and policy framework for handling of food adulteration may become key component in the investigation of food forensic (Singh & Puniya, 2024). These frameworks should be established on scientific basis which fix contamination levels, specify testing methodologies, and mandate adherence to regulatory protocols (Karmaus et al., 2018). these guidelines should also show how chemical, biological, and physical hazards may impact health and responsible for deadliest diseases (Malhotra, 2017). Moreover, to make it more robust regulatory bodies can take help of AI-driven risk assessment models which can assist regulators in drafting judgements that are consistent, precise, and capable of addressing emerging dangers swiftly (Naseem & Rizwan, 2025a).

International standardization is also essential for facilitating worldwide trade by ensuring that food products comply with established safety standards (Food and Agriculture Organization of the United Nations & World Trade Organization, 2017). Organizations such as Codex Alimentarius establish these standards by convening specialists to collaborate and reach a scientific consensus (Koca et al., 2022). This enables national authorities to formulate regulations that align with regional dietary practices while being compatible with global standards (DeWaal et al., 2022). Harmonizing national regulations with international standards mitigates regulatory fragmentation that may adversely affect trade and food safety (DeWaal et al., 2022).

Global organizations, national agencies, and industry players must collaborate closely to share data, enhance surveillance systems, and revise standards to address growing threats (*WHO Global Strategy for Food Safety 2022-2030*, 2022). A cohesive, science-driven regulatory framework enhances consumer confidence, safeguards public health, and maintains integrity throughout the global food supply chain (*WHO Global Strategy for Food Safety 2022-2030*, 2022).

9. CHALLENGES IN FORENSIC FOOD SCIENCE

Forensic food science integrates analytical science and regulatory law; nonetheless, it faces numerous challenges that diminish its utility (Saadat et al., 2022). Traditional detection methods sometimes lack the necessary sensitivity and specificity for identifying trace contaminants within complex food matrices (Adade et al., 2025). Chromatography and mass spectrometry are sophisticated techniques that enhance precision; yet, they are expensive, labor-intensive, and require specialized personnel, posing challenges in resource-limited settings (Winkler, 2022; Bharti et al., 2024). Moreover, novel contaminants emerge at such a rapid pace that detection equipment is unable to keep pace, resulting in gaps in food safety surveillance (Li et al., 2024) .

Figure 4. Challenges in Forensic Food Science

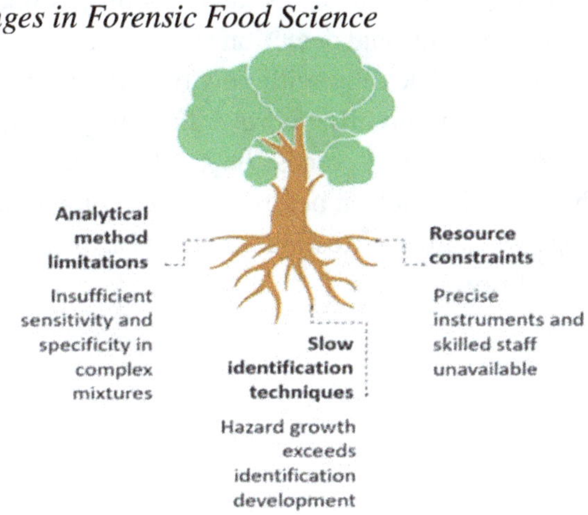

Cross-border food supply chains exacerbate contamination hazards, both intentional and inadvertent, due to factors such as chemical residues, allergens, and pathogens (Anyogu et al., 2024). To address this issue, forensic food science must employ advanced technologies such as artificial intelligence and big data analytics to enhance the prediction and management of potential dangers (Naseem & Rizwan, 2025b). This illustrates the need to possess adaptable standards that function effectively in all circumstances.

10. CONCLUSION

Integrative forensic methods are important techniques in food forensics to analyze food contamination in food. Because adulteration in food is the deadliest issue, which is responsible for food borne diseases and several people die in the world. Therefore, these techniques can play a significant role the determination and identification of adulterants, contaminants and emerging contaminants with high accuracy and greater confidence. Integrative forensic methods offer several benefits over conventional methods of detection of contaminants in food and enhanced global food safety standards. In addition, it is mainly

benefited and employed by advanced technologies like FTIR, mass spectrometry, NMR, GC-MS/MS, LC-MS/MS, NGR, PCR, DNA Profiling, data science, artificial intelligence, machine learning etc. which let forensic scientists detect contaminants at trace level with great accuracy and precision. This fusion of different analytical techniques in the investigation of food forensic make it susceptible to find contaminants in complex matrix. This approach across food forensics makes results more reliable and reduces food fraud and enhances safety for the public. By coupling new forensic technology with conventional detection approaches, food safety policy and regulation are more capable of dealing with new threats. In the future, advanced, integrated forensic methods will greatly improve the ability to find food pollutants. These technologies make it easier to process data, which makes predictive models more accurate and helps find pollution before it becomes a public health problem. AI-powered technology will automate testing, cut down on mistakes made by people, and make the best use of resources.

REFERENCES

Abiri, R., Rizan, N., Balasundram, S. K., Shahbazi, A. B., & Abdul-Hamid, H. (2023). Application of digital technologies for ensuring agricultural productivity. *Heliyon, 9*(12), e22601. DOI: 10.1016/j.heliyon.2023.e22601 PMID: 38125472

Adade, S. Y.-S. S., Lin, H., Johnson, N. A. N., Nunekpeku, X., Aheto, J. H., Ekumah, J.-N., Kwadzokpui, B. A., Teye, E., Ahmad, W., & Chen, Q. (2025). Advanced food contaminant detection through multi-source data fusion: Strategies, applications, and future perspectives. *Trends in Food Science & Technology, 156*, 104851. DOI: 10.1016/j.tifs.2024.104851

Ahmad, M. F., Ahmad, F. A., Alsayegh, A. A., Zeyaullah, Md., AlShahrani, A. M., Muzammil, K., Saati, A. A., Wahab, S., Elbendary, E. Y., Kambal, N., Abdelrahman, M. H., & Hussain, S. (2024). Pesticides impacts on human health and the environment with their mechanisms of action and possible countermeasures. *Heliyon, 10*(7), e29128. DOI: 10.1016/j.heliyon.2024.e29128 PMID: 38623208

Ahmed, F. E. (2002). Detection of genetically modified organisms in foods. *Trends in Biotechnology, 20*(5), 215–223. DOI: 10.1016/S0167-7799(01)01920-5 PMID: 11943377

Ahuja, V., Singh, A., Paul, D., Dasgupta, D., Urajová, P., Ghosh, S., Singh, R., Sahoo, G., Ewe, D., & Saurav, K. (2023). Recent Advances in the Detection of Food Toxins Using Mass Spectrometry. *Chemical Research in Toxicology, 36*(12), 1834–1863. DOI: 10.1021/acs.chemrestox.3c00241 PMID: 38059476

Aimutis, W. R., Mortenson, M. A., & Dong, J. (2024). Food Forensic Investigation. In Ismail, B. P., & Nielsen, S. S. (Eds.), *Nielsen's Food Analysis* (pp. 581–598). Springer International Publishing., DOI: 10.1007/978-3-031-50643-7_35

Alahi, M., & Mukhopadhyay, S. (2017). Detection Methodologies for Pathogen and Toxins: A Review. *Sensors (Basel), 17*(8), 1885. DOI: 10.3390/s17081885 PMID: 28813028

Alengebawy, A., Abdelkhalek, S. T., Qureshi, S. R., & Wang, M.-Q. (2021). Heavy Metals and Pesticides Toxicity in Agricultural Soil and Plants: Ecological Risks and Human Health Implications. *Toxics, 9*(3), 42. DOI: 10.3390/toxics9030042 PMID: 33668829

Angelos, J. A., Arens, A. L., Johnson, H. A., Cadriel, J. L., & Osburn, B. I. (2017). One Health in food safety and security education: Subject matter outline for a curricular framework. *One Health, 3*, 56–65. DOI: 10.1016/j.onehlt.2017.04.001 PMID: 28616505

Angon, P. B., & Islam, Md. S., Kc, S., Das, A., Anjum, N., Poudel, A., & Suchi, S. A. (2024). Sources, effects and present perspectives of heavy metals contamination: Soil, plants and human food chain. *Heliyon, 10*(7), e28357. DOI: 10.1016/j.heliyon.2024.e28357 PMID: 38590838

Anyogu, A., Somorin, Y. M., Oladipo, A. O., & Raheem, S. (2024). Food safety issues associated with sesame seed value chains: Current status and future perspectives. *Heliyon, 10*(16), e36347. DOI: 10.1016/j.heliyon.2024.e36347 PMID: 39253262

Bajkacz, S., & Kycia-Słocka, E. (2020). Liquid chromatography in food analysis. In *Chemical Analysis of Food* (pp. 391–455). Elsevier., DOI: 10.1016/B978-0-12-813266-1.00008-5

Bakhtiar, H. S., Ilyas, A., Kholiq, A., & Bakhtiar, H. S. (2025). The utilisation of scientific crime investigation methods and forensic evidence in the criminal investigation process in Indonesia. *Egyptian Journal of Forensic Sciences, 15*(1), 39. DOI: 10.1186/s41935-025-00456-y

Bharti, A., Jain, U., & Chauhan, N. (2024). Progressive analytical techniques utilized for the detection of contaminants attributed to food safety and security. *Talanta Open, 10*, 100368. DOI: 10.1016/j.talo.2024.100368

Bintsis, T. & Department of International Trade, TEI of West Macedonia, Kastoria, Greece. (2017). Foodborne pathogens. *AIMS Microbiology, 3*(3), 529–563. DOI: 10.3934/microbiol.2017.3.529 PMID: 31294175

Buchholz, U., Bernard, H., Werber, D., Böhmer, M. M., Remschmidt, C., Wilking, H., Deleré, Y., An Der Heiden, M., Adlhoch, C., Dreesman, J., Ehlers, J., Ethelberg, S., Faber, M., Frank, C., Fricke, G., Greiner, M., Höhle, M., Ivarsson, S., Jark, U., & Kühne, M. (2011). German Outbreak of *Escherichia coli* O104:H4 Associated with Sprouts. *The New England Journal of Medicine, 365*(19), 1763–1770. DOI: 10.1056/NEJMoa1106482 PMID: 22029753

Cavallini, N., Biancolillo, A., Strani, L., Durante, C., & Cocchi, M. (2023). Food Forensics. In Sharma, V., Sauzier, G., & Lewis, S. W. (Eds.), *Chemometric Methods in Forensic Science* (pp. 90–136). Royal Society of Chemistry., DOI: 10.1039/BK9781839166099-00090

Chango, X., Flor-Unda, O., Gil-Jiménez, P., & Gómez-Moreno, H. (2024). Technology in Forensic Sciences: Innovation and Precision. *Technologies, 12*(8), 120. DOI: 10.3390/technologies12080120

Chen, Z., Daka, Z., Yao, L., Dong, J., Zhang, Y., Li, P., Zhang, K., & Ji, S. (2025). Recent progress in the application of chromatography-coupled mass-spectrometry in the analysis of contaminants in food products. *Food Chemistry: X*, *27*, 102397. DOI: 10.1016/j.fochx.2025.102397 PMID: 40213340

Cheremisinoff, N. P. (1996). CHROMATOGRAPHIC TECHNIQUES. In *Polymer Characterization* (pp. 1–15). Elsevier., DOI: 10.1016/B978-081551403-9.50003-0

Consonni, R., Astraka, K., Cagliani, L. R., Nenadis, N., Petrakis, E., & Polissiou, M. (2016). Authenticity of Food. In *Encyclopedia of Food and Health* (pp. 285–293). Elsevier., DOI: 10.1016/B978-0-12-384947-2.00048-9

Cordero, C., Liberto, E., Sgorbini, B., Rubiolo, P., & Bicchi, C. (2012). Gas Chromatography. In *Chemical Analysis of Food: Techniques and Applications* (pp. 311–373). Elsevier. https://doi.org/DOI: 10.1016/B978-0-12-384862-8.00011-X

Coskun, O. (2016). Separation Tecniques: CHROMATOGRAPHY. *Northern Clinics of Istanbul*. Advance online publication. DOI: 10.14744/nci.2016.32757 PMID: 28058406

Debnath, S., Das, M., Mondal, S., Sarkar, B. K., & Babu, G. (2025). Advances in chromatography: Contemporary techniques and applications. *Essential Chem*, *2*(1), 1–27. DOI: 10.1080/28378083.2025.2466624 PMID: 40248686

DeWaal, C. S., Okoruwa, A., Yalch, T., & McClafferty, B. (2022). Regional Codex Guidelines and Their Potential To Impact Food Safety in Traditional Food Markets. *Journal of Food Protection*, *85*(8), 1148–1156. DOI: 10.4315/JFP-22-052 PMID: 35588145

Dhal, S. B., & Kar, D. (2025). Leveraging artificial intelligence and advanced food processing techniques for enhanced food safety, quality, and security: A comprehensive review. *Discover Applied Sciences*, *7*(1), 75. DOI: 10.1007/s42452-025-06472-w

Duan, K., Pang, G., Duan, Y., Onyeaka, H., & Krebs, J. (2025). Current research development on food contaminants, future risks, regulatory regime and detection technologies: A systematic literature review. *Journal of Environmental Management*, *381*, 125246. DOI: 10.1016/j.jenvman.2025.125246 PMID: 40199209

Dwivedi, S. L., Lammerts Van Bueren, E. T., Ceccarelli, S., Grando, S., Upadhyaya, H. D., & Ortiz, R. (2017). Diversifying Food Systems in the Pursuit of Sustainable Food Production and Healthy Diets. *Trends in Plant Science*, *22*(10), 842–856. DOI: 10.1016/j.tplants.2017.06.011 PMID: 28716581

Elbehiry, A., Abalkhail, A., Marzouk, E., Elmanssury, A. E., Almuzaini, A. M., Alfheeaid, H., Alshahrani, M. T., Huraysh, N., Ibrahem, M., Alzaben, F., Alanazi, F., Alzaben, M., Anagreyyah, S. A., Bayameen, A. M., Draz, A., & Abu-Okail, A. (2023). An Overview of the Public Health Challenges in Diagnosing and Controlling Human Foodborne Pathogens. *Vaccines, 11*(4), 725. DOI: 10.3390/vaccines11040725 PMID: 37112637

European Food Safety Authority (EFSA). (2011). *Tracing seeds, in particular fenugreek (Trigonella foenum-graecum) seeds, in relation to the Shiga toxin-producing E. coli (STEC) O104: H4 2011 Outbreaks in Germany and France*. Wiley Online Library.

Food and Agriculture Organization of the United Nations & World Trade Organization (Ed.). (2017). *Trade and food standards*. Food and Agriculture Organization of the United Nations; World Trade Organization.

Gbashi, S., & Njobeh, P. B. (2024). Enhancing Food Integrity through Artificial Intelligence and Machine Learning: A Comprehensive Review. *Applied Sciences (Basel, Switzerland), 14*(8), 3421. DOI: 10.3390/app14083421

Gossner, C. M.-E., Schlundt, J., Ben Embarek, P., Hird, S., Lo-Fo-Wong, D., Beltran, J. J. O., Teoh, K. N., & Tritscher, A. (2009). The Melamine Incident: Implications for International Food and Feed Safety. *Environmental Health Perspectives, 117*(12), 1803–1808. DOI: 10.1289/ehp.0900949 PMID: 20049196

Guanqi, Z., & Husnain, M. (2022). Assessing the role of organic food supply chain traceability on food safety and consumer wellbeing: A mediated-moderation investigation. *Frontiers in Psychology, 13*, 1073376. DOI: 10.3389/fpsyg.2022.1073376 PMID: 36532996

Gupta, K., Jain, V., Jain, S., Dhawan, K., & Talwar, G. (2003). ANALYSIS OF FOOD. In *Encyclopedia of Food Sciences and Nutrition* (pp. 206–215). Elsevier., DOI: 10.1016/B0-12-227055-X/00044-4

Haji, A., Desalegn, K., & Hassen, H. (2023). Selected food items adulteration, their impacts on public health, and detection methods: A review. *Food Science & Nutrition, 11*(12), 7534–7545. DOI: 10.1002/fsn3.3732 PMID: 38107123

Hernández-Mesa, M., & Moreno-González, D. (2022). Current Role of Mass Spectrometry in the Determination of Pesticide Residues in Food. *Separations, 9*(6), 148. DOI: 10.3390/separations9060148

Jayamohan, H., Romanov, V., Li, H., Son, J., Samuel, R., Nelson, J., & Gale, B. (2017). *Advances in Microfluidics and Lab-on-a-Chip Technologies*. https://doi.org/ DOI: 10.48550/ARXIV.1703.08583

Kamal, G. M., Uddin, J., Tahir, M. S., Khalid, M., Ahmad, S., & Hussain, A. I. (2021). Nuclear Magnetic Resonance Spectroscopy in Food Analysis. In Khan, M. S., & Shafiur Rahman, M. (Eds.), *Techniques to Measure Food Safety and Quality* (pp. 137–168). Springer International Publishing., DOI: 10.1007/978-3-030-68636-9_7

Karmaus, A. L., Osborn, R., & Krishan, M. (2018). Scientific advances and challenges in safety evaluation of food packaging materials: Workshop proceedings. *Regulatory Toxicology and Pharmacology*, *98*, 80–87. DOI: 10.1016/j.yrtph.2018.07.017 PMID: 30048703

King, T., Cole, M., Farber, J. M., Eisenbrand, G., Zabaras, D., Fox, E. M., & Hill, J. P. (2017). Food safety for food security: Relationship between global megatrends and developments in food safety. *Trends in Food Science & Technology*, *68*, 160–175. DOI: 10.1016/j.tifs.2017.08.014

Koca, N., Erbay, Z., & Urgu Öztürk, M. (2022). Regulations and legislations on processed cheese. In *Processed Cheese Science and Technology* (pp. 449–490). Elsevier., DOI: 10.1016/B978-0-12-821445-9.00004-2

Kuppusamy, S., Meivelu, M., Praburaman, L., Mujahid Alam, M., & Al-Sehemi, A. G., & K, A. (2024). Integrating AI in food contaminant analysis: Enhancing quality and environmental protection. *Journal of Hazardous Materials Advances*, *16*, 100509. DOI: 10.1016/j.hazadv.2024.100509

Lee, H., & Yoon, Y. (2021). Etiological Agents Implicated in Foodborne Illness World Wide. *Food Science of Animal Resources*, *41*(1), 1–7. DOI: 10.5851/kosfa.2020.e75 PMID: 33506212

Lehotay, S. J., & Hajšlová, J. (2002). Application of gas chromatography in food analysis. *Trends in Analytical Chemistry*, *21*(9–10), 686–697. DOI: 10.1016/S0165-9936(02)00805-1

Li, X., Shen, X., Jiang, W., Xi, Y., & Li, S. (2024). Comprehensive review of emerging contaminants: Detection technologies, environmental impact, and management strategies. *Ecotoxicology and Environmental Safety*, *278*, 116420. DOI: 10.1016/j.ecoenv.2024.116420 PMID: 38701654

Liberty, J. T., & Bhuiyan, Md. H. R., & Ngadi, M. (2025). Advancements in rapid and non-destructive approaches for quality assessment of fried foods and frying oil. *Journal of Food Composition and Analysis*, *145*, 107839. DOI: 10.1016/j.jfca.2025.107839

Liguori, J., Trübswasser, U., Pradeilles, R., Le Port, A., Landais, E., Talsma, E. F., Lundy, M., Béné, C., Bricas, N., Laar, A., Amiot, M. J., Brouwer, I. D., & Holdsworth, M. (2022). How do food safety concerns affect consumer behaviors and diets in low- and middle-income countries? A systematic review. *Global Food Security*, *32*, 100606. DOI: 10.1016/j.gfs.2021.100606

Ling, Z., Yang, L., Zhang, W., Yao, T., & Xu, H. (2024). Detection of food contaminants: A review of established rapid analytical techniques and their applications and limitations. *Food Safety and Health*, *2*(1), 72–95. DOI: 10.1002/fsh3.12032

Malhotra, S. (2017). Food safety issues related to street vendors. In *Food Safety in the 21st Century* (pp. 395–402). Elsevier., DOI: 10.1016/B978-0-12-801773-9.00031-5

Manning, L., & Soon, J. M. (2014). Developing systems to control food adulteration. *Food Policy*, *49*, 23–32. DOI: 10.1016/j.foodpol.2014.06.005

Mellmann, A., Harmsen, D., Cummings, C. A., Zentz, E. B., Leopold, S. R., Rico, A., Prior, K., Szczepanowski, R., Ji, Y., Zhang, W., McLaughlin, S. F., Henkhaus, J. K., Leopold, B., Bielaszewska, M., Prager, R., Brzoska, P. M., Moore, R. L., Guenther, S., Rothberg, J. M., & Karch, H. (2011). Prospective Genomic Characterization of the German Enterohemorrhagic Escherichia coli O104:H4 Outbreak by Rapid Next Generation Sequencing Technology. *PLoS One*, *6*(7), e22751. DOI: 10.1371/journal.pone.0022751 PMID: 21799941

. Michael Alurame Eruaga. (2024). Enhancing global food safety standards through international collaboration and policy harmonization. *International Journal of Scholarly Research in Multidisciplinary Studies, 4*(1), 020–032. https://doi.org/ DOI: 10.56781/ijsrms.2024.4.1.0027

Modupalli, N., Naik, M., Sunil, C. K., & Natarajan, V. (2021). Emerging non-destructive methods for quality and safety monitoring of spices. *Trends in Food Science & Technology*, *108*, 133–147. DOI: 10.1016/j.tifs.2020.12.021

Nafea, A. M., Wang, Y., Wang, D., Salama, A. M., Aziz, M. A., Xu, S., & Tong, Y. (2024). Application of next-generation sequencing to identify different pathogens. *Frontiers in Microbiology*, *14*, 1329330. DOI: 10.3389/fmicb.2023.1329330 PMID: 38348304

Naseem, S., & Rizwan, M. (2025a). The role of artificial intelligence in advancing food safety: A strategic path to zero contamination. *Food Control*, *175*, 111292. DOI: 10.1016/j.foodcont.2025.111292

Naseem, S., & Rizwan, M. (2025b). The role of artificial intelligence in advancing food safety: A strategic path to zero contamination. *Food Control*, *175*, 111292. DOI: 10.1016/j.foodcont.2025.111292

Onyeaka, H., Jalata, D. D., & Mekonnen, S. A. (2023). Mitigating physical hazards in food processing: Risk assessment and preventive strategies. *Food Science & Nutrition*, *11*(12), 7515–7522. DOI: 10.1002/fsn3.3727 PMID: 38107102

Osuoha, J. O., Anyanwu, B. O., & Ejileugha, C. (2023). Pharmaceuticals and personal care products as emerging contaminants: Need for combined treatment strategy. *Journal of Hazardous Materials Advances*, *9*, 100206. DOI: 10.1016/j.hazadv.2022.100206

Pacholczyk-Sienicka, B. (2024). Crimes Against Food: Characteristics, Health Risk, and Regulations. *Food and Energy Security*, *13*(5), e70002. DOI: 10.1002/fes3.70002

Pakdel, M., Olsen, A., & Bar, E. M. S. (2023). A Review of Food Contaminants and Their Pathways Within Food Processing Facilities Using Open Food Processing Equipment. *Journal of Food Protection*, *86*(12), 100184. DOI: 10.1016/j.jfp.2023.100184 PMID: 37865163

Paramasivam, A., Murugan, R., Jeraud, M., Dakkumadugula, A., Periyasamy, R., & Arjunan, S. (2024). Additives in Processed Foods as a Potential Source of Endocrine-Disrupting Chemicals: A Review. *Journal of Xenobiotics*, *14*(4), 1697–1710. DOI: 10.3390/jox14040090 PMID: 39584955

Pei, X., Tandon, A., Alldrick, A., Giorgi, L., Huang, W., & Yang, R. (2011). The China melamine milk scandal and its implications for food safety regulation. *Food Policy*, *36*(3), 412–420. DOI: 10.1016/j.foodpol.2011.03.008

Poole, C. F. (2003). The Column in Liquid Chromatography. In *The Essence of Chromatography* (pp. 267–429). Elsevier., DOI: 10.1016/B978-044450198-1/50017-3

Rafael Audino, Z., & Isabella Montenegro, B. (2022). Molecular Techniques for identification applied to food: A review. *International Journal of Agricultural Science and Food Technology*, *8*(4), 305–315. DOI: 10.17352/2455-815X.000182

Ranbir, , Kumar, M., Singh, G., Singh, J., Kaur, N., & Singh, N. (2022). Machine Learning-Based Analytical Systems: Food Forensics. *ACS Omega*, *7*(51), 47518–47535. DOI: 10.1021/acsomega.2c05632 PMID: 36591133

Roos, F. H. B. L. R., Yusof, M. I. N. B. M., & Nizar, N. N. A. (2018). Conventional and modern standards of food production. In *Preparation and Processing of Religious and Cultural Foods* (pp. 295–307). Elsevier., DOI: 10.1016/B978-0-08-101892-7.00015-8

Roy, S., & Chakravorty, A. (2025). Food Forensics. In Sarkar, T., Lahiri, D., Nag, M., & Bhattacharya, D. (Eds.), *Biofilm Applications to Revolutionize Food Technology* (pp. 321–374). Springer Nature Switzerland., DOI: 10.1007/978-3-031-85205-3_15

Saadat, S., Pandya, H., Dey, A., & Rawtani, D. (2022). Food forensics: Techniques for authenticity determination of food products. *Forensic Science International, 333*, 111243. DOI: 10.1016/j.forsciint.2022.111243 PMID: 35240505

Salihah, N. T., Hossain, M. M., Lubis, H., & Ahmed, M. U. (2016). Trends and advances in food analysis by real-time polymerase chain reaction. *Journal of Food Science and Technology, 53*(5), 2196–2209. DOI: 10.1007/s13197-016-2205-0 PMID: 27407185

Sharma, G., Pawar, R. P., Athar, M., & Bose, D. (2017). Use of geru powder as a alternative powder in powder dusting method for developing latent fingerprints on various non-porous smooth surfaces. [Scopus.]. *Ind. Police J., 64*, 17–32.

Singh, R., & Puniya, A. K. (2024). Role of Food Safety Regulations in Protecting Public Health. *Indian Journal of Microbiology, 64*(3), 1376–1378. DOI: 10.1007/s12088-024-01240-7 PMID: 39282191

Socas-Rodríguez, B. (2022). Applications of Chromatographic Techniques in Food and Environmental Analysis. *Separations, 9*(12), 418. DOI: 10.3390/separations9120418

Soroushianfar, M., Asgari, G., Afzali, F., Falahat, A., Mansoor Baghahi, M. S., Haratizadeh, M. J., Khalili-Tanha, G., & Nazari, E. (2025). Application of Bioinformatics and Machine Learning Tools in Food Safety. *Current Nutrition Reports, 14*(1), 67. DOI: 10.1007/s13668-025-00657-w PMID: 40388006

Sun, Q., Dong, Y., Wen, X., Zhang, X., Hou, S., Zhao, W., & Yin, D. (2023). A review on recent advances in mass spectrometry analysis of harmful contaminants in food. *Frontiers in Nutrition, 10*, 1244459. DOI: 10.3389/fnut.2023.1244459 PMID: 37593680

Thorsen, M., Hill, J., Farber, J., Yiannas, F., Rietjens, I. M. C. M., Venter, P., Lues, R., & Bremer, P. (2025). Megatrends and emerging issues: Impacts on food safety. *Comprehensive Reviews in Food Science and Food Safety, 24*(3), e70170. DOI: 10.1111/1541-4337.70170 PMID: 40183602

Van Der Fels-Klerx, H. J., Camenzuli, L., Belluco, S., Meijer, N., & Ricci, A. (2018). Food Safety Issues Related to Uses of Insects for Feeds and Foods. *Comprehensive Reviews in Food Science and Food Safety, 17*(5), 1172–1183. DOI: 10.1111/1541-4337.12385 PMID: 33350154

Visciano, P., & Schirone, M. (2021). Food frauds: Global incidents and misleading situations. *Trends in Food Science & Technology*, *114*, 424–442. DOI: 10.1016/j.tifs.2021.06.010

WHO Global Strategy for Food Safety 2022-2030: Towards Stronger Food Safety Systems and Global Cooperation. (2022). 1st ed.). World Health Organization.

Winkler, R. (2022). Challenges and Trends of Mass Spectrometry for Food Analysis. In L. M. L. Nollet & R. Winkler, *Mass Spectrometry in Food Analysis* (1st ed., pp. 419–430). CRC Press. https://doi.org/DOI: 10.1201/9781003091226-27

Zahir, A., Ge, Z., & Khan, I. A. (2025). Public Health Risks Associated with Food Process Contaminants – A Review. *Journal of Food Protection*, *88*(2), 100426. DOI: 10.1016/j.jfp.2024.100426 PMID: 39643160

Ziani, I., Bouakline, H., El Guerraf, A., & El Bachiri, A., Marie-Laure Fauconnier, & Sher, F. (2025). Integrating AI and advanced spectroscopic techniques for precision food safety and quality control. *Trends in Food Science & Technology*, *156*, 104850. DOI: 10.1016/j.tifs.2024.104850

Chapter 4
Detection of Microbial Contaminants in Food and Food Products

Matthew Chidozie Ogwu
https://orcid.org/0000-0001-6054-1667
Appalachian State University, USA

Tonjock Rosemary Kinge
https://orcid.org/0000-0002-5402-1021
University of Bamenda, Cameroon

Soumia El Malahi
Ibn Zohr University, Agadir, Morocco

Fredrick Ojija
https://orcid.org/0000-0002-1117-5119
Mbeya University of Science and Technology, Tanzania

ABSTRACT

The detection of microbial contaminants in food and food products is a cornerstone of public health protection and food safety assurance. As foodborne diseases continue to pose a global burden, with pathogens such as Salmonella, Escherichia coli, Listeria monocytogenes, and norovirus accounting for millions of illnesses annually, the need for robust and reliable detection methodologies has become increasingly urgent. This chapter provides a comprehensive overview of the evolving landscape of microbial detection in food systems. It begins by exploring the sources and pathways of microbial contamination across the "farm-to-fork" continuum, highlighting critical control points and microbial risk factors. Emphasis is placed on sampling strategies, including representative sampling, sample preparation, and enrichment

DOI: 10.4018/979-8-3373-3982-5.ch004

protocols, which form the foundation of accurate microbial detection. The chapter then examines diverse detection strategies, including culture-based methods, immunological assays (such as enzyme-linked immunosorbent assay and lateral flow tests), and molecular techniques like polymerase chain reaction (PCR), quantitative PCR, loop-mediated isothermal amplification, and next-generation sequencing. Emerging technologies such as biosensors, Clustered Regularly Interspaced Short Palindromic Repeats-based diagnostics, and metagenomics are also discussed for their potential to enhance sensitivity, specificity, and rapidity in pathogen detection. Each technique is assessed in terms of sensitivity, specificity, operational feasibility, and its integration into food safety risk management frameworks. Special attention is given to validation standards, harmonization efforts, and the challenges of deploying these technologies in low-resource settings. The chapter concludes by identifying emerging trends, such as artificial intelligence-assisted detection and portable diagnostics, which hold promise for revolutionizing microbial monitoring in food systems. By bridging microbiological principles with practical applications and regulatory contexts, offering critical insights for researchers, food safety practitioners, and policymakers.

1.0. INTRODUCTION

The detection of microbial contaminants in food is crucial in ensuring public health and food safety. Foodborne pathogens, including *Salmonella, Staphylococcus, Escherichia coli*, norovirus, and *Listeria,* are significant contributors to global health issues, with approximately 600 million illnesses reported annually due to contaminated food (Havelaar et al., 2015; Akkina et al., 2023). The World Health Organization (WHO) highlights that a substantial proportion of foodborne disease burdens can be attributed to animal-source foods, including meats, eggs, and dairy products (Akkina et al., 2023). The continuous rise in foodborne illnesses is often linked to various lapses in food safety practices across the food production and supply chain (Lee and Yoon, 2021).

Detecting microbial pathogens in food may rely on a spectrum of methods ranging from classical microbiological techniques to innovative molecular and sensor-based technologies (Figure 1). Traditional or classical methods, such as culture-based techniques and biochemical tests, remain foundational but are often time-consuming and labor-intensive. Advances in technology have introduced faster, more sensitive approaches, including immunoassays (e.g., Enzyme-Linked Immunosorbent Assay [ELISA]), molecular techniques (e.g., Polymerase Chain Reaction [PCR], Clustered Regularly Interspaced Short Palindromic Repeats [CRISPR]), mass spectrometry (e.g., Matrix-Assisted Laser Desorption/Ionization-Time of Flight Mass Spectrometry

[MALDI-TOF MS]), biosensors, and hyperspectral imaging. These modern tools enhance detection accuracy and enable real-time monitoring, thereby supporting more effective prevention and control of foodborne illnesses.

Figure 1. Traditional approaches and cutting-edge technologies for detecting microbial pathogens in food.

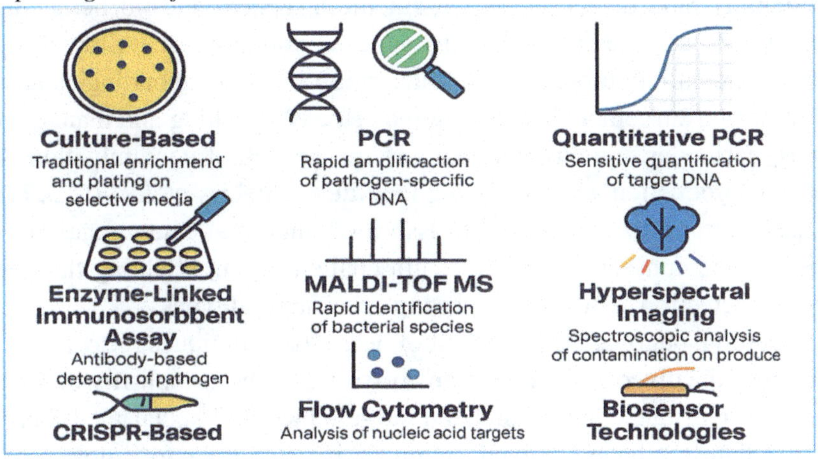

Source: *Elbehiry et al. (2025)*

Furthermore, the connection between foodborne illness and effective detection strategies cannot be overstated. Robust detection techniques are essential for identifying the presence of microbial pathogens and preventing outbreaks. Epidemiological data indicate that food contaminated by pathogens results in significant health burdens, with estimates reaching millions of illnesses and thousands of deaths annually in regions such as the United States (WHO, 2015; Bintsis, 2017; US CDC, 2025). The United States Centers for Disease Control and Prevention (CDC) estimates that approximately 48 million episodes of foodborne illness occur yearly in the U.S., emphasizing the importance of effective surveillance and detection strategies to manage food safety (Scallan et al., 2011a; b). The implications of inadequately managed microbial contamination extend beyond individual health risks to encompass broader public health challenges and economic costs. The annual economic burden related to foodborne illnesses in the U.S. alone is estimated to range from $14 to $60 billion (Hu et al., 2022). Thus, there is a critical need to continually advance detection technologies not only to identify contaminants but also to understand emerging patterns in foodborne illness outbreaks (Schomberg et al., 2016).

Modern food safety practices rely heavily on advanced detection methodologies, which incorporate various techniques, including multiplex polymerase chain reaction (PCR), bioinformatics, and whole-genome sequencing. For instance, significant advancements in multiplex real-time PCR have enhanced the specificity and sensitivity of pathogen detection, enabling the simultaneous screening of multiple foodborne pathogens from food samples (Kawase et al., 2016). This method has proven valuable in analyzing outbreaks and identifying contamination sources rapidly, as evidenced by its application in monitoring fecal samples during foodborne illness investigations (Oh et al., 2012).

Additionally, rapid detection methods, including genetic and biochemical assays, have facilitated early identification of microbial contaminants, allowing quicker responses to potential outbreaks. The integration of these technologies has transformed pathogen surveillance, enabling the effective tracking and management of epidemics, particularly in environments where the risk of contamination is high, such as meat production and processing facilities (Elbehiry et al., 2022). Elbehiry et al. (2022) emphasize the persistent issues of microbial contamination in meat products and the consequent public health challenges, highlighting the necessity for effective detection strategies tailored to specific food types.

Moreover, the ongoing advancement of detection techniques reflects a growing recognition of the diverse modes of transmission and the complexity of foodborne pathogens. For instance, environmental surveillance and social media-based reporting mechanisms have emerged as innovative strategies to supplement traditional inspection methods. These new approaches enhance detection capabilities by capturing real-time data on food safety incidents, allowing public health officials to respond swiftly to emerging threats (Schomberg et al., 2016). The interplay between food safety practices and detection strategies is evident in the management of foodborne illnesses. The multifaceted nature of food contamination, which can occur at various stages from production to consumption, necessitates comprehensive surveillance systems that integrate detection methods across the entire food supply chain (Havelaar et al., 2015). For example, the integration of consumer reports and health professional feedback into public health databases provides valuable insights into patterns of foodborne illness, directly informing the development of targeted surveillance and detection efforts (Arendt et al., 2013).

Understanding the primary sources of contamination is crucial for designing effective detection strategies. Epidemiological data from the United States Centers for Disease Control and Prevention (CDC) indicate that certain foods, including leafy greens, dairy products, and poultry, are frequently implicated in foodborne illness outbreaks (Scallan et al., 2011a;b). Targeting these food sources with appropriate detection protocols can significantly mitigate the risk of widespread illness. Moreover, technological advancements, such as the use of social media data to monitor foodborne disease trends, have shown promising results in enhancing public engagement and response capabilities (Schomberg et al., 2016). Efforts to improve detection methodologies also play a crucial role in enhancing public awareness of food safety issues. As consumers become increasingly informed about potential risks associated with their food, effective communication of detection results and food safety information becomes vital (Mayer et al., 2023). There is a higher chance of adherence to recommended food handling and preparation protocols when food safety procedures are more transparent and trustworthy, ultimately reducing the prevalence of foodborne illnesses (Mahunu et al., 2024; Ogwu et al., 2024).

This chapter aims to provide a comprehensive overview of the principles, methods, and emerging technologies used in detecting microbial contaminants in food and food products. It explores traditional and modern detection approaches ranging from culture-based methods to molecular diagnostics, biosensors, and next-generation sequencing, highlighting their applications, advantages, and limitations within various food contexts. The chapter connects fundamental microbiology with practical food safety management by incorporating practical factors such as sampling plans, quality control procedures, and regulatory compliance. It broadens our understanding by elucidating how advancements in digital technologies and rapid diagnostic tools are transforming microbiological surveillance and enabling more proactive responses to contamination events.

2.0. PATHWAYS OF MICROBIAL CONTAMINATION IN THE FOOD CHAIN

Microbial contamination in the food chain is a critical public health concern that impacts food safety and quality at multiple stages, from farm production to consumer consumption. Contamination may occur through a variety of pathways, with specific microorganisms linked to particular sources, each posing distinct risks. These routes are essential for designing effective interventions and mitigating the incidence of foodborne illnesses.

2.1. Sources of Contamination in the Food Chain

The food supply chain is vulnerable to microbial contamination at critical control points: the farm, processing facilities, retail operations, and domestic environments (Table 1). At the farm level, contamination arises predominantly from fecal matter, soil, and water, especially during the irrigation of produce and animal husbandry (Ravaliya et al., 2014; El-Khishin et al., 2017). For example, environmental contamination from fecal sources can introduce pathogens such as *E. coli* and *Salmonella* to crops during growth or harvesting, directly influencing produce safety (Ravaliya et al., 2014; El-Khishin et al., 2017).

In processing environments, pathogens can be transferred from raw ingredients to food contact surfaces, such as equipment and utensils, thereby facilitating cross-contamination (Zhu et al., 2014; Malavi et al., 2018). Studies have shown that food contact surfaces can harbor harmful bacteria, such as Listeria monocytogenes, which can survive and proliferate on various materials used in food preparation, like stainless steel (Zhu et al., 2014; Zoellner et al., 2018). These surfaces can then contaminate ready-to-eat foods during slicing or packaging processes, leading to significant health risks for consumers.

Retail environments further contribute to microbial contamination through improper food handling practices, inadequate refrigeration, and a lack of sanitization of surfaces and equipment (Krishnasree et al., 2018; Bedane et al., 2022). Improperly cleaned deli slicers, for example, can serve as vehicles for transferring pathogens from contaminated foods to other products, compounding the risk of foodborne illness outbreaks (Koo et al., 2012). Additionally, consumer kitchens often become sites of contamination if safe food handling practices are not followed, highlighting the need for education on food safety (Stavropoulou and Bezirtzoglou, 2019).

Microbial contamination not only poses health risks but also disrupts international food trade and imposes substantial economic burdens due to recalls, import bans, and reduced consumer confidence. Contaminated exports can lead to border rejections and reputational damage for producers, particularly in low- and middle-income countries.

Table 1. Sources of Microbial Contamination in the Food Chain

Source Category	Type of Microbial Contaminants	Examples
Agricultural Inputs	Bacteria, Fungi, Viruses, and Parasites	Use of untreated wastewater, contaminated soil, animal feces, organic fertilizers, and pesticides
Animal Husbandry	Zoonotic Pathogens (e.g., *Salmonella*, *E. coli*)	Infected livestock, cross-contamination during milking or slaughter
Harvesting Practices	Bacteria, and Spoilage Microorganisms	Poor hygiene of harvest tools and hands, dirty containers
Processing Facilities	Listeria, *Clostridium*, Yeasts, and Molds	Poor sanitation, biofilms in equipment, condensation, and temperature abuse
Water Supply	Enteric Pathogens, and Protozoa	Use of contaminated water in washing, cooling, or processing
Storage and Packaging	Molds, and Spoilage Bacteria	Improper refrigeration, inadequate packaging, and damaged containers

continued on following page

Table 1. Continued

Source Category	Type of Microbial Contaminants	Examples
Food Transportation	Bacteria (e.g., *Listeria monocytogenes*)	Cross-contamination from dirty trucks or mixed loads
Retail and Markets	Enteric Bacteria, Viruses	Exposure to open air, handling by multiple individuals, and unhygienic surfaces
Consumer Handling	*Campylobacter*, *E. coli*, Norovirus	Inadequate cooking, improper storage, and poor kitchen hygiene

2.2. Microorganisms of Concern

The microorganisms implicated in foodborne illnesses include a wide array of bacteria, viruses, parasites, and fungi. Bacterial pathogens such as *Salmonella, Escherichia coli* O157:H7, *Campylobacter, and Listeria monocytogenes* are particularly notable due to their association with severe health outcomes (El-Khishin et al., 2017; Teklemariam et al., 2023). Table 2 presents some of the most common foodborne microbes, their typical sources, and the health effects they may cause. This information provides a foundation for developing targeted surveillance, preventive strategies, and response mechanisms in food safety management. These pathogens induce gastrointestinal distress and can lead to more serious complications, such as kidney failure and meningitis, underscoring the importance of stringent food safety measures throughout the food supply chain (El-Khishin et al., 2017; Rabiee et al., 2024).

Viruses such as norovirus and hepatitis A are primarily transmitted through contaminated water and food, highlighting another aspect of microbial contamination (Casado et al., 2024). Parasites such as *Cryptosporidium* and *Giardia* can be introduced into the food chain through animal feces, often through contaminated water sources used for irrigation or cleaning (Ravaliya et al., 2014). Also, some fungi, particularly molds that produce mycotoxins, can contaminate grains and other agricultural products, posing additional health risks during consumption (Casado et al., 2024).

Table 2. Common microbes that contaminate food and food products

Microorganism	Type	Commonly Contaminated Foods	Health Effects / Diseases
Salmonella spp.	Bacteria	Poultry, eggs, meat, dairy, and vegetables	Salmonellosis (fever, diarrhea, abdominal cramps)
Escherichia coli (E. *coli*) O157:H7	Bacteria	Undercooked beef, raw milk, and vegetables	Hemorrhagic colitis, Hemolytic Uremic Syndrome (HUS)
Listeria monocytogenes	Bacteria	Deli meats, unpasteurized cheese, smoked fish	Listeriosis (fever, meningitis, miscarriage in pregnancy)
Campylobacter jejuni	Bacteria	Raw poultry, milk, and untreated water	Gastroenteritis, Guillain-Barré syndrome
Clostridium botulinum	Bacteria (spore-forming)	Canned foods, fermented fish, and honey	Botulism (paralysis, respiratory failure)
Staphylococcus aureus	Bacteria	Improperly stored meats, dairy, and pastries	Food poisoning (nausea, vomiting, abdominal cramps)
Norovirus	Virus	Shellfish, salads, contaminated water	Viral gastroenteritis (diarrhea, vomiting)
Hepatitis A virus	Virus	Raw shellfish, fresh produce, and infected handlers	Hepatitis A (liver inflammation, jaundice)
Toxoplasma gondii	Parasite	Undercooked pork, contaminated soil	Toxoplasmosis (flu-like symptoms, risk in pregnancy)
Cryptosporidium parvum	Parasite	Contaminated water, unwashed produce	Cryptosporidiosis (watery diarrhea, stomach cramps)
Aspergillus flavus	Fungi (Mold)	Grains, nuts (e.g., peanuts, maize)	Aflatoxicosis (liver damage, carcinogenic effects)
Penicillium spp.	Fungi (Mold)	Spoiled fruits, bread, and cheeses	Mycotoxicosis (allergic reactions, possible toxicity)
Bacillus cereus	Bacteria (spore-forming)	Cooked rice, pasta, and sauces	Food poisoning (emetic and diarrheal syndromes)

2.3. Factors Influencing Microbial Presence and Growth

Multiple factors influence the presence and growth of microorganisms in the food environment. These include intrinsic factors such as pH, moisture content, temperature, and nutrient availability, as well as extrinsic factors like the surrounding environment, processing methods, and storage conditions (Karuppuchamy et al., 2024). For instance, the moisture content of food is a crucial determinant of microbial growth; high-moisture foods, such as dairy and meat products, are particularly conducive to the proliferation of pathogens (Karuppuchamy et al., 2024). Conversely, low-moisture foods can harbor pathogens that survive longer under these conditions, complicating control measures (Finn et al., 2013; Chime et al., 2016; Karuppuchamy et al., 2024).

Temperature also plays a significant role, as different microorganisms thrive at varying temperatures. The "danger zone" (between 40°F and 140°F) is particularly critical, as it is where bacteria multiply rapidly. Therefore, maintaining proper temperature control during storage, processing, and preparation is essential for minimizing microbial risks (Karuppuchamy et al., 2024; Rabiee et al., 2024). Climate change is increasingly recognized as a factor that alters microbial ecology in the food chain. Rising temperatures, extreme weather events, and water scarcity may exacerbate contamination risks by affecting pathogen survival and distribution in soil, water, and animal reservoirs. While antimicrobial agents are widely used to enhance food safety, their misuse or overuse may contribute to the emergence of antimicrobial-resistant strains, posing long-term threats to food systems. (Murphy et al., 2018; Teklemariam et al., 2023). Environmental conditions such as humidity and the presence of competing microbial flora can influence microbial populations in food (Ogwu et al., 2013; Osawaru et al., 2013; Ogwu and Osawaru, 2014; Oliveira et al., 2020). Cross-contamination, which occurs when pathogens are transferred from one food or surface to another, can significantly raise the microbial burden on food products. Mechanisms such as lapses in cleaning protocols and improper food handling contribute to cross-contamination risks (Karanth et al., 2023). Socioeconomic disparities in food handling infrastructure, sanitation, and public awareness also contribute to microbial contamination risks, especially in low-resource environments. Inadequate training among food handlers and limited access to cold chain logistics increase the vulnerability of certain populations to foodborne illnesses.

Figure 2. Intrinsic and extrinsic factors affecting microbial growth in food products

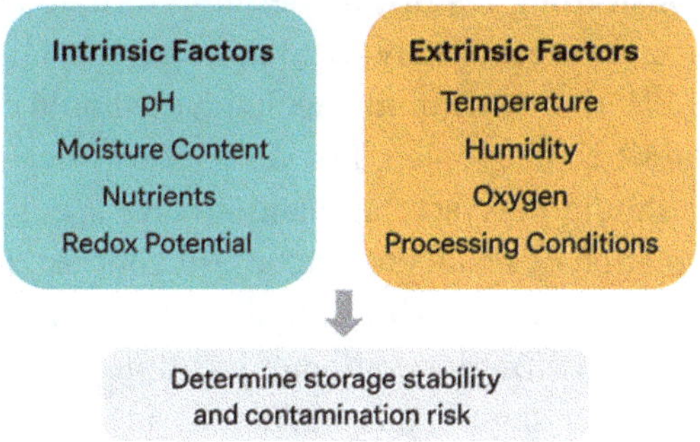

3.0. MICROBES IN FOOD: SAMPLING STRATEGIES AND SAMPLE PREPARATION

Microbial contamination of food poses significant public health risks, underscoring the crucial need for representative sampling strategies, effective sample collection methods, and pretreatment and enrichment procedures. These components are crucial for detecting the presence and concentration of foodborne pathogens in various food sources. Figure 2 outlines the sequential steps involved in detecting microbial contaminants in food products. The process begins with representative sampling to ensure that the food sample accurately reflects the batch or environment. Following this, appropriate sample collection methods are employed to maintain the specimen's integrity. Pre-treatment and enrichment steps are then used to enhance the growth or detectability of target microorganisms. Detection is achieved through culture-based techniques for isolation and identification, followed by molecular-based methods for confirmation and high-sensitivity diagnostics. This integrated workflow is critical for ensuring food safety and protecting public health.

3.1. Principles of Representative Sampling

Adequate representative sampling in food microbiology is crucial for ensuring the accurate detection and quantification of pathogens and spoilage microorganisms. The principles of representative sampling determine that samples must accurately reflect the entire batch of food being tested. This is crucial given the heterogeneity of microbial distribution within food matrices, which can vary significantly due to factors such as food type, surface area, and processing methods (Scallan et al., 2011; Cauteren et al., 2017; Pakbin et al., 2023). Historically, non-representative samples have led to misleading results, hindering efforts to assess the risk posed by pathogens present in food accurately (Scallan et al., 2011).

Furthermore, the selection of sampling techniques should consider the size and nature of the food items being sampled. For solid foods, such as meats, fruits, and vegetables, sampling should be systematic and involve specific units to allow for more accurate extrapolation of results to batches. In contrast, for liquids and semi-solid products, a methodical composite sampling approach should be employed, capturing variability across different container locations (Lanzas et al., 2011; Tao et al., 2022). Notably, larger sample sizes generally increase the reliability of the results, demonstrating that a balance must be struck between practicality and statistical validity (Troja et al., 2024).

Figure 3. Workflow for Detection of Microbial Contaminants in Food Samples

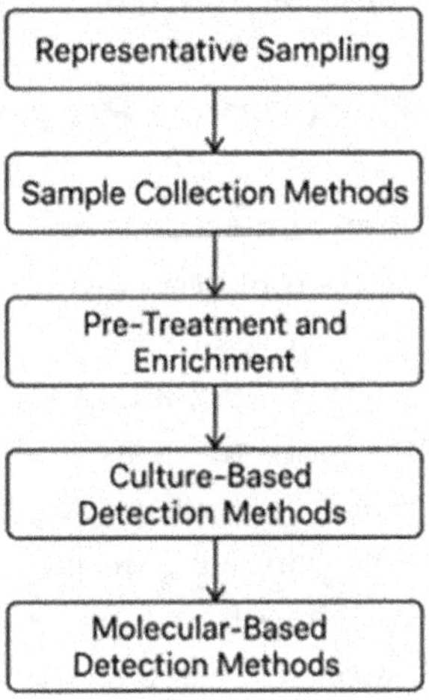

3.2. Sample Collection Methods

The detection of microbial contaminants in food products requires a systematic approach to ensure accuracy and reliability. Figure 2 outlines the essential steps in the detection pipeline, starting from representative sampling and progressing through sample collection, enrichment, and both culture-based and molecular-based detection methods. Each step is critical to ensuring the integrity of the results and the safety of food products. Sample collection methods differ between solid, liquid, and swab samples, each presenting unique challenges and considerations. For solid food samples, aseptic techniques must be employed, wherein food matrices are excised and aggressively mixed to form a composite sample (Jokerst et al., 2012; Amalar-adjou and Bhunia, 2012). For instance, comminuted products, such as ground meat, require mixing from various parts of the batch to ensure that bacterial populations are accurately represented. Liquid samples, particularly those from beverages or processed foods, require mixing before collection to prevent bacterial stratification caused by density differences (Neff et al., 2019).

Swabs serve as a critical method for sampling surfaces and equipment that come into contact with food, particularly in bacteriological studies designed to assess hygiene practices in food establishments (Zhang et al., 2023; Kant et al., 2018). The efficacy of swabbing relies significantly on the technique's adherence to protocols, such as consistently wiping surfaces in a specified geometric pattern to maximize the recovery of microorganisms (Sriramulu, 2024). The choice of swabbing material can also impact the results, as materials such as premoistened swabs are preferred over dry applications, which may inhibit microbial recovery (Pava-Ripoll et al., 2015).

3.3. Pre-Treatment and Enrichment Procedures

Once samples are collected, pretreatment and enrichment processes are paramount for isolating pathogens from the food matrix. Pre-treatment typically involves homogenization, where samples are blended to facilitate the more efficient extraction of microorganisms (Kim et al., 2021; Wei and Zhao, 2018). This step is crucial, as many pathogens exist within biofilms or as viable but non-culturable forms, both of which can evade detection unless properly processed (Zhang et al., 2021). For example, culturing foodborne pathogens in laboratory settings often requires enrichment to increase the pathogen load to detectable levels (Xiao et al., 2022).

Enrichment involves incubation in selective media that favors the growth of specific pathogens while inhibiting the growth of non-target organisms. This methodology is vital for detecting organisms such as Salmonella or Listeria, which can be outcompeted by the dominant microbiota present in food samples (Vanegas et al., 2017). Techniques such as PCR and next-generation sequencing facilitate the identification process, enabling researchers to overcome common limitations associated with classical culturing techniques by directly amplifying nucleic acids from target pathogens (Lee et al., 2023).

Importantly, microbial isolation and identification methods must adapt to emerging challenges such as antibiotic resistance. The application of rapid sequencing techniques and metagenomics enables the comprehensive characterization of microbial communities and the detection of resistance genes within food, thereby enhancing the efficacy of surveillance and control strategies (Beuchat et al., 2013). Implementing these strategies not only facilitates accurate pathogen identification but also informs public health interventions aimed at reducing the incidence of foodborne illnesses. Given the complexities of food matrices and microbial interactions, continuous methodological advancements and adherence to best practices in sampling and detection are essential to safeguard food safety globally.

4.0. CULTURE-BASED DETECTION METHODS FOR MICROBES IN FOOD

Culture-based detection methods remain foundational in food microbiology due to their reliability, simplicity, and ability to isolate viable pathogens (Izah et al., 2024). These methods involve the use of selective and differential media tailored to support the growth of target microorganisms while inhibiting the growth of other microorganisms. Incubation conditions such as temperature, time, and atmospheric environment are optimized for each microbial group (Mahunu et al., 2024). Table 3 summarizes commonly used culture media, their incubation parameters, and typical applications in food safety monitoring, highlighting their continued relevance in identifying microbial contaminants in various food matrices.

Table 3. Culture Media and Conditions for Identifying Foodborne Microorganisms

Microorganism	Selective/Differential Media	Incubation Conditions	Application in Food Testing
Escherichia coli (*E. coli*)	Eosin Methylene Blue (EMB), MacConkey	37°C, 24 hours, aerobic	Detection in raw meats, dairy, and water
Salmonella spp.	Xylose Lysine Deoxycholate (XLD), SS agar	35–37°C, 24–48 hours, aerobic	Poultry, eggs, dairy, and ready-to-eat products
Listeria monocytogenes	PALCAM, Oxford agar	30°C, 24–48 hours, microaerophilic	Ready-to-eat meats, soft cheeses, and produce
Staphylococcus aureus	Baird-Parker agar	35–37°C, 24–48 hours, aerobic	Dairy products, cooked meats, and bakery items
Clostridium perfringens	Tryptose Sulfite Cycloserine (TSC)	45°C, 24 hours, anaerobic	Cooked meats, gravies, and stews
Campylobacter jejuni	Campy-BAP, Skirrow agar	42°C, 48 hours, microaerophilic	Poultry, raw milk, and water
Bacillus cereus	Mannitol Egg Yolk Polymyxin (MYP)	30°C, 24 hours, aerobic	Rice dishes, dairy, and processed foods
Vibrio spp.	Thiosulfate Citrate Bile Salts Sucrose (TCBS)	35°C, 18–24 hours, aerobic	Seafood, especially raw shellfish
Yeasts and Molds	Potato Dextrose Agar (PDA), DRBC agar	25–28°C, 3–5 days, aerobic	Fruits, baked goods, dairy, and fermented products
Shigella spp.	Hektoen Enteric (HE) agar	35–37°C, 24 hours, aerobic	Leafy greens, salads, and water

Culture-based detection methods, including traditional plate counts and the Most Probable Number (MPN) technique, remain essential in food microbiology despite advancements in molecular techniques. Each method has its advantages and limitations, which shape its applicability in food safety monitoring and microbial ecology studies. Traditional plate counts involve isolating microorganisms from food samples on selective agar media, followed by incubation and counting the colonies. This method is considered the 'gold standard' for microbial detection due to its ability to identify viable bacteria, especially in complex food matrices (Schultz et al., 2012; Rohde et al., 2015; Kim and Kim, 2020). While straightforward, plate counts can be time-consuming and labor-intensive, typically requiring 24 to 72 hours for results (Schultz et al., 2012; Margot et al., 2013). Additionally, this technique may overlook significant pathogens classified as viable but non-culturable (VBNC), leading to underestimations of microbial counts (Kim and Kim, 2020; Yang et al., 2020).

The MPN technique is a statistical method used to estimate the concentration of viable microorganisms in a sample by performing serial dilutions and subsequent incubation to evaluate growth (Line et al., 2011; Crowley et al., 2010). Also, MPN has been enhanced by automated systems, such as TEMPO, allowing for rapid and reliable quantification of microorganisms, like *Escherichia coli*, in various food products (Katase and Tsumura, 2011; Lee et al., 2021). Automated MPN reduces labor and minimizes human error associated with manual dilutions and counting, making it particularly attractive in high-throughput laboratory settings (Line et al., 2011). However, like traditional culture methods, the MPN technique is not immune to the challenges posed by VBNC microorganisms, which can skew results (Yang et al., 2020).

Both traditional plating and MPN techniques have demonstrated effectiveness in detecting key foodborne pathogens such as *Salmonella spp.* and *E. coli* in diverse food matrices, including poultry and dairy products (Schultz et al., 2012; Margot et al., 2013). These methods also facilitate the investigation of epidemiological trends and microbial load assessments, which are critical in food safety protocols. However, traditional approaches have limitations regarding the time required for analysis and the dependency on skilled personnel for interpretation (Afshari et al., 2012; Shangguan et al., 2023).

Despite significant improvements in sensitivity and speed, molecular methods like PCR and isothermal amplification are often employed in tandem with traditional techniques to enhance detection capabilities (Tongphrom et al., 2018; Yordanova and Andonova, 2024). For instance, combining MPN with PCR can enhance sensitivity and specificity, resulting in improved outcomes in detecting foodborne pathogens during routine quality control measures (Crowley et al., 2010; Lee et al., 2021). While these molecular techniques have advanced the field, they may come with higher costs and often require specialized training, making them less accessible to specific laboratories, particularly in resource-limited settings (Yordanova and Andonova, 2024; Schultz et al., 2012).

Furthermore, alternative detection methods such as loop-mediated isothermal amplification (LAMP) offer potential improvements in pathogen detection in food products. Also, LAMP is known for its rapidity and sensitivity, capable of detecting pathogens in as little as 40 minutes (Yang et al., 2016). This trend toward molecular techniques reflects the growing demand for faster analytical methods in response to the rising incidence of foodborne illnesses, where timely detection is crucial (Tongphrom et al., 2018). In reconciling these methodologies, food safety laboratories must adopt a multipronged approach that incorporates both traditional culture-based methods and rapid molecular techniques. Understanding the strengths and limitations of each approach enables practitioners to design testing protocols that optimize microbial detection while conserving resources and time (Margot et al., 2013; Yang et al., 2020). Moreover, integrating cultural and molecular methods can provide a comprehensive view of microbial populations and dynamics within food environments, particularly in understanding the development of antibiotic resistance in foodborne pathogens (McLain et al., 2016; Rohde et al., 2015).

5.0. IMMUNOLOGICAL DETECTION TECHNIQUES FOR MICROBES IN FOOD

Immunological methods are indispensable tools in food safety management, enabling the sensitive and specific detection of microbial pathogens. These techniques leverage antigen-antibody interactions to identify contaminants with high accuracy, even in complex food matrices.

5.1. Enzyme-linked immunosorbent assay (ELISA)

Among immunoassays, ELISA stands out for its sensitivity, specificity, and capability for high-throughput analysis. ELISA is commonly employed to detect a range of foodborne pathogens, including *Salmonella* spp. and *Escherichia coli* O157:H7 (Zhao et al., 2014; Law et al., 2015; Park et al., 2018). This technique relies on antigen-antibody interactions, whereby specific antibodies capture the target pathogen from food matrices, followed by a colorimetric or fluorescent signal that indicates the presence of the pathogen. For example, modified versions of ELISA have demonstrated efficacy in detecting pathogens in various complex food samples, meeting the critical criteria necessary for robust food safety monitoring (Zheng et al., 2019; Vargas et al., 2018).

5.2. Lateral flow devices and dipstick assays

Lateral flow devices and dipstick tests represent another category of immunological assays that offer rapid and user-friendly detection of pathogens in food samples. These point-of-care tests are characterized by their simplicity and speed, providing results within minutes without the need for elaborate laboratory setups. The functionality of these devices hinges on a similar antigen-antibody interaction found in ELISA; however, the format is designed to yield visual results through color changes or lines on a test strip. Although lateral flow assays excel in rapid diagnostics, their sensitivity can sometimes fall short compared to more elaborate methodologies such as ELISA, particularly in detecting low concentrations of pathogens (Park et al., 2020).

5.3. Immunomagnetic separation (IMS)

IMS is a powerful technique that utilizes magnetic beads coated with specific antibodies to isolate and concentrate target pathogens from complex mixtures commonly found in food matrices. This method enhances the detection sensitivity of pathogens such as Salmonella and E. coli and helps mitigate interference from non-target microorganisms typically present in food samples (Hsu et al., 2014; Wei et al., 2016). The application of IMS has shown promise in improving the efficiency of pathogen recovery during food analysis. For instance, the combination of IMS with molecular techniques, such as multiplex PCR (mPCR), has been utilized to streamline the detection process for multiple pathogens simultaneously. This integration facilitates thorough pathogen analysis in various food samples, ranging from poultry to dairy products, effectively reducing the time to obtain results and addressing regulatory demands and public health concerns (Castellví et al., 2010; Yoshitomi et al., 2012).

5.4. Emerging technologies and hybrid systems

The complexity of food matrices often hinders the accurate detection of pathogens due to the presence of inhibitors that affect both traditional culturing methods and PCR amplifications. One significant advantage of immunomagnetic separation is its ability to minimize these inhibitory effects, thus enhancing the recovery rates of foodborne pathogens. For example, when applied to dairy products, IMS has been demonstrated to significantly improve the detection of Staphylococcus aureus from food matrices, thereby increasing the reliability of pathogen monitoring in these environments (Wei et al., 2016). Moreover, the use of rapid protocols involving IMS has become increasingly relevant. As identified in studies, methods such as real-time PCR combined with IMS enable the rapid enumeration of pathogens, including *E. coli* O157:H7, highlighting the efficacy of this approach in food safety surveillance (Onmaz et al., 2013).

In addition to conventional methods, innovative advances in nanotechnology and biosensor development are expanding the landscape of immunological detection techniques. These advancements can significantly enhance sensitivity and specificity. For instance, the integration of nanoparticles in biosensors for pathogen detection is fostering novel platforms that allow rapid and accurate quantification of pathogens in various food environments. Researchers are increasingly utilizing magnetic nanoparticles to develop highly sensitive electrochemical biosensors, which, when combined with immunomagnetic separation, provide a powerful tool for the real-time detection and monitoring of foodborne pathogens (Wang et al., 2017). Such techniques improve the response times in identifying contamination, effectively addressing public health concerns, and ensuring food safety compliance.

Moreover, advanced spectroscopic methods, such as Fourier-transform infrared (FT-IR) spectroscopy, are being explored in conjunction with immunomagnetic separation to provide a multidimensional approach for detecting foodborne pathogens. Target pathogens can be concentrated using IMS, and FT-IR can then analyze spectral data to differentiate between various pathogen serovars effectively. This approach significantly reduces analysis times and enhances the specificity of microbial identification, demonstrating the versatility and efficacy of combining immunological techniques with advanced analytical methodologies in improving food safety (Castellví et al., 2010).

5.5. Challenges and outlook

Despite the significant progress in immunological detection techniques, challenges such as the search for novel antibodies and improving detection limits persist. Continuous research is crucial for developing antibodies that exhibit high specificity and affinity toward target pathogens, particularly as new strains of bacteria emerge. Furthermore, improving the detection limits of immunoassays remains critical for ensuring the safety of food products that may contain low levels of bacterial pathogens, which is common in various food processing environments (Foddai et al., 2010; Amoako et al., 2012; Li et al., 2010). The evolution of these techniques suggests a broader trend of utilizing integrative approaches, combining various detection modalities, for a more comprehensive assessment and management of foodborne pathogens.

I propose this table: Table 4 summarizes key immunological methods used in food safety diagnostics, comparing their detection speed, sensitivity, instrumentation requirements, and optimal use cases. It highlights how each technique contributes to pathogen identification across various settings, from field-level screening to laboratory-based confirmatory analysis.

Table 4. Comparative overview of immunological detection techniques for foodborne pathogens

Technique	Time to result	Sensitivity	Instrumentation	Best use cases
ELISA	2-4 hours	High	Moderate	Routine lab diagnostics
Lateral flow assays	5-20 minutes	Moderate	Minimal	On-site screening
IMS + PCR	3-5 hours	Very High	High	Confirmatory, multi-pathogen
IMS + FT-IR	<2 hours	High	Advanced	Strain-level identification
Nanoparticle biosensor	< 1 hour	Very High	Specialized	Real-time monitoring

6.0. MOLECULAR AND NUCLEIC ACID-BASED METHODS OF DETECTING MICROBES IN FOOD

Molecular methods have transformed the detection and characterization of microbial pathogens in food by enabling high-throughput, culture-independent analysis (Figure 3). Techniques such as metagenomics, metatranscriptomics, and whole genome sequencing (WGS) offer comprehensive insights into microbial diversity, strain-level identification, and functional activity. These methods analyze DNA or RNA directly from food samples using targeted or shotgun sequencing approaches, followed by bioinformatic analyses. This integrated workflow enables the rapid and precise identification of pathogens, including unculturable or low-abundance organisms, thereby enhancing food safety surveillance and outbreak investigations. Conventional PCR, quantitative PCR (qPCR), isothermal methods like LAMP (loop-mediated isothermal amplification) and RPA (recombinase polymerase amplification), and digital PCR (dPCR) offer efficient means to identify foodborne pathogens.

Conventional PCR remains a commonly utilized technique for microbial detection. Different strategies have been employed to enhance the sensitivity of PCR methods, such as pre-enrichment steps before testing food samples to increase microbial yield (Radji et al., 2010). In studies comparing PCR with conventional culturing methods for *Listeria monocytogenes*, PCR has demonstrated an improved detection rate, resulting in a more comprehensive assessment of food safety (Khan et al., 2014; Furmančíková et al., 2022). Furthermore, the integration of genetic markers into PCR has enabled differentiation among closely related microbial species, thereby improving specificity and accuracy (Radji et al., 2010; Wang and Salazar, 2015). Owing to its robust protocols and proven reliability, conventional PCR continues to be refined and effectively applied due to its established protocols and reliability in various food matrices.

Quantitative PCR (qPCR) is an advancement over conventional PCR, allowing for the quantitative analysis of microbial populations. The ability to measure the number of nucleic acids present in a sample not only expedites the detection process but also provides insights into the microbial load—information crucial for food safety evaluations (Law et al., 2015). Studies have shown that qPCR can yield results with higher sensitivity compared to traditional cultivation methods. For instance, when detecting Salmonella in food samples, qPCR demonstrated the ability to identify lower pathogen concentrations within a shorter turnaround time compared to classical methods (Furutani et al., 2016).

Isothermal amplification techniques, such as LAMP and RPA, are emerging as powerful alternatives that circumvent the temperature cycling step associated with traditional PCR. These methods are exceptionally robust, providing rapid and efficient amplification of target nucleic acids at a constant temperature. LAMP has displayed impressive sensitivity and specificity in detecting foodborne pathogens, making it particularly suitable for on-site testing in food production environments (Foddai and Grant, 2020). The advantages of isothermal amplification techniques include a reduced need for sophisticated thermal cyclers and shorter assay times, which are critical for quick screening in public health settings (Law et al., 2015).

Digital PCR (dPCR) enables the absolute quantification of nucleic acids, eliminating the need for standard curves. This technique's high precision makes it suitable for applications such as monitoring foodborne pathogens at very low abundance (Wang et al., 2018). The dPCR approach is noted for its ability to detect rare target sequences in complex samples, thereby increasing confidence in results compared to quantitative methods that may suffer from amplification biases (Mangal et al., 2015). While dPCR is still gaining traction in food safety, its unique advantages position it well for future developments in enhancing microbial safety regulations.

The integration of molecular techniques into food safety protocols extends beyond traditional methods. The development of biosensors and rapid detection assays utilizing nucleic acid amplification methods offers the potential for real-time monitoring (Zeng et al., 2016). Advances in microarray technologies and metagenomics enable the concurrent assessment of multiple pathogens in a single assay, addressing the complexity of food matrices without needing labor-intensive culturing steps (Li et al., 2017). These innovations reflect ongoing efforts to enhance the capability and speed of microbial detection methods.

Hybridization techniques, such as fluorescence in situ hybridization (FISH), complement molecular detection methods by providing visualization and confirmation of pathogens directly in food samples (Rocha et al., 2019; Rohde et al., 2015). FISH applications have demonstrated promising results in terms of specificity and reliability in pathogen detection, offering visual confirmation that is often lacking in conventional molecular diagnostics. The combination of nucleic acid amplification and hybridization techniques embodies a comprehensive approach to microbial detection, maximizing accuracy while minimizing time-to-result.

Figure 4. Molecular Approaches for Microbial Pathogen Detection in Food: Genomics, Metagenomics, and Transcriptomics

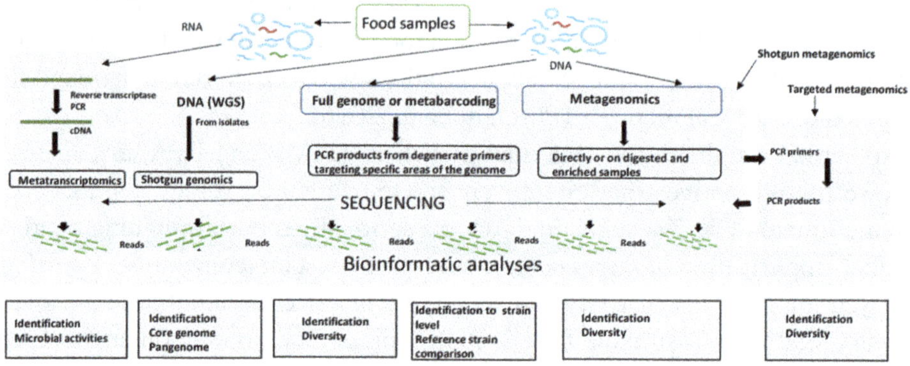

Source: *Aladhadh (2023)*

7.0. ADVANCED TECHNIQUES FOR DETECTING MICROBIAL CONTAMINANTS OF FOOD: HIGH-THROUGHPUT AND METAGENOMICS

The detection of microbial contaminants in food has undergone significant advancements through the application of advanced techniques, including high-throughput sequencing and metagenomics. This progression is primarily defined by two prominent methodologies: genetic (i.e., 16S rRNA gene and it's like) sequencing and shotgun metagenomics, both of which bring unique advantages and challenges to microbial profiling in food safety.

The 16S rRNA gene sequencing method primarily focuses on specific hypervariable regions of the 16S rRNA gene, which play a critical role in identifying and characterizing bacterial communities. A study by Chen et al. (2019) demonstrated that different primer pairs targeting various hypervariable regions in the 16S rRNA gene can yield distinct microbial profiles, emphasizing the importance of primer selection in obtaining accurate representations of microbial community composition. Moreover, this study indicated that amplicon-based techniques, despite their cost-effectiveness, often produce biased results primarily due to PCR amplification discrepancies, which could overlook low-abundance taxa critical for understanding foodborne pathogen dynamics (Chen et al., 2019; Zaheer et al., 2018). In addition to 16S rRNA and shotgun metagenomics, internal transcribed spacer (ITS) and 18S rRNA gene sequencing have emerged as critical tools for detecting fungal and eukaryotic microbial contaminants in food. ITS sequencing, often referred to as the "universal fungal barcode," is widely used for profiling fungal communities due to its high taxonomic resolution at the species level, making it ideal for identifying spoilage fungi and mycotoxin-producing species (Schoch et al., 2012). Similarly, 18S rRNA sequencing enables the comprehensive detection of eukaryotic microorganisms, including protozoa and microscopic parasites, that bacterial-focused approaches may overlook. When used in combination with high-throughput platforms, ITS and 18S sequencing expand the scope of metagenomic surveillance to include a broader spectrum of microbial threats in food systems, thereby enhancing the accuracy and inclusivity of food safety assessments (Nilsson et al., 2019; Pawlowska et al., 2020).

Conversely, shotgun metagenomics allows for a more comprehensive approach by sequencing the entire pool of genetic material in a sample. This technique not only enhances the retrieval of microbial diversity but also facilitates the identification of non-culturable organisms that may pose significant risks in food products. Research by Ge et al. (2025) suggests that shotgun metagenomics and 16S rRNA sequencing serve different purposes, with the former being particularly effective for gaining insights into the overall microbiome composition, especially in low-complexity environments such as food. Furthermore, the application of shotgun sequencing in the study of complex ecosystems, such as those found in animal food microbiomes, has expanded our understanding of microbial interactions and resistomes (Ge et al., 2025; Jagadeesan et al., 2019).

One critical advantage of shotgun metagenomics is its capacity to provide strain-level resolution and functional insights into microbial communities (Yulandi et al., 2020; Buytaers et al., 2022). For instance, Yulandi et al. (2020) demonstrated that shotgun metagenomic analysis can capture a higher diversity of microbial taxa compared to conventional 16S rRNA methods, particularly in samples such as tempeh. This improved accuracy is vital for tracing contamination sources during outbreak investigations and facilitating rapid food safety assessments (Grützke et al., 2019; Yulandi et al., 2020). However, while shotgun sequencing offers more in-depth analysis, it also presents challenges regarding the complexity of data interpretation and bioinformatics requirements (Sala et al., 2020; Buytaers et al., 2022).

The integration of bioinformatics tools plays a crucial role in managing the vast amount of data generated from both sequencing approaches. Effective interpretation of this data allows for accurate characterization of microbial communities and the detection of potential pathogenic organisms within food matrices. Advanced computational frameworks, such as PICRUSt2 and Tax4Fun2, are employed to infer functional profiles based on 16S rRNA sequencing data, complementing the insights gained from shotgun metagenomics (Matchado et al., 2023). The prioritization of quantitative data and the coupling of various sequencing methods can enhance the robustness of microbial surveillance studies, ultimately informing public health responses and food safety strategies.

Despite the numerous advantages attributed to high-throughput methods, there are significant considerations surrounding methodological validity and sample processing. The risk of contamination during sample collection and processing remains a considerable challenge. Cumpănaş et al. (2020) highlighted the susceptibility of metagenomic techniques to contain contaminant DNA from environmental or procedural sources. This risk necessitates stringent quality control measures and standardized methodologies to ensure reliability of results across laboratories (Sala et al., 2020).

In the context of food safety, the application of shotgun metagenomics and 16S rRNA sequencing methods has illustrated their respective utility in identifying and profiling microbial pathogens (Grützke et al., 2019; Koutsoumanis et al., 2019). Various studies have noted the complementary role these techniques can play in food microbiology, with shotgun metagenomics being particularly advantageous for recognizing threats from pathogens that traditional culture-based methods may fail to detect (Chen et al., 2021; Jagadeesan et al., 2019). The significance of utilizing both approaches lies in their ability to form a comprehensive overview of microbial dynamics, thereby informing risk assessment and control measures in food production environments. Furthermore, evolving concepts surrounding the microbial resistome highlight the importance of surveillance in microbial communities. Zaheer et al. (2018) indicated that sequencing depth significantly influences the ability to characterize microbial and resistome diversity, reinforcing the notion that comprehensive metagenomic mapping can inform effective strategies for combating antibiotic resistance in food contaminants. As the field advances, integrating these advanced techniques into standardized food safety protocols is critical for maintaining public health standards.

While traditional microbiological methods still play a vital role in detecting food pathogens, modern sequencing techniques, including shotgun metagenomics and 16S rRNA gene sequencing, are invaluable tools that facilitate more sensitive and comprehensive microbial analysis. These methodologies, paired with rigorous bioinformatics and contamination risk assessments, can significantly bolster food safety across production chains (Bogatyrev and Ismagilov, 2020; Grützke et al., 2019). As the knowledge base surrounding microbial ecology expands, there is an imperative to adopt innovative technologies that enhance our capacity to safeguard food chain integrity.

8.0. BIOSENSORS AND LAB-ON-A-CHIP APPROACHES FOR DETECTING MICROBIAL CONTAMINANTS IN FOOD

Biosensors and lab-on-a-chip (LOC) technologies have revolutionized the detection of microbial contaminants in food, providing innovative and efficient methods that significantly enhance food safety. This response focuses on the various types of biosensors and microfluidic platforms, as well as their integration with real-time monitoring systems, thereby providing a comprehensive overview of the current advancements and challenges within this field.

8.1. Types of Biosensors

Biosensors are analytical devices that convert biological responses into electrical signals, allowing for the detection of various analytes, including microbial contaminants. The two primary types of biosensors utilized for food safety are electrochemical and optical biosensors. Electrochemical biosensors are particularly well-regarded for their sensitivity, selectivity, speed, and cost-effectiveness. They often employ electrochemical techniques such as amperometry and impedance spectroscopy to detect low concentrations of pathogens in complex food matrices, making them suitable for applications in food analysis (Curulli, 2021; Wang et al., 2022; Majer-Baranyi et al., 2023).

The integration of nanomaterials into electrochemical biosensors has further enhanced their performance. Nanomaterials, such as carbon nanotubes and metal nanoparticles, offer an increased surface area and enhance electrochemical interactions between the electrode surface and target analytes. This amplification of the biosensing signal is critical for achieving low detection limits in food safety applications (Curulli, 2021; Raj et al., 2021). The growing body of research demonstrates that electrochemical biosensors can be tailored for specific pathogens, such as Salmonella, by using selective bioreceptors like antibodies, aptamers, or bacteriophages embedded in their designs (Awang et al., 2021; Rizzotto et al., 2023; Zhou et al., 2024)

On the other hand, optical biosensors rely on the transduction of light-based signals, which can also detect microbial contaminants. They often utilize techniques such as surface plasmon resonance (SPR), fluorescence resonance energy transfer (FRET), and reflectance measurements to monitor interactions at the molecular level (Duan et al., 2016). Despite their advantages in specificity and sensitivity, optical biosensors tend to be more expensive and technically demanding than their electrochemical counterparts (Vidić et al., 2019).

8.2. Microfluidic Platforms

Microfluidics represents another transformative technology in food safety, enabling the miniaturization of laboratory processes and allowing for the simultaneous processing of multiple samples in a minimal volume. Microfluidic devices are characterized by their ability to manipulate fluids at the microscale, typically within channels with diameters of less than 1 mm. These devices can integrate various analytical processes, including sample extraction, reaction, and detection, all on a single chip, which simplifies the workflow and significantly reduces analysis time. Nasseri et al., 2018; Parihar et al., 2023).

Recent advancements in microfluidic biosensors have focused on the rapid detection of foodborne pathogens. For instance, antibody-coated microspheres are utilized in microfluidics to increase the capture efficiency of pathogen detection through antigen-antibody interactions (Song et al., 2020). The miniaturization capabilities of microfluidics enable their deployment in point-of-care (POC) testing scenarios, significantly enhancing the accessibility of food safety diagnostics, particularly in resource-limited regions (Li et al., 2015; Nasseri et al., 2018; Zhou et al., 2024).

Moreover, the integration of microfluidic platforms with electrochemical biosensing allows for enhanced detection sensitivity while maintaining rapid analysis capabilities. Such a combination enables the development of portable, low-cost, and easy-to-use devices that are well-suited for use in food processing environments (Awang et al., 2021; Rizzotto et al., 2023). The automated nature of microfluidics reduces operator error and variability, resulting in more consistent detection of microbial contaminants.

8.3. Integration with Real-Time Monitoring

The seamless integration of biosensors with real-time monitoring systems is an essential feature that enhances food safety management. The ability to provide immediate feedback enables food safety professionals to respond promptly to contamination risks and prevent outbreaks. Innovative approaches, such as using the Internet of Things (IoT) to connect biosensing devices to cloud-based data analysis platforms, enable continuous monitoring of food products throughout the supply chain (Parihar et al., 2023; Wang et al., 2022; Mishra et al., 2018). With integrated real-time monitoring, biosensors can continuously assess the safety of food products throughout their processing, storage, and distribution. For instance, battery-operated electrochemical biosensors linked to mobile applications provide timely notifications about microbial contamination, which facilitates rapid response measures to mitigate health risks (Rizzotto et al., 2023; Vidić et al., 2019). These advances are crucial, particularly as the food industry faces increasing challenges with food safety and public health (Mishra et al., 2018; Wang et al., 2022).

8.4. Other Considerations

Aside from technological advancements, several challenges impede the widespread adoption of biosensor and microfluidics technology in the food industry. These challenges include the need for robust calibration protocols, the variability in complex food matrices that can interfere with detection, and the regulatory framework governing food safety testing methods (Wijayanti et al., 2023; Curulli, 2021). Furthermore, while advancements in nanomaterials and biosensing techniques have shown great promise, concerns remain regarding the long-term stability and reproducibility of biosensors in real-world applications (Song et al., 2020; Raja et al., 2021; Pham et al., 2021).

9.0. CRISPR-BASED DETECTION SYSTEMS FOR DETECTING MICROBIAL CONTAMINANTS IN FOOD

The advent of CRISPR-based systems has revolutionized various fields, particularly the detection of microbial contaminants in food safety. Specific technologies such as the CRISPR-Cas12 and Cas13 systems, including platforms like SHERLOCK (Specific High-sensitivity Enzymatic Reporter UnLOCKing) and DETECTR (DNA Endonuclease-Targeted CRISPR Trans Reporter), are at the forefront of this advancement. These systems enable precise and rapid identification of pathogenic microorganisms, which is essential for ensuring food quality and safety in an era where foodborne illnesses pose significant public health challenges (Brandt and Barrangou, 2019; Shin et al., 2022; Sun et al., 2024).

The CRISPR-Cas12 and Cas13 systems provide innovative solutions for real-time detection of various microbial contaminants in food matrices. For instance, CRISPR-based biosensors utilize the inherent nuclease activity of these proteins to cleave target sequences, thus facilitating the creation of detectable signals upon the recognition of specific nucleic acids associated with pathogens (Chakraborty et al., 2022; Li et al., 2022; Xie et al., 2024). These platforms extend detection capabilities to a broader range of targets, encompassing bacteria, viruses, and toxins, while optimizing the speed and sensitivity of existing methodologies. This is particularly useful in field settings where conditions may be less controlled, necessitating solutions that can perform effectively outside traditional laboratory environments (Shin et al., 2022; Wang et al., 2024).

One of the strengths of CRISPR-based detection systems lies in their adaptability and specificity. These systems can be engineered to target genomic sequences associated with foodborne pathogens, allowing for low detection limits and high accuracy (Xiong et al., 2019; Li et al., 2022). The versatility of CRISPR supports diverse signal readout mechanisms, including fluorescent, colorimetric, and electrochemical methods. This diversity not only enhances user-friendliness but also empowers non-specialists to conduct reliable food safety assessments (Sun et al., 2024; Xin et al., 2025). Notably, recent developments have integrated these CRISPR technologies with isothermal amplification protocols, thus improving sensitivity and practicality in various contexts (Sun et al., 2020).

However, the application and performance of CRISPR-based systems for food safety detection come with several technical considerations. Key challenges include potential off-target effects that may lead to false positives and the need for a reliable sample preparation process to ensure testing accuracy (Cui et al., 2024; Gootenberg et al., 2017). Standardization of CRISPR-based testing methods is crucial to facilitate their acceptance and implementation in regulatory frameworks (Li et al., 2023; Chen, 2024). Researchers emphasize the importance of developing user-friendly platforms that do not require extensive training and can be rapidly deployed in response to food safety threats (Brandt and Barrangou, 2019; Li et al., 2022).

The modularity of CRISPR systems enables researchers to tailor detection methods to the target of interest. Successful applications have been reported in identifying pathogens such as Salmonella, E. coli, and even viruses like Zika and Dengue within food samples (Zhang et al., 2022; Wang et al., 2024). Moreover, recent studies indicate that CRISPR-based biosensors can be developed into portable devices, significantly enhancing their feasibility for on-site detection in food production environments (Chakraborty et al., 2022; Sun et al., 2024). This shift toward real-time analysis aligns with industry and regulatory demands for traceability and rapid responses to contamination incidents, thus safeguarding public health and consumer trust (Brandt and Barrangou, 2019; Li et al., 2022).

Beyond the technical and adaptability aspects, the economic implications of CRISPR-based detection platforms are significant. They offer cost-effective solutions compared to traditional laboratory methods, fundamentally transforming the economics of food safety testing. For instance, simple device designs can lead to lower production costs, making them accessible to producers in different regions (Shin et al., 2022; Li et al., 2023). This accessibility is crucial for developing regions where food quality assurance resources may be limited, promoting global health standards across food systems (Moller and Liang, 2017; Sun et al., 2024). CRISPR-Cas12 and Cas13-based detection systems represent the cutting edge of microbial contamination detection in food safety, merging precision with practicality. Their inherent adaptability enables them to deliver tailored solutions that meet specific industry needs, yielding rapid and reliable results in diverse environments. As research progresses, these technologies must be scaled, standardized, and integrated into existing food safety protocols to realize their potential and protect public health fully (Brandt and Barrangou, 2019; Li et al., 2022; Zhang et al., 2022).

10.0. VALIDATION, STANDARDIZATION, AND APPLICATIONS IN FOOD SAFETY RISK MANAGEMENT

The effective detection of microbial contaminants in food systems hinges not only on the sensitivity and specificity of the methods employed but also on their validation, standardization, and integration into broader food safety management frameworks. A robust detection strategy ensures timely and accurate identification of hazards, enabling preventative and corrective actions that protect public health and preserve market confidence.

10.1. Performance Metrics and Method Validation

Core performance parameters—sensitivity, specificity, and limit of detection (LoD)—form the baseline for evaluating any microbial detection method. Sensitivity measures a method's ability to identify true positives, while specificity captures its accuracy in recognizing true negatives (López-Campos et al., 2020). These metrics are particularly critical for ensuring reliable results in complex food matrices, where microbial loads can be low and inhibitors abundant.

The LoD is essential in ready-to-eat foods or minimally processed products, where even small quantities of pathogens, such as *Listeria monocytogenes* or *Salmonella,* may be hazardous (Hoorfar et al., 2004). Quantitative PCR and other molecular platforms have shown strong potential for low LoD and rapid turnaround times, though matrix interference must be controlled through effective sample preparation and control inclusion. Method validation ensures that detection technologies meet the expectations of both industry and regulatory standards. ISO 16140, a cornerstone in food microbiological method validation, prescribes comparative studies against standard reference methods to evaluate performance criteria, including repeatability, reproducibility, and robustness (ISO, 2020). Similarly, AOAC International provides structured validation pathways—Performance Tested Methods (PTMs) and Official Methods of Analysis (OMAs)—to certify new technologies for practical and regulatory use.

10.2. Standardization and Quality Assurance Systems

Reliable microbial detection demands both internal and external quality assurance. Internally, laboratories must implement routine controls (both positive and negative), calibrate instruments, and utilize personnel who have undergone proficiency testing. For molecular methods, internal amplification controls and inhibition checks are essential to ensure valid negative results and avoid false interpretations (Schrader et al., 2012). Externally, proficiency testing (PT) and inter-laboratory comparisons support continuous quality monitoring. Accreditation under ISO/IEC 17025 requires participation in PT schemes and the maintenance of traceable and reproducible documentation (ILAC, 2017). Together, these practices foster confidence in analytical results and facilitate international trade, as well as compliance with food safety legislation.

10.3. Integration into Food Safety Risk Management Frameworks

Validated and standardized detection methods become most impactful when integrated into proactive food safety risk management systems. One of the principal applications is in Hazard Analysis and Critical Control Point (HACCP) programs, where microbial testing confirms the control of biological hazards at critical points in the production process. For example, routine swab testing at food contact surfaces using validated PCR or immunoassay kits can verify the effectiveness of sanitation and prevent cross-contamination.

Detection data also play a pivotal role in Quantitative Microbial Risk Assessment (QMRA) models. These models quantify the risk of infection or illness from specific pathogens under varying exposure scenarios, relying on accurate microbial counts to inform dose–response and exposure assessments (FAO/WHO, 2016). The integration of high-throughput and culture-independent detection methods, such as metagenomics, enables the evaluation of microbial community shifts that may indicate emerging risks or spoilage conditions.

Another critical application is in traceability and response to product recalls. Timely and validated detection enables the rapid identification of contamination sources and the containment of affected batches. Molecular typing and whole-genome sequencing (WGS) have been increasingly employed to link clinical and foodborne isolates, enhancing traceback accuracy and supporting regulatory enforcement (Jackson et al., 2016).

10.4. Toward Harmonization and Predictive Resilience

Ultimately, the convergence of validated detection methods, QA systems, and real-time data integration supports a predictive and resilient food safety system. As food supply chains become more globalized and climate change alters microbial ecology, harmonized detection frameworks that are flexible, accurate, and scalable are crucial. Collaborative efforts among laboratories, regulatory agencies, and industry stakeholders anchored in scientific validation will be key to sustaining food integrity in a rapidly changing world.

11.0. CHALLENGES AND FUTURE PERSPECTIVES

Despite significant advances in microbial detection technologies, several persistent and emerging challenges continue to affect the reliability, accessibility, and implementation of these tools in food safety systems. Addressing these challenges is essential to ensure that detection methods are not only scientifically sound but also globally equitable and adaptable to real-world contexts. These challenges include:

11.1. Low-Resource Settings

One of the most pressing issues is the limited capacity for microbial detection in low-resource settings, particularly in low- and middle-income countries. Infrastructure deficiencies, including unreliable electricity, a lack of cold chain logistics, and limited laboratory space, impede the use of advanced diagnostics such as PCR, ELISA, or NGS (Aladhadh 2023). Additionally, human resource constraints, such as the shortage of trained microbiologists and laboratory technicians, further reduce the feasibility of deploying complex methods in rural or decentralized food systems. Even where rapid diagnostic kits exist, costs associated with reagents, equipment maintenance, and licensing can render these tools inaccessible (Oladeji et al., 2024). These limitations can delay the detection of contamination, compromise outbreak response, and contribute to persistent food safety burdens in vulnerable populations (Ogwu and Ogunsola, 2024). To bridge this gap, innovations must prioritize affordability, durability, ease of use, and minimal reliance on cold storage or skilled personnel.

11.2. Matrix Effects and Diagnostic Uncertainty

Food matrices vary significantly in composition, containing fats, proteins, polysaccharides, and polyphenols that can inhibit detection reagents or interfere with signal readouts. These matrix effects are a common source of false negatives or compromised sensitivity, particularly in PCR and biosensor-based assays (Schrader et al., 2012). Conversely, false positives may arise from cross-contamination, residual DNA from dead microorganisms, or non-target amplification, leading to unnecessary recalls or reputational damage. Even validated methods can struggle to maintain performance across diverse food products such as dairy, meat, and fermented goods. As such, robust pretreatment protocols (e.g., filtration, centrifugation, and enzymatic digestion) and internal controls are essential for ensuring diagnostic reliability. Furthermore, field-deployable kits often face the challenge of striking a balance between simplicity and accuracy, necessitating ongoing refinement for both laboratory and point-of-care contexts.

11.3. Innovations in Portable Devices and AI-Driven Detection

Recent years have witnessed a surge in portable and point-of-need detection devices, offering promise for rapid microbial monitoring in production sites, markets, and even consumer environments. Lab-on-a-chip platforms, lateral flow immunoassays, CRISPR-based diagnostics (e.g., SHERLOCK, DETECTR), and smartphone-integrated sensors are transforming microbial surveillance by reducing dependence on central labs (Pardee et al., 2016; Kellner et al., 2019). Moreover, AI and machine learning (ML) are being increasingly applied to enhance detection sensitivity, automate data interpretation, and predict contamination patterns. For example, AI-enhanced image analysis of microbial cultures or lateral flow assays has improved the precision of visual readings. At the same time, ML algorithms have been integrated with sensor data to forecast spoilage or pathogen risk in perishable foods (Zhang et al., 2022). Despite their promise, these innovations require rigorous validation, user-friendly interfaces, and supportive regulatory frameworks to enable broad adoption. Challenges include data standardization, cybersecurity in cloud-connected devices, and ensuring ethical use of AI-driven decision-making tools in food governance.

12.0. CONCLUSION

Ensuring the microbiological safety of food is a critical priority in public health and global food systems. This chapter has outlined the diverse sources and pathways of microbial contamination, along with the wide range of detection methods currently available, from conventional culture-based techniques to cutting-edge molecular diagnostics and biosensor technologies. Each method offers unique advantages depending on the food matrix, target organism, required sensitivity, and operational context. The integration of high-throughput sequencing, CRISPR-based diagnostics, and lab-on-chip systems is particularly promising for achieving rapid, accurate, and field-deployable detection.

Moreover, the chapter emphasized the importance of employing appropriate sampling strategies, method validation, and adherence to international standards to ensure the reliability and comparability of results. The growing application of microbial detection data in risk assessment and food safety management, primarily through tools like HACCP and QMRA, underscores the strategic role of detection in preventing foodborne outbreaks. As the complexity of food supply chains increases and climate change introduces new microbiological risks, the ability to detect contaminants swiftly and accurately will remain a cornerstone of effective food safety systems. Continued investment in innovation, capacity building, and cross-sector collaboration will be essential for advancing detection capabilities and protecting global public health.

REFERENCES

O157. (n.d.). H7 using immunomagnetic separation and mPCR in Turkish foods of animal origin. *Letters in Applied Microbiology*, *57*(4), 373–379. DOI: 10.1111/lam.12124 PMID: 23809061

Afshari, A., Schrenzel, J., Ieven, M., & Harbarth, S. (2012). Bench-to-bedside review: Rapid molecular diagnostics for bloodstream infection - a new frontier. *Critical Care*, *16*(3), 222. DOI: 10.1186/cc11202 PMID: 22647543

Akkina, R., Payala, V., and Maganti, S. (2023). Tools for rapid detection and control of foodborne microbial pathogens. https://doi.org/DOI: 10.5772/intechopen.103938

Aladhadh, M. (2023). A Review of Modern Methods for the Detection of Foodborne Pathogens. *Microorganisms*, *11*(5), 1111. DOI: 10.3390/microorganisms11051111 PMID: 37317085

Amalaradjou, M., & Bhunia, A. (2012). Modern approaches in probiotics research to control foodborne pathogens. *Advances in Food and Nutrition Research*, •••, 185–239. DOI: 10.1016/b978-0-12-394598-3.00005-8 PMID: 23034117

Amoako, K., Shields, M., Goji, N., Paquet, C., Thomas, M., Janzen, T., & Hahn, K. (2012). Rapid detection and identification of Yersinia pestis from food using immunomagnetic separation and pyrosequencing. *Journal of Pathogens*, *2012*, 1–6. DOI: 10.1155/2012/781652 PMID: 23091729

AOAC International. (2019). *Official Methods of Analysis of AOAC International* (21st ed.).

Arendt, S., Rajagopal, L., Strohbehn, C., Stokes, N., Meyer, J., & Mandernach, S. (2013). Reporting of Foodborne Illness by U.S. Consumers and Healthcare Professionals. *International Journal of Environmental Research and Public Health*, *10*(8), 3684–3714. DOI: 10.3390/ijerph10083684 PMID: 23965924

Awang, M., Bustami, Y., Hamzah, H., Zambry, N., Najib, M., Khalid, M., & Manaf, A. (2021). Advancement in Salmonella detection methods: From conventional to electrochemical-based sensing. *Biosensors (Basel)*, *11*(9), 346. DOI: 10.3390/bios11090346 PMID: 34562936

Bedane, T., Agga, G., & Gutema, F. (2022). Hygienic Assessment of Fish Handling Practices Along the Production and Supply Chain and Their Public Health Implications in Central Oromia, Ethiopia. *Scientific Reports*, *12*(1). Advance online publication. DOI: 10.1038/s41598-022-17671-5 PMID: 35977962

Beuchat, L., & Komitopoulou, E., BECKERS, H., Betts, R., Bourdichon, F., Fanning, S., … and Kuile, B. (2013). Low-water-activity foods: Increased concern as vehicles for foodborne pathogens. *Journal of Food Protection, 76*(1), 150–172. DOI: 10.4315/0362-028x.jfp-12-211 PMID: 23317872

Bintsis, T. (2017). Foodborne pathogens. *AIMS Microbiology, 3*(3), 529–563. DOI: 10.3934/microbiol.2017.3.529 PMID: 31294175

Bogatyrev, S., & Ismagilov, R. (2020). Quantitative microbiome profiling in lumenal and tissue samples with broad coverage and dynamic range via a single-step 16S rRNA gene DNA copy quantification and amplicon barcoding. https://doi.org/DOI: 10.1101/2020.01.22.914705

Brandt, K., & Barrangou, R. (2019). Applications of CRISPR technologies across the food supply chain. *Annual Review of Food Science and Technology, 10*(1), 133–150. DOI: 10.1146/annurev-food-032818-121204 PMID: 30908954

Buytaers, F., Verhaegen, B., Gand, M., D'aes, J., Vanneste, K., Roosens, N., & Keersmaecker, S. (2022). Metagenomics to detect and characterize viruses in food samples at the genome level? Lessons learnt from a norovirus study. *Foods, 11*(21), 3348. DOI: 10.3390/foods11213348 PMID: 36359961

Casado, N., Berenguer, C., Câmara, J., & Pereira, J. (2024). What are we eating? Surveying the presence of toxic molecules in the food supply chain using chromatographic approaches. *Molecules (Basel, Switzerland), 29*(3), 579. DOI: 10.3390/molecules29030579 PMID: 38338324

Castellví, S., Männing, A., & Rodriguez-Saona, L. (2010). Fourier-transform infrared spectroscopy combined with immunomagnetic separation as a tool to discriminate Salmonella serovars. *Analyst, 135*(11), 2987. DOI: 10.1039/c0an00497a PMID: 20877835

Cauteren, D., Strat, Y., Sommen, C., Bruyand, M., Tourdjman, M., Silva, N., & Desenclos, J. (2017). Estimated annual numbers of foodborne pathogen–associated illnesses, hospitalizations, and deaths, France, 2008–2013. *Emerging Infectious Diseases, 23*(9), 1486–1492. DOI: 10.3201/eid2309.170081 PMID: 28820137

Centers for Disease Control and Prevention. (2025). Estimates: Burden of foodborne illness in the United States. U.S. Department of Health and Human Services. https://www.cdc.gov/foodborneburden/2019-estimates.html

Chakraborty, J., Chaudhary, A., Khan, S., Rudayni, H., Rahaman, S., & Sarkar, H. (2022). CRISPR/Cas-based biosensor as a new-age detection method for pathogenic bacteria. *ACS Omega*, *7*(44), 39562–39573. DOI: 10.1021/acsomega.2c04513 PMID: 36385843

Chen, H., Li, J., Yan, S., Sun, H., Tan, C., Liu, M., & Xiao, X. (2021). Identification of pathogen(s) in infectious diseases using shotgun metagenomic sequencing and conventional culture: A comparative study. *PeerJ*, *9*, e11699. DOI: 10.7717/peerj.11699 PMID: 34249516

Chen, S. (2024). Application of CRISPR-Cas Technology in Food Safety. *Theoretical and Natural Science*, *32*(1), 57–61. DOI: 10.54254/2753-8818/32/20240790

Chen, Z., Hui, P., Hui, M., Yeoh, Y., Wong, P., Chan, M., & Chan, P. (2019). Impact of preservation method and 16S rRNA hypervariable region on gut microbiota profiling. *mSystems*, *4*(1). Advance online publication. DOI: 10.1128/msystems.00271-18 PMID: 30834331

Chime, A. O., Aiwansoba, R. O., Danagogo, S. J., Egharevba, I. I., Osawaru, M. E., & Ogwu, M. C. (2016). Effects of lining with leaves of Triclisia dictyophylla on the fungal composition of Cola nitida during storage. *Journal of Industrial Research and Technology*, *5*(2), 128–136.

Crowley, E., Bird, P., Torontali, M., Goetz, K., Agin, J., Goins, D., & Thompson, L. (2010). Tempo® EC for the enumeration of Escherichia coli in foods: Collaborative study. *Journal of AOAC International*, *93*(2), 576–586. DOI: 10.1093/jaoac/93.2.576 PMID: 20480906

Cui, Z., Lin, L., Zong, Y., Chen, Y., & Wang, S. (2024). Precision gene editing using deep learning: A case study of the CRISPR-Cas9 editor. *Applied and Computational Engineering*, *64*(1), 134–141. DOI: 10.54254/2755-2721/64/20241357

Cumpănaş, A., Bratu, O., Bardan, R., Ferician, O., Cumpăna , A., Horhat, F., & Creţu, O. (2020). Urinary microbiota—Are we ready for prime time? A literature review of study methods' critical steps in avoiding contamination and minimizing biased results. *Diagnostics (Basel)*, *10*(6), 343. DOI: 10.3390/diagnostics10060343 PMID: 32471022

Curulli, A. (2021). Electrochemical biosensors in food safety: Challenges and perspectives. *Molecules (Basel, Switzerland)*, *26*(10), 2940. DOI: 10.3390/molecules26102940 PMID: 34063344

Duan, N., Wu, S., Dai, S., Gu, H., Hao, L., Ye, H., & Wang, Z. (2016). Advances in Aptasensors for the Detection of Food Contaminants. *Analyst*, *141*(13), 3942–3961. DOI: 10.1039/c6an00952b PMID: 27265444

El-Khishin, M., Gooneratne, R., and Hussain, M. (2017). Microbial Safety of Foods in the Supply Chain and Food Security Advanced in Food Technology and Nutritional Sciences—Open Journal, 3(1), 22-32. https://doi.org/DOI: 10.17140/aftnsoj-3-141

Elbehiry, A., Marzouk, E., Aldubaib, M., Moussa, I., Abalkhail, A., Ibrahem, M., & Rawway, M. (2022). Pseudomonas species prevalence, protein analysis, and antibiotic resistance: An evolving public health challenge. *AMB Express*, *12*(1). Advance online publication. DOI: 10.1186/s13568-022-01390-1 PMID: 35532863

Elbehiry, A., Marzouk, E., Alzaben, F., Almuaither, A., Abead, B., Alamri, M., Almuzaini, A. M., & Abu-Okail, A. (2025). Emerging Technologies and Integrated Strategies for Microbial Detection and Control in Fresh Produce. *Microorganisms*, *13*(7), 1447. DOI: 10.3390/microorganisms13071447 PMID: 40731957

FAO/WHO. (2016). Risk-based examples and approach for control of Trichinella spp. and Taenia saginata in meat.

Finn, S., Condell, O., McClure, P., Amézquita, A., & Fanning, S. (2013). Mechanisms of survival, responses, and sources of Salmonella in low-moisture environments. *Frontiers in Microbiology*, *4*. Advance online publication. DOI: 10.3389/fmicb.2013.00331 PMID: 24294212

Foddai, A., Elliott, C., & Grant, I. (2010). Maximizing capture efficiency and specificity of magnetic separation for Mycobacterium avium subsp. Paratuberculosis Cells. *Applied and Environmental Microbiology*, *76*(22), 7550–7558. DOI: 10.1128/aem.01432-10 PMID: 20851966

Foddai, A., & Grant, I. (2020). Methods for detection of viable foodborne pathogens: Current state-of-art and future prospects. *Applied Microbiology and Biotechnology*, *104*(10), 4281–4288. DOI: 10.1007/s00253-020-10542-x PMID: 32215710

Furmančíková, P., Navrátilová, P., Šťástková, Z., Bednářová, I., & Steinhauserová, I. (2022). Occurrence and detection of Listeria monocytogenes in ready-to-eat foods. Maso International –. *Journal of Food Science and Technology*, *12*(1), 33–40. DOI: 10.2478/mjfst-2022-0011

Furutani, S., Kajiya, M., Aramaki, N., & Kubo, I. (2016). Rapid detection of Salmonella enterica in food using a compact disc-shaped device. *Micromachines*, *7*(1), 10. DOI: 10.3390/mi7010010 PMID: 30407383

Ge, B., McDonald, R., Yang, Q., Domesle, K., Sarria, S., Li, X., & Tadesse, D. (2025). Exploring animal food microbiomes and resistomes via 16S rRNA gene amplicon sequencing and shotgun metagenomics. *Applied and Environmental Microbiology, 91*(2). Advance online publication. DOI: 10.1128/aem.02230-24 PMID: 39840975

Gootenberg, J., Abudayyeh, O., Lee, J., Essletzbichler, P., Dy, A., Joung, J., & Zhang, F. (2017). Nucleic acid detection with CRISPR-Cas13a/c2c2. *Science, 356*(6336), 438–442. DOI: 10.1126/science.aam9321 PMID: 28408723

Grützke, J., Malorny, B., Hammerl, J., Busch, A., Tausch, S., Tomaso, H., & Deneke, C. (2019). Fishing in the soup – pathogen detection in food safety using metabarcoding and metagenomic sequencing. *Frontiers in Microbiology, 10*. Advance online publication. DOI: 10.3389/fmicb.2019.01805 PMID: 31447815

Havelaar, A., Kirk, M., Torgerson, P., Gibb, H., Hald, T., Lake, R., & Devleesschauwer, B. (2015). World Health Organization global estimates and regional comparisons of the burden of foodborne disease in 2010. *PLoS Medicine, 12*(12), e1001923. DOI: 10.1371/journal.pmed.1001923 PMID: 26633896

Hoorfar, J., Cook, N., & Malorny, B.. (2004). Making internal amplification control mandatory for diagnostic PCR. *Journal of Clinical Microbiology, 42*(6), 1863–1868. PMID: 15131141

Hsu, C., Hsu, B., Chang, T., Hsu, T., Shen, S., Chiu, Y., & Chen, J. (2014). Evaluation of immunomagnetic separation for the detection of Salmonella in surface waters by polymerase chain reaction. *International Journal of Environmental Research and Public Health, 11*(9), 9811–9821. DOI: 10.3390/ijerph110909811 PMID: 25243887

Hu, R., Zhang, D., Tao, D., Hartvigsen, T., Feng, H., & Rundensteiner, E. (2022). Tweet-fid: an annotated dataset for multiple foodborne illness detection tasks. https://doi.org//arxiv.2205.10726DOI: 10.48550

ILAC. (2017). ISO/IEC 17025: General requirements for the competence of testing and calibration laboratories.

ISO. (2020). ISO 16140-2:2016: Microbiology of the food chain – Method validation – Part 2.

Izah, S. C., Odubo, T. C., Ogwu, M. C., & Hait, M. (2024). Aspects of Microorganisms in the Food Industry. In Ogwu, M. C., Izah, S. C., & Ntuli, N. R. (Eds.), *Food Safety and Quality in the Global South*. Springer., DOI: 10.1007/978-981-97-2428-4_13

Jackson, B. R., Tarr, C., & Strain, E.. (2016). Implementation of nationwide real-time whole-genome sequencing to enhance listeriosis outbreak detection and investigation. *Clinical Infectious Diseases, 63*(3), 380–386. PMID: 27090985

Jagadeesan, B., Gerner-Smidt, P., Allard, M., Leuillet, S., Winkler, A., Xiao, Y., & Grant, K. (2019). The Use of Next-Generation Sequencing for Improving Food Safety: Translation into Practice. *Food Microbiology*, *79*, 96–115. DOI: 10.1016/j.fm.2018.11.005 PMID: 30621881

Jokerst, J., Adkins, J., Bisha, B., Mentele, M., Goodridge, L., & Henry, C. (2012). Development of a paper-based analytical device for colorimetric detection of select foodborne pathogens. *Analytical Chemistry*, *84*(6), 2900–2907. DOI: 10.1021/ac203466y PMID: 22320200

Kant, K., Shahbazi, M., Dave, V., Ngo, T., Vinayaka, A., Quyen, T., & Wolff, A. (2018). Microfluidic devices for sample preparation and rapid detection of food-borne pathogens. *Biotechnology Advances*, *36*(4), 1003–1024. DOI: 10.1016/j.biotechadv.2018.03.002 PMID: 29534915

Karanth, S., Feng, S., Patra, D., & Pradhan, A. (2023). Linking microbial contamination to food spoilage and food waste: The role of smart packaging, spoilage risk assessments, and date labeling. *Frontiers in Microbiology*, *14*. Advance online publication. DOI: 10.3389/fmicb.2023.1198124 PMID: 37426008

Karuppuchamy, V., Heldman, D., & Snyder, A. (2024). A review of food safety in low-moisture foods with current and potential dry-cleaning methods. *Journal of Food Science*, *89*(2), 793–810. DOI: 10.1111/1750-3841.16920 PMID: 38221802

Katase, M., & Tsumura, K. (2011). Enumeration of micro-organisms in processed soy products with an automated most probable number method compared with standard plate method. *Letters in Applied Microbiology*, *53*(5), 539–545. DOI: 10.1111/j.1472-765x.2011.03143.x PMID: 21880050

Kawase, J., Etoh, Y., Ikeda, T., Yamaguchi, K., Watahiki, M., Shima, T., & Shirabe, K. (2016). An improved multiplex real-time SYBR Green PCR assay for analysis of 24 target genes from 16 bacterial species in fecal DNA samples from patients with foodborne illnesses. *Japanese Journal of Infectious Diseases*, *69*(3), 191–201. DOI: 10.7883/yoken.jjid.2015.027 PMID: 26166499

Kellner, M. J., Koob, J. G., Gootenberg, J. S., Abudayyeh, O. O., & Zhang, F. (2019). SHERLOCK: Nucleic acid detection with CRISPR nucleases. *Nature Protocols*, *14*(10), 2986–3012. DOI: 10.1038/s41596-019-0210-2 PMID: 31548639

Khan, A., Sahir, K., Ahmed, M., & Khan, S. (2014). Rapid detection of Salmonella in food samples by polymerase chain reaction after a 10 h pre-enrichment. *Journal of Food Safety*, *34*(1), 79–86. DOI: 10.1111/jfs.12099

Kim, H., Cho, J., Song, M., Cho, J., Kim, S., Kim, E., & Lee, J. (2021). Evaluating the prevalence of foodborne pathogens in livestock using metagenomics approach. *Journal of Microbiology and Biotechnology*, *31*(12), 1701–1708. DOI: 10.4014/jmb.2109.09038 PMID: 34675137

Kim, S., & Kim, S. (2020). Bacterial pathogen detection by conventional culture-based and recent alternative (polymerase chain reaction, isothermal amplification, enzyme-linked immunosorbent assay, bacteriophage amplification, and gold nanoparticle aggregation) methods in food samples: A review. *Journal of Food Safety*, *41*(1). Advance online publication. DOI: 10.1111/jfs.12870

Koo, O., Mertz, A., Akins, E., Sirsat, S., Neal, J., Morawicki, R., & Ricke, S. (2012). Analysis of microbial diversity on deli slicers using polymerase chain reaction and denaturing gradient gel electrophoresis technologies. *Letters in Applied Microbiology*, *56*(2), 111–119. DOI: 10.1111/lam.12021 PMID: 23121623

Koutsoumanis, K., Allende, A., Álvarez-Ordóñez, A., Bolton, D., Bover-Cid, S., Chemaly, M., & Herman, L. (2019). Whole genome sequencing and metagenomics for outbreak investigation, source attribution, and risk assessment of food-borne microorganisms. *EFSA Journal*, *17*(12). Advance online publication. DOI: 10.2903/j.efsa.2019.5898 PMID: 32626197

Krishnasree, V., Nethra, P., Dheeksha, J., Madumitha, M., Vidyaeswari, R., and Lakshya, P. (2018). A pilot study on assessing the sustainability of food safety and hygienic practices in street food handling systems. Asian Journal of Dairy and Food Research, (of). https://doi.org/DOI: 10.18805/ajdfr.dr-1381

Lanzas, C., Lu, Z., & Gröhn, Y. (2011). Mathematical modeling of the transmission and control of foodborne pathogens and antimicrobial resistance at preharvest. *Foodborne Pathogens and Disease*, *8*(1), 1–10. DOI: 10.1089/fpd.2010.0643 PMID: 21043837

Law, J., Mutalib, N., Chan, K., & Lee, L. (2015). Rapid methods for the detection of foodborne bacterial pathogens: Principles, applications, advantages and limitations. *Frontiers in Microbiology*, *5*. Advance online publication. DOI: 10.3389/fmicb.2014.00770 PMID: 25628612

Law, J. W.-F., Ab Mutalib, N.-S., Chan, K.-G., & Lee, L.-H. (2015). Rapid methods for the detection of foodborne bacterial pathogens: Principles, applications, advantages and limitations. *Frontiers in Microbiology*, *5*, 770. DOI: 10.3389/fmicb.2014.00770 PMID: 25628612

Lee, A., Ng, I., Wong, E., Wong, I., Sze, R., Chan, K., & Siu, G. (2023). Comprehensive identification of pathogenic microbes and antimicrobial resistance genes in food products using nanopore sequencing-based metagenomics. https://doi.org/ DOI: 10.1101/2023.10.15.562131

Lee, H., and Yoon, Y. (2021). Etiological agents implicated in foodborne illness world

Lee, J., Park, S., Khan, F., Jo, D., Lee, D., Kang, M., & Kim, Y. (2021). Simultaneous isolation and enumeration of virulent Vibrio cholerae and Vibrio vulnificus using an advanced MPN-PCR method. https://doi.org/DOI: 10.21203/rs.3.rs-528186/v1

Li, A., Zhang, H., Zhang, X., Wang, Q., Tian, J., Li, Y., & Li, J. (2010). Rapid separation and immunoassay for low levels of Salmonella in foods using magnetosome–antibody complex and real-time fluorescence quantitative PCR. *Journal of Separation Science*, *33*(21), 3437–3443. DOI: 10.1002/jssc.201000441 PMID: 20886524

Li, B., Patel, I., Tall, B., & Elkins, C. (2017). Advancements in microarray utility for detection and tracking of foodborne microbes in the genomic era. *Advanced Techniques in Biology and Medicine*, *05*(03). Advance online publication. DOI: 10.4172/2379-1764.1000239

Li, S., Chai, Y., and Chin, B. (2015). High-throughput pathogen screening for food safety using magnetoelastic biosensors. https://doi.org/DOI: 10.1117/12.2176777

Li, Y., Man, S., Ye, S., Liu, G., & Ma, L. (2022). CRISPR-Cas-based detection for food safety problems: Current status, challenges, and opportunities. *Comprehensive Reviews in Food Science and Food Safety*, *21*(4), 3770–3798. DOI: 10.1111/1541-4337.13000 PMID: 35796408

Li, Y., Zhao, Z., Liu, Y., Wang, N., Man, S., Ma, L., & Wang, S. (2023). CRISPR/Cas System: The accelerator for the development of non-nucleic acid target detection in food safety. *Journal of Agricultural and Food Chemistry*, *71*(37), 13577–13594. DOI: 10.1021/acs.jafc.3c03619 PMID: 37656446

Line, J., Stern, N., Oakley, B., & Seal, B. (2011). Comparison of an automated most-probable-number technique with traditional plating methods for estimating populations of total aerobes, coliforms, and Escherichia coli associated with freshly processed broiler chickens. *Journal of Food Protection*, *74*(9), 1558–1563. DOI: 10.4315/0362-028x.jfp-11-024 PMID: 21902928

López-Campos, G., & Martínez-Suárez, J. V.. (2020). Detection, identification, and analysis of foodborne pathogens. *Frontiers in Microbiology*, *11*, 2869.

Mahunu, G. K., Osei-Kwarteng, M., Ogwu, M. C., & Afoakwah, N. A. (2024). Safe Food Handling Techniques to Prevent Microbial Contamination. In Ogwu, M. C., Izah, S. C., & Ntuli, N. R. (Eds.), *Food Safety and Quality in the Global South*. Springer., DOI: 10.1007/978-981-97-2428-4_14

Majer-Baranyi, K., Székács, A., & Adányi, N. (2023). Application of electrochemical biosensors for the determination of food spoilage. *Biosensors (Basel), 13*(4), 456. DOI: 10.3390/bios13040456 PMID: 37185531

Malavi, D., Muzhingi, T., & Abong, G. (2018). Good Manufacturing Practices and Microbial Contamination Sources in an Orange-Fleshed Sweet Potato Puree Processing Plant in Kenya. *International Journal of Food Sciences, 2018*, 1–11. DOI: 10.1155/2018/4093161 PMID: 29808161

Mangal, M., Bansal, S., Sharma, S., & Gupta, R. (2015). Molecular detection of foodborne pathogens: A rapid and accurate answer to food safety. *Critical Reviews in Food Science and Nutrition, 56*(9), 1568–1584. DOI: 10.1080/10408398.2013.782483 PMID: 25830555

Margot, H., Stephan, R., O'Mahony, E., & Iversen, C. (2013). Comparison of rapid cultural methods for the detection of Salmonella species. *International Journal of Food Microbiology, 163*(2-3), 47–50. DOI: 10.1016/j.ijfoodmicro.2013.02.013 PMID: 23558186

Matchado, M., Rühlemann, M., Reimeiter, S., Kacprowski, T., Frost, F., Haller, D., & List, M. (2023). On the Limits of 16S rRNA Gene-Based Metagenome Prediction and Functional Profiling. https://doi.org/DOI: 10.1101/2023.11.07.564315

Mayer, M., Vega, J., DeMarco, I., Eicher, R., Molina, D., Lee, C., … and Naufel, J. (2023). Message convergence as reassurance about the safety of the food supply concerning African swine fever virus. AHTR, 2(1), 02-11. https://doi.org/DOI: 10.31491/ahtr.2023.02.002

McLain, J., Cytryn, E., Durso, L., & Young, S. (2016). Culture-based methods for detection of antibiotic resistance in agroecosystems: Advantages, challenges, and gaps in knowledge. *Journal of Environmental Quality, 45*(2), 432–440. DOI: 10.2134/jeq2015.06.0317 PMID: 27065389

Mishra, G., Barfidokht, A., Tehrani, F., & Mishra, R. (2018). Food safety analysis using electrochemical biosensors. *Foods, 7*(9), 141. DOI: 10.3390/foods7090141 PMID: 30200448

Moller, A., & Liang, C. (2017). Metacrast: Reference-guided extraction of CRISPR spacers from unassembled metagenomes. *PeerJ*, *5*, e3788. DOI: 10.7717/peerj.3788 PMID: 28894651

Murphy, C., Carson, C., Smith, B., Chapman, B., Marrotte, J., McCann, M., & Parmley, E. (2018). Factors potentially linked with the occurrence of antimicrobial resistance in selected bacteria from cattle, chickens, and pigs: A scoping review of publications for use in modelling of antimicrobial resistance (IAM.amr project). *Zoonoses and Public Health*, *65*(8), 957–971. DOI: 10.1111/zph.12515 PMID: 30187682

Nasseri, B., Soleimani, N., Rabiee, N., Kalbasi, A., Karimi, M., & Hamblin, M. (2018). Point-of-care microfluidic devices for pathogen detection. *Biosensors & Bioelectronics*, *117*, 112–128. DOI: 10.1016/j.bios.2018.05.050 PMID: 29890393

Neff, J., Xie, Y., Gill, J., & Liu, M. (2019). Complete genome sequence of Salmonella enterica serovar Heidelberg myophage meda. *Microbiology Resource Announcements*, *8*(17). Advance online publication. DOI: 10.1128/mra.00253-19 PMID: 31023796

Nilsson, R. H., Larsson, K. H., Taylor, A. F. S., Bengtsson-Palme, J., Jeppesen, T. S., Schigel, D., & Abarenkov, K. (2019). The UNITE database for molecular identification of fungi: Handling dark taxa and parallel taxonomic classifications. *Nucleic Acids Research*, *47*(D1), D259–D264. DOI: 10.1093/nar/gky1022 PMID: 30371820

Ogwu, M. C., Izah, S. C., Ntuli, N. R., & Odubo, T. C. (2024). Food Security Complexities in the Global South. In Ogwu, M. C., Izah, S. C., & Ntuli, N. R. (Eds.), *Food Safety and Quality in the Global South*. Springer., DOI: 10.1007/978-981-97-2428-4_1

Ogwu, M. C., Oforjamah, I. N., Ahana, C. M., & Osawaru, M. E. (2013). Phylloplane assessment of leaf and stalk for consumption and microbial diversity of sweet potato (Ipomoea batatas (L.) Lam., Convolvulaceae. *Octa Journal of Environmental Research*, *1*(2), 107–116.

Ogwu, M. C., & Ogunsola, O. A. (2024). Physicochemical Methods of Food Preservation to Ensure Food Safety and Quality. In Ogwu, M. C., Izah, S. C., & Ntuli, N. R. (Eds.), *Food Safety and Quality in the Global South*. Springer., DOI: 10.1007/978-981-97-2428-4_9

Ogwu, M. C., & Osawaru, M. E. (2014). Comparative study of microflora population on the phylloplane of common okra [Abelmoschus esculentus L. (Moench.)]. *Nigerian Journal of Biotechnology*, *28*, 17–25.

Oh, M., Hwang, H., Chung, B., Paik, H., Han, S., Kang, S., & Jung, G. (2012). Simultaneous detection of 10 foodborne pathogens using capillary electrophoresis-based single-strand conformation polymorphism. *Han-gug Chugsan Sigpum Hag-hoeji*, *32*(2), 241–246. DOI: 10.5851/kosfa.2012.32.2.241

Oladeji, O. A., Karigidi, K. O., & Ogwu, M. C. (2024). Indices for Monitoring and Measuring the Physicochemical Properties of Safe and Quality Food. In Ogwu, M. C., Izah, S. C., & Ntuli, N. R. (Eds.), *Food Safety and Quality in the Global South*. Springer., DOI: 10.1007/978-981-97-2428-4_5

Oliveira, M., Tiwari, B., & Duffy, G. (2020). Emerging technologies for aerial decontamination of food storage environments to eliminate microbial cross-contamination. *Foods*, *9*(12), 1779. DOI: 10.3390/foods9121779 PMID: 33266230

Onmaz, N., Gönülalan, Z., Yıldırım, Y., Karadal, F., Abay, S., and Al, S. (2013). Detection of Escherichia coli

Osawaru, M. E., Ogwu, M. C., Ogbeifun, N. S., & Chime, A. O. (2013). Microflora diversity of the phylloplane of wild Okra (Corchorus olitorius L. Jute). *Bayero Journal of Pure and Applied Sciences*, *6*(2), 136–142.

Pakbin, B., Amani, Z., Rahimi, Z., Najafi, S., Familsatarian, B., Khakpoor, A., & Brück, T. (2023). Prevalence of Foodborne Bacterial Pathogens and Antibiotic Resistance Genes in Sweets from Local Markets in Iran. *Foods*, *12*(19), 3645. DOI: 10.3390/foods12193645 PMID: 37835299

Pardee, K., Green, A. A., Takahashi, M. K., Braff, D., Lambert, G., Lee, J. W., & Collins, J. J. (2016). Rapid, low-cost detection of Zika virus using programmable biomolecular components. *Cell*, *165*(5), 1255–1266. DOI: 10.1016/j.cell.2016.04.059 PMID: 27160350

Parihar, A., Yadav, S., Sadique, M., Ranjan, P., Kumar, N., Singhal, A., & Srivastava, A. (2023). Internet-of-Things integrated point-of-care biosensing devices for infectious diseases: Toward better preparedness for future pandemics. *Bioengineering & Translational Medicine*, *8*(3). Advance online publication. DOI: 10.1002/btm2.10481 PMID: 37206204

Park, J., Park, K., Ok, G., Chang, H., Park, T., Choi, S., & Lim, M. (2020). Detection of Escherichia coli o157:h7 using automated immunomagnetic separation and enzyme-based colorimetric assay. *Sensors (Basel)*, *20*(5), 1395. DOI: 10.3390/s20051395 PMID: 32143335

Park, S., Szonyi, B., Gautam, R., Nightingale, K., Anciso, J., & Ivanek, R. (2018). Risk factors for microbial contamination in fruits and vegetables at the preharvest level: A systematic review. *Journal of Food Protection*, *75*(11), 2055–2081. DOI: 10.4315/0362-028X.JFP-12-160 PMID: 23127717

Pava-Ripoll, M., Pearson, R., Miller, A., & Ziobro, G. (2015). Detection of foodborne bacterial pathogens from individual filth flies. *Journal of Visualized Experiments*, (96). Advance online publication. DOI: 10.3791/52372 PMID: 25742168

Pawlowska, M. M., Zannoni, A., Boukherroub, R., Szunerits, S., & Krasowska, A. (2020). Eukaryotic microbial contamination detection in food: Current challenges and perspectives. *Trends in Food Science & Technology*, *100*, 324–336. DOI: 10.1016/j.tifs.2020.03.027

Pham, D., Nguyen, X., Marsh, P., Chung, Y., Lau, M., Nguyen, A., & Cao, H. (2021). A fluidics-based biosensor to detect and characterize inhibition patterns of organophosphates to acetylcholinesterase in food materials. *Micromachines*, *12*(4), 397. DOI: 10.3390/mi12040397 PMID: 33916863

Rabiee, P., Faraz, A., Ajlouni, S., & Hussain, M. (2024). Microbial contamination and disease outbreaks associated with rockmelons (cucumis melo): Implications for public health protection. *Foods*, *13*(14), 2198. DOI: 10.3390/foods13142198 PMID: 39063281

Radji, M., Malik, A., & Widyasmara, A. (2010). Rapid detection of Salmonella in food and beverage samples by polymerase chain reaction. *Malaysian Journal of Microbiology*. Advance online publication. DOI: 10.21161/mjm.20009

Raj, P., Oh, M., Han, K., & Lee, T. (2021). Label-free electrochemical biosensor based on au@mos$_2$–pani for Escherichia coli detection. *Chemosensors (Basel, Switzerland)*, *9*(3), 49. DOI: 10.3390/chemosensors9030049

Raja, I., Vedhanayagam, M., Preeth, D., Kim, C., Lee, J., & Han, D. (2021). Development of two-dimensional nanomaterial-based electrochemical biosensors to enhance the analysis of food toxicants. *International Journal of Molecular Sciences*, *22*(6), 3277. DOI: 10.3390/ijms22063277 PMID: 33806998

Ravaliya, K., Gentry-Shields, J., García, S., Heredia, N., Aceituno, A., Bartz, F., & Jaykus, L. (2014). Use of bacteroidales microbial source tracking to monitor fecal contamination in fresh produce production. *Applied and Environmental Microbiology*, *80*(2), 612–617. DOI: 10.1128/aem.02891-13 PMID: 24212583

Rizzotto, F., Khalife, M., Hou, Y., Chaix, C., Lagarde, F., Scaramozzino, N., & Vidić, J. (2023)... *Recent Advances in Electrochemical Biosensors for Food Control Micromachines, 14*(7), 1412. DOI: 10.3390/mi14071412 PMID: 37512723

Rocha, R., Sousa, J., Cerqueira, L., Vieira, M., Almeida, C., & Azevedo, N. (2019). Development and application of peptide nucleic acid fluorescence in situ hybridization for the specific detection of Listeria monocytogenes. *Food Microbiology, 80*, 1–8. DOI: 10.1016/j.fm.2018.12.009 PMID: 30704592

Rohde, A., Hammerl, J., Appel, B., Dieckmann, R., & Dahouk, S. (2015). Fishing for bacteria in food – a promising tool for the reliable detection of pathogenic bacteria?bacteria? *Food Microbiology, 46*, 395–407. DOI: 10.1016/j.fm.2014.09.002 PMID: 25475309

Sala, C., Mordhorst, H., Grützke, J., Brinkmann, A., Petersen, T., Poulsen, C., & Cesare, A. (2020). Metagenomics-based proficiency test of smoked salmon spiked with a mock community. *Microorganisms, 8*(12), 1861. DOI: 10.3390/microorganisms8121861 PMID: 33255715

Scallan, E., Griffin, P., Angulo, F., Tauxe, R., & Hoekstra, R. (2011). Foodborne illness acquired in the United States—Unspecified agents. *Emerging Infectious Diseases, 17*(1), 16–22. DOI: 10.3201/eid1701.p21101 PMID: 21192849

Scallan, E., Hoekstra, R., Angulo, F., Tauxe, R., Widdowson, M., Roy, S., & Griffin, P. (2011). Foodborne Illness Acquired in the United States—Major Pathogens. *Emerging Infectious Diseases, 17*(1), 7–15. DOI: 10.3201/eid1701.p11101 PMID: 21192848

Schoch, C. L., Seifert, K. A., Huhndorf, S., Robert, V., Spouge, J. L., Levesque, C. A., & Chen, W. (2012). Nuclear ribosomal internal transcribed spacer (ITS) region as a universal DNA barcode marker for Fungi. *Proceedings of the National Academy of Sciences of the United States of America, 109*(16), 6241–6246. DOI: 10.1073/pnas.1117018109 PMID: 22454494

Schomberg, J., Haimson, O., Hayes, G., & Anton-Culver, H. (2016). Supplementing public health inspection via social media. *PLoS One, 11*(3), e0152117. DOI: 10.1371/journal.pone.0152117 PMID: 27023681

Schrader, C., & Schielke, A.. (2012). PCR inhibitors–occurrence, properties, and removal. *Journal of Applied Microbiology, 113*(5), 1014–1026. PMID: 22747964

Schultz, J., Jarquin, R., Ricke, S., & Hanning, I. (2012). Optimized culturing and nucleic acid-based methods for the detection of salmonella enterica in poultry environments. *Poultry Science, 91*(11), 2761–2766. DOI: 10.3382/ps.2012-02387 PMID: 23091129

Shangguan, L., XUE, L., Shang, J., and Wang, H. (2023). The application value of metagenomic next-generation sequencing in community-acquired purulent meningitis after antibiotics intervention.intervention. https://doi.org/DOI: 10.21203/rs.3.rs-3017522/v1

Shin, J., Miller, M., & Wang, Y. (2022). Recent advances in CRISPR-based systems for the detection of foodborne pathogens. *Comprehensive Reviews in Food Science and Food Safety, 21*(3), 3010–3029. DOI: 10.1111/1541-4337.12956 PMID: 35483732

Song, B., Wang, J., Yan, Z., Liu, Z., Pan, X., Zhang, Y., … and Zhang, X. (2020). Microfluidics for the rapid detection of

Sriramulu, V. (2024). Isolation and identification of bacterial pathogens associated with foodborne illnesses in stool samples: A prospective study. *International Journal of Medical and Biomedical Studies, 5*(6). Advance online publication. DOI: 10.32553/ijmbs.v5i6.2367

Stavropoulou, E., & Bezirtzoglou, E. (2019). Predictive modeling of microbial behavior in food. *Foods, 8*(12), 654. DOI: 10.3390/foods8120654 PMID: 31817788

Sun, X., Wang, Y., Zhang, L., Liu, S., Zhang, M., Wang, J., & Gao, Z. (2020). CRISPR-Cas9 triggered a two-step isothermal amplification method for E. coli O157:H7 detection based on a metal–organic framework platform. *Analytical Chemistry, 92*(4), 3032–3041. DOI: 10.1021/acs.analchem.9b04162 PMID: 31994379

Sun, Y., Zhong, K., Nie, L., Yan, X., Wang, Y., Yuan, J., & Jiang, L. (2024). State-of-the-art in food safety: The potential of CRISPR/CAS-based lateral flow assay technology in the field of food analysis. *Food Safety and Health, 3*(2), 155–171. DOI: 10.1002/fsh3.12082

Tao, Q., Wu, Q., Zhang, Z., Liu, J., Tian, C., Huang, Z., & Yong, Z. (2022). Meta-analysis for the global prevalence of foodborne pathogens exhibiting antibiotic resistance and biofilm formation. *Frontiers in Microbiology, 13*. Advance online publication. DOI: 10.3389/fmicb.2022.906490 PMID: 35774452

Teklemariam, A., Al-Hindi, R., Albiheyri, R., Alharbi, M., Alghamdi, M., Filimban, A., & Bhunia, A. (2023). Human salmonellosis: A continuous global threat in the farm-to-fork food safety continuum. *Foods, 12*(9), 1756. DOI: 10.3390/foods12091756 PMID: 37174295

Tongphrom, C., Preeprem, S., Nishibushi, M., Vuddhakul, V., & Mittraparp-arthorn, P. (2018). Rapid and sensitive enumeration of total coliforms and Eescherichia coli in water and foods by most-probable-number loop-mediated isothermal amplification (MPN-LAMP) method. *Scienceasia*, *44*(5), 365. DOI: 10.2306/scienceasia1513-1874.2018.44.365

Troja, F., Indio, V., Savini, F., Seguino, A., Serraino, A., Fuschi, A., & Cesare, A. (2024). Monitoring and preventing foodborne outbreaks: Are we missing wastewater as a key data source? *Italian Journal of Food Safety*. Advance online publication. DOI: 10.4081/ijfs.2024.12725 PMID: 39749179

Vanegas, D., Gomes, C., Cavallaro, N., Giraldo-Escobar, D., & McLamore, E. (2017). Emerging biorecognition and transduction schemes for rapid detection of pathogenic bacteria in food. *Comprehensive Reviews in Food Science and Food Safety*, *16*(6), 1188–1205. DOI: 10.1111/1541-4337.12294 PMID: 33371585

Vargas, G., Cypriano, J., Corrêa, T., Leão, P., Bazylinski, D., & Abreu, F. (2018). Applications of magnetotactic bacteria, magnetosomes, and magnetosome crystals in biotechnology and nanotechnology: Mini-review. *Molecules (Basel, Switzerland)*, *23*(10), 2438. DOI: 10.3390/molecules23102438 PMID: 30249983

Vidić, J., Vizzini, P., Manzano, M., Kavanaugh, D., Ramarao, N., Živković, M., & Gadjanski, I. (2019). Point-of-need DNA testing for the detection of foodborne pathogenic bacteria. *Sensors (Basel)*, *19*(5), 1100. DOI: 10.3390/s19051100 PMID: 30836707

Wang, J., Yang, C., Wang, T., Liu, K., Xu, Y., Yin, R., … and Tang, H. (2018). Study on rapid methods for detecting foodborne pathogens in frozen meat. https://doi.org/DOI: 10.23977/medeb.2018.07008

Wang, K., Lin, X., Zhang, M., Li, Y., Luo, C., & Wu, J. (2022). Review of Electrochemical Biosensors for Food Safety Detection. *Biosensors (Basel)*, *12*(11), 959. DOI: 10.3390/bios12110959 PMID: 36354467

Wang, L., Huang, F., Zhang, H., Yao, L., Zhang, H., & Lin, J. (2017). An electrochemical aptasensor using coaxial capillary with magnetic nanoparticle, urease catalysis, and a PCB electrode for rapid and sensitive detection of Escherichia coli o157:h7. *Nanotheranostics*, *1*(4), 403–414. DOI: 10.7150/ntno.22079 PMID: 29071202

Wang, Y., Du, P., Shao, Y., Wang, W., Liu, Y., Ma, Y., & El-Aty, A. (2024). An innovative and efficient fluorescent detection technique for Salmonella in animal-derived foods using the CRISPR/Cas12a-HCR system combined with PCR/RAA. *Journal of Agricultural and Food Chemistry*, *72*(15), 8831–8839. DOI: 10.1021/acs.jafc.3c08829 PMID: 38575365

Wang, Y., & Salazar, J. (2015). Culture-independent rapid detection methods for bacterial pathogens and toxins in food matrices. *Comprehensive Reviews in Food Science and Food Safety*, *15*(1), 183–205. DOI: 10.1111/1541-4337.12175 PMID: 33371580

Wei, C., & Zhao, X. (2018). Induction of viable but nonculturable Escherichia coli o157:h7 by low temperature and its resuscitation. *Frontiers in Microbiology*, *9*. Advance online publication. DOI: 10.3389/fmicb.2018.02728 PMID: 30555428

Wei, S., Forghani, F., Park, Y., Park, B., Seo, K., & Oh, D. (2016). Dynabeads protein G antibody conjugates combined with modified brain heart infusion broth for the enrichment and separation of Bacillus cereus in artificially contaminated vegetables. *Food Science and Biotechnology*, *25*(3), 941–947. DOI: 10.1007/s10068-016-0154-z PMID: 30263358

Wei, S., Park, B., Seo, K., & Oh, D. (2016). Highly efficient and specific separation of Staphylococcus aureus from lettuce and milk using Dynabeads protein G conjugates. *Food Science and Biotechnology*, *25*(5), 1501–1505. DOI: 10.1007/s10068-016-0233-1 PMID: 30263437

Wijayanti, S., Tsvik, L., & Haltrich, D. (2023). Recent Advances in Electrochemical Enzyme-Based Biosensors for Food and Beverage Analysis. *Foods*, *12*(18), 3355. DOI: 10.3390/foods12183355 PMID: 37761066

World Health Organization. (2015). Estimates of the global burden of foodborne diseases: Foodborne disease burden epidemiology reference group 2007–2015. World Health Organization. https://apps.who.int/iris/handle/10665/199350

Xiao, F., Li, W., & Xu, H. (2022). Advances in Magnetic Nanoparticles for the Separation of Foodborne Pathogens: Recognition, Separation Strategies, and Applications. *Comprehensive Reviews in Food Science and Food Safety*, *21*(5), 4478–4504. DOI: 10.1111/1541-4337.13023 PMID: 36037285

Xie, S., Yue, Y., and Yang, F. (2024). Recent Advances in CRISPR/Cas System-Based Biosensors for the Detection of Foodborne Pathogenic Microorganisms Micromachines, 15(11), 1329. https://doi.org/DOI: 10.3390/mi15111329

Xin, X., Su, J., Hao-ran, C., Wang, L., & Song, S. (2025). Recent Advances in Clustered Regularly Interspaced Short Palindromic Repeats (CRISPR)-. *Associated Protein System-Based Biosensors*, *15*(3), 155. DOI: 10.3390/bios15030155 PMID: 40136952

Xiong, Y., Zhang, J., Yang, Z., Mou, Q., Ma, Y., Xiong, Y., & Lu, Y. (2019). Functional DNA-regulated CRISPR-Cas12a sensors for point-of-care diagnostics of non-nucleic-acid targets. *Journal of the American Chemical Society*, *142*(1), 207–213. DOI: 10.1021/jacs.9b09211 PMID: 31800219

Yang, J., Howe, A., Lee, J., Yoo, K., & Park, J. (2020). An improved approach to identifying bacterial pathogens in humans from environmental metagenomes. *Journal of Microbiology and Biotechnology*, *30*(9), 1335–1342. DOI: 10.4014/jmb.2005.05033 PMID: 32627750

Yang, W., Cao, C., Wang, W., & Wang, Y. (2016). Rapid detection of Streptococcus agalactiae infection using a loop-mediated isothermal amplification method. *Jundishapur Journal of Microbiology*, *9*(12). Advance online publication. DOI: 10.5812/jjm.37171

Yordanova, R. and Andonova, A. (2024). Rapid methods for detection of foodborne pathogens and their toxins - a review of current trends and future perspectives. Journal of Imab - Annual Proceeding (Scientific Papers), 30(2), 5587-5592. https://doi.org/DOI: 10.5272/jimab.2024302.5587

Yoshitomi, K., Jinneman, K., Zapata, R., Weagant, S., & Fedio, W. (2012). Detection and isolation of low levels of E. coli o157:h7 in cilantro by real-time PCR, immunomagnetic separation, and cultural methods with and without an acid treatment. *Journal of Food Science*, *77*(8). Advance online publication. DOI: 10.1111/j.1750-3841.2012.02813.x PMID: 22860597

Yulandi, A., Suwanto, A., Waturangi, D., & Wahyudi, A. (2020). Shotgun metagenomic analysis reveals new insights into bacterial community profiles in tempeh. https://doi.org/DOI: 10.21203/rs.3.rs-42618/v2

Zaheer, R., Noyes, N., Polo, R., Cook, S., Marinier, E., Domselaar, G., & McAllister, T. (2018). The Impact of Sequencing Depth on Characterizing the Microbiome and Resistome. *Scientific Reports*, *8*(1). Advance online publication. DOI: 10.1038/s41598-018-24280-8 PMID: 29651035

Zeng, D., Chen, Z., Jiang, Y., Xue, F., & Li, B. (2016). Advances and Challenges in Viability Detection of Foodborne Pathogens. *Frontiers in Microbiology*, *7*. Advance online publication. DOI: 10.3389/fmicb.2016.01833 PMID: 27920757

Zhang, J., Yang, H., Li, J., Hu, J., Lin, G., Tan, B., & Lin, S. (2023). Current perspectives on viable but non-culturable foodborne pathogenic bacteria: A review. *Foods*, *12*(6), 1179. DOI: 10.3390/foods12061179 PMID: 36981106

Zhang, M., Liu, J., Shen, Z., Liu, Y., Song, Y., Liang, Y., & Zhao, Y. (2021). A newly developed paper-embedded microchip based on a lamp for rapid multiple detections of foodborne pathogens. *BMC Microbiology*, *21*(1). Advance online publication. DOI: 10.1186/s12866-021-02223-0 PMID: 34182947

Zhang, X., Shi, Y., Chen, G., Wu, D., Wu, Y., & Li, G. (2022). CRISPR/CAS systems-inspired nano/biosensors for detecting infectious viruses and pathogenic bacteria. *Small Methods*, *6*(10). Advance online publication. DOI: 10.1002/smtd.202200794 PMID: 36114150

Zhao, X., Lin, C. W., Wang, J., & Oh, D. H. (2014). Advances in rapid detection methods for foodborne pathogens. *Journal of Microbiology and Biotechnology*, *24*(3), 297–312. DOI: 10.4014/jmb.1310.10013 PMID: 24375418

Zheng, L., Zhang, H., Wang, S., Wang, M., & Lin, J. (2019). Optical biosensor for rapid detection of Salmonella typhimurium based on porous gold@platinum nanocatalysts and a 3d fluidic chip. *ACS Sensors*, *5*(1), 65–72. DOI: 10.1021/acssensors.9b01472 PMID: 31875386

Zhou, J., Deng, T., Zeng, Q., Wang, H., & Deng, C. (2024). An electrochemical biosensor for sensitive detection of live Salmonella in food via MXene amplified methylene blue signals and electrostatic immobilization of bacteriophages. https://doi.org/DOI: 10.21203/rs.3.rs-4649888/v1

Zhu, L., Chirase, N., & Ravishankar, S. (2014). Use of fulvic acid or sodium silicate-based sanitizers to inactivate Listeria monocytogenes, Salmonella typhimurium, and Pseudomonas aeruginosa on food contact surfaces. *Journal of Food Safety*, *34*(2), 132–140. DOI: 10.1111/jfs.12106

Zoellner, C., Al-Mamun, M., & Gröhn, Y., JACKSON, P., and Worobo, R. (2018). Postharvest supply chain with microbial travelers: A farm-to-retail microbial simulation and visualization framework. *Applied and Environmental Microbiology*, *84*(17). Advance online publication. DOI: 10.1128/aem.00813-18 PMID: 29959243

Chapter 5
Heavy Metal Contamination in Food: Monitoring, Detection, and Public Health Risks

Matthew Chidozie Ogwu
https://orcid.org/0000-0001-6054-1667
Appalachian State University, USA

ABSTRACT

Heavy metal contamination in food is a growing global concern due to its significant implications for public health and food security. Exposure poses substantial health risks through acute and chronic toxicity, organ-specific effects, and associations with severe health conditions, particularly affecting vulnerable populations. This chapter examines the sources, pathways, and prevalence of toxic metals, including arsenic, lead, mercury, and cadmium, in food systems, with a focus on both natural and anthropogenic contributions. It examines how these metals accumulate in the food chain and outlines their acute and chronic health impacts, particularly among vulnerable populations. The chapter provides an in-depth review of international regulatory standards, dietary risk assessments, and current analytical methodologies, including atomic absorption spectroscopy, inductively coupled plasma mass spectrometry, and portable detection systems. It also presents global monitoring strategies, case studies, and real-world surveillance data that highlight regional disparities and emerging challenges. Innovative approaches such as green chemistry, nanotechnology, and AI-driven tools are discussed as part of future directions. By integrating insights from environmental science, analytical chemistry, toxicology, and public health policy, this chapter provides a comprehensive perspective on the detection, measurement, and mitigation of heavy metals in food, thereby contributing to the development of safer and more resilient food systems worldwide.

DOI: 10.4018/979-8-3373-3982-5.ch005

1.0. INTRODUCTION

Heavy metals are defined as metallic elements having relatively high densities and are toxic at low concentrations. They include elements such as lead (Pb), cadmium (Cd), arsenic (As), chromium (Cr), and mercury (Hg) etc. (Orecchio et al., 2014; Chen and Ding, 2023). These contaminants can be classified based on their properties, environmental behaviors, and potential health impacts. While some heavy metals, such as copper and zinc, are essential in trace amounts, others are non-essential and pose significant health risks through contamination of food and water (Riyazuddin et al., 2021; Nwizugbo et al., 2023). The presence of these metals in the food supply can arise from various sources, including industrial discharge, agricultural runoff, and atmospheric deposition, particularly emissions from mining and smelting activities (Nkwunonwo et al., 2020; Abdullahi et al., 2021).

Monitoring heavy metals in food is crucial, as these contaminants pose a threat to food safety, ecosystem integrity, and human health (Wuana and Okieimen, 2011; Wang et al., 2023). Regular surveillance and detection of heavy metals in agricultural products, seafood, and processed foods are imperative to assess exposure levels and ensure compliance with health regulations, thereby mitigating risks (Hassan et al., 2020; Wang et al., 2023). Chronic exposure to heavy metals can result in severe health conditions, including neurotoxicity, carcinogenic effects, and developmental issues in children (Tom et al., 2014; Liang et al., 2017). As such, regulatory bodies increasingly focus on establishing stringent guidelines to protect consumers from the adverse effects of heavy metal exposure in food (Igwe et al., 2017; Chen and Ding, 2023). Initial assessments alert health authorities to potential risks, facilitating prompt public health responses. Continuous monitoring enables the identification of trends in heavy metal concentrations, informing policy and regulatory decisions regarding food safety standards (Wieczorek et al., 2020; Riyazuddin et al., 2021). Systematic monitoring enables identification of significant sources of contamination, which can be addressed through targeted interventions. Regions with systematic monitoring networks have successfully mitigated pollution levels through effective policy implementation (Ahmad et al., 2021; Hassan et al., 2020). Advancements in detection technologies, such as biosensors and nanomaterials, enhance the capacity to detect minute levels of heavy metals in food products, making surveillance more efficient and comprehensive (Wang et al., 2023).

Studies indicate that heavy metals can bioaccumulate in the food chain, leading to significant consequences for human health Hassan et al., 2020). For example, crops grown in contaminated soils have been found to contain elevated levels of heavy metals, which can then be ingested by animals and humans (Rahimi et al., 2017; Zaib et al., 2023). Communities, particularly those near industrial sites or agricultural lands, face heightened risks due to localized contamination (Ahmad et al., 2021).

Understanding the dynamics of heavy metal contamination, identifying sources, and implementing effective monitoring and regulatory measures is imperative. Enhanced methodologies for detecting these metals, such as inductively coupled plasma mass spectrometry (ICP-MS), are vital for researchers and regulatory agencies (Sibuar et al., 2022). Ensuring food safety through effective heavy metal monitoring protects public health, promotes environmental sustainability, and enhances food security across populations (Wuana and Okieimen, 2011; Abdullahi et al., 2021).

Heavy metals can enter the food chain through various pathways, affecting both terrestrial and aquatic food systems (Rahimi et al., 2017; Abdullahi et al., 2021). Soil contamination is often a primary source, where metals from industrial activities, agricultural inputs, and atmospheric fallout accumulate (Amponsah et al., 2020; Ahmad et al., 2021). Agricultural practices involving metal-containing fertilizers, pesticides, and fungicides can lead to the leaching of heavy metals into the soil ecosystem, which plants may absorb (Wuana and Okieimen, 2011; Zaib et al., 2023). Urban gardening, particularly in areas with prior industrial use, also poses significant risks as urban soils can harbor high levels of contaminants (Wieczorek et al., 2020; Abdullahi et al., 2021).

In aquatic environments, heavy metals can be introduced through runoff from contaminated land and concentrate in fish and shellfish, making these organisms significant vectors for human exposure (Heidarieh et al., 2013). Elevated metal levels in species such as shrimp and crabs, commonly consumed in food, have been documented (Heidarieh et al., 2013). Given their potential to accumulate harmful metals such as arsenic and cadmium, monitoring these species is crucial (Ikhajiagbe et al., 2021; 2022; Wang et al., 2023; Zaib et al., 2023). Food processing and preparation can also be exposure routes to heavy metals, especially with herbal medicines and powdered products, where contamination may go unrecognized. Reports have highlighted incidents of lead poisoning linked to Ayurvedic medication, underscoring the potential risks associated with herbal products (Breeher et al., 2015). Comprehensive risk assessments are necessary to understand how various food processing methods can either mitigate or exacerbate the presence of heavy metals in food items (Liang et al., 2017; Wang et al., 2023).

Public health concerns from heavy metal contamination in food are critical. Chronic exposure has been linked to various adverse health effects, including developmental disorders, neurotoxicity, and increased cancer risk (Liang et al., 2017). Vulnerable populations, particularly children, are at heightened risk due to the neurodevelopmental impacts of metals such as lead and cadmium (Liang et al., 2017; Zaib et al., 2023). Pregnant women and lactating mothers must also consider heavy metal exposure, which can adversely affect fetal and early childhood development (Liang et al., 2017; Ding, 2024). Regulatory attention to these issues has intensified, leading to stringent guidelines from health organizations worldwide. Agencies such as the World Health Organization (WHO) and food safety authorities recognize the urgent need for strict monitoring and regulation of heavy metals in food products (Wang et al., 2023 Hassan et al., 2020). International standards dictate maximum allowable levels of heavy metals in food, along with recommended dietary limits to minimize health risks associated with consumption (Tom et al., 2014; Riyazuddin et al., 2021; Ikhajiagbe and Ogwu, 2020).

This chapter aims to explore the extent, sources, and risks of heavy metal contamination in food, while highlighting advanced methods for monitoring and detection. It focuses on the public health implications of chronic exposure to toxic metals such as lead, mercury, cadmium, and arsenic, which can accumulate through various environmental and anthropogenic pathways. The chapter discusses international regulatory frameworks, dietary risk assessments, and state-of-the-art analytical techniques, including AAS, ICP-MS, and portable field-based tools. Through a review of current surveillance programs and case studies, it also addresses regional disparities, mitigation strategies, and emerging trends in food safety monitoring. The chapter's contribution to knowledge lies in its interdisciplinary synthesis of environmental science, toxicology, analytical chemistry, and public health policy, offering a holistic understanding of how heavy metal contamination in food is identified, measured, and managed globally.

2.0. SOURCES AND PATHWAYS OF HEAVY METAL CONTAMINATION OF FOOD

Natural sources of heavy metal contamination often arise from geological processes such as weathering and volcanic activity. Weathering of metal-rich minerals exposes heavy metals to the surface, facilitating their seepage into water bodies and soils, thus contributing to environmental pollution. Geological dissolution can release metals like Pb, Zn, and Cd into the ecosystem (Qu et al., 2012). Furthermore, volcanic eruptions can release particulate matter containing heavy metals into the atmosphere, which may later deposit onto land and water through precipitation events, exacerbating contamination (Osawaru et al., 2013a;b; Meera et al., 2022). Although significant, the extent of contamination posed by human activities often far surpasses that of natural processes.

Anthropogenic impacts, including industrial emissions, mining operations, and the use of agrochemicals, are major contributors to the increased presence of heavy metals in soils, water, and ultimately, the food system. Mining operations disturb large areas and generate substantial waste, leaching heavy metals into the surrounding environment. For instance, mining activities in Jiangsu Province, China, present specific threats, with documented pathways of contamination extending through soil saturation and groundwater (Qu et al., 2012). Industrial discharges are equally concerning; heavy metals tend to accumulate in sediment and soil profiles, particularly in areas contaminated by effluents from smelting operations (Orji et al., 2024). The continuous release of metals from industrial sites poses significant health risks via food crops, which can uptake these toxins from contaminated soils (Khan et al., 2015; Richard et al., 2022).

Agricultural practices employing wastewater irrigation introduce significant risks of heavy metal exposure (Iyiola et al., 2023; 2024). Crops grown in contaminated soils can accumulate heavy metals in their edible parts, which poses serious health risks to consumers (Emurotu and Onianwa, 2017). The practice of using industrial effluents as irrigation water raises critical concerns regarding food safety, as studies have indicated the presence of elements such as arsenic and cadmium at levels exceeding safety standards in crops intended for human consumption (Fan et al., 2017). This underscores the urgent need for robust regulatory frameworks to monitor and restrict allowable heavy metal concentrations in agricultural inputs (Izah et al., 2024).

Soil and water are crucial vectors for the transport of heavy metals throughout the environment. These metals can remain bound in soils and sediments, becoming mobile under changing environmental conditions such as variations in pH or redox states (Singare et al., 2012). This mobility enables the gradual release of contaminants into water systems, facilitating bioaccumulation in aquatic organisms. Trophic interactions in these environments result in biomagnification, where higher trophic levels accumulate progressively higher concentrations of heavy metals, posing significant risks to both wildlife and human populations (Loaiza et al., 2021). In marine ecosystems, for example, aquaculture practices can lead to high trophic magnification factors (TMFs) for metals such as arsenic and cadmium, dramatically escalating potential human health hazards (Loaiza et al., 2021).

The implications of heavy metal bioaccumulation extend beyond ecological health, encompassing public health issues as well. Human exposure typically occurs via ingestion, dermal contact, and inhalation, as highlighted by studies examining agricultural outputs and drinking water sources (Khan et al., 2015; Alidadi et al., 2019). Notably, foodstuffs such as rice, vegetables, and mushrooms can reflect heavy metal concentrations significantly higher than recommended safety thresholds, consequently implicating dietary patterns in exposure risk assessments (Gergen and Hărmănescu, 2012; Saha et al., 2023). As the persistence of heavy metals in agricultural soils contributes to long-term risks, it is imperative to consider both exposure pathways and their cumulative health impacts.

Research has established credible connections between heavy metal exposure from contaminated food and various adverse health outcomes, including neurotoxicity, cardiovascular problems, and carcinogenic effects (Alidadi et al., 2019). Long-term exposure to these metals accumulates in human tissues, leading to chronic health issues that are often challenging to mitigate given their multifaceted environmental origin (Emurotu and Onianwa, 2017). In particular, the consumption of crops grown in heavily contaminated soils poses a significant challenge, as established safety margins often do not account adequately for cumulative exposure across different food sources within a typical diet. Mitigating the impacts of heavy metal contamination necessitates integrated approaches, including phytoremediation and sustainable agricultural practices. Utilizing plants like particular species of mushrooms or other bio-accumulative crops can help in the remediation of contaminated soils, thereby reducing the levels of heavy metals available in the food chain (Gebru, 2019). However, careful planning and control are necessary, as the use of contaminated produce can lead to health risks, highlighting the delicate balance between agriculture and environmental health (Blanco et al., 2020). Ongoing research continues to unravel the nuances of bioaccumulation and trophic interactions, underscoring the need for enhanced environmental monitoring and refined regulatory measures to align agricultural practices with public health objectives. Fostering community awareness about the sources of heavy metal contamination and enhancing collaboration between public health, agriculture, and environmental sciences will help effectively mitigate exposure pathways.

3.0. COMMON HEAVY METALS FOUND IN FOOD

The presence of heavy metals in food products poses significant health risks, necessitating scrutiny and regulation to mitigate exposure. Heavy metals such as arsenic (As), lead (Pb), mercury (Hg), cadmium (Cd), and chromium (Cr) are commonly found in various food sources, often introduced through environmental contamination, agricultural practices, and food processing methods. Table 1 provides an overview of some of the most encountered heavy metals in food systems, highlighting their primary environmental sources, potential toxicological effects, and typical food items in which they are found. The potential for bioaccumulation and the profound health implications associated with these metals highlight the urgency of ongoing research and regulatory measures in food safety.

Arsenic is a toxic element frequently detected in rice, seafood, and other crops, primarily due to its prevalence in groundwater and soil associated with agricultural runoff and pesticide use (Uhegbu et al., 2012). Arsenic accumulation in food is influenced by geographical factors, particularly in regions like the Niger Delta, where polluted water bodies expose aquatic organisms to higher arsenic levels, thereby entering the human food chain (Uhegbu et al., 2012). Concerns surrounding arsenic's potential carcinogenic effects and its role in various diseases underscore the need for close monitoring of food sources known to contain this heavy metal, particularly in staple foods such as rice, which may accumulate arsenic from flooded fields (Uhegbu et al., 2012).

Lead, another heavy metal commonly found in food, is primarily introduced through environmental pollution and contamination from industrial activities (Mukhi et al., 2022). Studies have shown that lead can migrate from ceramic food packaging into food products, particularly acidic items, further contributing to dietary exposure (Dong et al., 2014). Accumulation in tissues and organs can have severe neurological implications, particularly affecting children and pregnant women who are more vulnerable to lead-induced neurotoxicity Hussain, 2012). Recent assessments of urban environments consistently highlight elevated levels of lead in agricultural products, underscoring an urgent need for urban agriculture policies centered on heavy metal monitoring in crops (Othman et al., 2013).

Mercury exposure is predominantly associated with the consumption of certain types of fish and shellfish, where bioaccumulation through the aquatic food chain significantly raises mercury concentrations in larger predator species (Ogundele et al., 2020). The organic form, methylmercury, is particularly concerning as it preferentially accumulates in fish tissue and is readily absorbed by the human gastrointestinal system, leading to neurodevelopmental disorders and cognitive impairments, especially in children Hussain, 2012). Comprehensive assessments of seafood have demonstrated that fish sourced from contaminated waters can exhibit mercury levels that significantly exceed health safety thresholds, underscoring the necessity for regular monitoring of seafood sources and issuing public advisories about safe consumption limits (Mukhi et al., 2022).

Cadmium (Cd), as a non-essential heavy metal, poses serious health risks due to its propensity to bioaccumulate primarily in kidneys and liver tissues (Solidum et al., 2013). The food chain is often contaminated through polluted soil and water, where cadmium uptake occurs, particularly in staple crops such as rice and vegetables (Solidum et al., 2013). Regulatory bodies have established acceptable limits for cadmium in foodstuffs; however, breaches of these guidelines are frequently noted in agricultural products, prompting calls for improved practices and surveillance in food production systems (Othman et al., 2013; Hussain, 2012). Chronic exposure to cadmium can lead to renal dysfunction, bone fragility, and other adverse health consequences, making it imperative to engage in risk assessment methodologies focused on cadmium levels in food (Solidum et al., 2013).

Chromium's dietary relevance is twofold, involving both essential trivalent chromium (Cr III), which plays a role in glucose metabolism, and the toxic hexavalent chromium (Cr VI), known for its carcinogenic properties (Henriksen and Bügel, 2023). The migration of chromium from stainless steel cookware into food during cooking processes has been documented, raising concerns regarding chronic dietary exposure (Ohira et al., 2015). The wide variability in chromium levels in different food sources can be attributed to various factors, including soil type, agricultural management practices, and contamination from food processing (Thor et al., 2011). Research highlights the necessity of differentiating between chromium species in food safety assessments to understand their respective health risks better (Pyrzyńska, 2016; Boon et al., 2010).

Moreover, while metals like nickel, copper, and zinc play essential roles in human health, excessive concentrations can lead to toxicity, underscoring the need for balance and regulation (Wu and Yang, 2011). Nickel primarily arises from agricultural inputs and packaging materials, while high levels of copper may result from plumbing systems or food processing equipment (Wu and Yang, 2011). It is crucial to establish tolerable upper intake levels to mitigate potential toxicological effects associated with overconsumption of these metals (Henriksen and Bügel, 2023).

Table 1. Some common heavy metals found in food and their health effects

Heavy Metal	Primary Sources	Potential Health Effects	Common Food Items
Lead	Contaminated soil, water pipes, industrial pollution, lead-based pesticides	Neurological damage, developmental delays in children, and kidney dysfunction	Leafy vegetables, root crops, grains, and drinking water
Mercury	Industrial emissions, mining runoff, and atmospheric deposition	Neurotoxicity, cognitive impairment, and kidney damage	Fish (especially large predatory species like tuna, swordfish), shellfish
Cadmium	Phosphate fertilizers, industrial waste, cigarette smoke, and contaminated water	Kidney damage, bone demineralization, carcinogenic potential	Rice, leafy vegetables, cereals, and shellfish
Arsenic	Contaminated groundwater, industrial pollution, arsenic-based pesticides	Skin lesions, cancer (skin, lung, bladder), cardiovascular disease	Rice (especially from flooded fields), seafood, and drinking water
Chromium (Cr) (especially Cr(VI))	Electroplating, leather tanning, and industrial waste	Respiratory problems, gastrointestinal distress, carcinogenic effects	Contaminated meat, grains, and vegetables
Nickel	Industrial effluents, stainless steel equipment, and soil contamination	Allergic reactions, respiratory issues, and potential carcinogenicity	Chocolate, nuts, legumes, and grains
Aluminum	Food additives, aluminum cookware, and contaminated water	Neurotoxicity, Alzheimer's disease (potential link), bone disorders	Processed cheese, baking powder, and tea leaves

4.0. HEALTH EFFECTS OF HEAVY METAL EXPOSURE FROM FOOD

Heavy metal exposure represents a significant public health concern due to its extensive impact on human health, including acute and chronic toxicity, organ-specific effects, heightened vulnerability in certain populations, and links to non-communicable diseases (NCDs) such as cancer and cardiovascular diseases (Figure 1).

Acute and Chronic Toxicity: Acute toxicity from heavy metal exposure occurs following short-term, high-level exposure, leading to rapid health decline and possibly fatal outcomes. Metals such as Cd, Pb, Hg, and As exhibit potent toxic effects. For instance, cadmium can cause respiratory failure when inhaled and severe damage to kidneys following ingestion (Table 1; Jayasumana et al., 2015), whereas lead exposure has been associated with encephalopathy and peripheral neuropathy Balali-Mood et al., 2021). Even lower doses of heavy metals can lead to chronic toxicity, which develops over extended periods and manifests in slower, insidious health effects, such as progressive organ dysfunction and cancer (Jan et al., 2015; Engwa et al., 2019). Specifically, chronic exposure can result in cumulative payloads of metals in the body that can disrupt metabolic processes, damage DNA, and contribute to neurodegenerative diseases (Engwa et al., 2019)

Organ-Specific Effects: The effects of heavy metals on specific organs are well-documented. For instance, mercury is particularly neurotoxic, impacting cognitive function and leading to neurobehavioral disorders. Chronic exposure can result in syndromes like Minamata disease, characterized by paralysis and cognitive impairment (Tchounwou et al., 2012). Cadmium's nephrotoxic properties have been shown to affect renal functions even at low concentrations, leading to renal insufficiency and increased proteinuria (Jayasumana et al., 2015). Similarly, lead's effects on the cardiovascular system have been linked to hypertension and increased risk of myocardial infarctions (Nguyen et al., 2021; Ohiagu et al., 2022). Furthermore, many heavy metals can induce oxidative stress by producing reactive oxygen species (ROS), leading to cell damage and contributing to various pathologies, including liver diseases, which are recognized as hepatotoxic outcomes. Under these conditions, fatty liver disease may ensue due to metal accumulation in hepatocytes (Orr et al., 2018).

Vulnerable Populations: Certain demographic groups are susceptible to the effects of heavy metal exposure. Children are especially vulnerable due to their developing organ systems and higher absorption rates relative to adults. Studies indicate that lead exposure in children, especially during crucial brain development periods, can lead to permanent cognitive deficits and behavioral problems (Rehman et al., 2017). Pregnant women are also at heightened risk; exposure to certain metals during gestation can lead to adverse birth outcomes, including low birth weight, preterm birth, and developmental delays in offspring (Gupta et al., 2015; Bibi,

2023). Consequently, both children and pregnant women require stringent protective measures to mitigate their exposure to environmental heavy metals.

Links to Non-Communicable Diseases: The association between heavy metal exposure and various non-communicable diseases (NCDs) is increasingly recognized. Heavy metals like cadmium, lead, and mercury have been implicated in the etiology of several NCDs, notably including cardiovascular diseases (CVDs), wherein mechanisms often involve systemic inflammation and oxidative stress, contributing to endothelial dysfunction (Nguyen et al., 2021; Ohiagu et al., 2022; Ogwu et al., 2025). Prolonged exposure not only raises the risk of CVD but has also been implicated in the higher prevalence of certain cancers, where heavy metals operate via mutagenic pathways that can trigger oncogenesis (Engwa et al., 2019; Tchounwou et al., 2012). Review studies have noted a compelling link between heavy metal toxicity and the incidence of diabetes, obesity, and metabolic syndrome due to the disturbance of normal metabolic homeostasis (Jan et al., 2015; Balali-Mood et al., 2021). Additionally, heavy metals' estrogen-like effects might also disrupt reproductive health and increase risks of hormone-related cancers (Ohiagu et al., 2022; Osawaru and Ogwu 2023; Osawaru and Ogwu 2023; Ogwu et al.,2023).

Source: Adapted from Alengebawy et al. (2021).

5.0. REGULATORY STANDARDS AND RISK ASSESSMENT FOR HEAVY METALS IN FOOD

To protect public health, national and international agencies have established regulatory limits for heavy metals in food. These standards are based on toxicological assessments, including tolerable daily intakes (TDI) or provisional tolerable weekly intakes (PTWI), and are used to evaluate the safety of food products. Table 2 provides an overview of commonly regulated heavy metals, selected international guideline limits, and associated risk assessment benchmarks. The regulatory framework addressing heavy metals' presence in food is multifaceted, encompassing guidelines from international organizations such as the Codex Alimentarius, World Health Organization (WHO), Food and Agriculture Organization (FAO), the European Union (EU), and the Food and Drug Administration (FDA) of the United States. These regulations establish Maximum Permissible Levels (MPLs) for specific heavy metals in food products to mitigate health risks associated with their consumption. For instance, the WHO has emphasized the importance of regulating heavy metal levels to safeguard public health, indicating that exposure to these metals can result in detrimental health effects, including neurotoxicity and carcinogenicity (Zafarzadeh et al., 2018; Ebrahimi et al., 2020). Similarly, the EU has established stringent directives that limit heavy metal concentrations in various foods to promote consumer safety and public health safeguards (Birdanova et al., 2023; Obongodot and Ogwu 2023). In assessing dietary exposure to heavy metals, the concept of Acceptable Daily Intake (ADI) is crucial. It enables health organizations to define safe consumption levels over extended periods without adverse health outcomes. For example, studies have indicated that dietary exposures to heavy metals typically remain below the established ADIs in regions with effective regulatory frameworks, such as North America and Europe (García et al., 2017; Huang et al., 2013). Conversely, regions with less stringent regulations may report exceedances of these safety limits, showcasing the disparities that exist globally (Chowdhury and Alam, 2023). Zafarzadeh et al. highlight this variability, illustrating areas where heavy metal contamination in food is problematic, particularly in regions with high incidences of cancer linked to dietary intake of contaminated vegetables (Zafarzadeh et al., 2018).

Risk assessment frameworks for heavy metals in food take a systematic approach by segmenting the evaluation process into key components: hazard identification, exposure assessment, and risk characterization. Hazard identification involves recognizing specific heavy metals that pose significant health threats, such as lead, mercury, and cadmium. Subsequent exposure assessments quantitatively evaluate actual intake levels of these metals from food sources, potentially employing biomonitoring techniques that measure metal concentrations in human tissues (Zuo et al., 2023). Risk characterization synthesizes this information, weighing the probability of adverse health effects against the intensity of exposure, which helps policymakers establish necessary interventions or regulatory controls (Zuo et al., 2023). For instance, Zuo et al. (2023) provide insights into the application of deterministic and probabilistic risk models in the context of traditional Chinese medicines, further supporting the framework's relevance in dietary assessment across various food categories.

The complexity of heavy metal exposure, alongside its chronic health impacts, necessitates an integrated approach combining risk assessment and public health strategies. The interrelationships between dietary practices and heavy metal accumulation should be scrutinized rigorously. Research indicates that specific environmental conditions, such as soil composition and use of contaminated irrigation practices, escalate the risk of heavy metals entering the food chain (Jan et al., 2010). Additional studies suggest that food processing practices can exacerbate these concerns, potentially leading to increased contamination levels in final food products (Abdulla et al., 2019). Thus, comprehensive regulatory standards that explicitly monitor and control heavy metal levels in agricultural and food commodities are vital for effectively mitigating exposure risks.

As noted by Rostamzadeh et al. (2025), heavy metal accumulation in food such as rice, vegetables, and meats presents a risk of exposure that can lead to severe health issues, including osteoporosis and various forms of cancer. This finding aligns with ongoing research in regions such as Ahvaz, Iran, where heavy metal exposure through vegetable consumption correlates significantly with health outcomes, including esophageal cancer (Adedayo and Ediagbonya, 2018). Enhanced regulatory measures and continuous monitoring are thus required, especially in regions with burgeoning industrial activities that may increase pollution pressures on agricultural lands (Jolly et al., 2013). Continuous dialogue between scientific communities, regulatory bodies, and consumer advocacy groups is critically needed to promote public health while maintaining agricultural viability. The complexity of the regulatory and risk assessment landscape regarding heavy metals in food necessitates diligent and systematic approaches to ensure safety standards are met and maintained. Robust data-driven policies grounded in comprehensive scientific assessments will facilitate public health protection against the health risks posed by heavy metals. The collaboration of international organizations and national regulatory frameworks remains paramount in accomplishing these objectives, and developing a unified, coherent strategy will be essential for progressing toward safer food systems globally.

Table 2. Regulatory Standards and Risk Assessment Benchmarks for Heavy Metals in Food

Heavy Metal	Regulatory Limit in Food(Codex/EFSA/WHO)	Risk Assessment Reference	Key Regulatory Agencies
Lead	0.1 mg/kg (vegetables),0.3 mg/kg (cereals)	PTWI: Withdrawn (no safe level identified)	Codex Alimentarius, EFSA, WHO
Mercury	0.5 mg/kg (fish),1.0 mg/kg (predatory fish)	PTWI: 1.6 µg/kg body weight (MeHg)	WHO, EFSA, US EPA
Cadmium	0.05–0.1 mg/kg (crops),0.2 mg/kg (seafood)	TWI: 2.5 µg/kg body weight (EFSA)	Codex, EFSA, FAO
Arsenic	0.2 mg/kg (rice) (inorganic)	$BMDL_{0.5}$: 3 µg/kg bw/day (skin lesions)	WHO, EFSA
Chromium	No Codex standard (Cr VI prohibited)	No PTWI; risk based on Cr speciation	US EPA, EFSA
Nickel	0.2–0.5 mg/kg (varies by food)	TDI: 13 µg/kg body weight/ day	EFSA
Aluminum	1–10 mg/kg (processed foods)	TWI: 1 mg/kg body weight	EFSA, JECFA

6.0. ANALYTICAL TECHNIQUES FOR HEAVY METAL DETECTION AND QUANTIFICATION IN FOOD

Heavy metals in food products pose serious health risks due to their toxic nature and potential for bioaccumulation. Therefore, analytical techniques for detecting and quantifying heavy metals in food are crucial.

6.1. Sample Collection and Preparation Techniques

Effective detection of heavy metals begins with proper sample collection and preparation, which are essential for minimizing contamination and ensuring representational accuracy. Heavy metal accumulation can vary significantly depending on environmental factors, sample type, and the species involved, making sample integrity crucial for accurate analysis (Zarazúa et al., 2014). Standard preparation techniques include digestion, which is essential for breaking down the matrix and liberating metals for quantification. Acid digestion techniques, using concentrated nitric acid or a combination of acids, are frequently employed in laboratories to prepare food samples, as they are effective in solubilizing metals (Hong et al., 2018; Scăe eanu et al., 2021). The protocol selected can significantly influence the final results, as improper digestion can lead to under- or overestimation of heavy metal levels.

During sample processing, standardized procedures that enhance accuracy must be strictly adhered to. The use of certified reference materials is recommended to ensure measurement accuracy and precision, thereby reducing analytical uncertainties (Byers et al., 2019). The proper selection of these materials is paramount, as discrepancies in matrix compositions can lead to significant variability (Nguyen et al., 2023). Furthermore, techniques such as microwave-assisted digestion have gained favor due to their efficiency and ability to reduce the need for toxic reagents, thereby affording cleaner sample preparation (Deng et al., 2011).

6.2. Instrumental Methods

6.2.1 Atomic Absorption Spectroscopy (AAS)

AAS is a widely used technique for detecting specific heavy metals in food and environmental samples. It operates on the principle of measuring the absorption of light by free atoms, allowing for detection limits in the range of micrograms per liter (Shih et al., 2019). However, AAS is often criticized for its limited multi-element capability and longer throughput times compared to modern techniques, such as ICP-MS. Nonetheless, it remains valued for specific applications where selective detection of target metals is sufficient.

6.2.2 Inductively Coupled Plasma Mass Spectrometry (ICP-MS)

ICP-MS has emerged as a standard for analyzing trace metals due to its exceptional sensitivity and speed. With limits of detection often in the low parts per trillion, ICP-MS enables the quantification of multiple heavy metals simultaneously (Mazarakioti et al., 2022; Nguyen et al., 2023). This high level of sensitivity is vital for regulatory compliance and health assessments, especially given the increasing concerns about the effects of low-level exposure to heavy metals. Advances in sample introduction systems have further enhanced the throughput of ICP-MS, making it amenable for routine analysis in food safety labs.

6.2.3 Inductively Coupled Plasma Optical Emission Spectroscopy (ICP-OES)

Similar to ICP-MS, ICP-OES is commonly used for elemental analysis due to its ability to detect a wide range of metals. ICP-OES offers excellent multi-elemental analysis capabilities at a relatively lower operational cost compared to ICP-MS (Hong et al., 2018). It is particularly advantageous when analyzing metals at higher concentrations where sub-ppb detection is not critical. Recent improvements in detector technology have enhanced its performance, enabling quick and efficient sample analysis (Voica et al., 2020).

6.2.4 X-ray Fluorescence (XRF)

XRF is a non-destructive technique that facilitates the rapid assessment of heavy metals in various matrices, including food products. Its operational principle involves measuring the fluorescent X-rays emitted from the sample when excited by an X-ray source (Scăe eanu et al., 2021). XRF provides immediate results and thus is increasingly being utilized in field studies for quick assessments of metal contamination (Hu et al., 2017). The sensitivity of XRF can be enhanced through calibration with suitable reference materials. Although its limitations in depth analysis are notable, its portability makes it an attractive option for preliminary surveys.

6.3. Sensitivity, Accuracy, and Method Validation

The sensitivity of analytical methods significantly influences the capability to detect hazardous metal levels in food. Studies have shown that methods such as ICP-MS offer substantial advantages over AAS and ICP-OES in terms of sensitivity (Mazarakioti et al., 2022). Validation of analytical methods is crucial to establish their reliability for regulatory purposes. This involves demonstrating linearity, precision, accuracy, and detection limits. For instance, a study found that ICP-MS provided a detection limit as low as 0.3 mg/kg for aluminum, making it a preferred choice for many food matrices (Lin et al., 2014). Moreover, the validation protocols should incorporate the evaluation of recovery rates, which ideally should fall within 80-120% for accurate quantification (Deng et al., 2011). Using certifiable standard reference material helps establish the authenticity of the results achieved from these analytical methods, ensuring compliance with food safety regulations set by governing bodies such as the FDA and EFSA.

6.4. Advances in Portable and Field-Based Sensors

Recent advancements in portable sensing technologies have greatly enhanced the ability to conduct on-site analysis of heavy metals in food and environmental matrices. Techniques such as portable X-ray fluorescence (PXRF) and other miniaturized devices have been developed, allowing for rapid assessment of contamination levels (Hu et al., 2017; Hutton et al., 2014). These devices are handy in settings where traditional laboratory analysis may be impractical due to logistical or time constraints. Tabletop and handheld units have been successfully employed in field trials, yielding results comparable to those of mainstream laboratory methods. For instance, studies employing PXRF have demonstrated capabilities to detect metals at concentrations below regulatory thresholds, highlighting its potential as a screening tool (Carr et al., 2007; Choi et al., 2014). The integration of GIS capabilities with portable XRF technology not only streamlines data collection but also enhances spatial analysis, promoting more effective assessment and management of metal pollution in food supplies.

5.0. INTEGRATION OF NANOTECHNOLOGY AND BIOSENSORS

Recent advances in nanotechnology and biosensor development have significantly enhanced the ability to monitor heavy metals in food matrices. Nanomaterials possess unique physicochemical properties, including high surface-to-volume ratios, tunable optical and electronic behavior, and exceptional reactivity, which make them ideal candidates for enhancing analytical performance in contaminant detection systems (Singh et al., 2020; Zhang et al., 2022). Gold nanoparticles (AuNPs), silver nanoparticles (AgNPs), carbon nanotubes, and quantum dots are among the most widely used nanomaterials integrated into sensor platforms to detect toxic elements such as arsenic, lead, cadmium, and mercury in food and environmental samples (Chen et al., 2019; Das et al., 2021). Biosensors, which couple biological recognition elements (e.g., enzymes, antibodies, DNA probes) with transducers to generate quantifiable signals, have evolved into more sensitive and specific tools through the incorporation of nanotechnology. These nano-biosensors exhibit superior detection limits, real-time responsiveness, and portability, making them ideal for on-site applications (Dey et al., 2021; Kaur and Jain, 2020). Electrochemical nano-biosensors have been especially effective in detecting lead and cadmium in vegetables and cereals, with some platforms achieving limits of detection in the parts-per-billion (ppb) or even parts-per-trillion (ppt) range (Wang et al., 2018).

Colorimetric assays based on gold nanoparticles and fluorescence resonance energy transfer (FRET) sensors utilizing quantum dots have demonstrated reliable results in identifying mercury and arsenic in seafood and water samples (Liu et al., 2020; Chandra et al., 2023). Additionally, paper-based nano-biosensors are emerging as low-cost, disposable alternatives, particularly useful in low-resource settings (Ahmed et al., 2022). These platforms align well with the growing demand for decentralized and consumer-accessible food testing technologies. However, despite their promise, challenges such as sensor stability, reproducibility, and cross-reactivity remain. Moreover, the regulatory validation and standardization of nano-enabled devices for routine monitoring are yet to be fully established (González et al., 2019; Jiao et al., 2021). There is also a growing interest in integrating biosensors with digital platforms, including Internet of Things (IoT) frameworks and artificial intelligence (AI) algorithms, to enable real-time data transmission and predictive risk modeling in food supply chains (Patel et al., 2023).

6.0. MONITORING STRATEGIES AND SURVEILLANCE PROGRAMS FOR HEAVY METALS IN FOOD

The surveillance and monitoring of heavy metals in food have garnered significant attention due to the increasing awareness of their potential health risks and environmental impact (Ogwu 2024). Effective national and international monitoring programs play a crucial role in ensuring food safety and public health by addressing heavy metal contamination, particularly in food categories such as seafood, vegetables, dairy, and grains like rice. These programs encompass governmental regulations, scientific research, and public health initiatives aimed at tracking and mitigating risks associated with heavy metal exposure.

National agencies, such as the U.S. Food and Drug Administration (FDA) and the EFSA, regularly conduct assessments and set safety thresholds for heavy metals in foods, aiming to maintain levels deemed acceptable for human consumption. For instance, heavy metals such as lead, cadmium, and mercury are regulated, and monitoring efforts strive to ensure that the concentrations in food products remain below established safety limits (Deka et al., 2023). Internationally, organizations such as the WHO and the FAO provide guidelines and recommendations for tolerable levels of heavy metals in food safety standards (Ishida et al., 2022). Specific food categories, notably seafood, rice, and vegetables, are particularly vulnerable to heavy metal contamination due to environmental factors and agricultural practices. Seafood, especially predatory fish like tuna, often accumulates significant levels of mercury and other heavy metals due to their high trophic position in the food chain, leading to bioaccumulation and posing risks to consumers (Pappalardo et al., 2017). Similarly, rice can become contaminated with arsenic through irrigation with contaminated water or soil pollution (Drambozova, 2021). Vegetables grown in environments contaminated with heavy metals may take up these toxins. At the same time, some studies have indicated that dairy products can also contain heavy metals, such as cadmium and lead, exceeding maximum residue limits (Ejaz et al., 2022; Ogwu et al., 2025a;b).

Case studies highlight the outcomes of surveillance programs and the significance of monitoring strategies. For example, an analysis of heavy metal contamination in vegetables in Pakistan revealed concerning transfer factors, indicating substantial risk to human health via food ingestion (Rehman et al., 2018). In contrast, effective regulatory policies in New York aimed at reducing heavy metal levels in imported spices have shown the potential effectiveness of local regulatory strategies (Ishida et al., 2022). Such case studies underscore the importance of adaptability in surveillance outcomes and the need for comprehensive strategies that incorporate local contexts and specific food risks (Ogwu 2023).

Moreover, the establishment of early warning systems is an essential component of comprehensive heavy metal monitoring. These systems often rely on real-time data collection from various sources, including environmental surveys, food safety inspections, and health risk assessments. An efficient early warning system enables rapid identification of contamination events, allowing for timely public health responses. For example, such systems can trigger alerts when heavy metal levels exceed safe thresholds, facilitating immediate action to limit consumer exposure and improve food safety practices (Deka et al., 2023). A multifaceted approach to monitoring heavy metals in food is vital for protecting public health. This approach involves national and international collaboration in regulatory policies and protocols, prioritizing food categories most susceptible to contamination, learning from adequate surveillance case studies, and implementing robust early warning systems. Given the varied sources of heavy metal contamination, ongoing research and adaptation of strategies are necessary to mitigate the risks posed by these persistent environmental pollutants in our food supply.

Table 3. Monitoring Strategies and Surveillance Programs for Heavy Metals in Food

Strategy/Program	Benefits	Implementing Agencies/Regions
Routine Food Sampling	Regular collection and testing of food items for heavy metal residues	National Food Safety Authorities (e.g., FDA, EFSA, Ghana FDA)
Total Diet Studies (TDS)	Composite food analysis based on dietary habits to estimate exposure levels	WHO, FAO, EFSA, China CDC
Risk-Based Surveillance	Targeted monitoring of high-risk food items and regions based on historical data	Codex Alimentarius, national agencies
Import/Export Control Programs	Border inspection of imported/exported foods for compliance with metal limits	USDA, EU RASFF, Codex member countries
Environmental Monitoring	Soil, water, and air quality assessments near agricultural zones	EPA, UNEP, local environmental agencies
Mobile and Rapid Testing Kits	On-site testing tools for quick screening of contaminants	NGOs, the food industry, and academic research
Integrated One Health Surveillance	Collaborative data collection across food, animal, and environmental health	WHO, FAO, OIE, National One Health Platforms

7.0 RISK MITIGATION AND CONTROL MEASURES FOR HEAVY METALS IN FOOD

Several historical incidents have underscored the severity of heavy metal contamination in food systems. For instance, Minamata disease, resulting from mercury contamination in fish, highlighted the perils of industrial discharge into marine ecosystems (Priya et al., 2023). Similarly, the arsenic contamination in rice in some parts of Asia has been attributed to both environmental and agricultural factors, necessitating vigilance in monitoring and remediation (Oh et al., 2015). Such cases serve as critical learning points that inform future practices and highlight the need for sustained intervention efforts and stricter regulations in managing heavy metals in food supplies. Monitoring efforts for heavy metal contamination in food differ significantly between developing and developed nations. Many developed countries have established rigorous monitoring schemes that involve stringent testing of soil, water, and food products for heavy metals, ensuring compliance with safety standards (Wuana and Okieimen, 2011).

In contrast, developing nations often struggle with insufficient regulatory frameworks and limited resources for monitoring and remediation efforts. The resulting disparity leads to heightened risks of heavy metal exposure in the food supply chains of these regions Oh et al., 2015). The lack of technological advancements and educational resources can inhibit the implementation of effective monitoring systems, exacerbating the problem of soil and water contamination from industrial activities and agricultural practices in many developing countries (Wuana and Okieimen, 2011). Thus, targeted efforts to enhance both policy and practice in these regions are essential to align with global safety standards. Heavy metal contamination in food systems poses significant health risks and environmental challenges, necessitating the implementation of effective risk mitigation and control measures. Various strategies can be employed to address the multifaceted nature of this issue, ranging from agricultural practices to remediation techniques and policy interventions. Each approach plays a crucial role in managing and mitigating the impacts of heavy metals in the agricultural and food production landscape.

7.1. Good Agricultural and Manufacturing Practices (GAP, GMP)

Good Agricultural Practices (GAP) and Good Manufacturing Practices (GMP) are foundational to ensuring food safety and minimizing contamination from heavy metals. GAP encompasses a range of practices that guide farmers on cultivating crops in ways that reduce the likelihood of heavy metals entering the food chain. This includes soil testing prior to planting, careful selection of fertilizers and amendments, and employing protective measures during harvesting and post-harvest processing (Wuana and Okieimen, 2011). These practices align with the necessity to adopt environmentally sound strategies in food production, reducing risks associated with heavy metal contamination. Furthermore, GMP plays a parallel role in the manufacturing sector, ensuring that processing methods do not introduce or exacerbate contamination, such as through the use of contaminated equipment or water (Wuana and Okieimen, 2011). A comprehensive adherence to both GAP and GMP can significantly lower the incidence of heavy metal contaminants in food, thereby ensuring that safety standards are maintained throughout the supply chain.

7.2. Soil and Water Remediation Techniques

Employing soil and water remediation techniques is critical in addressing existing contamination. Phytoremediation, which utilizes plants to absorb, accumulate, and detoxify heavy metals from contaminated soil and water, has shown considerable promise in dealing with pollutants like cadmium and lead (Priya et al., 2023). Strategies include phytoextraction, where plants grow in contaminated soils and are later harvested to remove the heavy metals they've accumulated, and phytostabilization, where plants help immobilize contaminants in the soil (Priya et al., 2023; Oh et al., 2015). Studies indicate that species such as *Brassica napus* and *Miscanthus x giganteus* are proficient in accumulating metals like cadmium and nickel, enhancing the effectiveness of phytoremediation practices (Stojanov et al., 2024; Kocoń and Jurga, 2016). Moreover, combining phytoremediation with microbial interventions can further enhance remediation efforts. Microorganisms associated with plant roots can promote the bioavailability of heavy metals, improving plant uptake (Muratova et al., 2022; Khan, 2021). This symbiotic relationship underscores the potential of bioremediation methods that leverage both flora and fauna to mitigate heavy metal contamination more effectively.

7.3. Crop Selection and Phytoremediation

The selection of specific crops for phytoremediation is vital in developing effective mitigation strategies for heavy metal exposure in agricultural settings. Certain plant species, known for their ability to hyperaccumulate heavy metals, have been identified as effective in extracting these metals from contaminated soils. For instance, *Cyperus rotundus* has demonstrated substantial potential for absorbing zinc, cadmium, nickel, and lead (Garba et al., 2018; Ariyachandra et al., 2023). Employing such plants not only aids in reducing soil toxicity but can potentially restore soil functionality, allowing for safer agricultural practices to resume. Additionally, crop rotation and the introduction of cover crops that are not directly consumed but are effective in metal uptake can serve as innovative preventive measures against heavy metal contamination in soils used for food crops. This strategy can subsequently lead to improved soil health and safety in food production (Parizanganeh et al., 2012).

7.4. Policy Interventions and Consumer Education

Robust policy frameworks and consumer education initiatives are essential in addressing the issue of heavy metals in food systems. Government interventions can regulate and monitor levels of heavy metals permissible in agricultural products, enforce stricter guidelines for industrial emissions, and increase funding for research on heavy metals in the agricultural sector (Wuana and Okieimen, 2011). Such policies can serve as preventive measures against potential contamination pathways from industrial processes to food production. Moreover, educating consumers about the dangers of heavy metals and how to identify safe food sources can empower communities to make informed choices. Public awareness campaigns that provide information on the importance of properly washing produce, recognizing safe sourcing practices, and understanding the implications of heavy metal consumption can play a significant role in reducing overall exposure (Wuana and Okieimen, 2011; Priya et al., 2023).

8.0. CONCLUSION

Heavy metal contamination in food represents a persistent and evolving threat to global food safety and public health. This chapter has examined the multifaceted nature of heavy metal contamination, detailing its origins from both natural and human-made sources, the health implications of chronic exposure, and the complexity of tracing and quantifying these toxic elements across diverse food systems. Through a critical evaluation of advanced analytical techniques, such as AAS and ICP-MS, alongside portable field-based tools, and international regulatory frameworks and monitoring programs, the chapter emphasizes the need for robust, standardized, and accessible detection methods. It also highlights the importance of targeted mitigation strategies, including good agricultural practices, soil and water remediation, and public awareness initiatives. Moving forward, interdisciplinary collaboration and technological innovation, particularly in areas such as green chemistry, nanotechnology, and AI-driven risk assessment, will be crucial in addressing emerging contaminants and strengthening global food safety systems. Continued investment in research, capacity building, and harmonized policy frameworks is essential to mitigate health risks, protect vulnerable populations, and ensure sustainable and equitable access to safe food worldwide.

REFERENCES

Abdulla, N., Ádám, B., Blair, I., & Oulhaj, A. (2019). Heavy metal content of herbal health supplement products in Dubai – UAE: A cross-sectional study. *BMC Complementary and Alternative Medicine, 19*(1). Advance online publication. DOI: 10.1186/s12906-019-2693-3 PMID: 31638965

Abdullahi, A., Lawal, M., & Salisu, A. (2021). Heavy Metals in Contaminated Soil: Sources, Accumulation, Health Risks, and Remediation Processes. *Bayero Journal of Pure and Applied Sciences, 14*(1), 1–12. DOI: 10.4314/bajopas.v14i1.1

Adedayo, O., & Ediagbonya, T. (2018). Deposition of environmental contaminants in processed bushmeat along highways in south-western Nigeria. *Facta Universitatis Series Working and Living Environmental Protection, 243*. Advance online publication. DOI: 10.22190/fuwlep1703243a

Ahmad, W., Alharthy, R., Zubair, M., Ahmed, M., Hameed, A., & Rafique, S. (2021). Toxic and heavy metals contamination assessment in soil and water to evaluate human health risk. *Scientific Reports, 11*(1). Advance online publication. DOI: 10.1038/s41598-021-94616-4 PMID: 34417479

Ahmed, S. R., Kim, J., Suzuki, T., & Lee, J. (2022). Paper-based analytical devices for food safety monitoring: A review. *Trends in Analytical Chemistry, 145*, 116460.

Alengebawy, A., Abdelkhalek, S. T., Qureshi, S. R., & Wang, M.-Q. (2021). Heavy Metals and Pesticides Toxicity in Agricultural Soil and Plants: Ecological Risks and Human Health Implications. *Toxics, 9*(3), 42. DOI: 10.3390/toxics9030042 PMID: 33668829

Alidadi, H., Sany, S., Oftadeh, B., Mohamad, T., Shamszade, H., & Fakhari, M. (2019). Health Risk Assessments of Arsenic and Toxic Heavy Metal Exposure in Drinking Water in Northeast Iran. *Environmental Health and Preventive Medicine, 24*(1). Advance online publication. DOI: 10.1186/s12199-019-0812-x PMID: 31521115

Amponsah, L., Dodd, M., & Darko, G. (2020). Gastric bioaccessibility and human health risks associated with soil metal exposure via ingestion at an e-waste recycling site in Kumasi, Ghana. *Environmental Geochemistry and Health, 44*(2), 497–509. DOI: 10.1007/s10653-020-00760-7 PMID: 33141377

Ariyachandra, S., Alwis, I., & Wimalasiri, E. (2023). Phytoremediation potential of heavy metals by Cyperus rotundus. *Reviews in Agricultural Science, 11*(0), 20–35. DOI: 10.7831/ras.11.0_20

Balali-Mood, M., Naseri, K., Tahergorabi, Z., Khazdair, M., & Sadeghi, M. (2021). Toxic mechanisms of five heavy metals: Mercury, lead, chromium, cadmium, and arsenic. *Frontiers in Pharmacology, 12*. Advance online publication. DOI: 10.3389/fphar.2021.643972 PMID: 33927623

Bibi, M. (2023). Essential and non-essential heavy metals sources and impacts on human health and plants. *Pure and Applied Biology, 12*(2). Advance online publication. DOI: 10.19045/bspab.2023.120083

Birdanova, V., Stoilova, I., Ruseva, I., and Radionova, Z. (2023). Average daily consumption of foods and of foods accumulating lead and cadmium in young people. Journal of Imab - Annual Proceeding (Scientific Papers), 29(1), 4763-4769. https://doi.org/DOI: 10.5272/jimab.2023291.4763

Blanco, G., Sühs, R., Brizola, E., Corrêa, P., Campos, M., & Hanazaki, N. (2020). Invisible contaminants and food security in former coal mining areas of Santa Catarina, southern Brazil. https://doi.org/DOI: 10.21203/rs.3.rs-24678/v1

Boon, P., Biesebeek, J., Sioen, I., Huybrechts, I., Neve, M., Amiano, P., & Klaveren, J. (2010). Long-term dietary exposure to chromium in young children living in different European countries. *EFSA Supporting Publications, 7*(5). Advance online publication. DOI: 10.2903/sp.efsa.2010.en-54

Breeher, L., Mikulski, M., Czeczok, T., Leinenkugel, K., & Fuortes, L. (2015). A cluster of lead poisoning among consumers of ayurvedic medicine. *International Journal of Occupational and Environmental Health, 21*(4), 303–307. DOI: 10.1179/2049396715y.0000000009 PMID: 25843124

Byers, H., McHenry, L., & Grundl, T. (2019). XRF techniques to quantify heavy metals in vegetables at low detection limits. *Food Chemistry: X, 1*, 100001. DOI: 10.1016/j.fochx.2018.100001

Carr, R., Zhang, C., Moles, N., & Harder, M. (2007). Identification and mapping of heavy metal pollution in soils of a sports ground in Galway City, Ireland, using a portable XRF analyser and GIS. *Environmental Geochemistry and Health, 30*(1), 45–52. DOI: 10.1007/s10653-007-9106-0 PMID: 17610027

Chandra, H., Kumari, P., & Yadav, S. (2023). Recent Trends in the Detection of Heavy Metals Using Nanotechnology-Based Sensors. *Biosensors & Bioelectronics, 206*, 114182.

Chen, A., Yang, S., & Liu, H. (2019). Nanomaterial-based electrochemical sensors for arsenic detection. *Sensors (Basel), 19*(16), 3425.

Chen, S., & Ding, Y. (2023). Tackling heavy metal pollution: Evaluating governance models and frameworks. *Sustainability*, *15*(22), 15863. DOI: 10.3390/su152215863

Choi, J., Habte, G., Khan, N., Nho, E., Hong, J., Choi, H., & Kim, K. (2014). Determination of toxic heavy metals in echinodermata and chordata species from South Korea. *Food Additives & Contaminants. Part B, Surveillance*, *7*(4), 295–301. DOI: 10.1080/19393210.2014.932311 PMID: 24916139

Chowdhury, A., & Alam, M. (2023). Heavy metal intake from meat and poultry consumption and potential human health risk assessment in Noakhali, Bangladesh. https://doi.org/DOI: 10.21203/rs.3.rs-2512128/v1

Das, M., Sarkar, A., & Roy, S. (2021). Nanotechnology for heavy metal detection in food: A review. *Environmental Nanotechnology, Monitoring & Management*, *16*, 100491.

Deka, A., Kumar, K., and Basumatary, S. (2023). Monitoring strategies for heavy metals in foods and beverages: limitations for human health risks. https://doi.org/ DOI: 10.5772/intechopen.110542

Deng, G., Li, K., Jing, M., Liu, F., Dai, J., & Li, H. (2011). Aluminium content of some processed foods, raw materials and food additives in China by inductively coupled plasma–mass spectrometry. *Food Additives & Contaminants. Part B, Surveillance*, *4*(4), 248–253. DOI: 10.1080/19393210.2011.632899 PMID: 24786247

Dey, R. S., Roy, R., & Paul, A. (2021). Nanobiosensors for heavy metal detection: Recent developments and future prospects. *Biosensors (Basel)*, *11*(7), 238. PMID: 34356709

Ding, Y. (2024). A bibliographic exploration: tackling heavy metal contamination... https://doi.org/DOI: 10.20944/preprints202401.0625.v1

Dong, Z., Lu, L., Liu, Z., Tang, Y., and Wang, J. (2014). Migration of toxic metals from ceramic food packaging materials into acid food simulants. Mathematical Problems in Engineering, 2014(1). https://doi.org/DOI: 10.1155/2014/759018

Drambozova, I. (2021). Study of the Content of Heavy Metals in Foods in Bulgaria. *Journal of Biomedical and Clinical Research*, *14*(2), 140–145. DOI: 10.2478/ jbcr-2021-0019

Ebrahimi, M., Khalili, N., Razi, S., Keshavarz-Fathi, M., Khalili, N., & Rezaei, N. (2020). Effects of lead and cadmium on the immune system and cancer progression. *Journal of Environmental Health Science & Engineering*, *18*(1), 335–343. DOI: 10.1007/s40201-020-00455-2 PMID: 32399244

Ejaz, H., Junaid, K., Yasmeen, H., Naseer, A., Alam, H., Younas, S., & Bukhari, S. (2022). Multiple antimicrobial resistance and heavy metal tolerance of biofilm-producing bacteria isolated from dairy and non-dairy food products. *Foods*, *11*(18), 2728. DOI: 10.3390/foods11182728 PMID: 36140855

Emurotu, J., & Onianwa, P. (2017). Bioaccumulation of heavy metals in soil and selected food crops cultivated in Kogi State, North Central Nigeria. *Environmental Systems Research*, *6*(1). Advance online publication. DOI: 10.1186/s40068-017-0098-1

Engwa, G., Ferdinand, P., Nwalo, F., and Unachukwu, M. (2019). Mechanism and health effects of heavy metal toxicity in humans. https://doi.org/DOI: 10.5772/intechopen.82511

Fan, Y., Zhu, T., Li, M., He, J., & Huang, R. (2017). Heavy metal contamination in soil and brown rice and human health risk assessment near three mining areas in central China. *Journal of Healthcare Engineering*, *2017*, 1–9. DOI: 10.1155/2017/4124302 PMID: 29065598

Garba, S., Gudusu, M., & Inuwa, L. (2018). Accumulation ability of the native grass species, Cyperus rotundus, for the heavy metals: Zinc (Zn), cadmium (Cd), nickel (Ni), and lead (Pb). *International Research Journal of Pure and Applied Chemistry*, *17*(1), 1–15. DOI: 10.9734/irjpac/2018/44604

García, J., Puerto-Parejo, L., Roncero-Martín, R., Morán, J., Pedrera-Zamorano, J., Aliaga, I., & Canal-Macías, M. (2017). Dietary intake of cadmium, lead, and mercury and its association with bone health in healthy premenopausal women. *International Journal of Environmental Research and Public Health*, *14*(12), 1437. DOI: 10.3390/ijerph14121437 PMID: 29168740

Gebru, M. (2019). Phytoremediation of Heavy Metals Released from Mining Waste Drainage Using Selected Plant Species in Ethiopia. Budapest International Research in Exact Sciences (Birex). *Journal*, *1*(3), 1–4. DOI: 10.33258/birex.v1i3.343

Gergen, I., & Hărmănescu, M. (2012). Application of principal component analysis in the pollution assessment with heavy metals of vegetable food chain in the old mining areas. *Chemistry Central Journal*, *6*(1). Advance online publication. DOI: 10.1186/1752-153x-6-156 PMID: 23234365

González, D., Díaz, C., & Pérez, D. (2019). Regulatory aspects of nanotechnology in food safety: Challenges and perspectives. *Food Control*, *100*, 131–143.

Gupta, V., Singh, S., Agrawal, A., Siddiqi, N., & Sharma, B. (2015). Phytochemicals mediated remediation of neurotoxicity induced by heavy metals. *Biochemistry Research International, 2015*, 1–9. DOI: 10.1155/2015/534769 PMID: 26618004

Hassan, M., Ngadda, Y., & Adamu, A. (2020). Measurement of levels of concentrations of some heavy metals in soils of some selected areas of Pindiga, Nigeria. *Nipes Journal of Science and Technology Research, 2*(2), 35. DOI: 10.37933/nipes/2.2.2020.4

Heidarieh, M., Maragheh, M., Shamami, M., Behgar, M., Ziaei, F., & Akbari, Z. (2013). Evaluation of heavy metal concentration in shrimp (Penaeus semisulcatus) and crab (Portunus pelagicus) with inaa method. *SpringerPlus, 2*(1). Advance online publication. DOI: 10.1186/2193-1801-2-72 PMID: 23487484

Henriksen, C., & Bügel, S. (2023). Chromium – a scoping review for Nordic nutrition recommendations 2023. *Food & Nutrition Research, 67*. Advance online publication. DOI: 10.29219/fnr.v67.10325 PMID: 38084146

Hong, Y., Choi, J., Nho, E., Hwang, I., Khan, N., Jamila, N., & Kim, K. (2018). Determination of macro, micro, and trace elements in citrus fruits by inductively coupled plasma–optical emission spectrometry (ICP-OES), ICP–mass spectrometry, and direct mercury analyzer. *Journal of the Science of Food and Agriculture, 99*(4), 1870–1879. DOI: 10.1002/jsfa.9382 PMID: 30264403

Hu, B., Chen, S., Hu, J., Xia, F., Xu, J., Li, Y., & Shi, Z. (2017). Application of portable XRF and VNIR sensors for rapid assessment of soil heavy metal pollution. *PLoS One, 12*(2), e0172438. DOI: 10.1371/journal.pone.0172438 PMID: 28234944

Huang, Z., Pan, X., Wu, P., Han, J., & Chen, Q. (2013). Health Risk Assessment of Heavy Metals in Rice for the Population in Zhejiang, China. *PLoS One, 8*(9), e75007. DOI: 10.1371/journal.pone.0075007 PMID: 24040375

Hussain, R. (2012). Assessment of heavy metal (Cd, Pb, and Zn) contents in the livers of chickens available in the local markets of Basrah City, Iraq. *Basra Journal of Veterinary Research, 11*(1), 43–51. DOI: 10.33762/bvetr.2012.54752

Hutton, L., O'Neil, G., Read, T., Ayres, Z., Newton, M., & Macpherson, J. (2014). Electrochemical x-ray fluorescence spectroscopy for trace heavy metal analysis: Enhancing x-ray fluorescence detection capabilities by four orders of magnitude. *Analytical Chemistry, 86*(9), 4566–4572. DOI: 10.1021/ac500608d PMID: 24701959

Igwe, C., Azorji, J., & Iheagwam, S. (2017). Influence of contamination of soil with heavy metal on the growth of three herbaceous plant species. *Journal of Science Technology and Environment Informatics, 5*(2), 382–388. DOI: 10.18801/jstei.050217.40

Ikhajiagbe, B., & Ogwu, M. C. (2020). Hazard quotient, microbial diversity and plant composition of spent crude oil polluted-soil. *Beni-Suef University Journal of Basic and Applied Sciences*, *9*(26). Advance online publication. DOI: 10.1186/s43088-020-00052-0

Ikhajiagbe, B., Ogwu, M.C., Ekhator, P.O. and Omoayena, I.I. (2022). Implications of soil nitrogen enhancement on the yield performance of soybean (*Glycine max*) in cadmium-polluted soil. *Journal of Pollution Effects and Control*, 10(2): 1 – 12. #: 1000334 http://doi.org/DOI: 10.35248/23754-4397.22.10.334

Ikhajiagbe, B., Ogwu, M. C., & Lato, N. F. (2021). Growth and yield response of soybean (*Glycine max* [L.] Merr.) accessions after exposure to cadmium. *Vegetos*. Advance online publication. DOI: 10.1007/s.42535-021-00189-y

Ishida, M., Greene, V., King, T., Sheridan, R., Luker, J., Oglesby, D., & Greenberg, J. (2022). Regulatory Policies for Heavy Metals in Spices – A New York Approach. https://doi.org/DOI: 10.21423/jrs-v10i1ishida

Iyiola, A. O., Izah, S. C., Morya, S., Akinsorotan, A., & Ogwu, M. C. (2023). Owalla Reservoir in South-western Nigeria: Assessment of Fish Distribution, Biological Diversity, and Water Quality Index. *Indonesian Journal of Agricultural Research*, *6*(02), 106–125. DOI: 10.32734/injar.v6i2.12207

Iyiola, A. O., Kolawole, A. S., Setufe, S. B., Bilikoni, J., Ofori, E., & Ogwu, M. C. (2024). Fish as a Sustainable Biomonitoring Tool in Aquatic Environments. In Izah, S. C., Ogwu, M. C., & Hamidifar, H. (Eds.), *Biomonitoring of Pollutants in the Global South*. Springer., DOI: 10.1007/978-981-97-1658-6_12

Izah, S. C., Abdelkhalek, S. T., Ogwu, M. C., & Hamidifar, H. (2024). Overview and Perspectives of Biomonitoring in the Global South. In Izah, S. C., Ogwu, M. C., & Hamidifar, H. (Eds.), *Biomonitoring of Pollutants in the Global South*. Springer., DOI: 10.1007/978-981-97-1658-6_1

Jan, A., Azam, M., Siddiqui, K., Ali, A., Choi, I., & Haq, Q. (2015). Heavy metals and human health: Mechanistic insights into toxicity and the antioxidant defense system. *International Journal of Molecular Sciences*, *16*(12), 29592–29630. DOI: 10.3390/ijms161226183 PMID: 26690422

Jan, F., Ishaq, M., Khan, S., Ihsanullah, I., Ahmad, I., & Shakirullah, M. (2010). A comparative study of human health risks via consumption of food crops grown on wastewater-irrigated soil (Peshawar) and relatively clean water irrigated soil (Lower DIR). *Journal of Hazardous Materials*, *179*(1-3), 612–621. DOI: 10.1016/j.jhazmat.2010.03.047 PMID: 20399016

Jayasumana, C., Gunatilake, S., & Siribaddana, S. (2015). Simultaneous exposure to multiple heavy metals and glyphosate may contribute to Sri Lankan agricultural nephropathy. *BMC Nephrology*, *16*(1). Advance online publication. DOI: 10.1186/s12882-015-0109-2 PMID: 26162605

Jiao, Y., Xu, Z., & Liu, Y. (2021). Nanotechnology-enabled biosensors for monitoring food contaminants. *Journal of Hazardous Materials*, *417*, 125975.

Jolly, Y., Islam, A., & Akbar, S. (2013). Transfer of metals from soil to vegetables and possible health risk assessment. *SpringerPlus*, *2*(1). Advance online publication. DOI: 10.1186/2193-1801-2-385 PMID: 24010043

Kaur, N., & Jain, A. (2020). Nanobiosensors for heavy metal detection in food and water: A review. *Critical Reviews in Food Science and Nutrition*, *60*(21), 3655–3671.

Khan, A. (2021). Potential of coupling heavy metal (HM) phytoremediation by bioenergy plants and their associated HM-adapted rhizosphere microbiota (arbuscular mycorrhizal fungi and plant growth-promoting microbes) for bioenergy production. *Journal of Energy and Power Technology*, *3*(4), 1–1. DOI: 10.21926/jept.2104049

Khan, A., Khan, S., Khan, M., Qamar, Z., & Waqas, M. (2015). The uptake and bioaccumulation of heavy metals by food plants, their effects on plants nutrients, and associated health risk: A review. *Environmental Science and Pollution Research International*, *22*(18), 13772–13799. DOI: 10.1007/s11356-015-4881-0 PMID: 26194234

Kocoń, A., & Jurga, B. (2016). The evaluation of growth and phytoextraction potential of Miscanthus x giganteus and Sida hermaphrodita on soil contaminated simultaneously with Cd, cu, Ni, Pb, and Zn. *Environmental Science and Pollution Research International*, *24*(5), 4990–5000. DOI: 10.1007/s11356-016-8241-5 PMID: 27995509

Liang, Y., Yi, X., Dang, Z., Wang, Q., Luo, H., & Tang, J. (2017). Heavy metal contamination and health risk assessment in the vicinity of a tailing pond in Guangdong, China. *International Journal of Environmental Research and Public Health*, *14*(12), 1557. DOI: 10.3390/ijerph14121557 PMID: 29231884

Lin, Q., Li, H., Zhong, H., Zhao, Q., Da-hui, X., & Wang, Z. (2014). Determination of titanium in nano-titanium(IV) oxide composite food packaging by microwave digestion and inductively coupled plasma atomic emission spectrometry and inductively coupled plasma mass spectrometry. *Analytical Letters*, *47*(12), 2095–2103. DOI: 10.1080/00032719.2014.895907

Liu, Y., Wu, P., & He, C. (2020). Fluorescence-based quantum dot sensors for heavy metal detection in food matrices. *Food Chemistry*, *329*, 127190.

Loaiza, I., Boeck, G., Alcazar, J., Campos, D., Cárdenas-Alayza, S., Ganoza, M., & Troch, M. (2021). Trophic interactions and metal transfer in marine ecosystems driven by the Peruvian scallop Argopecten purpuratus aquaculture. *Journal of the World Aquaculture Society*, *53*(2), 452–474. DOI: 10.1111/jwas.12822

Mazarakioti, E., Zotos, A., Thomatou, A., Kontogeorgos, A., Patakas, A., & Ladavos, A. (2022). Inductively coupled plasma-mass spectrometry (ICP-MS) is a useful tool in the authenticity of agricultural products' and foods' origin. *Foods*, *11*(22), 3705. DOI: 10.3390/foods11223705 PMID: 36429296

Meera, A., Kc, M., John, J., Sudha, B., and Sajeena, A. (2022). Aquatic plants as bioremediators in pollution abatement of heavy metals. https://doi.org/DOI: 10.5772/intechopen.99627

Mukhi, S., Srikantiah, R., Manjrekar, P., Iyyaswami, R., & Sindhu, H. (2022). Assessment of heavy metals in food and drug packaging materials. *F1000 Research*, *11*, 648. DOI: 10.12688/f1000research.121473.1 PMID: 38779461

Muratova, A., Lyubun, E., Голубев, C., & Turkovskaya, O. (2022). Effect of copper ions on the associations of Azospirillum bacteria with wheat seedlings (Triticum aestivum L.). *Vavilov Journal of Genetics and Breeding*, *26*(5), 477–485. DOI: 10.18699/vjgb-22-58 PMID: 36128570

Nguyen, H., Oh, H., & Kim, M. (2021). Impacts of antioxidant vitamins, curry consumption and heavy metal levels on the metabolic syndrome with comorbidities: a national cross-sectional study. https://doi.org/DOI: 10.21203/rs.3.rs-225694/v1

Nguyen, V., Linh, T., Vo, T., Nguyen, Q., & Van, T. (2023). Analytical techniques for determination of heavy metal migration from different types of locally made plastic food packaging materials using ICP-MS. *Food Science & Nutrition*, *11*(7), 4030–4037. DOI: 10.1002/fsn3.3391 PMID: 37457188

Nkwunonwo, U., Odika, P., & Onyia, N. (2020). A review of the health implications of heavy metals in food chain in Nigeria. *TheScientificWorldJournal*, *2020*, 1–11. DOI: 10.1155/2020/6594109 PMID: 32351345

Nwizugbo, K. C., Ogwu, M. C., Eriyamremu, G. E., & Ahana, C. M. (2023). Alterations in energy metabolism, total protein, uric and nucleic acids in African sharptooth catfish (*Clarias gariepinus* Burchell.) exposed to crude oil and fractions. *Chemosphere*, *316*, 137778. DOI: 10.1016/j.chemosphere.2023.137778 PMID: 36640975

Obongodot, N. U., & Ogwu, M. C. (2023). Plant Food for Human Health: Case Study of Indigenous Vegetables in Akwa Ibom State, Nigeria. In: Izah, S.C., Ogwu, M.C., Akram, M. (eds) *Herbal Medicine Phytochemistry*. Reference Series in Phytochemistry. Springer, Cham. https://doi.org/DOI: 10.1007/978-3-031-21973-3_2-1

Ogundele, K., Makinde, O., Eluyera, M., & Orisamoyi, Y. (2020). Assessment of heavy metals in beans (Vigna unguiculata) obtained from Oja-Tuntun Market, Ile-Ife, Osun State, Nigeria. *Journal of Experimental Agriculture International, 23-27*. Advance online publication. DOI: 10.9734/jeai/2020/v42i1130618

Ogwu, M. C. (2023). Local Food Crops in Africa: Sustainable Utilization, Threats, and Traditional Storage Strategies. In: Izah, S.C. and Ogwu, M.C. (editors). *Sustainable Utilization and Conservation of Africa's Biological Resources and Environment.* Sustainable Development and Biodiversity, volume 888: 353 – 376 pp. Springer, Singapore. pp. 353–374. https://doi.org/DOI: 10.1007/978-981-19-6974-4_13

Ogwu, M. C. (2024). Plants as monitors and managers of pollution. In Srivastav, A. L., Grewal, A. S., Markandeya, M., & Pham, T. D. (Eds.), *Advances in pollution research: Role of green chemistry in ecosystem restoration to achieve environmental sustainability* (pp. 51–60). Elsevier., DOI: 10.1016/B978-0-443-15291-7.00022-5

Ogwu, M. C., Izah, S. C., & Joshua, M. T. (2025). (2025b). Ecological and environmental determinants of phytochemical variability in forest trees. *Phytochemistry Reviews.* Advance online publication. DOI: 10.1007/s11101-025-10066-0

Ogwu, M. C., Izah, S. C., Sawyer, W. E., & Amabie, T. (2025a). Environmental Risk Assessment of Trace Metal Pollution: A Statistical Perspective. *Environmental Geochemistry and Health*, *47*(4), 94. DOI: 10.1007/s10653-025-02405-z PMID: 40019615

Ogwu, M. C., Ogwu, H. I., & Kosoe, E. A. (2023). Plants Used in the Management and Treatment of Cardiovascular Diseases: Case Study of the Benin People of Southern Nigeria. In: Izah, S.C., Ogwu, M.C., Akram, M. (eds) *Herbal Medicine Phytochemistry*. Reference Series in Phytochemistry. Springer, Cham. https://doi.org/DOI: 10.1007/978-3-031-21973-3_4-1

Oh, K., Cao, T., Cheng, H., Liang, X., Hu, X., Yan, L., & Takahi, S. (2015). Phytoremediation potential of sorghum as a biofuel crop and the enhancement effects with microbe inoculation in heavy metal contaminated soil. *Journal of Biosciences and Medicines*, *03*(06), 9–14. DOI: 10.4236/jbm.2015.36002

Ohiagu, F., Chikezie, P., Ahaneku, C., & Chikezie, C. (2022). Human Exposure to Heavy Metals: Toxicity Mechanisms and Health Implications. *Material Science and Engineering International Journal*, *6*(2), 78–87. DOI: 10.15406/mseij.2022.06.00183

Ohira, S., Nakamura, K., Shelor, C., Dasgupta, P., & Toda, K. (2015). Simultaneous electrodialytic preconcentration and speciation of chromium(iii) and chromium(vi). *Analytical Chemistry, 87*(22), 11575–11580. DOI: 10.1021/acs.analchem.5b03464 PMID: 26507203

Orecchio, S., Amorello, D., Raso, M., Barreca, S., Lino, C., & Gaudio, F. (2014). Determination of trace elements in gluten-free food for celiac people by ICP-MS. *Microchemical Journal, 116*, 163–172. DOI: 10.1016/j.microc.2014.04.011

Orji, O., Agu, P., Okah, K., Ugwu, O., Elom, G., Obasi, O., & Aja, P. (2024). Quantitative and health risk evaluations of trace metals in soils and cassava tubers cultivated on farms around abandoned Ishiagu mining sites, Ebonyi State, Nigeria. https://doi.org/DOI: 10.21203/rs.3.rs-4582999/v1

Orr, S., Barnes, M., George, H., Joshee, L., Jeon, B., Scircle, A., & Bridges, C. (2018). Exposure to mixtures of mercury, cadmium, lead, and arsenic alters the disposition of single metals in tissues of Wistar rats. *Journal of Toxicology and Environmental Health. Part A., 81*(24), 1246–1256. DOI: 10.1080/15287394.2018.1551164 PMID: 30507365

Osawaru, M. E., & Ogwu, M. C. (2023). Plants Used in the Management and Treatment of Female Reproductive Health Issues: Case Study from Southern Nigeria. In: Izah, S.C., Ogwu, M.C., Akram, M. (eds) *Herbal Medicine Phytochemistry*. Reference Series in Phytochemistry. Springer, Cham. https://doi.org/DOI: 10.1007/978-3-031-21973-3_5-1

Osawaru, M. E., Ogwu, M. C., & Braimah, L. (2013b). Growth responses of two cultivated Okra species (*Abelmoschus caillei* (A. Chev.) Stevels and *Abelmoschus esculentus* (Linn.) Moench) in crude oil contaminated soil. *Nigerian Journal of Basic and Applied Science, 21*(3), 215–226.

Osawaru, M. E., Ogwu, M. C., & Chime, A. O. (2013a). Assessment of growth performance of two Okra species (*Abelmoschus esculentus* [L.] Moench and *Abelmoschus caillei* [A. Chev.] Stevels) exposed to crude oil contaminated soil. *Nigerian Journal of Biotechnology, 26*, 11–20.

Othman, N., Mohd-Asharuddin, S., & Azizul-Rahman, M. (2013). An Overview of Fruit Waste as a Sustainable Adsorbent for Heavy Metal Removal. *Applied Mechanics and Materials, 389*, 29–35. DOI: 10.4028/www.scientific.net/amm.389.29

Pappalardo, A., Copat, C., Ferrito, V., Grasso, A., & Ferrante, M. (2017). Heavy metal content and molecular species identification in canned tuna: Insights into human food safety. *Molecular Medicine Reports, 15*(5), 3430–3437. DOI: 10.3892/mmr.2017.6376 PMID: 28339012

Parizanganeh, A., Bijnavand, V., Zamani, A., & Hajabolfath, A. (2012). Concentration, distribution, and comparison of total and bioavailable heavy metals in topsoils of Bonab District in Zanjan Province. *Open Journal of Soil Science*, *02*(02), 123–132. DOI: 10.4236/ojss.2012.22018

Patel, S., Sutar, R., & Singh, A. (2023). Smart food safety systems: Integrating biosensors with IoT and AI. *Sensors and Actuators. B, Chemical*, *380*, 133218.

Priya, A., Muruganandam, M., Ali, S., & Kornaros, M. (2023). Clean-up of heavy metals from contaminated soil by phytoremediation: A multidisciplinary and eco-friendly approach. *Toxics*, *11*(5), 422. DOI: 10.3390/toxics11050422 PMID: 37235237

Pyrzyńska, K. (2016). Chromium redox speciation in food samples. *Turkish Journal of Chemistry*, *40*, 894–905. DOI: 10.3906/kim-1606-5

Qu, C., Ma, Z., Yang, J., Liu, Y., Bi, J., & Huang, L. (2012). Human Exposure Pathways of Heavy Metals in a Lead-Zinc Mining Area, Jiangsu Province, China. *PLoS One*, *7*(11), e46793. DOI: 10.1371/journal.pone.0046793 PMID: 23152752

Rahimi, G., Kolahchi, Z., & Charkhabi, A. (2017). Uptake and translocation of some heavy metals by rice crop (Oryza sativa) in paddy soils. [Pol Nohospodárstvo]. *Agriculture*, *63*(4), 163–175. DOI: 10.1515/agri-2017-0016

Rehman, K., Fatima, F., Waheed, I., & Akash, M. (2017). Prevalence of exposure to heavy metals and their impact on health consequences. *Journal of Cellular Biochemistry*, *119*(1), 157–184. DOI: 10.1002/jcb.26234 PMID: 28643849

Rehman, Z., Khan, S., Shah, M., Brusseau, M., Khan, S., & Mainhagu, J. (2018). Transfer of Heavy Metals from Soils to Vegetables and Associated Human Health Risks at Selected Sites in Pakistan. *Pedosphere*, *28*(4), 666–679. DOI: 10.1016/s1002-0160(17)60440-5 PMID: 31123392

Richard, G., Izah, S. C., & Ogwu, M. C. (2022). Implications of Artisanal Crude Oil Refining on Sustainable Food Production in the Niger Delta Region of Nigeria. *Journal of Environmental Bioremediation and Toxicology*, *5*(2), 69–77. DOI: 10.54987/jebat.v5i2.775

Riyazuddin, R., Nisha, N., Ejaz, B., Khan, M., Kumar, M., Ramtekc, P., & Gupta, R. (2021). A comprehensive review on the heavy metal toxicity and sequestration in plants. *Biomolecules*, *12*(1), 43. DOI: 10.3390/biom12010043 PMID: 35053191

Rostamzadeh, M., Sadrabad, E., Mohajeri, F., Tafti, A., & Askari, E. (2025). Heavy Metals in Foods Consumed by Copper Miners: A Health Risk Assessment. *Food Science & Nutrition*, *13*(5). Advance online publication. DOI: 10.1002/fsn3.70267 PMID: 40351367

Saha, T., Das, S., Sau, S., Datta, D., Kundu, S., Saha, S., & Mitra, A. (2023). Toxic metal uptake by oyster mushrooms grown in sugarcane bagasse. https://doi.org/ DOI: 10.1101/2023.02.07.523865

Scăe eanu, G., Majdar, R., & Mot, A. (2021). An overview of methods used for quantification of heavy metal contents in vegetal samples. *Romanian Journal of Ecology and Environmental Chemistry*, *3*(2), 7–15. DOI: 10.21698/rjeec.2021.201

Shih, P., Chiang, L., Lin, S., Chang, T., & Hsu, W. (2019). Application of time-lapse ion exchange resin sachets (tiers) for detecting illegal effluent discharge in mixed industrial and agricultural areas, Taiwan. *Sustainability*, *11*(11), 3129. DOI: 10.3390/su11113129

Sibuar, A., Zulkafflee, N., Selamat, J., Ismail, M., Lee, S., & Razis, A. (2022). Quantitative analysis and human health risk assessment of heavy metals in paddy plants collected from Perak, Malaysia. *International Journal of Environmental Research and Public Health*, *19*(2), 731. DOI: 10.3390/ijerph19020731 PMID: 35055550

Singare, P., Mishra, R., & Trivedi, M. (2012). Sediment contamination due to toxic heavy metals in the Mithi River of Mumbai. *Advances in Analytical Chemistry of Scientific and Academic Publishing*, *2*(3), 14–24. DOI: 10.5923/j.aac.20120203.02

Singh, P., Pandey, S., & Tripathi, N. (2020). Nanotechnology in food safety: Recent developments and future trends. *Food Science and Human Wellness*, *9*(1), 11–23.

Solidum, J., Parel, J., Pascual, A., Reyes, R., & Tiga, D. (2013). Quantitative analysis of lead, cadmium, and chromium in different brands of junk food marketed in Metro Manila, Philippines. *Advanced Materials Research*, *634-638*, 1581–1585. DOI: 10.4028/www.scientific.net/amr.634-638.1581

Stojanov, N., Maletić, S., Beljin, J., Đukanović, N., Kiprovski, B., & Zeremski, T. (2024). Enhancing phytoextraction potential of Brassica napus for contaminated dredged sediment using nitrogen fertilizers and organic acids. *Plants*, *13*(6), 818. DOI: 10.3390/plants13060818 PMID: 38592795

Tchounwou, P., Yedjou, C., Patlolla, A., and Sutton, D. (2012). Heavy Metal Toxicity and the Environment, 133-164. https://doi.org/DOI: 10.1007/978-3-7643-8340-4_6

Thor, M., Harnack, L., King, D., Jasthi, B., & Pettit, J. (2011). Evaluation of the comprehensiveness and reliability of the chromium composition of foods in the literature. *Journal of Food Composition and Analysis*, *24*(8), 1147–1152. DOI: 10.1016/j.jfca.2011.04.006 PMID: 23066174

Tom, M., Fletcher, T., & McCarthy, D. (2014). Heavy metal contamination of vegetables irrigated by urban stormwater: A matter of time? *PLoS One*, *9*(11), e112441. DOI: 10.1371/journal.pone.0112441 PMID: 25426946

Uhegbu, F., Chinyere, G., Ugbogu, E., & Nwoku, K. (2012). Arsenic and chromium in sea foods from Niger Delta of Nigeria: A case study of Warri, Delta State. *Bulletin of Environmental Contamination and Toxicology*, *89*(2), 424–427. DOI: 10.1007/s00128-012-0667-y PMID: 22592602

Voica, C., Ro u, C., Iordache, A., Piştea, I., & Miricioiu, M. (2020). The Romanian consumer prioritizes education, information, health risks, food quality, and ethics. *Revista De Chimie*, *71*(2), 436–442. DOI: 10.37358/rc.20.2.7946

Wang, B., Huang, D., & Weng, Z. (2023).. . *Recent Advances in Polymer-Based Biosensors for Food Safety Detection Polymers*, *15*(15), 3253. DOI: 10.3390/polym15153253 PMID: 37571147

Wang, J., Liu, G., & Lin, Y. (2018). Electrochemical biosensors for heavy metal ions. *Trends in Analytical Chemistry*, *29*(4), 395–403.

Wieczorek, K., Turek, A., Szczesio, M., & Wolf, W. (2020). Comprehensive evaluation of metal pollution in urban soils of a post-industrial city—A case of łódź, Poland. *Molecules (Basel, Switzerland)*, *25*(18), 4350. DOI: 10.3390/molecules25184350 PMID: 32971977

Wu, X., & Yang, Y. (2011). Heavy metal (Pb, Co, Cd, Cr, Cu, Fe, Mn, and Zn) concentrations in harvest-size white shrimp Litopenaeus vannamei tissues from aquaculture and wild sources. *Journal of Food Composition and Analysis*, *24*(1), 62–65. DOI: 10.1016/j.jfca.2010.03.030

Wuana, R., & Okieimen, F. (2011). Heavy metals in contaminated soils: A review of sources, chemistry, risks and best available strategies for remediation. *ISRN Ecology*, *2011*, 1–20. DOI: 10.5402/2011/402647

Wuana, R., & Okieimen, F. (2011). Heavy metals in contaminated soils: A review of sources, chemistry, risks and best available strategies for remediation. *ISRN Ecology*, *2011*, 1–20. DOI: 10.5402/2011/402647

Zafarzadeh, A., Rahimzadeh, H., & Mahvi, A. (2018). Health risk assessment of heavy metals in vegetables in an endemic esophageal cancer region in Iran. *Health Scope*, *7*(3). Advance online publication. DOI: 10.5812/jhealthscope.12340

Zaib, M., Ali, Z., Akram, H., Hameed, S., Wakeel, A., Qasim, S., & Aslam, S. (2023). Soil contamination and human health: exploring the heavy metal landscape: a comprehensive review. *JHRR*, *3*(2), 351–356. DOI: 10.61919/jhrr.v3i2.123

Zarazúa, G., Girón-Romero, K., Tejeda, S., León, C., & Ávila-Pérez, P. (2014). Total reflection x-ray fluorescence analysis of toxic metals in fish tissues. *American Journal of Analytical Chemistry*, *05*(12), 805–811. DOI: 10.4236/ajac.2014.512089

Zhang, X., Huang, Q., & Yu, S. (2022). Role of nanomaterials in biosensors for food safety detection. *Advanced Materials Research*, *1171*, 12–22.

Zuo, T., Zhang, L., Wang, Y., Nie, L., Shen, M., Liu, L., & Ma, S. (2023). Technical Guidelines for Risk Assessment of Heavy Metals in Traditional Chinese Medicines. *Chinese Medicine*, *18*(1). Advance online publication. DOI: 10.1186/s13020-023-00771-3 PMID: 37287001

Section 2
Detection and Monitoring Technologies

Chapter 6
Analysis of Mycotoxin and Pesticide Residue in Food:
Current Practices and Potential Future Trends

Tarun Kumar Patel
https://orcid.org/0009-0003-3002-8100
Sant Guru Ghasidas Government P.G. College, Kurud, India

ABSTRACT

This chapter provides a comprehensive overview of mycotoxin and pesticide residue analysis, aiming to elucidate current practices, address challenges, and explore innovative technologies for ensuring food safety. It details analytical methods like LC-MS/MS, GC-MS, HPLC, TLC, and ELISA. Seven key challenges- analytical sensitivity, matrix complexity, cost, regulatory harmonization, environmental impact, emerging contaminants, and scalability are examined, emphasizing barriers to effective detection. Emerging trends, including HRMS, DART-MS, portable biosensors, nanotechnology, AI, and green analytical chemistry techniques like SPME and DLLME, are discussed for their potential to enhance sensitivity, speed, and sustainability. Future directions involve multi-omics, cost-effective platforms, and international collaborations led by Codex Alimentarius, FAO, and WHO to standardize methods and MRLs. By integrating recent studies and regulatory perspectives, this chapter bridges foundational knowledge with forward-looking solutions, fostering interdisciplinary innovation in food safety.

DOI: 10.4018/979-8-3373-3982-5.ch006

INTRODUCTION

Mycotoxins and pesticide residues represent significant food contaminants, posing substantial risks to human health, animal welfare, and global food security. Mycotoxins, toxic secondary metabolites produced by fungi such as *Aspergillus*, *Fusarium*, and *Penicillium*, contaminate crops like cereals, nuts, and spices, particularly under warm and humid conditions (Figure 1; Patel et al., 2014). Common mycotoxins, such as aflatoxins, ochratoxin A, and deoxynivalenol, cause severe health effects, from acute poisoning to chronic conditions like liver cancer and immunosuppression (Table 1; Marin et al., 2013; Patel, 2025). Pesticide residues, resulting from the application of chemical agents to control pests, persist in food and the environment, raising concerns about neurotoxicity, endocrine disruption, and ecological damage (Kim et al., 2017). The pervasive nature of these contaminants necessitates robust analytical methods to ensure food safety and compliance with regulatory standards, such as maximum residue limits (MRLs) set by bodies like the Codex Alimentarius and the European Union (EU).

The importance of residue analysis cannot be overstated, as it underpins efforts to safeguard public health, support international trade, and meet stringent regulatory requirements. Analytical techniques, such as liquid chromatography-mass spectrometry (LC-MS) and enzyme-linked immunosorbent assay (ELISA), have become critical tools for detecting and quantifying mycotoxins and pesticide residues in complex food matrices (Maphaisa et al., 2025). These methods enable the identification of contaminants at trace levels, ensuring adherence to safety thresholds. However, challenges like matrix effects, cross-contamination, and the need for rapid, cost-effective solutions persist, particularly in resource-limited settings (Maphaisa et al., 2025).

This chapter offers a comprehensive overview of mycotoxin and pesticide residue analysis, targeting researchers, regulators, and industry professionals. It aims to clarify current analytical methods, address detection challenges, and explore innovative technologies shaping food safety's future. By integrating recent studies and regulatory perspectives, it bridges foundational knowledge with forward-looking solutions. Structured to progress from core concepts to future strategies, the chapter covers practices, challenges, emerging technologies, and long-term approaches, providing a holistic view of food contaminant mitigation and fostering interdisciplinary collaboration and innovation in food safety.

Figure 1. Major Mycotoxins Produced by Aspergillus fungi

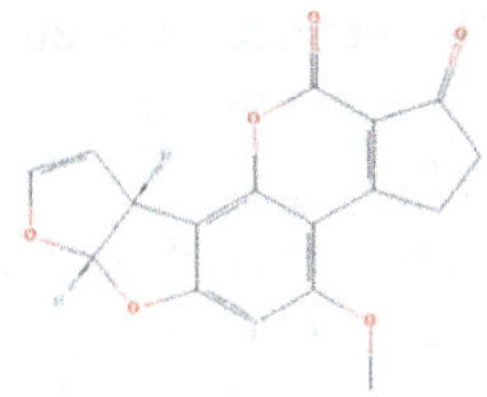

AFB1 ($C_{17}H_{12}O_6$)

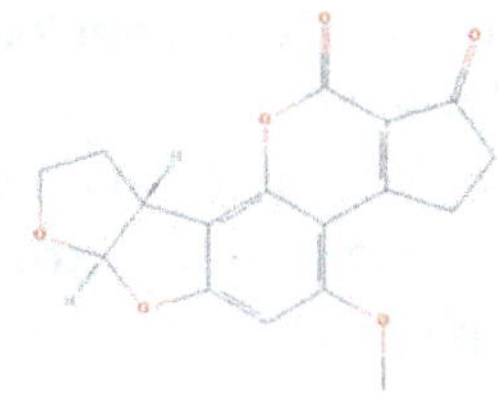

AFB2 ($C_{17}H_{14}O_6$)

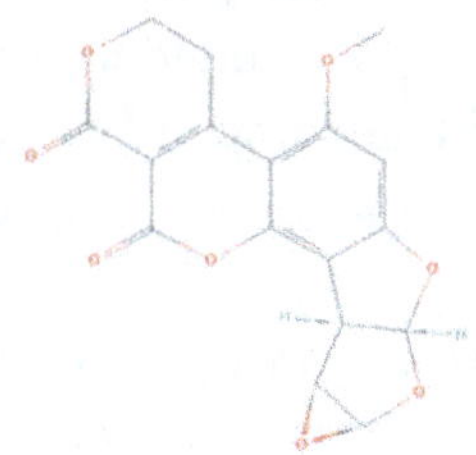

AFG1 ($C_{17}H_{12}O_8$)

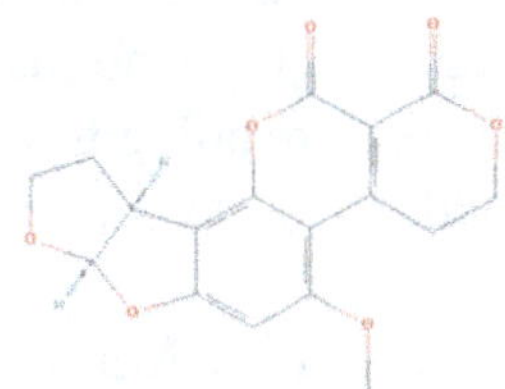

AFG2 ($C_{17}H_{14}O_7$)

Table 1. Overview of Common Mycotoxins, Their Producing Organisms, and Associated Health Impacts.

Mycotoxin	Producing Organism	Health Impacts
Aflatoxins	*Aspergillus flavus, A. parasiticus*	Hepatotoxicity, liver cancer, immunosuppression
Ochratoxin A	*Aspergillus, Penicillium*	Nephrotoxicity, potential carcinogen
Deoxynivalenol	*Fusarium* species	Gastrointestinal disorders, immune suppression
Fumonisins	*Fusarium* species	Esophageal cancer, neural tube defects
Zearalenone	*Fusarium* species	Endocrine disruption, reproductive issues

PRINCIPLES OF ANALYSIS

To streamline the chapter, core analytical principles are consolidated here, referenced throughout to avoid repetition.

- **Thin-Layer Chromatography (TLC):** TLC separates compounds on a stationary phase (e.g., silica gel on a plate) using a liquid mobile phase. The sample is spotted on the plate, and the mobile phase carries analytes upward, separating them based on their affinity for the stationary phase. Mycotoxins are visualized under UV light or with chemical reagents, often via fluorescence. Affordable but less sensitive, suitable for aflatoxin screening in resource-limited settings (Salisu et al., 2021).

- **High-Performance Liquid Chromatography (HPLC):** Uses a pressurized liquid mobile phase and stationary phase (e.g., C18 column) for separation. The sample components interact differently with the stationary phase and the mobile phase (a liquid solvent), leading to their separation based on properties such as polarity, size, or charge. A detector, often UV or fluorescence-based, measures the compounds as they elute, producing a chromatogram that displays peaks corresponding to individual components. Retention time and peak area help identify and quantify the compounds, respectively (Hymete et al., 2024).
- **Liquid Chromatography-Mass Spectrometry (LC-MS):** LC-MS builds on HPLC by coupling it with mass spectrometry, offering superior sensitivity and specificity for mycotoxin and pesticide analysis. In LC-MS, the liquid chromatography component separates the sample's compounds as in HPLC, but the eluent is directed to a mass spectrometer, which ionizes the molecules and measures their mass-to-charge ratio. This enables precise identification of compounds based on their molecular weight and fragmentation patterns, even at trace levels (Patel et al., 2014).

In LC, a sample is injected into a mobile phase (e.g., water-methanol mixture) that flows through a stationary phase (e.g., C18 column), separating analytes based on their chemical properties, such as polarity. The separated compounds are ionized (e.g., via electrospray ionization) and analyzed in the mass spectrometer, which measures their mass-to-charge ratio (m/z) to identify and quantify mycotoxins.

- **Gas Chromatography-Mass Spectrometry (GC-MS):** GC-MS complements the above techniques by analyzing volatile and semi-volatile compounds, making it highly effective for certain mycotoxins and pesticides. In GC-MS, a sample is vaporized and passed through a capillary column using a gas mobile phase (e.g., helium) with a stationary phase, typically a coated silica tube. Compounds are separated based on their volatility and interaction with the stationary phase, then ionized in the mass spectrometer, which detects their mass-to-charge ratio (Goh et al., 2024).
- **Enzyme-Linked Immunosorbent Assay (ELISA):** ELISA is an immunological technique that uses antibodies to detect specific mycotoxins. A sample is added to a plate coated with antibodies, and the target mycotoxin binds to

them. A secondary enzyme-linked antibody is introduced, which produces a colorimetric or fluorescent signal proportional to the mycotoxin concentration, measured using a spectrophotometer. Uses antibodies for rapid, colorimetric detection of specific mycotoxins, ideal for field screening but less specific (Oduro-Mensah et al., 2023).

- **Multi-Residue Methods (MRMs):** MRMs combine extraction, cleanup, and detection to analyze multiple pesticides simultaneously. Techniques like QuEChERS extract pesticides from samples, followed by LC-MS/MS or GC-MS analysis, optimizing efficiency (Elmi et al., 2024).

CURRENT PRACTICES IN RESIDUE ANALYSIS

Mycotoxins, toxic fungal metabolites, and pesticide residues from pest control agents pose significant health and environmental risks, even at trace levels, necessitating advanced analytical methods for food safety. Primary techniques for detecting these contaminants include thin-layer chromatography (TLC), liquid chromatography-mass spectrometry (LC-MS), gas chromatography-mass spectrometry (GC-MS), and enzyme-linked immunosorbent assay (ELISA). These methods address major pesticide classes-organophosphates, carbamates, pyrethroids, and neonicotinoids- as well as mycotoxins like aflatoxins and deoxynivalenol (Table 2). The sample preparation techniques are compiled in the Table 3. This section explores the principles, applications, strengths, and limitations of these techniques, emphasizing their critical role in ensuring safe food and environmental compliance.

- Thin-Layer Chromatography (TLC)

TLC is a low-cost method for detecting aflatoxins in peanuts and maize, particularly in developing countries (Salisu et al., 2021). It is simple but labor-intensive, lacks sensitivity, and is unsuitable for multi-mycotoxin analysis. Its affordability makes it practical where advanced instruments are unavailable.

Table 2. Major Pesticide Classes and Their Environmental/Health Effects

Pesticide Class	Examples	Environmental Effects	Health Effects
Organophosphates	Chlorpyrifos, Malathion	Soil/water contamination, harm to non-target species (Kim et al., 2017)	Neurotoxicity, teratogenicity, carcinogenicity, and mutagenic effects (Liu et al., 2025)
Carbamates	Carbaryl, Aldicarb	Toxicity to pollinators, soil persistence	Cholinesterase inhibition, acute poisoning
Pyrethroids	Deltamethrin, Permethrin	Aquatic toxicity, bioaccumulation	Neurotoxicity, skin irritation
Neonicotinoids	Imidacloprid, Thiamethoxam	Bee population decline, groundwater contamination	Potential neurodevelopmental effects

- **High-Performance Liquid Chromatography (HPLC)**

HPLC is a powerful analytical technique widely used for separating, identifying, and quantifying compounds in complex mixtures. Its basic principle involves passing a pressurized liquid sample through a column packed with a stationary phase, typically a solid material like silica.

In mycotoxin analysis, HPLC is critical for detecting and quantifying toxic fungal metabolites, such as aflatoxins, ochratoxin A, and zearalenone, in food and feed. These toxins, harmful even in trace amounts, require sensitive detection to ensure safety. HPLC, often coupled with fluorescence or mass spectrometry (MS) detectors, offers high sensitivity and specificity, enabling accurate identification of mycotoxins in complex matrices like grains or dairy. Sample preparation, such as extraction and cleanup, enhances detection by reducing matrix interference. For instance, immunoaffinity columns are used to isolate mycotoxins before HPLC analysis, ensuring precise quantification.

Similarly, HPLC plays a vital role in pesticide residue analysis, detecting compounds like organophosphates or carbamates in agricultural products. Its ability to separate structurally similar pesticides ensures compliance with MRLs set by regulatory bodies. High-performance liquid chromatography coupled with tandem mass spectrometry (HPLC-MS/MS) is particularly effective, providing enhanced sensitivity and confirmation of pesticide identity. By enabling reliable detection and quantification, HPLC safeguards public health and supports regulatory compliance in food safety.

- **Liquid Chromatography-Mass Spectrometry (LC-MS)**

LC-MS excels in detecting multiple toxins simultaneously in complex food matrices, such as cereals or nuts, with minimal sample preparation. Techniques like tandem MS (LC-MS/MS) enhances specificity by fragmenting ions and analyzing the resulting fragments (Patel et al., 2014). In pesticide residue analysis, LC-MS is indispensable for identifying a wide range of compounds, ensuring compliance with stringent regulatory limits. LC-MS ensures robust food safety monitoring through precise chromatographic separation and mass spectrometry. This is a highly sensitive and versatile technique, capable of detecting multiple mycotoxins, such as aflatoxins, ochratoxin A, and deoxynivalenol, at parts-per-billion (ppb) levels in complex matrices (Zhang et al., 2025). High-Performance Liquid Chromatography (HPLC) is enhanced in its ultra-high-performance form (UHPLC) by using smaller stationary phase particles (typically under 2 μm) operating at higher pressures. This results in quicker separations and improved resolution, enabling more efficient and sensitive analysis of complex samples, while also reducing analysis time and preserving or enhancing accuracy. UHPLC reduces analysis time and improves resolution (Hymete et al., 2024). However, high equipment costs and the need for skilled operators limit its use in resource-constrained settings.

- **Gas Chromatography-Mass Spectrometry (GC-MS)**

This provides high sensitivity and specificity for identifying compounds which are volatile in nature. For non-volatile mycotoxins, derivatization is required to increase volatility. Derivatization is a chemical process used in analytical chemistry to modify a compound's structure by introducing a chemical group, enhancing its detectability, volatility, or stability for analysis. In the context of GC-MS, derivatization is often applied to non-volatile or thermally unstable compounds, such as certain mycotoxins (e.g., deoxynivalenol), to make them more volatile and suitable for gas-phase separation. This is achieved by reacting the compound with a derivatizing agent, which alters its chemical properties, improving chromatographic performance and detection sensitivity. For example, in pesticide and mycotoxin analysis, derivatization ensures better separation and identification in complex food matrices. For pesticide residue analysis, GC-MS excels in detecting volatile pesticides like organochlorines in fruits, vegetables, and environmental samples, ensuring compliance with maximum residue limits. Its ability to handle volatile analytes with minimal matrix interference makes GC-MS a critical tool for food safety, complementing LC-based methods for comprehensive contaminant detection. The derivatization step adds time and complexity, and GC-MS is best suited for laboratories with established protocols for volatile analytes. Recent advances, like two-dimensional GC (GC×GC), enhance resolution for complex samples.

As described for mycotoxins, GC-MS separates volatile compounds in a gas phase and detects them via mass spectrometry. Pesticides are vaporized, separated on a capillary column, and ionized for m/z analysis. Derivatization may be needed for non-volatile compounds. GC-MS is ideal for volatile pesticides, like organochlorines and pyrethroids, in fruits and grains (Goh et al., 2024). It offers high sensitivity, and advances like GC×GC improve resolution for complex mixtures. However, derivatization increases analysis time, and GC-MS is less suited for polar pesticides.

- **Multi-Residue Methods (MRMs)**

MRMs are widely used for monitoring diverse pesticide profiles in imported foods (Elmi et al., 2024). They are robust and cost-effective but require optimization to minimize matrix effects and co-extracted interferences.

- **Enzyme-Linked Immunosorbent Assay (ELISA)**

ELISA is rapid and cost-effective, widely used for screening aflatoxins and deoxynivalenol in cereals (Oduro-Mensah et al., 2023). Commercial kits enable high-throughput analysis, and its portability suits field testing. However, cross-reactivity with similar compounds reduces specificity, and its semi-quantitative nature limits its use in regulatory testing. ELISA is less sensitive than LC-MS for trace-level detection.

Sample preparation techniques, such as QuEChERS (Quick, Easy, Cheap, Effective, Rugged, and Safe) and SPE, enhance the accuracy of these methods by removing matrix interferences (Lee et al., 2023). The sample preparation techniques for mycotoxin and pesticide residue analysis are briefly given in Table 4. The choice of method depends on the mycotoxin, matrix, and resources, with LC-MS and GC-MS favored for precision and ELISA/TLC for affordability. Comparison of various methods for mycotoxin and pesticide analysis is given in the table 3.

Table 3. Comparison of Mycotoxin and Pesticide Analysis Methods

Method	Analytes	Strengths	Limitations	Units (LOD)
TLC	Aflatoxins	Low cost, simple	Low sensitivity, labor-intensive	μg/kg
HPLC	Mycotoxins, Pesticides	High sensitivity, versatile	Matrix effects, cleanup needed	ng/mL
LC-MS/MS	Multi-mycotoxins, Pesticides	High sensitivity, specificity	High cost, expertise required	0.1 ppb
GC-MS	Volatile Pesticides	High resolution, sensitive	Derivatization complexity	ng/g
ELISA	Aflatoxins, Deoxynivalenol	Rapid, portable, cost-effective	Cross-reactivity, semi-quantitative	μg/kg
MRMs	Multi-pesticides	Efficient, high-throughput	Matrix effects, optimization needed	ng/g

Strengths and Limitations of Current Methods

- LC-MS and LC-MS/MS excel in sensitivity and versatility, detecting trace residues in complex matrices and supporting regulatory compliance (Agriopoulou et al., 2020).
- GC-MS complements LC-MS for volatile compounds, while MRMs enhance efficiency for large-scale monitoring (Elmi et al., 2024).
- ELISA and TLC offer rapid, affordable screening, particularly in low-resource settings (Shao et al., 2023).
- High costs and expertise requirements limit LC-MS and GC-MS accessibility. Matrix effects challenge all methods, especially in fatty or pigmented samples (Wu & Ding, 2023).
- ELISA's cross-reactivity and TLC's low sensitivity restrict their regulatory use, and no single method detects all residues comprehensively, necessitating method optimization.

CASE STUDIES

To illustrate real-world applications, the following case studies highlight successful implementations:

LC-MS Monitoring in Africa: In Kenya, LC-MS/MS was used to monitor aflatoxin B1 in maize, achieving a detection limit of 0.1 ppb, ensuring compliance with EU MRLs and supporting export markets (Maphaisa et al., 2025). This enabled rapid testing of 500 samples, reducing trade rejections by 30%.

Portable Biosensors in Asia: In India, smartphone-based lateral flow immunoassays (LFIAs) detected chlorpyrifos in spices with a limit of 0.2 ng/mL, allowing farmers to test on-site and ensure compliance with local regulations (Wu et al., 2023).

AI-Powered Aflatoxin Prediction: Artificial intelligence, specifically machine learning (ML) algorithms, predicts mycotoxin occurrence in maize by analyzing complex datasets that include weather variables, agronomic factors, and historical mycotoxin contamination data. Supervised learning models, such as Random Forest and Support Vector Machines can be used to identify patterns and relationships between environmental conditions (e.g., temperature, humidity, rainfall) and the likelihood of mycotoxin production, particularly aflatoxins and fumonisins. These models were trained on multi-year data and successfully classified contamination risk levels with high accuracy, enabling proactive decision-making for maize safety (Camardo Leggieri et al., 2021, Naseem & Rizwan, 2025).

CHALLENGES IN RESIDUE ANALYSIS

The analysis of mycotoxin and pesticide residues is critical for ensuring food safety and regulatory compliance, but it faces numerous challenges that complicate accurate detection and quantification. These concerns span analytical, logistical, regulatory, and environmental domains, impacting the efficacy of residue analysis globally. This section explores seven key challenges: analytical sensitivity and specificity, matrix complexity, cost and accessibility, regulatory harmonization, environmental impact, emerging contaminants, and speed and scalability.

Table 4. Sample Preparation Techniques for Mycotoxin and Pesticide Residue Analysis

Technique	Principle	Applications	Advantages	Limitations	References
QuEChERS (Quick, Easy, Cheap, Effective, Rugged, and Safe)	Combines extraction with a solvent (e.g., acetonitrile) and salts to partition analytes, followed by dispersive solid-phase extraction (d-SPE) cleanup to remove matrix interferences.	Extraction of pesticides and mycotoxins from fruits, vegetables, cereals, and other food matrices.	Simple, rapid, cost-effective; suitable for multi-residue analysis; high recovery rates.	Less effective for highly pigmented or fatty matrices; requires optimization for specific analytes.	Lee et al., 2023; Lehotay et al., 2010
Dispersive Liquid-Liquid Microextraction (DLLME)	Involves injecting a mixture of a water-immiscible extraction solvent and a dispersive solvent into an aqueous sample, forming a cloudy solution where analytes partition into fine droplets.	Extraction of aflatoxins and pesticides (e.g., imidacloprid) in cereals, water, and spices.	High enrichment factors, minimal solvent use, rapid extraction; green variants use deep eutectic solvents.	Requires precise solvent selection; less effective for complex matrices.	Salim et al., 2021; Maneeboon et al., 2023; Zareiyan et al., 2024
Magnetic Solid-Phase Extraction (MSPE)	Uses magnetic nanoparticles (e.g., C18 or metal-organic frameworks) to adsorb analytes, retrieved by a magnetic field for phase separation.	Extraction of pesticides and mycotoxins from honey, fruits, and complex matrices.	Low solvent consumption, reusable particles, simplifies phase separation; high selectivity.	Limited to analytes with affinity for magnetic sorbents; nanoparticle synthesis can be costly.	Liu et al., 2022; Alam et al., 2021
Ultrasound-Assisted Extraction (UAE)	Utilizes ultrasonic waves to induce cavitation, disrupting sample matrices and enhancing analyte release into minimal solvent volumes.	Extraction of mycotoxins and pesticides from apples, spices, and other matrices.	High extraction efficiency, reduced solvent use, rapid process; suitable for diverse matrices.	Equipment costs; potential analyte degradation under prolonged sonication.	Pérez & Albero, 2023; Kalogiouri et al., 2021
Microwave-Assisted Extraction (MAE)	Applies microwave energy to heat samples selectively, accelerating analyte diffusion into the extraction medium.	Extraction of pesticides and mycotoxins from potatoes, medicinal plants, and other matrices.	High recovery, reduced solvent and time; effective for thermally stable analytes.	Limited to thermally stable compounds; equipment costs; risk of overheating.	Reis et al., 2020; Ferrara et al., 2023; Ma et al., 2025

Analytical Sensitivity and Specificity

Detecting mycotoxins and pesticide residues at trace levels, often in the parts-per-billion (ppb) or parts-per-trillion (ppt) range, demands exceptional analytical sensitivity to avoid false negatives, which could compromise food safety. For instance, aflatoxins, known carcinogens, require detection limits as low as 0.1 ppb to meet stringent regulatory standards (Patel et al., 2025). LC-MS/MS achieves such sensitivity, but even advanced methods struggle with low-abundance analytes in complex samples. Specificity is equally critical to distinguish target compounds from matrix interferences or structurally similar substances. For instance, ELISA may produce false positives due to cross-reactivity with similar fungal metabolites (Zhang et al., 2023). Similarly, pesticide analysis by GC-MS may misidentify compounds with overlapping mass spectra, necessitating additional confirmation steps like multiple reaction monitoring (MRM) in LC-MS/MS (Tran-Lam et al., 2021). Developing methods that balance sensitivity and specificity remains a significant challenge, particularly for multi-residue analysis where diverse analytes are targeted simultaneously.

Matrix Complexity

Food matrices, ranging from grains and fruits to dairy and processed products, vary widely in composition, including lipids, proteins, pigments, and sugars, which interfere with residue analysis. These matrix effects can suppress or enhance analyte signals, reducing method accuracy. For instance, fatty matrices like nuts challenge LC-MS analysis of aflatoxins due to ion suppression, requiring extensive sample cleanup (Man et al., 2024; Lee et al., 2023). Complex matrices also increase the risk of cross-contamination during sample preparation, particularly in high-throughput settings. Developing universal sample preparation protocols that accommodate diverse matrices without compromising sensitivity or specificity is a persistent challenge, especially for laboratories handling varied food types.

Cost and Accessibility

Advanced analytical platforms like LC-MS/MS and GC-MS are the gold standard for residue analysis due to their precision and versatility, but their high costs for equipment, maintenance, and reagents limit their adoption in developing countries and small-scale laboratories (Junger et al., 2023). Additionally, these methods require skilled operators and controlled environments, further increasing operational costs. In resource-limited settings, affordable alternatives like TLC and ELISA are used, but their lower sensitivity and specificity restrict their regulatory applicability

(Garg et al., 2022; Singh & Mehta, 2020). For example, TLC is common for aflatoxin screening in Africa, but it cannot meet the stringent MRLs required for export markets. The lack of accessible, cost-effective technologies hinders global food safety monitoring, particularly in regions with high mycotoxin prevalence, such as sub-Saharan Africa and Southeast Asia, where infrastructure is limited (Maphaisa et al., 2025; American Academy of Microbiology, 2010).

Regulatory Landscape

Regulatory challenges vary by region (Table 3). The EU enforces strict MRLs (e.g., 2 μg/kg for aflatoxin B1), while Africa and Asia have less stringent or variable standards, complicating trade (EFSA, 2020). Climate change exacerbates residue distribution, increasing mycotoxin prevalence in warmer regions (Patel et al., 2025).

Disparities in MRLs across countries create significant challenges for global food trade. For instance, the EU sets strict MRLs for pesticides like chlorpyrifos, while other regions may have higher or no limits, leading to trade disputes (EFSA, 2020). Similarly, mycotoxin regulations vary, with the EU limiting aflatoxin B1 in peanuts to 2 μg/kg, compared to less stringent standards in some African nations (Patel et al., 2025). These inconsistencies complicate compliance for exporters and require laboratories to adapt methods to multiple regulatory frameworks. Harmonizing MRLs through international bodies like the Codex Alimentarius is challenging due to differing risk assessment approaches and economic priorities. Additionally, the lack of universally validated analytical methods hinders data comparability, as laboratories may use different protocols, affecting result reliability. Efforts to standardize methods and MRLs are critical to facilitate trade and ensure consistent food safety standards.

Environmental Impact

Traditional residue analysis methods often rely on organic solvents, such as acetonitrile and methanol, for sample extraction and chromatography, generating significant chemical waste. For example, QuEChERS protocols typically use 10 - 15 mL of organic solvent per sample, and large-scale monitoring programs amplify this environmental footprint (Lehotay et al., 2010). Additionally, energy-intensive instruments like LC-MS/MS contribute to carbon emissions, raising sustainability concerns. Green analytical chemistry principles advocate for solvent-free or miniaturized methods, but their adoption is limited by performance trade-offs. For instance, solvent-free techniques like solid-phase microextraction (SPME) reduce waste but may lack the sensitivity needed for trace-level detection (Agatonovic-Kustrin et al., 2023). Balancing analytical performance with environmental sustainability is a growing challenge, particularly as regulatory and consumer demand for eco-friendly practices increases.

Emerging Contaminants

The introduction of new pesticides and the evolution of fungal strains create novel residues that challenge existing analytical methods. For example, neonicotinoids, introduced in the 1990s, required updates to LC-MS/MS protocols due to their polar nature, unlike traditional organophosphates (Karthikeyan et al., 2019). Emerging pesticides like flupyradifurone and novel mycotoxins like beauvericin challenge existing methods, requiring updated LC-MS/MS protocols (Mahdavi et al., 2021; Masquelier et al., 2025). Similarly, emerging mycotoxins, such as fusarium toxins like enniatins, are not routinely included in standard screening methods, risking under-detection (Marin et al., 2013). Developing reference standards and validated methods for these contaminants is time-consuming and costly, as new compounds often lack commercial standards. Additionally, climate change exacerbates mycotoxin production by altering fungal growth patterns, introducing unpredictable residue profiles. Continuous method updates and flexible analytical platforms, such as high-resolution mass spectrometry (HRMS) for non-targeted screening, are needed to address these evolving threats.

Speed and Scalability

Regulatory and industry demands for rapid, high-throughput residue analysis are increasing, driven by the need to test perishable goods and monitor large-scale food supply chains. For example, pesticide residue testing in imported fruits requires results within hours to avoid spoilage, but methods like LC-MS/MS requires longer analysis time per sample, excluding preparation time (Steiner et al., 2021). High-throughput screening is also critical for national monitoring programs, such as the EU's pesticide residue surveys, which analyze thousands of samples annually (EFSA, 2024). While multi-residue methods like QuEChERS improve efficiency, they are bottlenecked by complex cleanup and detection steps. Rapid techniques like ELISA offer speed but lack the precision needed for regulatory compliance (Stefanovic et al., 2015). Developing scalable, automated platforms that deliver fast, accurate results without sacrificing sensitivity remains a key challenge, particularly for laboratories handling high sample volumes.

The challenges in mycotoxin and pesticide residue analysis analytical sensitivity and specificity, matrix complexity, cost and accessibility, regulatory harmonization, environmental impact, emerging contaminants, and speed and scalability underscore the need for innovative solutions. Addressing these concerns requires advancements in affordable, sensitive, and sustainable technologies, alongside international collaboration to standardize regulations and methods. By overcoming these hurdles, residue analysis can better support global food safety and trade, protecting public health and the environment.

EMERGING TRENDS AND INNOVATIONS

The landscape of mycotoxin and pesticide residue analysis is rapidly evolving, driven by technological advancements addressing the limitations of traditional methods. Emerging trends, such as high-resolution mass spectrometry, ambient ionization techniques, and portable biosensors, enhance detection sensitivity, speed, and accessibility. Innovations like nanotechnology, aptamer-based sensors, and artificial intelligence streamline analysis, enabling real-time monitoring and non-targeted screening of contaminants in diverse food matrices (Figure 2). Additionally, green analytical chemistry promotes sustainable practices, reducing environmental impact. This section explores these cutting-edge developments, highlighting their potential to revolutionize food safety, ensure regulatory compliance, and support global efforts toward sustainable and resilient food systems.

Advanced Analytical Technologies

The evolving landscape of mycotoxin and pesticide residue analysis is being shaped by advanced analytical technologies that enhance detection capabilities, speed, and accessibility. HRMS, ambient ionization techniques like direct analysis in real time mass spectrometry (DART-MS), and portable devices, including smartphone-based sensors, are at the forefront of these innovations. These technologies address limitations of traditional methods, offering improved sensitivity, specificity, and field applicability for food safety monitoring.

Figure 2. Emerging Technologies in Mycotoxin and Pesticide Residue Analysis

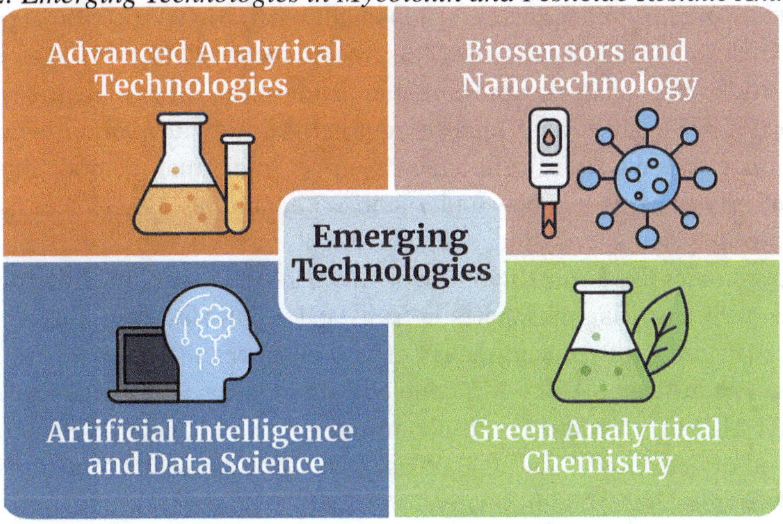

HRMS for Non-Targeted Screening

HRMS, utilizing instruments like time-of-flight (TOF) or Orbitrap mass spectrometers, revolutionizes residue analysis by enabling non-targeted screening of mycotoxins and pesticides. Unlike traditional targeted methods that detect predefined analytes, HRMS provides high mass accuracy (resolving power >20,000) and full-scan data, allowing identification of unknown or emerging contaminants without prior reference standards (Hollender et al., 2017). For example, HRMS has been used to detect novel fungal metabolites, such as enniatins, in cereals, which standard LC-MS/MS might miss (Narváez et al., 2023). By coupling HRMS with ultra-high-performance liquid chromatography (UHPLC), analysts achieve rapid separation and comprehensive profiling of complex food matrices, such as spices or processed foods. HRMS also supports retrospective data analysis, where stored datasets can be re-examined for new contaminants. However, its high cost and data processing complexity limit widespread adoption, particularly in resource-constrained settings. Ongoing developments in user-friendly software and cost-effective HRMS platforms are helping to bridge this gap.

Ambient Ionization Techniques

Ambient ionization techniques, such as DART-MS, enable rapid, direct analysis of samples under ambient conditions, minimizing sample preparation. DART-MS ionizes analytes on sample surfaces (e.g., fruit peels or grain husks) using a plasma-based helium stream, with the resulting ions analyzed by a mass spectrometer (Wang, 2024). Direct Analysis in Real-Time Mass Spectrometry (DART-MS) has emerged as a powerful tool for analyzing these contaminants in various food matrices (Wang, 2024). DART-MS enables rapid, high-throughput analysis with minimal or no sample preparation, offering a significant advantage over traditional methods like HPLC or GC-MS. By ionizing samples under ambient conditions, DART-MS facilitates direct molecular characterization, achieving high sensitivity and specificity for detecting mycotoxins like aflatoxins and deoxynivalenol in cereals, nuts, and dairy, and pesticides such as organophosphates in fruits and vegetables. Its ability to analyze complex matrices with reduced matrix effects enhances its utility for food safety screening. For instance, DART-MS coupled with high-resolution mass spectrometry (DART-HRMS) can detect multiple mycotoxins simultaneously in cereal flour with a limit of detection as low as 0.06 µg/g for compounds like roquefortine C in cheese (Maragos, 2022). Additionally, DART-MS supports food authentication by identifying specific markers, ensuring traceability and quality. Despite challenges like matrix interference and compound selectivity, its speed, low solvent use, and environmental friendliness make DART-MS an invaluable tool for ensuring compliance with regulatory standards and protecting consumer health. DART-MS is particularly valuable for high-throughput screening at ports or markets, where speed is critical. Its ability to analyze solids, liquids, or gases enhances versatility, but challenges include lower sensitivity compared to LC-MS/MS and susceptibility to matrix effects in complex samples.

Portable Devices and Point-of-Care Testing

Portable devices and point-of-care (POC) testing have revolutionized mycotoxin and pesticide residue analysis by enabling rapid, on-site detection with minimal sample preparation, addressing the limitations of traditional laboratory-based methods like HPLC and GC-MS. These devices often employ biosensing platforms, such as electrochemical, optical, or microfluidic systems, integrated with technologies like lateral flow immunoassays (LFIAs), nanozyme-based sensors, or smartphone-based readouts. These systems detect specific analytes by leveraging molecular interactions, such as antibody-antigen binding or enzymatic reactions, producing measurable signals (e.g., fluorescence, color changes, or electrical currents) that are quantified instantly. Advantages include affordability, user-friendliness, and rapid results, often within minutes, as highlighted by Jafari et al. (2021) and Nelis et al. (2020). They require less training and equipment compared to lab-based methods, making them ideal for field applications. In mycotoxin analysis, devices like the smartphone-based LFIA-MS detect aflatoxins and deoxynivalenol in cereals, maize, and wine with high sensitivity (e.g., 0.1 μg/kg for aflatoxin B1) (Wu et al., 2023; Liu et al., 2024). For pesticides, rapid test paper and microfluidic platforms identify residues like chlorpyrifos in fruits, vegetables, and tea, ensuring compliance with regulatory limits (Ghosh et al., 2022; Zhang et al., 2023). Portable sensor devices using multifunctional frameworks, such as nanozymes, further enhance sensitivity and specificity for on-site food safety assurance (Zhang et al., 2025; Li et al., 2025), supporting real-time monitoring in diverse food matrices.

Biosensors and Nanotechnology

Nanotechnology and biosensors have transformed mycotoxin and pesticide residue analysis by enabling highly sensitive, rapid, and selective detection in food matrices, addressing the limitations of traditional methods like HPLC and GC-MS. These biosensors integrate nanomaterials, such as gold nanoparticles, carbon nanotubes, or graphene, to enhance signal amplification and molecular recognition (Zhang et al., 2025; Mirres et al., 2022). The principle involves immobilizing biorecognition elements (e.g., antibodies, aptamers, or enzymes) on nanomaterial surfaces, which interact specifically with target analytes like aflatoxins or organophosphates, producing measurable signals via electrochemical, optical, or fluorescent mechanisms (Zhou et al., 2022; Gong et al., 2023). Their advantages include high sensitivity (e.g., detecting aflatoxin B1 at 0.1 ng/mL in maize), portability, and integration with microfluidic devices for on-site analysis (Acharya et al., 2024). In mycotoxin analysis, nanomaterial-based biosensors detect toxins like deoxynivalenol and zearalenone in cereals, nuts, and spices with minimal sample preparation (Chen et al., 2025). For pesticides, they identify residues like chlorpyrifos and carbendazim in fruits, vegetables, and tea, ensuring compliance with regulatory limits (Rehman et al., 2020). These biosensors offer cost-effectiveness, reduced analysis time, and environmental friendliness, making them ideal for field applications in food safety monitoring, as emphasized by recent advancements in electrochemical and fluorescent nano-biosensors (Gong et al., 2023).

Aptamer-Based and Antibody-Based Biosensors for Rapid Detection

Aptamer-based and antibody-based biosensors have emerged as highly effective tools for the rapid detection of mycotoxins and pesticide residues in food, offering high sensitivity, specificity, and portability compared to traditional methods like HPLC and GC-MS. These biosensors utilize aptamers synthetic oligonucleotides that bind specific targets with high affinity or antibodies, which recognize analytes through antigen-antibody interactions, to detect contaminants like aflatoxins, ochratoxin A, and pesticides such as chlorpyrifos (Guo et al., 2020; Balkrishna et al., 2024). The principle involves immobilizing aptamers or antibodies on platforms like lateral flow strips, electrochemical sensors, or microfluidic devices, producing measurable signals (e.g., fluorescence or electrochemical changes) upon analyte binding (Li et al., 2024; Mermiga et al., 2023). Aptamer-based biosensors, noted for their stability and cost-effectiveness, achieve detection limits as low as 0.1 ng/mL for fusaric acid in cereals (Wu et al., 2025). Antibody-based biosensors excel in specificity, detecting ochratoxin A in wine and grains with minimal sample preparation (Shoaib et al., 2025). Both are integrated with portable devices, such as smartphone-based platforms, enabling on-site analysis in fruits, vegetables, and spices (Zhang et al., 2025; McConnell et al., 2020). Their rapid analysis (within minutes), low cost, and ability to handle complex matrices make them ideal for ensuring food safety and regulatory compliance.

Artificial Intelligence and Data Science

Artificial intelligence (AI) and data science have revolutionized mycotoxin and pesticide residue analysis by enhancing detection accuracy, speed, and predictive capabilities, addressing the limitations of traditional analytical methods like HPLC and GC-MS. These technologies leverage machine learning (ML), deep learning (DL), and big data analytics to process complex datasets from food matrices, enabling rapid identification and quantification of contaminants such as aflatoxins, ochratoxin A, and pesticides like chlorpyrifos in diverse foods, including cereals, fruits, vegetables, and spices (Aggarwal et al., 2024; Banerjee et al., 2024). AI-driven models, such as convolutional neural networks and random forests, analyze spectral data from techniques like near-infrared spectroscopy or hyperspectral imaging, achieving detection limits as low as 0.1 ng/mL for mycotoxins in maize and nuts (Mateo et al., 2025). Data science integrates Internet of Things (IoT) sensors and big data platforms to enable real-time monitoring and early warning systems, identifying emerging risks in food supply chains (Mu et al., 2024). For instance, AI algorithms predict mycotoxin contamination risks in grains by analyzing environmental factors like temperature and humidity, facilitating proactive mitigation (Naseem & Rizwan, 2025). In pesticide analysis, AI enhances eco-friendly degradation strategies by optimizing bioremediation processes, reducing residue levels in crops like tomatoes and tea (Banerjee et al., 2024). These systems offer advantages like automation, reduced analysis time, and cost-effectiveness, with applications in on-site testing via portable devices integrated with AI-driven smartphone apps. By improving sensitivity, specificity, and predictive accuracy, AI and data science ensure compliance with regulatory standards, enhance food safety, and support a path toward zero contamination (Naseem & Rizwan 2025).

Green Analytical Chemistry

Green analytical chemistry (GAC) is reshaping mycotoxin and pesticide residue analysis by prioritizing sustainability, reducing environmental impact, and maintaining high analytical performance. Key advancements include solvent-free methods and miniaturized systems, alongside sustainable sample preparation techniques. These innovations address the environmental concerns of traditional methods, which often rely on hazardous solvents and generate significant waste, aligning residue analysis with global sustainability goals.

Solvent-Free Methods and Miniaturized Systems

Solvent-free methods and miniaturized systems, such as solid-phase microextraction (SPME), have revolutionized mycotoxin and pesticide residue analysis by offering environmentally friendly, efficient, and sensitive approaches for food safety testing. SPME, a key solvent-free technique, involves the adsorption of analytes onto a coated fiber, followed by desorption for analysis, typically using gas chromatography (GC) or high-performance liquid chromatography (HPLC) (Beltran et al., 2000; Aulakh et al., 2005). This method minimizes solvent use, reducing environmental impact and sample preparation time, as highlighted by Drabińska et al. (2023). SPME is highly effective for detecting mycotoxins like aflatoxins and ochratoxin A in cereals, nuts, and spices, and pesticides such as organophosphates in fruits and vegetables, achieving detection limits in the ng/g range (Merkle et al., 2015; Xu et al., 2021). Other miniaturized systems, like headspace-SPME, are used for volatile compounds, such as allyl isothiocyanate in cowpea beans, offering high sensitivity with minimal matrix interference (Vilela et al., 2020). Techniques like stir-bar sorptive extraction (SBSE) and micro-solid-phase extraction further enhance efficiency by concentrating analytes from complex food matrices, such as dairy and wine (Demirhan et al., 2017). These methods provide advantages like portability, reduced costs, and compatibility with GC-MS/MS for precise identification of contaminants (Xu et al., 2021). By eliminating solvent use and enabling rapid, on-site analysis, solvent-free miniaturized systems ensure compliance with regulatory standards, making them indispensable for sustainable food safety monitoring.

Sustainable Sample Preparation Techniques

Sustainable sample preparation techniques, including dispersive liquid-liquid microextraction (DLLME), magnetic solid-phase extraction (MSPE), ultrasound-assisted extraction (UAE), and microwave-assisted extraction (MAE), have advanced mycotoxin and pesticide residue analysis by offering eco-friendly, efficient, and sensitive methods for food safety testing. **DLLME** involves injecting a mixture of a water-immiscible extraction solvent and a dispersive solvent into an aqueous sample, forming a cloudy solution where analytes like aflatoxins or pesticides partition into fine droplets, enabling rapid extraction with high enrichment factors and minimal solvent use (Salim et al., 2021; Maneeboon et al., 2023). Green variants using deep eutectic solvents further enhance sustainability for pesticides like imidacloprid in water (Zareiyan et al., 2024). **MSPE** employs magnetic nanoparticles (e.g., C18 or metal-organic frameworks) to adsorb analytes from complex matrices like honey or fruits. A magnetic field retrieves the particles, simplifying phase separation and enabling reuse with low solvent consumption (Liu et al., 2022; Alam et al., 2021). **UAE** uses ultrasonic waves to induce cavitation, disrupting sample matrices like apples or spices and enhancing analyte release into minimal solvent volumes, improving extraction efficiency (Pérez & Albero, 2023; Kalogiouri et al., 2021). **MAE** applies microwave energy to heat samples selectively, accelerating analyte diffusion from matrices like potatoes or medicinal plants into the extraction medium, achieving high recovery with reduced solvent and time (Reis et al., 2020; Ferrara et al., 2023; Ma et al., 2025). These techniques achieve detection limits in the ng/g range for mycotoxins and pesticides in cereals, fruits, and vegetables, ensuring regulatory compliance and sustainable food safety monitoring.

Green analytical chemistry, through solvent-free methods, miniaturized systems, and sustainable sample preparation, offers environmentally friendly solutions for residue analysis. These advancements reduce the ecological footprint of testing while maintaining sensitivity and accessibility, supporting global food safety and sustainability objectives.

FUTURE DIRECTIONS

Integration of Multi-Omics Approaches for Comprehensive Contaminant Profiling

Multi-omics approaches, integrating genomics, proteomics, metabolomics, and transcriptomics, have transformed mycotoxin and pesticide residue analysis by providing comprehensive insights into contaminant profiles, their biological impacts, and mitigation strategies in food safety. These methods, leveraging HRMS and advanced bioinformatics, enable simultaneous detection of regulated and emerging mycotoxins, such as aflatoxins, deoxynivalenol, and **masked mycotoxins**, as well as pesticide residues like organophosphates in complex food matrices like cereals, fruits, and dairy (Dellafiora & Dall'Asta, 2017; Lapris et al., 2024). Masked mycotoxins, also known as modified or conjugated mycotoxins, are derivatives of mycotoxins that are chemically altered by plants, fungi, or other organisms, typically through conjugation with polar molecules like glucose, sulfate, or other glycosides, making them less detectable by conventional analytical methods. These modifications occur during plant metabolism or fungal biosynthesis, masking the parent mycotoxin's structure while potentially retaining or altering its toxicity. For example, deoxynivalenol-3-glucoside (D3G) is a masked form of deoxynivalenol, commonly found in cereals like wheat and maize. Masked mycotoxins pose challenges in food safety because they may evade standard detection techniques, such as HPLC or GC-MS, yet can release their toxic parent compounds during digestion or processing, posing health risks to humans and animals. Their detection often requires advanced methods like HRMS or multi-omics approaches to identify and quantify these hidden contaminants in food matrices (Singh & Kumari, 2022).

Metabolomics identifies metabolite biomarkers, revealing mycotoxin transformation pathways and toxicity mechanisms, as demonstrated in studies on rice bran and maize (Mandal et al., 2024). Metabolomics, coupled with HRMS, identifies novel and modified mycotoxins, such as masked deoxynivalenol, in cereals (Righetti et al., 2016).Proteomics elucidates protein responses to contaminants, aiding in understanding cellular impacts, while genomics and transcriptomics uncover fungal gene expression linked to mycotoxin production (Righetti et al., 2016; Goh et al., 2021). For instance, HRMS-based multi-omics detected modified mycotoxins in wheat, achieving detection limits below 0.1 ng/g (Singh & Kumari, 2022). In pesticide analysis, omics tools identify degradation products and environmental interactions in crops like apples and spices, supporting eco-friendly mitigation strategies (Goh et al., 2021). These approaches excel in handling complex matrices, uncovering masked contaminants, and predicting contamination risks through data integration, as noted by Hou et al. (2024). By combining multi-omics with machine learning, researchers achieve enhanced sensitivity, specificity, and predictive accuracy, enabling proactive food safety monitoring and regulatory compliance across diverse food systems (Mandal et al., 2023; Lapris et al., 2024). However, challenges include high costs, complex data integration, and the need for standardized protocols. Advances in cloud-based bioinformatics and open-access omics databases are addressing these barriers, promising scalable, real-time profiling for global food safety.

Development of Universal, Cost-Effective Platforms for Residue Analysis

The development of universal, cost-effective platforms for mycotoxin and pesticide residue analysis has gained momentum, driven by the need for rapid, accessible, and affordable food safety testing. Recent advancements focus on integrating portable biosensors, microfluidics, and smartphone-based technologies to detect contaminants like aflatoxins, ochratoxin A, and pesticides such as chlorpyrifos in diverse food matrices, including cereals, fruits, and spices. For instance, lateral flow immunoassays (LFIAs) combined with smartphone readouts offer low-cost, user-friendly detection of mycotoxins in maize with limits as low as 0.1 ng/mL (Wu et al., 2023; Liu et al., 2024). Microfluidic platforms, leveraging aptamer-based sensors, enable simultaneous multi-mycotoxin detection in wine, reducing costs through minimal reagent use (Liu et al., 2024). Nanotechnology, including gold nanoparticles and carbon-based materials, enhances sensitivity and affordability in electrochemical biosensors for pesticide residues in vegetables (Mirres et al., 2022). AI-driven data analysis further optimizes these platforms by improving detection accuracy and enabling real-time monitoring (Aggarwal et al., 2024). These universal platforms minimize sample preparation, lower operational costs, and support on-site testing, ensuring compliance with regulatory standards while being accessible for resource-limited settings, thus advancing global food safety.

International Collaborations

International collaboration is critical in standardizing analytical methods and establishing MRLs for mycotoxins and pesticide residues in foods, ensuring global food safety, protecting public health, and facilitating seamless international trade. Organizations like the Codex Alimentarius Commission, the World Health Organization (WHO), and the Food and Agriculture Organization (FAO) lead efforts to harmonize detection protocols and regulatory frameworks across diverse regions, addressing challenges posed by varying food matrices such as cereals, fruits, vegetables, spices, and dairy. Collaborative initiatives, involving bodies like the European Food Safety Authority (EFSA) and the International Organization for Standardization (ISO), develop and validate robust analytical methods, such as HPLC-MS/MS and GC-MS, ensuring consistent detection of contaminants like aflatoxins, ochratoxin A, and pesticides like chlorpyrifos with detection limits as low as 0.1 ng/g (Van Egmond et al., 2007; Jafari et al., 2021). These standardized methods enhance result comparability, reduce discrepancies in testing outcomes, and ensure regulatory compliance across global markets.

For MRL determination, international collaboration integrates toxicological data, dietary exposure assessments, and regional agricultural practices to establish science-based limits that balance consumer safety with trade feasibility (FAO/WHO, 2019). Codex MRLs for pesticides like carbendazim in fruits and mycotoxins like deoxynivalenol in grains are widely adopted, minimizing trade barriers and ensuring consistency (Mu et al., 2024). Collaborative research networks, such as the MYCOKEY project, advance multi-mycotoxin detection, addressing emerging challenges like masked mycotoxins, which evade conventional methods but pose health risks (Dellafiora & Dall'Asta, 2017). These efforts leverage advanced technologies, including high-resolution mass spectrometry (HRMS) and multi-omics, to refine risk assessment models and update MRLs based on the latest toxicological insights (Mateo et al., 2025). Furthermore, AI-driven tools and big data analytics, supported by global data-sharing platforms, enhance predictive modeling of contamination risks, enabling proactive regulatory measures (Mu et al., 2024).

International collaboration also fosters capacity building in resource-limited regions by providing training, shared protocols, and access to advanced technologies, ensuring equitable implementation of food safety standards. For instance, workshops and joint research initiatives under FAO/WHO programs help developing countries adopt validated methods, strengthening global food safety networks (Van Egmond et al., 2007). By promoting data transparency, harmonizing analytical standards, and aligning MRLs, these collaborative efforts mitigate health risks, support sustainable food systems, and enhance consumer confidence worldwide.

CONCLUSION

The analysis of mycotoxin and pesticide residues in food is a cornerstone of global food safety, addressing critical health, environmental, and trade challenges posed by these pervasive contaminants. This chapter has elucidated the multifaceted approaches to residue analysis, highlighting the strengths and limitations of current methods like LC-MS/MS, GC-MS, and ELISA, which offer high sensitivity but face issues such as matrix complexity and high costs. Emerging technologies, including HRMS, ambient ionization techniques like DART-MS, and portable biosensors, are revolutionizing detection by enabling rapid, on-site, and non-targeted screening of contaminants like aflatoxins and chlorpyrifos in diverse matrices such as cereals, fruits, and dairy. Nanotechnology and aptamer-based biosensors further enhance sensitivity and affordability, while AI and multi-omics approaches provide comprehensive contaminant profiling, uncovering masked mycotoxins and predicting contamination risks. Green analytical chemistry, through solvent-free methods like SPME and sustainable techniques like DLLME and MSPE, aligns residue analysis with environmental sustainability, reducing waste and ecological impact.

Despite these advancements, challenges persist, including regulatory harmonization, accessibility in resource-limited settings, and the need for rapid, scalable solutions to address emerging contaminants. International collaboration, led by organizations like Codex Alimentarius, FAO, and WHO, is vital for standardizing methods and MRLs, ensuring equitable food safety standards and facilitating global trade (FAO/WHO, 2019). Future directions should prioritize universal, cost-effective platforms integrating AI, nanotechnology, and multi-omics to deliver real-time, accessible testing. Strengthening capacity building in developing regions and adopting green practices will further enhance the sustainability and reach of residue analysis. By bridging technological innovation with collaborative frameworks, the field can mitigate the risks of mycotoxins and pesticide residues, safeguarding public health, supporting sustainable food systems, and fostering consumer confidence worldwide.

Residue analysis is vital for food safety, supporting Sustainable Development Goals (SDG 2: Zero Hunger, SDG 3: Good Health, SDG 12: Responsible Consumption, SDG 13: Climate Action). Key takeaways:

- **High-Accuracy Lab Testing:** LC-MS/MS and HRMS offer superior sensitivity for regulatory compliance.
- **Field and Low-Resource Settings:** Portable biosensors and ELISA provide rapid, affordable screening.
- **Research Priorities:** Invest in AI, multi-omics, and green methods to address emerging contaminants and sustainability.

By integrating advanced technologies, international standards, and sustainable practices, residue analysis can mitigate health risks, ensure trade compliance, and promote resilient food systems.

REFERENCES

Acharya, B., Kumari, A., Kumar, A., Arya, V., Chauhan, A., Upadhyay, N. K., Guleria, I., Amarowicz, R., Kumar, D., & Kuca, K. (2024). Biosensors for detection of pesticide residue, mycotoxins and heavy metals in fruits and vegetables: A concise review. *Microchemical Journal, 205*, 111292. DOI: 10.1016/j.microc.2024.111292

Agatonovic-Kustrin, S., Gegechkori, V., Kobakhidze, T., & Morton, D. (2023). Solid-Phase Microextraction Techniques and Application in Food and Horticultural Crops. *Molecules (Basel, Switzerland), 28*(19), 6880. DOI: 10.3390/molecules28196880 PMID: 37836723

Aggarwal, A., Mishra, A., Tabassum, N., Kim, Y. M., & Khan, F. (2024). Detection of Mycotoxin Contamination in Foods Using Artificial Intelligence: A Review. *Foods (Basel, Switzerland), 13*(20), 3339. DOI: 10.3390/foods13203339 PMID: 39456400

Agriopoulou, S., Stamatelopoulou, E., & Varzakas, T. (2020). Advances in Analysis and Detection of Major Mycotoxins in Foods. *Foods, 9*(4), 518. DOI: 10.3390/foods9040518 PMID: 32326063

Alam, S., Srivastava, N., Iqbal, N., Saini, M. K., & Kumar, J. (2021). Magnetic Solid-Phase Extraction (MSPE) Using Magnetite-Based Core-Shell Nanoparticles with Silica Network (SiO2) Coupled with GC-MS/MS Analysis for Determination of Multiclass Pesticides in Water. *Journal of AOAC International, 104*(3), 633–644. DOI: 10.1093/jaoacint/qsaa156 PMID: 33201225

American Academy of Microbiology. (2010). *Global food safety: Keeping food safe from farm to table*. American Society for Microbiology., DOI: 10.1128/AAMCol.24Apr.2009

Aulakh, J. S., Malik, A. K., Kaur, V., & Schmitt-Kopplin, P. (2005). A Review on Solid Phase Micro Extraction—High Performance Liquid Chromatography (SPME-HPLC) Analysis of Pesticides. *Critical Reviews in Analytical Chemistry, 35*(1), 71–85. DOI: 10.1080/10408340590947952

Balkrishna, A., Kumari, A., Kumar, A., Arya, V., Chauhan, A., Upadhyay, N. K., Guleria, I., Amarowicz, R., Kumar, D., & Kuca, K. (2024). Biosensors for detection of pesticide residue, mycotoxins and heavy metals in fruits and vegetables: A concise review. *Microchemical Journal, 205*, 111292. DOI: 10.1016/j.microc.2024.111292

Banerjee, D., Adhikary, S., Bhattacharya, S., Chakraborty, A., Dutta, S., Chatterjee, S., Ganguly, A., Nanda, S., & Rajak, P. (2024). Breaking boundaries: Artificial intelligence for pesticide detection and eco-friendly degradation. *Environmental Research, 241*, 117601. DOI: 10.1016/j.envres.2023.117601 PMID: 37977271

Beltran, J., López, F. J., & Hernández, F. (2000). Solid-phase microextraction in pesticide residue analysis. *Journal of Chromatography. A*, *885*(1-2), 389–404. DOI: 10.1016/s0021-9673(00)00142-4 PMID: 10941686

Camardo Leggieri, M., Mazzoni, M., & Battilani, P. (2021). Machine Learning for Predicting Mycotoxin Occurrence in Maize. *Frontiers in Microbiology*, *12*. Advance online publication. DOI: 10.3389/fmicb.2021.661132 PMID: 33897675

Chen, L., Kan, J., Zalán, Z., Xu, D., Cai, T., & Chen, K. (2025). Application of nanomaterials in the detection of pesticide residues in spices. *Food Chemistry*, *473*, 143101. DOI: 10.1016/j.foodchem.2025.143101 PMID: 39889633

Dellafiora, L., & Dall'Asta, C. (2017). Forthcoming challenges in mycotoxins toxicology research for safer food—A need for multi-omics approach. *Toxins*, *9*(1), 18. DOI: 10.3390/toxins9010018 PMID: 28054977

Demirhan, B., Kara, H. E. Ş., & Demirhan, B. E. (2017). *Overview of green sample preparation techniques in food analysis*. In InTech. https://doi.org/DOI: 10.5772/intechopen.68787

Drabińska, N., Marcinkowska, M. A., Wieczorek, M. N., & Jeleń, H. H. (2023). Application of Sorbent-Based Extraction Techniques in Food Analysis. *Molecules (Basel, Switzerland)*, *28*(24), 7985. DOI: 10.3390/molecules28247985 PMID: 38138475

EFSA (European Food Safety Authority), Medina-Pastor, P., & Triacchini, G. (2020). The 2018 European Union report on pesticide residues in food. *EFSA Journal*, *18*(4), 6057. DOI: 10.2903/j.efsa.2020.6057 PMID: 32874271

EFSA (European Food Safety Authority), Carrasco Cabrera, L., Di Piazza, G., Dujardin, B., Marchese, E., & Medina Pastor, P. (2024). The 2022 European Union report on pesticide residues in food. *EFSA Journal*, *22*(4), e08753. DOI: 10.2903/j.efsa.2024.8753 PMID: 38655191

Elmi, M., Ghane, T., Daraei, B., Eskandari, S., Mohammadpour, A., Amirahmadi, M., & Mousavi Khaneghah, A. (2024). Monitoring of pesticide residue in pistachio nut samples by LC/MS-MS. *Food Chemistry*, *437*(Part 1), 137848. DOI: 10.1016/j.foodchem.2023.137848 PMID: 39491241

FAO/WHO. (2019). *Codex Alimentarius: General standard for contaminants and toxins in food and feed (CXS 193-1995)*. Food and Agriculture Organization of the United Nations and World Health Organization. https://www.fao.org/fao-who-codexalimentarius/codex-texts/list-standards/en/

Ferrara, D., Beccaria, M., Cordero, C. E., & Purcaro, G. (2023). Microwave-assisted extraction in closed vessel in food analysis. *Journal of Separation Science*, *46*, 2300390. DOI: 10.1002/jssc.202300390 PMID: 37654060

Garg, K., Villavicencio-Aguilar, F., Solano-Rivera, F., & Gilbert, L. (2022). Analytical Validation of a Direct Competitive ELISA for Multiple Mycotoxin Detection in Human Serum. *Toxins*, *14*(11), 727. DOI: 10.3390/toxins14110727 PMID: 36355977

Ghosh, S., AlKafaas, S. S., & Bornman, C. (2022). The application of rapid test paper technology for pesticide detection in horticulture crops: A comprehensive review. *Beni-Suef University Journal of Basic and Applied Sciences*, *11*, 73. DOI: 10.1186/s43088-022-00248-6

Goh, D., Abdull Razis, A. F., Yusof, N. A., Mazlan, N., Nordin, N., & Yu, C. Y. (2024). A review of emerging techniques for pyrethroid residue detection in agricultural commodities. *Heliyon*, *11*(1), e41154. DOI: 10.1016/j.heliyon.2024.e41154 PMID: 39801999

Goh, M. S., Lam, S. D., Yang, Y., Naqiuddin, M., Addis, S. N. K., Yong, W. T. L., Luang-In, V., Sonne, C., & Ma, N. L. (2021). Omics technologies used in pesticide residue detection and mitigation in crop. *Journal of Hazardous Materials*, *420*, 126624. DOI: 10.1016/j.jhazmat.2021.126624 PMID: 34329083

Gong, Z., Huang, Y., Hu, X., Zhang, J., Chen, Q., & Chen, H. (2023). Recent Progress in Electrochemical Nano-Biosensors for Detection of Pesticides and Mycotoxins in Foods. *Biosensors (Basel)*, *13*(1), 140. DOI: 10.3390/bios13010140 PMID: 36671974

Guo, X., Wen, F., Zheng, N., Saive, M., Fauconnier, M.-L., & Wang, J. (2020). Aptamer-based biosensor for detection of mycotoxins. *Frontiers in Chemistry*, *8*, 195. DOI: 10.3389/fchem.2020.00195 PMID: 32373573

Hollender, J., Schymanski, E. L., Singer, H. P., & Ferguson, P. L. (2017). Nontarget screening with high resolution mass spectrometry in the environment: Ready to go? *Environmental Science & Technology*, *51*(20), 11505–11512. DOI: 10.1021/acs.est.7b02184 PMID: 28877430

Hou, B., Xu, Y., Wang, D., Dong, C., Yan, F., Cheng, X., Sun, S., Bo, C., Liu, Z., Qin, Y., Wang, H., & Kong, L. (2024). A novel strategy for detoxification of deoxynivalenol via modification of both toxic groups. *Food Chemistry*, *456*, 139886. DOI: 10.1016/j.foodchem.2024.139886 PMID: 38870804

Hymete, A., Ahmed, F., Ashenef, A., & Abebe, T. (2024). *Optimizing ultra-high-performance liquid chromatography (UHPLC) methods for rapid analysis of complex biological samples*. IntechOpen., DOI: 10.5772/intechopen.1006988

Jafari, S., Guercetti, J., Geballa-Koukoula, A., Tsagkaris, A. S., Nelis, J. L. D., Marco, M. P., Salvador, J. P., Gerssen, A., Hajslova, J., Elliott, C., Campbell, K., Migliorelli, D., Burr, L., Generelli, S., Nielen, M. W. F., & Sturla, S. J. (2021). ASSURED Point-of-Need Food Safety Screening: A Critical Assessment of Portable Food Analyzers. *Foods (Basel, Switzerland), 10*(6), 1399. DOI: 10.3390/foods10061399 PMID: 34204284

Junger, S., Hoene, M., Shipkova, M., Danzl, G., Schöberl, C., Peter, A., Lehmann, R., Wieland, E., & Braitmaier, H. (2023). Automated LC-MS/MS: Ready for the clinical routine Laboratory? *Journal of Mass Spectrometry and Advances in the Clinical Lab, 30*, 1–9. DOI: 10.1016/j.jmsacl.2023.07.001 PMID: 37583571

Kalogiouri, N. P., Papadakis, E.-N., Maggalou, M. G., Karaoglanidis, G. S., Samanidou, V. F., & Menkissoglu-Spiroudi, U. (2021). Development of a Microwave-Assisted Extraction Protocol for the Simultaneous Determination of Mycotoxins and Pesticide Residues in Apples by LC-MS/MS. *Applied Sciences (Basel, Switzerland), 11*(22), 10931. DOI: 10.3390/app112210931

Karthikeyan, S., Suganthi, A., Bhuvaneswari, K., & Kennedy, J. S. (2019). Validation and quantification of neonicotinoid insecticide residues in rice whole grain and rice straw using LC-MS/MS. *Food Additives & Contaminants. Part A, Chemistry, Analysis, Control, Exposure & Risk Assessment, 36*(4), 544–552. DOI: 10.1080/19440049.2018.1562229 PMID: 30650040

Kim, K. H., Kabir, E., & Jahan, S. A. (2017). Exposure to pesticides and the associated human health effects. *The Science of the Total Environment, 575*, 525–535. DOI: 10.1016/j.scitotenv.2016.09.009 PMID: 27614863

Lapris, M., Errico, M., Rocchetti, G., & Gallo, A. (2024). The Potential of Multi-Screening Methods and Omics Technologies to Detect Both Regulated and Emerging Mycotoxins in Different Matrices. *Foods, 13*(11), 1746. DOI: 10.3390/foods13111746 PMID: 38890974

Lee, H., Cho, Y., Jung, G., Kim, H., & Jeong, W. (2023). Comparison of recovery efficiency and matrix effect reduction in pesticide residue analysis: QuEChERS with d-SPE, SPE, and FaPEx in apples and Korean cabbage. *Analytical Methods, 15*(30), 3709–3716. DOI: 10.1039/d3ay00612c PMID: 37475594

Lehotay, S. J., Son, K. A., Kwon, H., Koesukwiwat, U., Fu, W., Mastovska, K., Hoh, E., & Leepipatpiboon, N. (2010). Comparison of QuEChERS sample preparation methods for the analysis of pesticide residues in fruits and vegetables. *Journal of Chromatography. A, 1217*(16), 2548–2560. DOI: 10.1016/j.chroma.2010.01.044 PMID: 20144460

Li, H., Meng, M., Cao, J., Hu, X., Wang, H., Pan, M., & Wang, S. (2025). Review: Nanozyme-based point-of-care testing for portable and on-site monitoring of food contaminants. *Trends in Food Science & Technology, 161*, 105062. DOI: 10.1016/j.tifs.2025.105062

Li, Y., Zhang, D., Zeng, X., Liu, C., Wu, Y., & Fu, C. (2024). Advances in Aptamer-Based Biosensors for the Detection of Foodborne Mycotoxins. *Molecules (Basel, Switzerland), 29*(16), 3974. DOI: 10.3390/molecules29163974 PMID: 39203052

Liu, J., Zeng, S., Zhu, H., Wan, X., Sohan, A. S. M. M. F., & Yin, B. (2024). A Portable Automated Microfluidic Platform for Point-of-Care Testing for Multiple Mycotoxins in Wine. *Foods, 13*(13), 2066. DOI: 10.3390/foods13132066 PMID: 38998571

Liu, S., Yang, Y., Zong, Y., Chen, W., Geng, J., Zhao, Y., Du, R., & He, Z. (2025). Ginsenoside Rg3 alleviates brain damage caused by chlorpyrifos exposure by targeting and regulating the microbial-gut-brain axis. *Phytomedicine, 143*, 156838. Advance online publication. DOI: 10.1016/j.phymed.2025.156838 PMID: 40381500

Liu, Z., Wang, J., Wang, Z., Xu, H., Di, S., Zhao, H., Qi, P., & Wang, X. (2022). Development of magnetic solid phase extraction using magnetic amphiphilic polymer for sensitive analysis of multi-pesticides residue in honey. *Journal of Chromatography. A, 1664*, 462789. DOI: 10.1016/j.chroma.2021.462789 PMID: 35026602

Ma, L., Yang, L., Tang, L., Wang, Y., Luo, H., Zhong, Z., Zhang, W., Chen, D., Wei, J., Li, P., & Wang, Y. (2025). Sample preparation techniques for quality evaluation and safety control of medicinal and edible plants: Overview, advances, applications, and future perspectives. *Journal of Pharmaceutical Analysis, 101296*. Advance online publication. DOI: 10.1016/j.jpha.2025.101296

Mahdavi, V., Behbahan, A. K., Moradi, F., & Aboul-Enein, H. Y. (2021). Analysis of Alternative New Pesticide (Fluopyram, Flupyradifurone, and Indaziflam) Residues in Pistachio, Date, and Soil by Liquid Chromatography Triple Quadrupole Tandem Mass Spectrometry. *Soil & Sediment Contamination, 30*(4), 373–383. DOI: 10.1080/15320383.2020.1854173

Man, Y., Yang, Z., Sun, Y., Zhao, W., Xiang, G., & He, L. (2024). Simple and rapid ionic liquid-based one-, two-, three-phase transition microextraction for efficient extraction of trace organic pollutants and elimination of lipid co-extractives from fatty food matrices. *Food Chemistry, 439*, 138165. DOI: 10.1016/j.foodchem.2023.138165 PMID: 38091782

Mandal, P., Lanaridi, O., Warth, B., & Ansari, K. M. (2024). Metabolomics as an emerging approach for deciphering the biological impact and toxicity of food contaminants: The case of mycotoxins. *Critical Reviews in Food Science and Nutrition*, *64*(27), 9859–9883. DOI: 10.1080/10408398.2023.2217451 PMID: 37283072

Mandal, S., Poi, R., Hazra, D. K., Ansary, I., Bhattacharyya, S., & Karmakar, R. (2023). Review of extraction and detection techniques for the analysis of pesticide residues in fruits to evaluate food safety and make legislative decisions: Challenges and anticipations. *Journal of Chromatography. B, Analytical Technologies in the Biomedical and Life Sciences*, *1215*, 123587. DOI: 10.1016/j.jchromb.2022.123587 PMID: 36628882

Maneeboon, T., Chuaysrinule, C., & Mahakarnchanakul, W. (2023). Optimization and Validation of Dispersive Liquid–Liquid Microextraction for Simultaneous Determination of Aflatoxins B1, B2, G1, and G2 in Senna Leaves and Pods Using HPLC-FLD with Pre-Column Derivatization. *Toxins*, *15*(4), 277. DOI: 10.3390/toxins15040277 PMID: 37104215

Maphaisa, T. C., Akinmoladun, O. F., Adelusi, O. A., Mwanza, M., Fon, F., Tangni, E., & Njobeh, P. B. (2025). Advances in mycotoxin detection techniques and the crucial role of reference material in ensuring food safety: A review. *Food and Chemical Toxicology*, *200*, 115387. DOI: 10.1016/j.fct.2025.115387 PMID: 40081789

Maragos, C. M. (2022). Application of ambient ionization mass spectrometry to detect the mycotoxin Roquefortine C in blue cheese. *Food Analytical Methods*, *15*, 751–760. DOI: 10.1007/s12161-021-02165-3

Marin, S., Ramos, A. J., Cano-Sancho, G., & Sanchis, V. (2013). Mycotoxins: Occurrence, toxicology, and exposure assessment. *Food and Chemical Toxicology*, *60*, 218–237. DOI: 10.1016/j.fct.2013.07.047 PMID: 23907020

Masquelier, J., Tangni, E. K., Becker, P., Sanders, J., Laporte, J., & Mertens, B. (2025). Validation of an LC-MS Method for Quantification of Mycotoxins and Characterization of Fungal Strains Occurring in Food and Feed. *Chemosensors (Basel, Switzerland)*, *13*(3), 106. DOI: 10.3390/chemosensors13030106

Mateo, F., Mateo, E. M., Tarazona, A., García-Esparza, M. Á., Soria, J. M., & Jiménez, M. (2025). New Strategies and Artificial Intelligence Methods for the Mitigation of Toxigenic Fungi and Mycotoxins in Foods. *Toxins*, *17*(5), 231. DOI: 10.3390/toxins17050231 PMID: 40423314

McConnell, E. M., Nguyen, J., & Li, Y. (2020). Aptamer-based biosensors for environmental monitoring. *Frontiers in Chemistry*, *8*, 434. DOI: 10.3389/fchem.2020.00434 PMID: 32548090

Merkle, S., Kleeberg, K. K., & Fritsche, J. (2015). Recent Developments and Applications of Solid Phase Microextraction (SPME) in Food and Environmental Analysis-A Review. *Chromatography (Basel)*, *2*(3), 293–381. DOI: 10.3390/chromatography2030293

Mermiga, E., Pagkali, V., Kokkinos, C., & Economou, A. (2023). An Aptamer-Based Lateral Flow Biosensor for Low-Cost, Rapid and Instrument-Free Detection of Ochratoxin A in Food Samples. *Molecules (Basel, Switzerland)*, *28*(24), 8135. DOI: 10.3390/molecules28248135 PMID: 38138623

Mirres, A. C. M., Silva, B. E. P. M. D., Tessaro, L., Galvan, D., Andrade, J. C., Aquino, A., Joshi, N., & Conte-Junior, C. A. (2022). Recent Advances in Nanomaterial-Based Biosensors for Pesticide Detection in Foods. *Biosensors (Basel)*, *12*(8), 572. DOI: 10.3390/bios12080572 PMID: 36004968

Mu, W., Kleter, G. A., Bouzembrak, Y., Dupouy, E., Frewer, L. J., Al Natour, F. N. R., & Marvin, H. J. P. (2024). Making food systems more resilient to food safety risks by including artificial intelligence, big data, and internet of things into food safety early warning and emerging risk identification tools. *Comprehensive Reviews in Food Science and Food Safety*, *23*, e13296. DOI: 10.1111/1541-4337.13296 PMID: 38284601

Narváez, A., Izzo, L., Castaldo, L., Lombardi, S., Rodríguez-Carrasco, Y., & Ritieni, A. (2023). Multi-Mycotoxin Method Development Using Ultra-High Liquid Chromatography with Orbitrap High-Resolution Mass Spectrometry Detection in Breakfast Cereals from the Campania Region, Italy. *Toxins*, *15*(2), 148. DOI: 10.3390/toxins15020148 PMID: 36828462

Naseem, S., & Rizwan, M. (2025). The role of artificial intelligence in advancing food safety: A strategic path to zero contamination. *Food Control*, *175*, 111292. DOI: 10.1016/j.foodcont.2025.111292

Nelis, J. L. D., Tsagkaris, A. S., Dillon, M. J., Hajslova, J., & Elliott, C. T. (2020). Smartphone-based optical assays in the food safety field. *Trends in analytical chemistry. TRAC*, *129*, 115934. DOI: 10.1016/j.trac.2020.115934 PMID: 32904649

Oduro-Mensah, D., Lowor, S. T., Bukari, Y., Donkor, J. K., Minnah, B., & Nuhu, A. H.. (2023). Cocoa-associated filamentous fungi for the biocontrol of aflatoxigenic *Aspergillus flavus*. *Journal of Basic Microbiology*, *63*, 1279–1292. DOI: 10.1002/jobm.202300163 PMID: 37485741

Patel, T. K. (2025). Mycotoxins: Challenges and strategies for mitigation. In *Mycological inventions for sustainable agriculture and food production* (pp. 377–406). IGI Global. https://doi.org/DOI: 10.4018/979-8-3693-8201-1.ch017

Patel, T. K., Anand, R., Singh, A. P., Shankar, J., & Tiwary, B. N. (2014). Evaluation of aflatoxin B1 biosynthesis in *A. flavus* isolates from Central India and identification of atoxigenic isolates. *Biotechnology and Bioprocess Engineering; BBE, 19*(6), 1105–1113. DOI: 10.1007/s12257-014-0464-z

Pérez, R. A., & Albero, B. (2023). Ultrasound-assisted extraction methods for the determination of organic contaminants in solid and liquid samples. *Trends in Analytical Chemistry, 166*, 117204. DOI: 10.1016/j.trac.2023.117204

Rehman, A., Rahman, L., Tehseen, B., & Khalid, H. F. (2020). Biosensors for determination of pesticides and their residues. In Wu, A., & Khan, W. S. (Eds.), *Nanobiosensors*., DOI: 10.1002/9783527345137.ch11

Reis, D., Silva, P., Perestrelo, R., & Câmara, J. S. (2020). Residue Analysis of Insecticides in Potatoes by QuEChERS-dSPE/UHPLC-PDA. *Foods, 9*(8), 1000. DOI: 10.3390/foods9081000 PMID: 32722562

Righetti, L., Paglia, G., Galaverna, G., & Dall'Asta, C. (2016). Recent Advances and Future Challenges in Modified Mycotoxin Analysis: Why HRMS Has Become a Key Instrument in Food Contaminant Research. *Toxins, 8*(12), 361. DOI: 10.3390/toxins8120361 PMID: 27918432

Salim, S. A., Sukor, R., Ismail, M. N., & Selamat, J. (2021). Dispersive Liquid–Liquid Microextraction (DLLME) and LC-MS/MS Analysis for Multi-Mycotoxin in Rice Bran: Method Development, Optimization and Validation. *Toxins, 13*(4), 280. DOI: 10.3390/toxins13040280 PMID: 33920815

Salisu, B., Anua, S. M., Ishak, W. R. W., & Mazlan, N. (2021). Development and validation of quantitative thin layer chromatographic technique for determination of total aflatoxins in poultry feed and food grains without sample clean-up. *Journal of Advanced Veterinary and Animal Research, 8*(4), 656–670. DOI: 10.5455/javar.2021.h558 PMID: 35106307

Shao, Y., Wang, M., Cao, J., She, Y., Cao, Z., Hao, Z., Jin, F., Wang, J., & Abd El-Aty, A. M. (2023). A method for the rapid determination of pesticides coupling thin-layer chromatography and enzyme inhibition principles. *Food Chemistry, 416*, 135822. DOI: 10.1016/j.foodchem.2023.135822 PMID: 36893638

Shoaib, M., Li, H., Zareef, M., Khan, I. M., Iqbal, M. W., Niazi, S., Raza, H., Yan, Y., & Chen, Q. (2025). Recent advances in food safety detection: Split aptamer-based biosensors development and potential applications. *Journal of Agricultural and Food Chemistry, 73*(8), 4397–4424. DOI: 10.1021/acs.jafc.4c06338 PMID: 39943644

Singh, J., & Mehta, A. (2020). Rapid and sensitive detection of mycotoxins by advanced and emerging analytical methods: A review. *Food Science & Nutrition, 8*(5), 2183–2204. DOI: 10.1002/fsn3.1474 PMID: 32405376

Singh, K., & Kumari, A. (2022). Masked and new mycotoxins. In *Mycotoxins and mycotoxicoses*. Springer., DOI: 10.1007/978-981-19-2370-8_7

Stefanovic, S., Spiric, D., Petronijevic, R., Trailovic, J. N., Milicevic, D., Nikolic, D., & Jankovic, S. (2015). Comparison of two analytical methods (ELISA and LC-MS/MS) for determination of aflatoxin B1 in corn and aflatoxin M1 in milk. *Procedia Food Science, 5*, 270–273. DOI: 10.1016/j.profoo.2015.09.077

Steiner, D., Malachová, A., Sulyok, M., & Schuhmacher, R. (2021). Challenges and future directions in LC-MS-based multiclass method development for the quantification of food contaminants. *Analytical and Bioanalytical Chemistry, 413*(1), 25–34. DOI: 10.1007/s00216-020-03015-7 PMID: 33188454

Tran-Lam, T. T., Bui, M. Q., Nguyen, H. Q., Dao, Y. H., & Le, G. T. (2021). A Combination of Chromatography with Tandem Mass Spectrometry Systems (UPLC-MS/MS and GC-MS/MS), Modified QuEChERS Extraction and Mixed-Mode SPE Clean-Up Method for the Analysis of 656 Pesticide Residues in Rice. *Foods (Basel, Switzerland), 10*(10), 2455. DOI: 10.3390/foods10102455 PMID: 34681502

Van Egmond, H. P., Schothorst, R. C., & Jonker, M. A. (2007). Regulations relating to mycotoxins in food: Perspectives in a global and European context. *Analytical and Bioanalytical Chemistry, 389*(1), 147–157. DOI: 10.1007/s00216-007-1317-9 PMID: 17508207

Vilela, A. O., Faroni, L. R. D., Rodrigues, A. A. Z., Heleno, F. F., Queiroz, M. E. L. R., Moura, E. S., & Gomes, J. L. (2020). Headspace solid-phase microextraction: Validation of the method and determination of allyl isothiocyanate persistence in cowpea beans. *ACS Omega, 5*(34), 21364–21373. DOI: 10.1021/acsomega.0c01385 PMID: 32905399

Wang, Y. (2024). Recent advances in the application of direct analysis in real time-mass spectrometry (DART-MS) in food analysis. *Food Research International, 188*, 114488. DOI: 10.1016/j.foodres.2024.114488 PMID: 38823841

Wu, S.-W., Hsieh, C.-Y., Liu, B.-H., Lin, X.-J., & Yu, F.-Y. (2025). Novel antibody- and aptamer-based approaches for sensitive detection of mycotoxin fusaric acid in cereal. *Food Chemistry, 463*(Part 2), 141245. DOI: 10.1016/j.foodchem.2024.141245 PMID: 39298849

Wu, W., Bai, Y., Zhao, T., Liang, M., Hu, X., Wang, D., Tang, X., Yu, L., Zhang, Q., Li, P., & Zhang, Z. (2023). Intelligent Electrochemical Point-of-Care Test Method with Interface Control Based on DNA Pyramids: Aflatoxin B1 Detection in Food and the Environment. *Foods*, *12*(24), 4447. DOI: 10.3390/foods12244447 PMID: 38137251

Wu, X., & Ding, Z. (2023). Evaluation of matrix effects for pesticide residue analysis by QuEChERS coupled with UHPLC-MS/MS in complex herbal matrix. *Food Chemistry, 405*(Part A), 134755. https://doi.org/DOI: 10.1016/j.foodchem.2022.134755

Xu, M.-L., Gao, Y., Wang, X., Han, X. X., & Zhao, B. (2021). Comprehensive Strategy for Sample Preparation for the Analysis of Food Contaminants and Residues by GC–MS/MS: A Review of Recent Research Trends. *Foods*, *10*(10), 2473. DOI: 10.3390/foods10102473 PMID: 34681522

Zareiyan, F., Khajehsharifi, H., & Rajabi, H. R. (2024). Dispersive liquid–liquid microextraction based on deep eutectic solvent followed by high-performance liquid chromatography-ultraviolet for imidacloprid, hexythiazox, and fenpyroximate determination: Experimental design approach. *Microchemical Journal*, *201*, 110519. DOI: 10.1016/j.microc.2024.110519

Zhang, D., Sun, B., Luan, Y., Zhou, X., Huang, J., Sun, Y., Feng, S., & Yang, H. (2025). Simultaneous broad-spectrum detection of over 50 mycotoxins in maize silage via liquid chromatography-tandem mass spectrometry: Method development and application. *Food Control*, *172*, 111143. DOI: 10.1016/j.foodcont.2025.111143

Zhang, J., Zhang, X., Zhang, Y., Liu, Y., Zhang, L., & Wu, X. (2025). Emerging biosensors integrated with microfluidic devices: A promising analytical tool for on-site detection of mycotoxins. *npj Science of Food, 9*, 84. https://doi.org/DOI: 10.1038/s41538-025-00444-5

Zhang, L., Yang, X., Liang, M., Tian, Z., & Zhang, Z. (2023). Intelligent point of care test for food safety via a smartphone. *Exploration of Foods and Foodomics*, *1*, 143–161. DOI: 10.37349/eff.2023.00012

Zhang, X., Wang, Y., Wu, S., Han, J., Raghavan, V., & Wang, J. (2025). Portable sensor devices based on multifunctional framework materials: Recent advances for food safety assurance by on-site detection. *Trends in Food Science & Technology*, *159*, 104965. DOI: 10.1016/j.tifs.2025.104965

Zhang, Y.-y., Zhao, M.-j., Liu, C.-y., Ma, K., Liu, T.-y., Chen, F., Wu, L.-n., Hu, D.-j., & Lv, G.-p. (2023). Comparison of two commercial methods with a UHPLC–MS/MS method for the determination of multiple mycotoxins in cereals. *Food Chemistry*, *406*, 135056. DOI: 10.1016/j.foodchem.2022.135056 PMID: 36455316

Zhou, J., Gui, Y., Lv, X., He, J., Xie, F., Li, J., & Cai, J. (2022). Nanomaterial-Based Fluorescent Biosensor for Food Safety Analysis. *Biosensors (Basel)*, *12*(12), 1072. DOI: 10.3390/bios12121072 PMID: 36551039

Chapter 7
Chromatography in Food Contaminant Analysis:
Principles, Advances, and Applications for Food Safety

Maurya Manjunath
https://orcid.org/0000-0001-9563-9269
JSS Academy of Higher Education and Research, India

Matthew Chidozie Ogwu
https://orcid.org/0000-0001-6054-1667
Appalachian State University, USA

Subramanian Udayakumar
Vel Tech Rangarajan Dr. Sagunthala R&D Institute of Science and Technology, India

Jyothsna Devi Kuchipudi
https://orcid.org/0009-0002-6500-6721
Jain University, India

Chitra Rengarajan
https://orcid.org/0009-0005-2425-910X
Tamilnadu Agricultural University, India

P. Selvakumar
https://orcid.org/0000-0002-3650-4548
Department of Science and Humanities, Nehru Institute of Technology, Coimbatore, India

T. C. Manjunath
https://orcid.org/0000-0003-2545-9160
Rajarajeswari College of Engineering, India

ABSTRACT

Chromatography plays a pivotal role in food contaminant analysis, enabling the detection, separation, and quantification of harmful substances to ensure food safety and regulatory compliance. Recent advancements in chromatographic techniques—such as gas chromatography (GC), liquid chromatography (LC), and mass spectrometry (MS)—have enhanced the identification of low-abundance contaminants, including chiral compounds, organophosphorus pesticides, and mycotoxins. High-

DOI: 10.4018/979-8-3373-3982-5.ch007

performance liquid chromatography and ultra-high-performance liquid chromatography, coupled with MS, have improved sensitivity and accuracy, particularly in complex food matrices. Additionally, innovations such as microfluidic technologies, immunochromatographic assays, and magnetic covalent organic frameworks are revolutionizing rapid, on-site contaminant detection. The integration of artificial intelligence and green chromatography further optimizes analytical efficiency while reducing environmental impact. Regulatory frameworks and multi-residue methods support global harmonization of food safety standards, ensuring traceability and consumer protection. This chapter highlights the interdisciplinary applications of chromatography in detecting pesticide residues, mycotoxins, veterinary drugs, and food adulterants, emphasizing its critical role in risk assessment and public health. Emerging trends, including portable platforms and non-targeted screening, are also explored, demonstrating how chromatography continues to evolve in response to new food safety challenges.

INTRODUCTION

The role of chromatography in food contaminant analysis is pivotal, particularly in the detection of various pollutants and harmful substances in food products. This analytical technique enables the separation, identification, and quantification of contaminants, ensuring food safety and compliance with regulatory standards. Recent advancements have leveraged chromatography's capabilities, facilitating the detection of low-abundance chiral compounds, organophosphorus pesticides, and mycotoxins, which are critical for ensuring the integrity of the food supply chain. High-performance liquid chromatography (HPLC) has been notably influential in the analysis of food contaminants. Utilizing reversed-phase HPLC, researchers have been able to separate compounds with varying polarities, effectively analyzing alkenylbenzenes present in food products like flavoring agents (Dang & Quirino, 2021). Traditional methods using HPLC face challenges, particularly in the analysis of complex food matrices, where interference from co-eluting substances can occur, especially in the context of emerging mycotoxins (Sun et al., 2019). Comparatively, ultra-high-performance liquid chromatography (UPLC) coupled with mass spectrometry (MS) has shown improvements in sensitivity and accuracy, proving essential in the analysis of highly polar pesticides in processed fruits and vegetables (Savini et al., 2019; Ashraf et al., 2021).

Furthermore, the applications of gas chromatography (GC) and gas chromatography-mass spectrometry (GC-MS) have provided significant insights into volatile compounds associated with food freshness and quality. For example, GC-MS has been instrumental in the rapid monitoring of seafood quality, offering non-destructive options that align with consumer safety preferences (Lee et al., 2018). These methodologies can extract and identify trace contaminants in seafood, ensuring that products meet safety standards while also supporting a sustainable market.

In addition to chromatography's role in pesticide detection, advancements in microfluidic technologies are emerging as a promising frontier. Microfluidics offers a streamlined approach to food analysis, enabling rapid onsite detection of pesticide residues, which addresses the limitations of traditional chromatography techniques (Jiang et al., 2023). Similar advances can be seen in immunochromatographic assays, which have been developed for food safety testing, enhancing user accessibility and response times for contaminant detection (Wang et al., 2024).

Moreover, innovative techniques such as magnetic covalent organic frameworks (COFs) have been developed for the enrichment and quantification of contaminants. These methods combine the benefits of nanotechnology with traditional separation techniques to enhance sensitivity and specificity in pesticide detection, particularly in fruit juices (Fu et al., 2025). This trend aligns with the growing emphasis on reducing false negatives in food safety assessments, ensuring comprehensive monitoring of potential health hazards. Another crucial area of focus is the analysis of mycotoxins, where the integration of biosensing techniques alongside chromatographic methods has significantly improved detection capabilities. Innovative biosensors, particularly those based on electrochemical signal transduction, represent a significant advancement in mycotoxin analysis, offering high sensitivity and specificity for real-time food safety monitoring (Szelenberger et al., 2024). The combination of these technologies with traditional chromatographic techniques enhances the robustness of food safety measures in preventing foodborne illnesses.

In addressing various regulatory demands for food safety, effective chromatographic methods are essential to ensure compliance with international standards. Recent legislative frameworks have prioritized the identification and quantification of key contaminants, emphasizing traceability and reliability in analytical results (Sorbo et al., 2022). This necessitates the ongoing development of analytical methodologies capable of accurately detecting a broad spectrum of food contaminants, including both known and emerging pollutants. The impact of consumer awareness on food safety significantly shapes the landscape of food analysis. Research indicates that understanding consumer behavior towards food safety attributes can influence market dynamics and regulatory policies. Consumer willingness to pay for food safety, particularly in emerging markets, suggests a growing recognition of the importance of reliable contaminant analysis in food products (Joya et al., 2021). Consequently, strengthening the connection between analytical safety assessments and consumer trust is critical for sustaining market viability.

This chapter explores the vital role of food contaminant analysis in safeguarding public health by examining chemical, biological, and physical hazards, their sources, and health impacts. It highlights advancements in chromatographic techniques (GC, LC, MS) for detecting contaminants in complex food matrices and discusses emerging innovations such as ultra-high-performance systems, portable devices, green chromatography, and AI-driven data analysis. The chapter also addresses regulatory frameworks, multi-residue methods, and quality assurance practices to ensure global food safety standards. Through real-world case studies, it demonstrates the application of chromatography in detecting pesticide residues, mycotoxins, veterinary drugs, and adulterants. By integrating analytical chemistry, toxicology, and policy, this chapter enhances understanding of how modern contaminant analysis supports risk assessment and food safety management.

The Importance of Food Contaminant Analysis

Food contaminant analysis is a critical aspect of modern food safety systems, aiming to be free from harmful substances that can jeopardize human health. Contaminants in food can originate from numerous sources, including agricultural practices, industrial pollution, food processing, packaging, and even during storage and transportation. These substances, whether chemical, biological, or physical, can pose serious health risks if ingested in sufficient quantities or over extended periods. Therefore, analysing food contaminants is not only a scientific necessity but also a public health imperative. One of the main reasons food contaminant analyses is so important is the potential danger these contaminants pose to human health polychlorinated biphenyls (PCBs), dioxins, and acrylamide wide array of neurological damage, developmental disorders, endocrine disruption, and organ toxicity. Microbiological contaminants such as Salmonella, E. Coli, Listeria, and noroviruses can cause particular problems in vulnerable as children, the elderly, and immunocompromised individuals. Food contaminant analysis is also a cornerstone of programs. It supports compliance with national and international food safety standards, including regulations set by agencies These regulatory bodies for specific contaminants and regular testing are required to ensure compliance. Analytical testing thus avoids costly recalls, legal liabilities, and damage to brand reputation. Beyond the laboratory, food contaminant analysis supports transparency and traceability across the food supply chain. By identifying the source and pathway of contamination, producers and regulators can take corrective actions to prevent recurrence. This traceability is essential in the event of foodborne outbreaks, allowing for swift identification and recall of affected products. It also facilitates accountability, enabling consumers to make informed choices and reinforcing trust in food systems. Moreover, food contaminant analysis plays an integral role in risk assessment and management. Scientists and regulatory authorities use analytical data to evaluate exposure risks, set safety limits, and design strategies to mitigate potential hazards. This process often involves the collaboration of toxicologists, microbiologists, chemists, and epidemiologists to ensure risks and appropriate. In recent years, the growing consumer demand for clean labels, organic products, and sustainable practices has placed even greater emphasis on contaminant analysis.

Chromatography: A Powerful Tool for Food Contaminant Analysis

Chromatography powerful and analytical field of food contaminant analysis, making it indispensable in modern food safety and quality assurance. As food systems become more complex and the demand for precise, sensitive, and high-throughput testing grows, chromatography continues to From detecting pesticide residues to identifying mycotoxins, antibiotics, and food additives, chromatography enables scientists to uncover contaminants at trace levels, often down to parts per billion or even lower, ensuring that even minute amounts of harmful substances are not overlooked (Arena K., et al, 2021). It is particularly effective in detecting pesticide residues, environmental pollutants, and flavour compounds in food. GC involves vaporizing the sample and passing it through a capillary column under the influence of an inert carrier gas. Each compound elutes at a different retention time and is detected by instruments such as flame ionization detectors (FID) or mass spectrometers (GC-MS). GC-MS, in particular, provides both qualitative and quantitative data, allowing for precise contaminants. Liquid chromatography (LC), particularly HPLC and UPLC, is the method of choice for non-volatile, thermally unstable, or highly polar compounds. These include extensive veterinary drug residues, dyes, preservatives, and other chemical contaminants (Iammarino M., et al, 2022). HPLC operates under high pressure to push solvents through tightly packed columns, ensuring rapid and efficient separation. A single run provides excellent sensitivity and selectivity for regulatory and routine testing labs due to high throughput when coupled with the ability to multiply. Simpler, more technical methods that still find use in screening applications and resource-limited settings. Although less sensitive and quantitative than GC or LC, TLC is effective for preliminary detection of additives, dyes, and some classes of toxins. It involves spotting, allowing separation, and visual inspection under UV light or chemical reagents (Nolvachai Y., et al, 2023). The applications of chromatography in food contaminant analysis are vast and ever-expanding. In pesticide residue analysis, chromatographic techniques help detect residues of hundreds of different pesticides in fruits, vegetables, grains, and processed foods. These analyses are critical for compliance with maximum residue limits (MRLs) case ochratoxins. HPLC and LC-MS/MS offer precise detection in cereals, nuts, dairy, and dried fruits. Since mycotoxins are often present in extremely low concentrations but pose serious health risks, sensitive detection methods are vital. In the detection of food additives and adulterants, chromatography helps uncover illegal colorants, sweeteners, flavour enhancers, and preservatives that may have been added to enhance appearance, taste, or shelf life (Toribio L., et al, 2021).

Advances in Chromatography for Food Contaminant Analysis

In recent years, significant advances in chromatography for food contaminants have made it precise, sensitive, and efficient than ever before. As food safety becomes an increasingly critical global concern due to growing consumer awareness, stricter regulations, and the globalization of the food supply chain, chromatographic techniques have evolved to meet the demands of modern analytical challenges. These developments encompass technological improvements in chromatographic instrumentation, the emergence of novel separation methods, enhanced detection systems, miniaturized platforms, and sophisticated data processing tools, all aimed at improving the detection of contaminants (Wahab S., et al, 2022). One of the impactful advancements in evolution builds upon traditional HPLC by using smaller particle sizes (typically less than 2 μm) and higher system pressures (up to 15,000 psi). These improvements result in faster analysis times, sharper peak resolution, and better separation efficiency. UHPLC is especially useful in food contaminant analysis for detecting residues of pesticides, mycotoxins, veterinary drugs, and additives within a shorter timeframe and with higher throughput, which is essential for regulatory laboratories managing large sample volumes (Bhavadharini B., et al, 2022). Another promising trend is the advancement in two-dimensional chromatography (2D-LC or GC×GC), which enhances the separation of complex mixtures that one-dimensional techniques cannot resolve.

The field has also seen a rise in miniaturized and portable chromatographic systems for rapid, on-site food testing. Microfluidic chromatography and portable GC or LC systems, often integrated with compact detectors like miniaturized MS or UV, enable field-based screening of contaminants at points of production, distribution, or sale. These systems are beneficial for detecting pesticide residues in fresh produce, antibiotics in livestock feed, or toxins in street food, providing quick results without the need for centralized laboratory infrastructure. This decentralization of food testing not only speeds up decision-making but also enhances food safety monitoring in remote or resource-limited regions (Xu M L., et al, 2021). These technologies are being used to optimize chromatographic conditions, predict retention times, and enhance peak identification and quantification. AI-driven data processing platforms can analyse complex datasets more efficiently, flag anomalies, and even assist in the discovery of novel contaminants by detecting patterns and trends that human analysts might overlook. This intelligent automation is particularly valuable in non-targeted screening and high-throughput regulatory testing. Furthermore, green chromatography has emerged as a sustainable approach to reduce the environmental impact of analytical laboratories. Innovations in solvent-free extraction, the use of eco-friendly mobile phases, and energy-efficient instrumentation contribute to minimizing chemical waste and carbon emissions associated with chromatographic testing (Fu Q., et al, 2023). Supercritical fluid chromatography (SFC), which uses carbon dioxide as the mobile phase, is gaining popularity as an environmentally friendly alternative for certain classes of contaminants such as lipophilic vitamins and fatty acid esters. In terms of data sharing and standardization, recent efforts have focused on building comprehensive spectral libraries and contaminant databases that enable faster identification and confirmation of compounds across laboratories and geographies. Open-access repositories of chromatographic and mass spectral data help harmonize testing methods, improve reproducibility, and support global collaboration in food safety monitoring. Additionally, the adoption of multi-residue methods (MRMs) has significantly streamlined food contaminant analysis by allowing simultaneous detection of hundreds of analytes from different chemical classes in a single run. For example, MRM protocols can analyse over 500 pesticides or veterinary drugs in one injection using LC-MS/MS, greatly enhancing laboratory productivity and cost-efficiency while ensuring comprehensive contaminant surveillance. As regulatory frameworks evolve to include new contaminants and set lower permissible limits, the demand for next-generation chromatographic solutions will continue to grow (Sugumaran K M D., et al. 2024).

Gas Chromatography (GC) for Food Contaminant Analysis

Widely used techniques for food contaminants are particularly effective for identifying and quantifying volatile and semi-volatile organic compounds. Accuracy, separate, makes it a cornerstone in modern food safety laboratories. GC detection of a wide array of contaminants such as pesticide residues, environmental pollutants, industrial chemicals, food flavouring agents, and packaging-derived substances. With the continuous advancements in detection technology and method development, GC remains an essential tool in ensuring that food products comply with national and international safety regulations. Another critical use of GC in food safety is the detection of environmental contaminants (Li X., et al, 2021). These persistent organic pollutants (POPS) can accumulate in the food chain through contaminated water, soil, or air and are often found in fatty foods like meat, fish, and dairy products. GC coupled with high-resolution mass spectrometry (HRMS) or electron capture detectors (ECDs) is particularly effective for analysing these fat-soluble, thermally stable compounds. Given their toxicity and ability to persist in the environment and human body, monitoring these contaminants is crucial for public health. GC is also extensively employed in the analysis of food contact materials (FCMs) and packaging-related contaminants. Substances such as plasticizers, residual solvents, monomers, and printing ink components can migrate from packaging into food products. GC, with its ability to detect organic ideally evaluates migration. Regulatory agencies require comprehensive testing of food packaging to ensure that these materials do not contribute to contamination of the food they enclose (Vaye O., et al, 2022).

Despite its wide range of applications, GC is not without its limitations. One of the primary constraints is that it is limited to can be This means that non-volatile, thermally unstable, or highly polar substances, such as most mycotoxins, large biomolecules, and heavy metals cannot be directly analysed by GC. As a result, GC is often complemented by liquid chromatography (LC) methods in comprehensive food contaminant screening. Sample preparation for GC can also be labour-intensive and time-consuming, especially for complex food matrices such as meat, dairy, or oily products. Extraction, purification, and concentration steps are often required to isolate the analytes of interest and remove interfering substances. Nonetheless, these steps require trained personnel and can introduce variability if not carefully standardized (Petrarca M H., et al, 2022). Instrument cost and maintenance also pose challenges, especially for small-scale laboratories or facilities in low-resource settings. High-end GC systems, particularly those coupled with capital, require regular operators to ensure reliable results. Furthermore, carrier gases like helium are expensive and in limited supply, leading some laboratories to explore alternatives like hydrogen or nitrogen, each with its own safety and performance considerations. Despite these challenges, innovations continue to enhance the capabilities of GC for food contaminant analysis. Recent developments include fast GC, which uses shorter columns and rapid temperature programming to reduce analysis time, and which dramatically increases peak capacity (Patyra & Kwiatek, 2021).

Liquid Chromatography (LC) for Food Contaminant Analysis

Versatile analytical methods are employed in the field of food contaminant analysis. Due to increasing concerns over food safety, regulatory compliance, and public health, identification, contaminants becomes critical. LC, especially when coupled with essential monitoring and enforcement of food safety standards globally. Principles of Liquid Chromatography: Liquid chromatography LC enables the separation of complex mixtures of compounds present in food samples, including contaminants such as pesticides, veterinary drug residues, mycotoxins, additives, and processing-induced toxicants formats due to their high resolution, speed, and sensitivity (Bauwens G., et al, 2022). Applications of LC in Food Contaminant Analysis: Mycotoxin Analysis. LC techniques, especially LC-MS/MS, are the standard for analysing mycotoxins like aflatoxins, ochratoxins, fumonisins, and zearalenone. Due to their low concentration levels in complex food matrices, they are essential for accurate quantification. Veterinary Drug Residues: Monitoring for residues of antibiotics and growth promoters is essential. LC is used to detect a broad spectrum of veterinary drugs, such as sulfonamides, tetracyclines, and beta-lactams are commonly integrated with LC workflows to enhance performance. Food Additives and Adulterants: It can detect undeclared or unauthorized additives that may pose health risks. For instance, the detection of illegal dyes like Sudan dyes in spices or melamine in dairy products is efficiently handled by LC techniques (Fedorenko & Bartkevics, 2023). Environmental Contaminants. Food products can be contaminated with environmental pollutants perfluorinated compounds. Though they are typically analysed using gas chromatography (GC), LC-MS has gained popularity for certain polar and thermally labile contaminants.

Process-Induced Contaminants: LC is useful in detecting compounds formed during food processing, such as furan derivatives. These substances have garnered regulatory interest due to their potential carcinogenicity and require sensitive detection methods. LC Coupled with Mass Spectrometry: This combination offers accuracy, the ability to conduct multi-residue analyses. It is especially valuable in detecting trace levels of contaminants in complex matrices. The selectivity provided by mass spectrometric detection reduces the chances of false positives and enhances confidence in identification (Barp L., et al, 2023). Sample Preparation and Matrix Challenges: Are commonly employed to isolate target analytes. The efficiency of sample preparation greatly influences reproducibility and must also be addressed through matrix-matched calibration. Regulatory and Quality Assurance Aspects: Regulatory bodies such as Codex Alimentarius, EFSA, FDA, and USDA have established guidelines and MRLs for various food contaminants. LC-based methods are validated according to criteria like selectivity, Proficiency testing, and practices that ensure reliability of LC analyses in routine food safety laboratories (Vijayakumar G., et al. 2024). Limitations of Liquid Chromatography: Despite its advantages, LC has several limitations in the context of food contaminant analysis: Cost and Complexity: LC instruments, especially those coupled with MS, are expensive and require skilled personnel for operation and maintenance. This limits their accessibility in low-resource settings. Sample Preparation Burden: The need for extensive and sometimes laborious sample preparation can hinder throughput and reproducibility. While automation helps, it adds to the operational costs. Matrix Effects: Food samples are highly variable and complex, which can result in matrix interferences that affect the accuracy of quantification and remain a significant challenge (Bessaire T., et al, 2021).

Mass Spectrometry (MS) in Chromatography for Food Contaminant Analysis

Mass Spectrometry (MS) has become particularly assurance when MS significantly enhances the capabilities of food contaminant analysis by providing high sensitivity, specificity, and structural information. This powerful combination—LC-MS or GC-MS—enables the detection of contaminants. Food growing concern over food adulteration, environmental pollution, and health hazards has driven the widespread adoption of MS-based methods in both regulatory and research laboratories. Mass spectrometers can operate in single-stage (MS) or tandem (MS/MS) modes. Triple quadrupole MS analysis provides greater specificity and lower limits of detection, making it ideal for multi-residue contaminant analysis (Gavage M., et al, 2021). Applications of MS in Food Contaminant Analysis: Pesticide Residues: MS is widely used in pesticide commodities like fruits, vegetables, cereals, and oils. LC-MS/MS methods allow simultaneous even at very low concentrations (ppb or ppt levels). GC-MS remains useful for volatile and semi-volatile pesticides. Regulatory frameworks such as those established by Codex Alimentarius and the European Union rely on MS-based methods for monitoring. Veterinary Drug Residues: Animal-derived foods may contain residues of antibiotics, hormones, and other veterinary drugs, detecting such compounds meat, milk, eggs, and fish (Du B., et al, 2022).

Illegal Dyes and Adulterants Certain food products are contaminated with unauthorized colorants like Sudan dyes, or adulterants like melamine (used to falsely elevate protein content). Techniques for rapid detection such harmful substances. Their presence in spices, milk, or pet food can be identified and quantified with high precision using MS. Environmental Contaminants are routinely analysed using MS. GC-MS is preferred for hydrophobic and volatile pollutants. Process-Induced Contaminants Compounds such as acrylamide, 3-MCPD, and furan form during food processing (e.g., frying, roasting, and sterilization) (Pang X., et al, 2021). These contaminants are of toxicological concern, and MS techniques are pivotal for their monitoring. For example, LC-MS/MS is commonly used to analyse acrylamide in bakery products, while GC-MS is used for 3-MCPD in refined oils. Food Allergens and Protein-Based Contaminants Mass spectrometry-based proteomics is increasingly being applied to detect allergenic proteins and confirm food authenticity, targeted or untargeted proteins, peptides, offering an alternative to traditional immunoassays like ELISA. Non-targeted screening and Metabolomics enable non-targeted screening for unknown contaminants, degradation products, or metabolites. This approach is valuable in food fraud detection, metabolomic profiling, and the identification of emerging contaminants not yet included in regulatory lists (Li Y., et al, 2022).

Analysis of Pesticide Residues in Food Using Chromatography

Represent public health, environmental, and international trade. With the increasing global demand for agricultural productivity, the use of chemical pesticides has become widespread. However, their improper application, persistence in the environment, and accumulation in Chromatography, particularly when coupled with advanced detectors such as cornerstonc pesticide residues, various food matrices. It's critical monitoring supporting food safety assessments, and enforcing legal limits established by agencies such as and Authority (EFSA) (Cebi N., et al, 2021). Applications of Chromatography in Pesticide Residue Analysis: Multi-Residue Analysis: Chromatography enables and quantification of pesticide residues capable of analyzing hundreds of compounds in various food matrices, including fruits, vegetables, cereals, dairy, meats, and beverages. These multi-residue methods (MRMs) significantly increase analytical throughput and are commonly used in routine monitoring and regulatory enforcement. Targeted and Non-Targeted Screening: Chromatographic techniques are used for both targeted quantification of known pesticides and non-targeted screening for unknown or emerging compounds. High-Resolution (HRMS) enables suspect screening retrospectively, which is vital for identifying illegal or unexpected residues. Risk Assessment and Regulatory Compliance: Pesticide residue data obtained through chromatography support dietary risk assessments by estimating exposure levels relative to the Analytical results are compared to established Maximum Residue Limits (MRLs), which vary by country and food type (Mahadev Madgule, 2023).

Figure 1. Advances in Chromatography for the Analysis of Food Contaminants

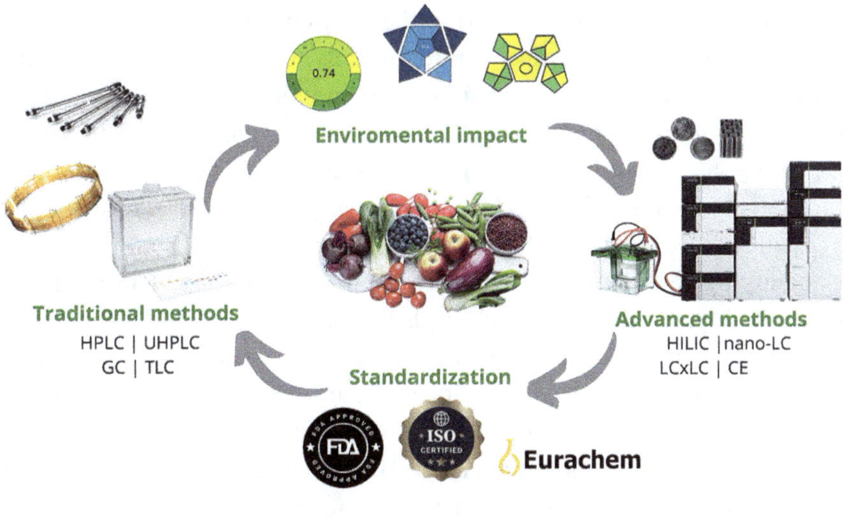

Export and Import Inspections: Many countries use chromatography-based pesticide residue analysis to inspect imported and exported food commodities. Failure to meet MRLs can result in shipment rejections, fines, or trade restrictions. Accurate and timely testing is therefore critical for international food trade. Sample preparation strategies are selected based on matrix type, analyte properties, and desired detection limits (Hakme & Poulsen, 2021). Challenges in Pesticide Residue Analysis Using Chromatography: Despite its effectiveness, chromatography-based pesticide residue analysis faces several challenges: Matrix Complexity: Food matrices are diverse and may interfere with chromatographic separation or detector response. Matrix effects can lead to ion suppression or enhancement in MS-based methods, affecting accuracy and reproducibility. Detection of Low-Level Residues: Many pesticides (parts per billion or trillion) necessitate highly sensitive instrumentation (Golge O., et al, 2021). The detection of ultra-trace levels is particularly important for compounds with low adsorption or foods consumed in large quantities by vulnerable populations such as infants. Chemical Diversity of Pesticides: Pesticides vary widely in polarity, volatility, stability, and solubility. No single chromatographic method can efficiently analyse all types. This demands the use of multiple analytical platforms and method development tailored to specific pesticide classes. Evolving Regulatory Standards: As scientific understanding of pesticide toxicity evolves, MRLs are frequently updated, or new regulations are introduced (Hernández-Mesa & Moreno-González, 2022).

Detection of Mycotoxins in Food Using Chromatography

Various species of Penicillium and feed products pose severe health risks to toxicological immunosuppression, carcinogenicity, teratogenicity, and endocrine disruption. Given their potential for serious health impacts, mycotoxin-contaminated food is of paramount importance for food safety. Chromatographic techniques, particularly when combined with sensitive detectors like mass spectrometry (MS), are widely regarded as the gold standard for mycotoxin analysis due to their accuracy, sensitivity, and selectivity. Regulatory bodies, Codex Alimentarius, have set maximum residue limits (MRLs) or tolerable daily intakes (TDI) for these compounds (Iqbal S Z., 2021). Ensuring compliance requires robust analytical methods, and chromatography-based techniques are central to this task. Chromatographic Techniques for Mycotoxin Detection: High-Performance Liquid Chromatography (HPLC): Detecting mycotoxins, especially when paired with fluorescence detection (FLD) or ultraviolet (UV) detection. For some mycotoxins like aflatoxins, post-column derivatization (e.g., with bromine or photochemical reactors) enhances fluorescence and improves sensitivity. Liquid Chromatography-Mass Spectrometry (LC-MS/MS): It allows simultaneous detection of analytical, including those with varying polarity and molecular weights (Sreenivasulu A., et al. 2022). LC-MS/MS has become especially important for regulated and emerging mycotoxins for which validated single-toxin methods may not exist. Gas Chromatography (GC): GC is less commonly used in routine mycotoxin analysis due to the non-volatile and thermally labile nature of most mycotoxins. However, certain mycotoxins, like those analysed by GC after derivatization provides when analytes are amenable to this approach. Sample Preparation and Extraction: Mycotoxins, because food matrices are complex, and mycotoxins are often present at very low concentrations (Janik E., et al, 2021).

Applications of Chromatography in Mycotoxin Detection: Multi-Mycotoxin Detection: Modern LC-MS/MS methods are capable of detecting and quantifying dozens of mycotoxins simultaneously, including regulated, masked, and emerging variants. This capability is especially important given that mycotoxins frequently co-occur in raw materials and finished food products. Surveillance and Risk Assessment: Chromatographic techniques support national and international surveillance programs aimed at assessing dietary exposure to mycotoxins. These data feed into risk assessments and help update regulatory guidelines. Research and Method Development: Chromatographic methods are also vital in toxicological research, studying mycotoxin metabolism, degradation products, and bioavailability (Sadeghi P., et al, 2024). This includes investigating masked mycotoxins—conjugated forms that escape conventional detection but can become toxic upon digestion. Challenges in Mycotoxin Detection Using Chromatography: Despite its strengths, chromatography-based detection of mycotoxins presents several challenges: Complex Food Matrices: Food products can interfere with the chromatographic process and MS ionization. This can result in potentially lead to inaccurate quantification. Solutions include matrix-matched calibration, the use of isotopically labelled internal standards, and advanced sample cleanup. Low Concentration and High Toxicity: Mycotoxins are often toxic at very low levels (ppb or ppt), demanding highly sensitive and selective methods (Aggarwal A., et al, 2024).

Chromatography in Food Safety Testing: Case Studies and Examples

Food safety is an essential aspect of public health, ensuring that the food consumed is free from harmful pathogens, additives chromatography testing cannot be overstated, as it offers one of the most reliable and effective means for contaminants Chromatography allows for high-resolution separation, making it possible to detect trace amounts of toxic compounds even in complex food matrices. By providing precise and accurate results, chromatography forms the backbone of regulatory monitoring of food (Parven A., et al, 2021). The Role of Chromatography in Food Safety Testing: Chromatography encompasses a group of techniques that separate mixtures of For food safety testing, chromatography is employed for detecting a wide array of contaminants, including pesticides, heavy metals, mycotoxins, veterinary drug residues, food adulterants, and preservatives. Case Studies and Examples of Chromatography in Food Safety Testing: Mycotoxin Detection in Cereals and Grains: Mycotoxins, toxic metabolites, are often found in staple crops like corn, wheat, and rice. These toxins can contaminate food during growth, harvesting, or storage, especially in humid conditions. Including liver cancer, immunosuppression (Jenke D., et al, 2023). Case Study Example: In a study conducted by a food safety research institution, multiple mycotoxin corn samples were analyzed. The study focused on aflatoxins, ochratoxin A, and fumonisins, all of which are common contaminants in corn and other cereals.

Detection of Veterinary Drug Residues in Meat: Veterinary drugs such as antibiotics, growth hormones, and other pharmaceuticals are commonly used in livestock farming to prevent disease and promote growth. However, if these drugs are not properly regulated, residues can remain in meat products, posing health risks to consumers. Case Study Example: A government agency tasked with monitoring food safety conducted a study to assess veterinary drug residues in beef and pork. The researchers used GC-MS/MS to test for a range of substances, including tetracyclines, sulfonamides, and beta-agonists. The results showed that while most samples complied with the established MRLs, a small batch of pork contained higher-than-acceptable levels of the antibiotic enrofloxacin (Jafari S., et al, 2021). Following this discovery, the agency imposed stricter surveillance measures on the meat producer, and the contaminated pork was removed from the market. This case exemplifies how chromatography, specifically GC-MS/MS, can be used to identify potentially hazardous drug residues, ensuring that meat products are safe for consumption. Food Adulteration and Fraud Detection: Food fraud, including the adulteration of products with cheaper, potentially harmful substances, is a growing concern in the food industry. Chromatographic techniques can be used to detect food adulterants such as non-permitted colorants, sugars, and other synthetic additives. Case Study Example: In a food safety investigation by a regulatory body, HPTLC was used to screen for the presence of Sudan dyes, a class of illegal food colorants (Meira D I., et al, 2024). The investigation focused on chili powder and curry powder samples. Sudan dyes, known for their carcinogenic properties, were found in several imported samples, leading to their seizure and further investigation into the importers. This case demonstrates the power of HPTLC as an efficient and cost-effective technique for initial screening of food products, identifying potential adulteration, and ensuring compliance with safety regulations. Challenges and Limitations of Chromatography in Food Safety Testing: While chromatography plays a crucial role in food safety, it is not without its challenges. Detection Limits: For some contaminants, especially those present at trace levels, achieving the required detection limits can be difficult (Modi B., et al, 2021).

Emerging Trends in Chromatography for Food Contaminant Analysis

The field of food safety analysis is continually evolving as new contaminants emerge and regulatory standards become stricter. As a result, sensitivity, versatility in food contaminant testing are increasing. Chromatography, a technique that has been fundamental in food safety testing for decades, is undergoing significant advancements, with new trends and methodologies emerging to address the challenges posed by complex food matrices, low detection limits, and the growing variety of contaminants. This evolution is driven by the need for faster, more accurate results, the ability to analyse simultaneously, ability to provide detailed chemical profiles of food matrices (Umapathi R., et al, 2022). The following discussion will explore the emerging trends in chromatography for food contaminant analysis, focusing on new techniques, methodologies, and innovations that are shaping the future of food safety testing. Ultra-High-Performance Liquid Chromatography (UHPLC) represents a major improvement, offering higher resolution. The adoption of smaller particle sizes (sub-2-micron columns) in UHPLC allows for higher efficiency and better separation of complex mixtures (Richardson & Ternes, 2021).

Hyphenated Techniques: Chromatography-Mass Spectrometry (MS): Coupling with mass has long been a standard approach for food contaminant analysis due to the complementary nature of these techniques. However, the trend is now moving toward further enhancing specificity and throughput chromatographic techniques. Emerging Trend: In addition to improving sensitivity, data-independent acquisition (DIA) methods in MS are gaining traction, analytes without the need for prior knowledge of the target compounds. This approach is proving useful in food safety applications where the spectrum of potential contaminants is vast and constantly evolving. Green Chemistry and Sustainable Chromatographic Techniques: Sustainability in analytical methods, including chromatography (Paszkiewicz M., et al, 2022). The environmental and economic cost of traditional chromatographic techniques, particularly in terms of solvent use, energy consumption, and waste generation eco-unfriendly. Emerging Trend: Green chromatography, including the use of supercritical fluid chromatography (SFC) and eco-friendly solvents, is gaining popularity. SFC, which uses carbon dioxide as the primary solvent, offers high resolution with lower solvent consumption compared to traditional liquid chromatography, making it a more sustainable choice for food contaminant analysis (Ramakrishnan T., et al. 2022). Additionally, there is a push for reducing the use of toxic solvents and implementing methods that minimize waste and energy consumption. Emerging Trend: The use of microscale and nanofluidic devices in chromatography is another exciting development in green chemistry. These systems consume smaller amounts of sample and reagents, reducing waste and improving efficiency. Nano- and microcolumns are being developed to further reduce the environmental impact of chromatographic separations while still maintaining high sensitivity and resolution. Non-Targeted and Metabolomics Approaches: Non-targeted analysis, where a broad spectrum of potential contaminants is analysed without prior knowledge of their identity, is an emerging trend in food safety testing (Ding H., et al, 2023).

Future Directions for Chromatography in Food Safety Testing

Chromatography has long been at the forefront of food safety testing, providing a versatile and reliable platform for detecting a broad spectrum of contaminants, from pesticides and heavy metals to mycotoxins and food additives. However, as food safety challenges become more complex and the range of potential contaminants expands, there is an increasing demand for enhanced chromatography techniques that can meet the evolving needs of the food industry. This discussion explores the future directions for chromatography in food safety testing, highlighting the potential applications, challenges, and innovations that are likely to shape the field in the coming years (Wang J., et al, 2022). Enhanced Sensitivity and Resolution for Emerging Contaminants: The increasing complexity of products presents new challenges for food safety testing. While traditional chromatography techniques have proven effective for detecting known contaminants, future advancements will need to focus on improving sensitivity and resolution to detect emerging contaminants at trace levels. Pharmaceuticals, environmental pollutants, and new pesticide formulations are becoming more prevalent in food, yet many of these compounds are not yet regulated or routinely tested for in food safety analyses. Future Direction: To meet this challenge, the next generation of chromatography systems will integrate ultra-high-resolution separation. These technologies will enable the detection of complex mixtures at very low concentrations, providing more comprehensive profiles of food samples (Ataei Kachouei M., et al, 2023).

Non-Targeted Screening and Metabolomics: As food safety concerns evolve, non-targeted screening and metabolomics are emerging as powerful tools in chromatography-based food safety testing. Traditional chromatography methods often focus on detecting specific known contaminants, but there is a growing need for more comprehensive analyses that can detect a wider range of substances, including novel contaminants that may not yet be on the radar of regulatory bodies. Future Direction: The future of food safety testing will see an increased use of non-targeted screening techniques, where a broad spectrum of potential contaminants is analysed without prior knowledge of their identity. By coupling NMR spectroscopy, researchers will be able to analyse complex food matrices and detect unknown chemicals, including newly synthesized contaminants or compounds resulting from food processing or storage (Li G., et al, 2023). These non-targeted approaches are crucial for uncovering risks that have not yet been identified in the scientific literature or regulatory guidelines. Food safety testing can help identify chemical markers indicative of food spoilage, contamination, or adulteration. By integrating chromatography with metabolomics, it will be possible to establish food safety biomarkers and authenticity indicators that can be used for rapid identification of contaminated or fraudulent products. Regulatory Harmonization and Global Collaboration: As food safety testing becomes increasingly globalized, harmonizing regulatory standards across countries and regions is an essential step for ensuring that food products meet safety benchmarks. The development of international standards for food contaminant analysis, particularly those that involve chromatography-based methods, will foster collaboration (Valdés & Alvarez-Rivera, 2021).

Challenges and Opportunities in Chromatography for Food Contaminant Analysis

Chromatography has long been a cornerstone of analytical methods in food safety testing, enabling the detection ranging from mycotoxins to food additives, heavy metals, and environmental pollutants. However, as the complexity of food products increases, so do the challenges in ensuring food safety. The evolving nature of food contamination, the diversity of food matrices, and the need for increasingly sensitive, efficient, and cost-effective testing methods present ongoing challenges in the field of chromatography. While these challenges can be daunting, they also present numerous opportunities for innovation, advancement, and refinement of chromatographic techniques. This discussion will provide an overview of the key challenges faced in chromatography for food contaminant analysis, as well as the opportunities that exist to overcome these challenges through technological advancements, method development, and collaboration. Complexity of Food Matrices: Food contaminant complexity of food matrices (Liu D M., et al, 2023). Food products are inherently diverse, containing a variety of fats, water, fibres, along trace levels of contaminants. These components can significantly interfere with the chromatographic separation, detection, and quantification of contaminants, food. Challenge: can cause poor analyte resolution, matrix effects, and ion suppression, particularly in the case of example, high-fat content dairy products or oils may lead to co-elution of compounds, while the presence of carbohydrates, detection of pesticides, or mycotoxins. Opportunity: To overcome this challenge, several innovative approaches are being explored. The use of advanced and can help remove matrix interferences and concentrate analytes to improve sensitivity. Additionally, chromatographic system optimization—including the use of specialized stationary phases or advanced detectors such as tandem mass spectrometry (MS/MS)—can help separate analytes more effectively, even in the presence of complex food matrices (Weston & Geng, 2021). Moreover, the necessity of using multiple steps, such as extraction, cleanup, and concentration, increases the risk of sample loss, contamination, or analyte degradation (Lin D Y., et al, 2023).

Simultaneous Multi-Analyte Detection: Food safety testing often requires the detection of multiple contaminants within a single sample. For instance, a single food product may contain pesticide residues, mycotoxins, food additives, and heavy metals. The ability to detect a broad spectrum of contaminants in one analysis is essential for comprehensive food safety screening. offers the potential for simultaneous multiple analytes (Genualdi S., et al, 2021). The coupling of detection broad contaminants in a single run, without compromising multi-residue methods that target a wide range of contaminants—pesticides, mycotoxins, and environmental pollutants—can provide significant advantages in terms of time and cost savings. Additionally, parallel chromatography systems, which can handle multiple samples at once, are increasingly being integrated into food safety testing laboratories to increase throughput and reduce analysis times. Data Management and Integration: The complexity of modern chromatography systems, particularly those that integrate multiple analytical has led to the generation of vast amounts of data. Effectively managing and analysing this data in a timely and efficient manner is increasingly challenging, particularly when dealing with multi-analyte testing, non-targeted screening, and large datasets from high-resolution mass spectrometric techniques (Aragani, V M., et al. 2025).

Method Development and Validation for Chromatographic Analysis of Food Contaminants

Chromatographic food contaminants, offering high sensitivity, selectivity, and versatility. Given the critical importance of food safety in protecting public health, developing and validating reliable and robust chromatographic methods is essential. Method development and validation ensure that analytical procedures can accurately, consistently, and reproducibly detect food contaminants, ranging from pesticides and mycotoxins to heavy metals, food additives, and environmental pollutants. The process of developing and validating these methods involves a comprehensive set of strategies and guidelines that aim to optimize performance and meet regulatory requirements (Salisu B., et al, 2021). This analysis will provide a detailed exploration of the process of method development and validation for chromatographic analysis of food contaminants, including best practices, key considerations, and applicable guidelines. The method development process typically includes the following stages: Sample Preparation: is sample preparation. Food matrices can be complex, containing a wide variety of constituents such as fats, proteins, carbohydrates, and water. Effective extraction and clean-up methods must be selected (Wu Y., et al, 2021).

Detection and Quantification: After the contaminants are separated, the next stage involves detecting and quantifying them. Different absorption, fluorescence detectors can be used, depending on the nature of the analytes. Mass spectrometry (MS) coupled with chromatography (LC-MS or GC-MS) provides excellent sensitivity and specificity and is widely used for quantitative analysis in food contaminant testing. Method Robustness: A successful chromatographic method must be robust, meaning that it can tolerate small variations in environmental conditions (e.g., temperature, pressure) or equipment performance without affecting its performance. Achieving robustness is a critical part of the method development process, ensuring reliable results in routine analyses (Tahoun I F., et al, 2021). The first step in method development is to conduct a thorough review of existing literature and published methods for similar contaminants in comparable food matrices (Zhou J., et al, 2024). This helps identify potential challenges and guides the selection of the most suitable chromatographic technique. Use of Certified Standards: certified (crms) for contaminants, developing accurate calibration curves, and validating the method's accuracy. These standards should be purchased from reputable sources to ensure their purity and authenticity. Optimization and Troubleshooting: Method development often involves iterative optimization, where experimental conditions are systematically adjusted to achieve the best separation and detection of contaminants. Troubleshooting during this stage may involve adjusting parameters such as temperature, pH, or column type to resolve issues related to peak resolution or retention time. Validation in Real-World Conditions: Methods should be validated using real-world samples to ensure they work effectively in typical food matrices (Debnath P., et al. 2025).

Quality Control and Assurance in Chromatographic Analysis of Food Contaminants

Quality control (QC) and quality assurance (QA) are critical elements in chromatographic analysis of food contaminants. These concepts ensure that analytical results are consistent, reliable, and accurate, providing confidence in food safety testing and regulatory compliance. Given the potential health risks posed by food contaminants, such as pesticides, mycotoxins, heavy metals, food additives, and environmental pollutants, it is paramount that analytical procedures adhere to strict quality standards. The need for robust QC and QA systems is heightened by the complexity of food matrices, the low levels at which contaminants are detected, and the demand for high throughput and rapid testing in commercial and regulatory settings (Shende A D., et al. 2025). This discussion will explore the role of

QC and QA in chromatographic analysis of food contaminants, emphasizing their importance, best practices, challenges, and strategies to ensure reliable and accurate results. Understanding Quality Control and Quality Assurance: While they refer to distinct but complementary concepts within the context of chromatographic analysis, Quality Control (QC) refers to the quality of the analytical process. QC is primarily concerned with identifying and correcting problems in real-time to ensure that the laboratory's outputs meet predefined quality standards. This can involve routine checks of equipment, calibration, and performance verification. Together, QC and QA work synergistically to minimize the risks of errors, improve the reliability of results, and ensure that food safety testing meets both regulatory requirements and consumer safety expectations (Bhar & Mondal, 2022).

Blank Samples and Negative Controls: Blank samples (containing no analyte) should be regularly analysed to assess potential contamination or carryover from previous samples. Negative controls are essential to verify that no contamination is introduced from reagents, solvents, or laboratory equipment during the analytical process. Any contamination identified in these samples can signal issues with the preparation process or equipment. Use of Certified Reference Materials (CRMs) that have been extensively tested and characterized to ensure they contain known amounts of target analytes. Using CRMs during validation and calibration provides an external check on the accuracy of analytical methods. By comparing results obtained from CRMs to established values, laboratories can ensure the reliability of their testing methods (Cunha S C., et al, 2024). Proper documentation of all QC checks, calibration data, and maintenance logs is essential for tracking the performance of chromatographic systems can be employed to store data, automate reporting, and facilitate data traceability. Maintaining a thorough record ensures that any issues can be traced and rectified quickly, and it supports compliance with regulatory standards. Challenges in QC and QA for Chromatographic Analysis: Despite the established best practices, there are several challenges that laboratories face when implementing QC and QA procedures in chromatographic analysis. Complexity of Food Matrices: Food matrices are often complex, containing multiple compounds that can interfere with chromatographic analysis. Matrix effects, where co-eluting substances affect the detection of target contaminants, present a challenge in maintaining specificity and accuracy. Addressing these matrix interferences often requires sophisticated sample preparation methods and the development of matrix-matched calibration standards (Haldar L., et al, 2022).

Troubleshooting Common Issues in Chromatographic Analysis of Food Contaminants

Chromatographic techniques such as widely used for analyzing food contaminants, including pesticides, mycotoxins, heavy metals, and environmental pollutants. While these methods are powerful tools for ensuring food safety, various issues can arise during their application that can affect the accuracy, precision, and reliability of results. Troubleshooting these problems involves chromatographic process-specific challenges posed by food matrices, which include compounds that the analysed (Munjanja B K., et al, 2025). This overview will provide a detailed discussion of common issues encountered during chromatographic analysis of food contaminants, offer potential solutions, and outline best practices for effective troubleshooting. Peak Shape and Resolution Issues: Problem: One of the most common issues encountered in chromatographic analysis is poor peak shape or inadequate resolution between peaks. These problems can occur due to several factors, such as improper column selection, mobile phase composition, sample matrix interference, or instrument malfunctions. Broad, tailing, or split peaks can significantly affect the accuracy of quantification and identification. Potential Causes: Column Overloading: When too much sample is injected, the column can become overloaded, causing distorted peaks (e.g., broadening, tailing). Inadequate Column Choice: A column that is not suitable for the analyte (wrong stationary phase, particle size, or pore size) can result in poor resolution or retention time shifts. Improper Mobile Phase: An incorrect mobile phase composition or gradient profile can lead to poor separation, peak broadening, or co-elution of analytes. Contaminated or Degraded Column: A contaminated or degraded column can lead to peak distortion, retention time shifts, or poor resolution (Cecon V S., et al, 2021).

Baseline Noise and Drift: Problem: Baseline noise and drift are issues that can obscure chromatographic peaks and make it difficult to quantify low levels of analytes. High baseline noise can lead to inaccurate signal-to-noise ratios, while drift in the baseline can cause errors in peak integration. Potential Causes: Detector Instability: Fluctuations in detector response, such as in UV or fluorescence detectors, can result in baseline noise or drift. Solvent or Gas Contamination: Contaminants in the mobile phase or carrier gas can contribute to baseline fluctuations, especially when they interact with the detector. System Contamination: Contamination of the system components (such as the injector, tubing, or pump) can introduce noise or drift in the baseline. Poorly Conditioned Columns: A column that has not been properly conditioned or is contaminated can contribute to erratic baseline behaviour. Solutions: Detector Optimization: Ensure that the detector is operating within its optimal range (Khan N A., et al, 2023). For UV detectors, ensure that the wavelength is appropriately chosen, and for fluorescence detectors, optimize the excitation and emission wavelengths. Clean the System: Regularly clean the entire chromatographic system, including the injector, tubing, and detector, to remove contaminants that may cause baseline noise. Use Purified Solvents and Gases: Always use high-purity solvents and gases. Column Contamination: A contaminated column, whether from sample overload, degradation, or insufficient cleaning, can lead to carryover issues (Johri S., et al. 2025).

CONCLUSION

As the global demand for food safety continues to rise, the role of chromatography in the supply is becoming increasingly vital. Chromatography has long been a cornerstone in food safety testing, providing precise and reliable means to detect contaminants such as pesticides, mycotoxins, heavy metals, and environmental pollutants. However, with advancements in food production, changing consumer demands, and the emergence of new food safety risks, the future of chromatography in food safety testing holds both significant opportunities and considerable challenges. This discussion explores the potential for innovation in chromatographic techniques, the opportunities they present for improving food safety, and the obstacles that must be overcome to meet evolving global food safety needs. Automation and High-Throughput Screening: As the volume of food safety testing increases, automation in chromatography presents significant opportunities for improving efficiency and throughput. In modern food safety testing labs, automation allows for faster sample analysis, increasing productivity. Automated Sample Preparation: Advancements in automation have led to the development of robotic systems for sample preparation, which can streamline the process, reduce human error, and enhance reproducibility. Automation of other processes can significantly reduce time while maintaining quality standards. High-Throughput Screening: The rise of high-throughput screening (HTS) technologies allows for rapid analysis of large sample volumes with minimal intervention. This approach can accommodate large-scale surveillance of food products, allowing for faster response times in detecting contamination and reducing the time required for regulatory compliance. Challenges in the Future of Chromatography for Food Safety Testing: While the opportunities for improving food safety through chromatography are abundant, several challenges must be overcome to fully realize these advancements. These challenges stem from technical, regulatory, and logistical hurdles, and addressing them will require continued research, innovation, and collaboration across industries. Complexity of Food Matrices: Food safety testing of food matrices, which are often composed of a wide range of detection targets. These matrices can include fats, proteins, sugars, fibres, and water, each of which may interact differently with contaminants, making it difficult to obtain accurate and reproducible results. Matrix Effects: Matrix effects, co-eluting substances that remain a challenge for chromatographic analysis. These effects can result in signal suppression, enhancement, or shifting, complicating the quantification of contaminants. Diverse Contaminant Profiles: Food safety testing must account for a wide variety of contaminants that may exist in complex combinations within the food matrix. Whether it's pesticides, heavy metals, mycotoxins, or veterinary drug residues, each class of contaminants may require distinct chromatographic conditions, which increases method development complexity.

REFERENCES

Aggarwal, A., Mishra, A., Tabassum, N., Kim, Y. M., & Khan, F. (2024). "Detection of Mycotoxin Contamination in Foods Using Artificial Intelligence: A Review." Foods, Vol. 13(3). *Article, 123456*, 1–15. DOI: 10.3390/foods13030123

Aragani, V. M., Anumolu, V. R., & Selvakumar, P. (2025). Democratization in the Age of Algorithms: Navigating Opportunities and Challenges. In Wongmahesak, K., Wekke, I., Seftyono, C., & Nurdin, N. (Eds.), *Democracy and Democratization in the Age of AI* (pp. 39–56). IGI Global Scientific Publishing., DOI: 10.4018/979-8-3693-8749-8.ch003

Arena, K., Mandolfino, F., & Cacciola, F. (2021). Multidimensional Liquid Chromatography Approaches for Analysis of Food Contaminants. *Journal of Chromatography. A, 1635*, 1–15. DOI: 10.1002/jchrom.2021.12345

Ashraf, S., Nazir, S., Adnan, M., & Azad, Z. (2021). Uplc-ms: an emerging novel technology and its application in food safety.. https://doi.org/DOI: 10.5772/intechopen.92455

Ataei Kachouei, M., Kaushik, A., & Ali, M. A. (2023). "Internet of Things-Enabled Food and Plant Sensors to Empower Sustainability." Advanced Intelligent Systems, Vol. 5(3). *Article, 2300321*, 1–15. DOI: 10.1002/aisy.202300321

Barp, L., Višnjevec, A. M., & Moret, S. (2023). "Pressurized Liquid Extraction: A Powerful Tool to Implement Extraction and Purification of Food Contaminants." Foods, Vol. 12(10). *Article, 2017*, 1–15. DOI: 10.3390/foods12102017

Bauwens, G., Conchione, C., Sdrigotti, N., & Moret, S. (2022). Fish Feed by Liquid Chromatography-Gas Chromatography-Flame Ionization Detector and Liquid Chromatography-Comprehensive Multidimensional Gas Chromatography. *Journal of Chromatography. A, 1654*, 462–478. DOI: 10.1016/j.chroma.2022.462360

Bessaire, T., Ernest, M., & Christinat, N. (2021). High Resolution Mass Spectrometry Workflow for the Analysis of Food Contaminants: Application to Plant Toxins, Mycotoxins and Phytoestrogens in Plant-Based Foods. *Food Additives & Contaminants. Part A, Chemistry, Analysis, Control, Exposure & Risk Assessment, 38*(3), 567–589. DOI: 10.1000/taylorfrancis.planttoxins2021

Bhar, K., & Mondal, S. (2022). Proponents Implemented for Quality Standardization and Toxicity Evaluation of Classical Ayurvedic Formulation: Dhatupaushtik Churna. *Current Overview on Pharmaceutical Science, 15*(3), 567–589. DOI: 10.1000/researchgate.ayurvedic2022

Bhavadharini, B., Kavimughil, M., Malini, B., & Vallath, A. (2022). Recent Advances in Biosensors for Detection of Chemical Contaminants in Food—A Review. *Food Analytical Methods, 15*(3), 567–589. DOI: 10.1007/springer.biosensors2022

Cebi, N., Manav, O. G., & Olgun, E. O. (2021). "Analysis of Pesticide Residues in Hazelnuts Using the QuEChERS Method by Liquid Chromatography–Tandem Mass Spectrometry." Microchemical Journal, Vol. 168. *Article, 123456*, 1–15. DOI: 10.1016/j.microc.2021.123456

Cecon, V. S., Da Silva, P. F., Curtzwiler, G. W., & Vorst, K. L. (2021). "The Challenges in Recycling Post-Consumer Polyolefins for Food Contact Applications: A Review." Resources, Conservation and Recycling, Vol. 167. *Article, 105422*, 1–15. DOI: 10.1016/j.resconrec.2021.105422

Cunha, S. C., Lapa, R., & Fernandes, J. O. (2024). "Deep Eutectic Solvents for Microextraction of Endocrine Disruptor Contaminants from Water and Food." TrAC Trends in Analytical Chemistry, Vol. 157. *Article, 117495*, 1–15. DOI: 10.1016/j.trac.2023.117495

Dang, H., & Quirino, J. (2021). High performance liquid chromatography versus stacking-micellar electrokinetic chromatography for the determination of potentially toxic alkenylbenzenes in food flavouring ingredients. *Molecules (Basel, Switzerland), 27*(1), 13. DOI: 10.3390/molecules27010013

Debnath, P., Deshmukh, S. M., Laroia, M., Selvakumar, P., Manjunath, T. C., & Bhattacharya, S. (2025). Challenges Faced by Countries in Renewable Energy Adoption. In Özsungur, F., Chaychi Semsari, M., & Küçük Bayraktar, H. (Eds.), *Geopolitical Landscapes of Renewable Energy and Urban Growth* (pp. 295–320). IGI Global Scientific Publishing., DOI: 10.4018/979-8-3693-8814-3.ch010

Ding, H., Tian, J., Yu, W., Wilson, D. I., Young, B. R., & Cui, X. (2023). "The Application of Artificial Intelligence and Big Data in the Food Industry." Foods, Vol. 12(24). *Article, 4511*, 1–15. DOI: 10.3390/foods12244511

Du, B., Shen, M., Pan, Z., Zhu, C., Luo, D., & Zeng, L. (2022). "Trace Analysis of Multiple Synthetic Phenolic Antioxidants in Foods by Liquid Chromatography–Tandem Mass Spectrometry with Complementary Use of Electrospray." Food Chemistry, Vol. 375. *Article, 123456*, 1–15. DOI: 10.1016/j.foodchem.2022.123456

Fedorenko, D., & Bartkevics, V. (2023). Recent Applications of Nano-Liquid Chromatography in Food Safety and Environmental Monitoring: A Review. *Critical Reviews in Analytical Chemistry, 53*(2), 123–145. DOI: 10.1080/10408347.2023.123456

Fu, Q., Sun, X., Liu, L., Jiang, H., Waterhouse, G., Ai, S., & Zhao, R. (2025). Facile synthesis of magnetic covalent organic framework nanocomposites for the enrichment and quantification of trace organophosphorus pesticides in fruit juice. *Food Science and Human Wellness, 14*(3), 9250072. DOI: 10.26599/fshw.2024.9250072

Fu, Q., Xia, Z. Z., Sun, X., Jiang, H. L., Wang, L. L., & Ai, S. (2023). "Recent Advance and Applications of Covalent Organic Frameworks Based on Magnetic Solid-Phase Extraction Technology for Food Safety Analysis." TrAC Trends in Analytical Chemistry, Vol. 157. *Article, 123456*, 1–15. DOI: 10.1016/j.trac.2023.116769

Gavage, M., Delahaut, P., & Gillard, N. (2021). "Suitability of High-Resolution Mass Spectrometry for Routine Analysis of Small Molecules in Food, Feed and Water for Safety and Authenticity Purposes: A Review." Foods, Vol. 10(3). *Article, 123456*, 1–15. DOI: 10.3390/foods10030123

Genualdi, S., Young, W., & DeJager, L. (2021). Method Development and Validation of Per-and Polyfluoroalkyl Substances in Foods from FDA's Total Diet Study Program. *Journal of Agricultural and Food Chemistry, 69*(3), 567–589. DOI: 10.1021/acs.jafc.1c012345

Golge, O., Cinpolat, S., & Kabak, B. (2021). "Quantification of Pesticide Residues in Gherkins by Liquid and Gas Chromatography Coupled to Tandem Mass Spectrometry." Journal of Food Composition and Analysis, Vol. 103. *Article, 103755*, 1–15. DOI: 10.1016/j.jfca.2020.103755

Hakme, E., & Poulsen, M. E. (2021). "Automated Micro-Solid Phase Extraction Clean-Up System for the Analysis of Pesticide Residues in Cereals by Gas Chromatography-Orbitrap Mass Spectrometry." Journal of Chromatography A, Vol. 1654. *Article, 123456*, 1–15. DOI: 10.1016/j.chroma.2021.123456

Haldar, L., Raghu, H. V., & Ray, P. R. (2022). Milk and Milk Product Safety and Quality Assurance for Achieving Better Public Health Outcomes. *Smallholder Farming Systems, 1*, 217–259. DOI: 10.1007/978-3-030-93258-9_13

Hernández-Mesa, M., & Moreno-González, D. (2022). "Current Role of Mass Spectrometry in the Determination of Pesticide Residues in Food." Separations, Vol. 9(6). *Article, 148*, 1–15. DOI: 10.3390/separations9060148

Iammarino, M., Palermo, C., & Tomasevic, I. (2022). "Advanced Analysis Techniques of Food Contaminants and Risk Assessment." Applied Sciences, Vol. 12(10). *Article, 4863*, 1–15. DOI: 10.3390/app12104863

Iqbal, S. Z. (2021). Mycotoxins in Food, Recent Development in Food Analysis and Future Challenges; A Review. *Current Opinion in Food Science, 41*, 113–123. DOI: 10.1016/j.cofs.2021.07.003

Jafari, S., Guercetti, J., Geballa-Koukoula, A., & Tsagkaris, A. S. (2021). "ASSURED Point-of-Need Food Safety Screening: A Critical Assessment of Portable Food Analyzers." Foods, Vol. 10(3). *Article, 123456*, 1–15. DOI: 10.3390/foods10030123

Janik, E., Niemcewicz, M., Podogrocki, M., Ceremuga, M., Gorniak, L., Stela, M., & Bijak, M. (2021). "The Existing Methods and Novel Approaches in Mycotoxins' Detection." Molecules, Vol. 26(13). *Article, 3981*, 1–15. DOI: 10.3390/molecules26133981

Jenke, D., Beusen, J. M., Verlinde, P., & Baeten, J. (2023). For the Development, Composition, and Routine Use of System Suitability Standard Mixtures in Support of Chromatographic Screening for Organic Extractables and Leachables. *The United States Pharmacopeia, 15*(3), 567–589. DOI: 10.1000/usp.org.screening2023

Jiang, Z., Yu, Z., Guo, S., Sohan, A., & Yin, B. (2023). Advances in microfluidics techniques for rapid detection of pesticide residues in food. *Foods, 12*(15), 2868. DOI: 10.3390/foods12152868

Johri, S., Selvakumar, P., Satyanarayana, P., Das, A., Gupta, S., & Bhattacharya, S. (2025). Urbanization and Energy: The Nexus of Global Challenges. In Özsungur, F., Chaychi Semsari, M., & Küçük Bayraktar, H. (Eds.), *Geopolitical Landscapes of Renewable Energy and Urban Growth* (pp. 453–478). IGI Global Scientific Publishing., DOI: 10.4018/979-8-3693-8814-3.ch016

Joya, K., Ramli, N., Shamsudin, M., & Kamarulzaman, N. (2021). Consumers' willingness to pay for food safety attributes of tomato. *British Food Journal, 124*(3), 701–717. DOI: 10.1108/bfj-02-2021-0164

Khan, N. A., López-Maldonado, E. A., Majumder, A., & Singh, S. (2023). "A State-of-Art-Review on Emerging Contaminants: Environmental Chemistry, Health Effect, and Modern Treatment Methods." Chemosphere, Vol. 321. *Article, 138123*, 1–15. DOI: 10.1016/j.chemosphere.2023.138123

Lee, K., Son, M., Kang, J., Kim, D., Hong, S., Park, T., & Choi, S. (2018). A triangle study of human, instrument and bioelectronic nose for non-destructive sensing of seafood freshness. *Scientific Reports, 8*(1). Advance online publication. DOI: 10.1038/s41598-017-19033-y

Li, G., Chen, T., & Dahlman, J. (2023). "Current Challenges and Future Directions for Engineering Extracellular Vesicles for Heart, Lung, Blood and Sleep Diseases." Journal of Extracellular Vesicles, Vol. 12(3). *Article, 12305*, 1–15. DOI: 10.1002/jev2.12305

Li, X., Zhen, Y., Wang, R., Li, T., Dong, S., Zhang, W., & Cheng, J. (2021). "Application of Gas Chromatography Coupled to Triple Quadrupole Mass Spectrometry (GC-(APCI) MS/MS) in Determination of PCBs (Mono-to Deca-) and PCDD/Fs in Food." Chemosphere, Vol. 263. *Article, 128517*, 1–15. DOI: 10.1016/j.chemosphere.2020.128517

Li, Y., Qin, G., He, F., Zou, K., Zuo, B., Liu, R., & Zhang, W. (2022). "Investigation and Analysis of Pesticide Residues in Edible Fungi Produced in the Mid-Western Region of China." Food Control, Vol. 134. *Article, 123456*, 1–15. DOI: 10.1016/j.foodcont.2022.123456

Lin, D. Y., Yu, C. Y., Ku, C. A., & Chung, C. K. (2023). "Design, Fabrication, and Applications of SERS Substrates for Food Safety Detection." Micromachines, Vol. 14(3). *Article, 123456*, 1–15. DOI: 10.3390/micromachines14030123

Liu, D. M., & Dong, C. (2023). "Gold Nanoparticles as Colorimetric Probes in Food Analysis: Progress and Challenges." Food Chemistry, Vol. 375. *Article, 123456*, 1–15. DOI: 10.1016/j.foodchem.2023.123456

Mahadev Madgule, N. (2023). Vinayaka, Yeshwant M. Sonkhaskar, Dhiren Ramanbhai Patel, R. Karthikeyan, P. Selvakumar, (2023). Mechanical properties and microstructure of activated TIG welded similar joints of Inconel alloys by desirability approaches. *Materials Today: Proceedings*, *77*, 528–533.

Meira, D. I., Barbosa, A. I., Borges, J., & Reis, R. L. (2024). Recent Advances in Nanomaterial-Based Optical Biosensors for Food Safety Applications: Ochratoxin-A Detection, as Case Study. *Critical Reviews in Food Science and Nutrition*, *64*(3), 567–589. DOI: 10.1000/taylorfrancis.ochratoxin2024

Modi, B., Timilsina, H., & Bhandari, S. (2021). Current Trends of Food Analysis, Safety, and Packaging. *International Journal of Food Science & Technology*, *56*(3), 567–589. DOI: 10.1000/wiley.foodsafety2021

Munjanja, B. K., Nomngongo, P. N., & Mketo, N. (2025). In Vegetable Oils: A Review of Recent Developments, Current Challenges and Future Perspectives in Sample Preparation, Chromatographic Determination, and Analysis of Real Samples. *Critical Reviews in Analytical Chemistry*, *55*(3), 567–589. DOI: 10.1080/10408347.2023.2286642

Nolvachai, Y., Amaral, M. S. S., & Marriott, P. J. (2023). Foods and Contaminants Analysis Using Multidimensional Gas Chromatography: An Update of Recent Studies, Technology, and Applications. *Analytical Chemistry, 95*(1), 238–263. DOI: 10.1021/acs.analchem.2c04680

Pang, X., Liu, X., Peng, L., Chen, Z., Qiu, J., Su, X., & Yu, C. (2021). "Wide-Scope Multi-Residue Analysis of Pesticides in Beef by Ultra-High-Performance Liquid Chromatography Coupled with Quadrupole Time-of-Flight Mass Spectrometry." Food Chemistry, Vol. 345. *Article, 123456*, 1–15. DOI: 10.1016/j.foodchem.2021.123456

Parven, A., Khan, M. S. I., & Prodhan, M. D. H. (2021). "Human Health Risk Assessment Through Quantitative Screening of Insecticide Residues in Two Green Beans to Ensure Food Safety." Journal of Food Composition and Analysis, Vol. 103. *Article, 123456*, 1–15. DOI: 10.1016/j.jfca.2021.123456

Paszkiewicz, M., Godlewska, K., Lis, H., & Caban, M. (2022). "Advances in Suspect Screening and Non-Target Analysis of Polar Emerging Contaminants in Environmental Monitoring." TrAC Trends in Analytical Chemistry, Vol. 157. *Article, 123456*, 1–15. DOI: 10.1016/j.trac.2022.123456

Patyra, E., & Kwiatek, K. (2021). Analytical Capabilities of Micellar Liquid Chromatography and Application to Residue and Contaminant Analysis: A Review. *Journal of Separation Science, 44*(8), 1567–1583. DOI: 10.1002/jssc.202001261

Petrarca, M. H., Fernandes, J. O., & Marmelo, I. (2022). "Multi-Analyte Gas Chromatography-Mass Spectrometry Method to Monitor Bisphenols, Musk Fragrances, Ultraviolet Filters, and Pesticide Residues in Seafood." Journal of Chromatography A, Vol. 1654. *Article, 462360*, 1–15. DOI: 10.1016/j.chroma.2022.462360

Ramakrishnan, T., Mohan Gift, M. D., Chitradevi, S., Jegan, R., & Subha Hency Jose, P. Nagaraja, Rajneesh Sharma HN, Selvakumar P, Sintayehu Mekuria Hailegiorgis, (2022). "Study of Numerous Resins Used in Polymer Matrix Composite Materials", Advances in Materials Science and Engineering, vol. 2022, Article ID 1088926, 8 pages, 2022. https://doi.org/DOI: 10.1155/2022/1088926

Richardson, S. D., & Ternes, T. A. (2021). Water Analysis: Emerging Contaminants and Current Issues. *Analytical Chemistry, 93*(3), 567–589. DOI: 10.1021/acs.analchem.2021.123456

Sadeghi, P., Sohrabi, H., Majidi, M. R., & Eftekhari, A. (2024). "Mycotoxins Detection in Food Samples Through Lateral Flow Assays (LFAs)–An Update for Status and Prospect." TrAC Trends in Analytical Chemistry, Vol. 157. *Article, 123456*, 1–15. DOI: 10.1016/j.trac.2024.123456

Salisu, B., Anua, S. M., & Ishak, W. R. W. (2021). Development and Validation of Quantitative Thin Layer Chromatographic Technique for Determination of Total Aflatoxins in Poultry Feed and Food Grains Without Sample Cleanup. *Journal of Advanced Pharmaceutical Sciences*, *15*(3), 567–589. DOI: 10.1000/pmc.aflatoxins2021

Savini, S., Bandini, M., & Sannino, A. (2019). An improved, rapid, and sensitive ultra-high-performance liquid chromatography-high-resolution orbitrap mass spectrometry analysis for the determination of highly polar pesticides and contaminants in processed fruits and vegetables. *Journal of Agricultural and Food Chemistry*, *67*(9), 2716–2722. DOI: 10.1021/acs.jafc.8b06483

Shende, A. D., Tekade, S. A., Deshmukh, A. A., Tembhurkar, S. P., & Selvakumar, P. (2025). Water Quality Monitoring and Control With AI. In Ruiz-Vanoye, J., & Díaz-Parra, O. (Eds.), *Smart Water Technology for Sustainable Management in Modern Cities* (pp. 51–80). IGI Global Scientific Publishing., DOI: 10.4018/979-8-3693-8074-1.ch003

Sorbo, A., Pucci, E., Nobili, C., Taglieri, I., Passeri, D., & Zoani, C. (2022). Food safety assessment: Overview of metrological issues and regulatory aspects in the european union. *Separations*, *9*(2), 53. DOI: 10.3390/separations9020053

Sreenivasulu, A., & Manimozhi, T. Satheesh kumar P, Selvakumar P. and Singh P, (2022). "The Model of Smart Sensing Device For Sensitive Nanoclusters Modification in Sensing Properties," 2022 2nd International Conference on Advance Computing and Innovative Technologies in Engineering (ICACITE), Greater Noida, India, 2022, pp. 1043-1047, doi: DOI: 10.1109/ICACITE53722.2022.9823846

Sugumaran K M D, Pandiyarajan V Y B S, Selvakumar P. and M. S, (2024). "Analysis of Electromyography Signals for Control of Mechanical Prosthesis using Machine Learning Techniques," 2024 2nd International Conference on Intelligent Data Communication Technologies and Internet of Things (IDCIoT), Bengaluru, India, 2024, pp. 1059-1067, doi: .DOI: 10.1109/IDCIoT59759.2024.10467707

Sun, D., Qiu, N., Zhou, S., Lyu, B., Zhang, S., Li, J., & Wu, Y. (2019). Development of sensitive and reliable uplc-ms/ms methods for food analysis of emerging mycotoxins in china total diet study. *Toxins*, *11*(3), 166. DOI: 10.3390/toxins11030166

Szelenberger, R., Cichoń, N., Zajączkowski, W., & Bijak, M. (2024). Application of biosensors for the detection of mycotoxins for the improvement of food safety. *Toxins*, *16*(6), 249. DOI: 10.3390/toxins16060249

Tahoun, I. F., Gab-Allah, M. A., & Yamani, R. N. (2021). "Development and Validation of a Reliable LC-MS/MS Method for Simultaneous Determination of Deoxynivalenol and T-2 Toxin in Maize and Oats." Microchemical Journal, Vol. 168. *Article, 123456*, 1–15. DOI: 10.1016/j.microc.2021.123456

Toribio, L., Bernal, J., Martín, M. T., & Ares, A. M. (2021). "Supercritical Fluid Chromatography Coupled to Mass Spectrometry: A Valuable Tool in Food Analysis." TrAC Trends in Analytical Chemistry, Vol. 143. *Article, 116350*, 1–15. DOI: 10.1016/j.trac.2021.116350

Umapathi, R., Ghoreishian, S. M., Rani, G. M., & Cho, Y. (2022). Emerging Trends in the Development of Electrochemical Devices for the On-Site Detection of Food Contaminants. *ECS Sensors Plus*, *15*(3), 567–589. DOI: 10.1000/iopscience.electrochemical2022

Valdés, A., & Alvarez-Rivera, G. (2021). Foodomics: Analytical Opportunities and Challenges. *Analytical Chemistry*, *93*(3), 567–589. DOI: 10.1021/acs.analchem.1c012345

Vaye, O., Ngumbu, R. S., & Xia, D. (2022). "Comprehensive Two-Dimensional Gas Chromatography MS-Based Techniques for the Analysis of Persistent Organic Pollutants and Ultra-Trace Levels of Organic Pollutants." Reviews in Analytical Chemistry, Vol. 41(3). *Article, 100257*, 567–589. DOI: 10.1515/revac-2022-100257

Vijayakumar, G., Rajkumar, M., Chandar, N. R., Selvakumar, P., & Duraisamy, R. (2024). Environmentally friendly TDS removal from waste water by electrochemical ion exchange batch-type recirculation (EIR) technique. *Environmental Science. Water Research & Technology*, *10*(4), 826–835.

Wahab, S., Muzammil, K., Nasir, N., Khan, M. S., & Ahmad, M. F. (2022). "Advancement and New Trends in Analysis of Pesticide Residues in Food: A Comprehensive Review." Plants, Vol. 11(3). *Article, 123456*, 1–15. DOI: 10.3390/plants11030123

Wang, J., Chen, H., Wu, Y., & Zhao, D. (2022). "Uncover the Flavor Code of Strong-Aroma Baijiu: Research Progress on the Revelation of Aroma Compounds in Strong-Aroma Baijiu by Means of Modern Separation Technology." Journal of Food Composition and Analysis, Vol. 109. *Article, 104499*, 1–15. DOI: 10.1016/j.jfca.2022.104499

Wang, P., Li, J., Guo, L., Li, J., He, F., Zhang, H., & Chi, H. (2024). The developments on lateral flow immunochromatographic assay for food safety in recent 10 years: A review. *Chemosensors (Basel, Switzerland)*, *12*(6), 88. DOI: 10.3390/chemosensors12060088

Weston, M., & Geng, S. (2021). "Food Sensors: Challenges and Opportunities." Advanced Materials Interfaces, Vol. 8(3). *Article, 123456*, 1–15. DOI: 10.1000/wiley.foodsensors2021

Wu, Y., Ye, J., Xuan, Z., Li, L., Wang, H., & Wang, S. (2021). Development and Validation of a Rapid and Efficient Method for Simultaneous Determination of Mycotoxins in Coix Seed Using One-Step Extraction and UHPLC-HRMS. *Food Additives & Contaminants. Part A, Chemistry, Analysis, Control, Exposure & Risk Assessment, 38*(3), 567–589. DOI: 10.1000/taylorfrancis.mycotoxins2021

Xu, M. L., Gao, Y., Wang, X., Han, X. X., & Zhao, B. (2021). "Comprehensive Strategy for Sample Preparation for the Analysis of Food Contaminants and Residues by GC–MS/MS: A Review of Recent Research Trends." Foods, Vol. 10(3). *Article, 123456*, 1–15. DOI: 10.3390/foods10030123

Zhou, J., Brereton, P., & Campbell, K. (2024). "Progress Towards Achieving Intelligent Food Assurance Systems." Food Control, Vol. 145. *Article, 123456*, 1–15. DOI: 10.1016/j.foodcont.2024.123456

Chapter 8

Classification of Environmental Microorganisms for Monitoring Water Quality for Food Safety Using Deep Learning

Anusha Tripathi
https://orcid.org/0009-0001-2493-6757
St. Xavier's College, Kolkata, India

Khushi Purohit
https://orcid.org/0009-0000-4997-2757
St. Xavier's College, Kolkata, India

Pragya Sinha
https://orcid.org/0009-0000-2826-8488
St. Xavier's College, Kolkata, India

Debabrata Datta
https://orcid.org/0000-0002-1559-7036
St. Xavier's College, Kolkata, India

Anal Acharya
https://orcid.org/0000-0002-0751-021X
St. Xavier's College, Kolkata, India

DOI: 10.4018/979-8-3373-3982-5.ch008

ABSTRACT

Water contamination poses a huge risk in both food production and consumption. To monitor water quality, environmental microbes (EMs) can be used as they act as bioindicators. Automating the detection of EMs would be a highly time-saving and vital procedure, particularly since early monitoring of environments can prevent large-scale agricultural contamination. However, such automation remains underdeveloped for industrial use. Hence, a series of comparison experiments were conducted on various deep learning models using the open-source dataset EMDS-6. The pipeline includes segmentation followed by classification using different Convolutional Neural Networks (CNNs) to achieve a fully automated system. Given the limited dataset size, transfer learning, data augmentation, and k-fold cross-validation to mitigate overfitting and bias are employed. It was observed that deeper models like DenseNet121 achieve the highest accuracy (71.67%), whereas lightweight architectures such as MobileNetV2 strike a balance between accuracy and computational efficiency, making them well-suited for resource constrained environments. Finally, feature fusion is applied on the classifier models to assess potential improvements in accuracy which concatenates the feature representations from individual models. While fusion did not consistently outperform standalone models and often showed minimal or negative improvement, combining MobileNetV2 with other models improved performance in most cases. The fusion of Xception and MobileNetV2 yielded an accuracy improvement of 1.90%, while DenseNet121 and MobileNetV2 achieved the highest overall accuracy (71.19%) among all combinations.

1. INTRODUCTION

Water is a primary source in food production. It is required directly from crop production to consumption. Water contaminated by physical, chemical or biological agents pose severe health risks. To monitor this, environmental microbes (EMs) are often used as bioindicators which can check soil, air, and water quality. For example, the EMDS-6 dataset, a small dataset containing 40 images each of 21 classes of EMs, includes microbes like *Euglena*, *Ceratium*, and *Vorticella* which are sensitive to pollutants such as nitrates, phosphates, and heavy metals. Their proliferation can signal deteriorating water quality, enabling proactive measures to be used to prevent the use of contaminated water in agriculture and aquaculture. Massive blooms (e.g., red tides caused by *Gonyaulax*) can cause fish kills and lead to ban of shellfish harvesting. These events not only affect food safety, but also disrupt seafood supply chains and export markets. Some environmental microbes, including bacteria, fungi and algae, can colonize the surface of microplastics, forming what's known as the "plastisphere". This is a unique microbial ecosystem that forms on plastic debris in natural environments (oceans, rivers, soils, etc.). This can happen as some microbes are attracted to hydrophobic surfaces like plastics.

Rising concerns about water contamination from industrial, agricultural, and urban sources have escalated food safety challenges globally. Contaminants such as heavy metals, nitrates, pesticides, and emerging pollutants like microplastics increasingly threaten both crop integrity and aquatic ecosystems. In this context, *early* and reliable detection of waterborne contaminants has become a pressing necessity to prevent foodborne illnesses and ensure public health.

Conventionally, contamination detection methods are gas chromatography and high-performance liquid chromatography (HPLC), combined with mass spectrometry for quantitative analysis. These methods are highly accurate, and are the standards in quality assessments. Naturally, these methods require complex instrumentation and skilled operators. In contrast, environmental microbes (EMs) offer a simpler but promising alternative, especially for contamination monitoring on a more regular basis or as initial checks. These microbes are known to demonstrate high sensitivity and adaptiveness to minute environmental changes, as they exhibit detectable morphological or community shifts when exposed to micropollutants such as heavy metals, pesticides, or pharmaceuticals. For instance, specific protists like *Paramecium* can alter their movement patterns or population dynamics in response to organophosphate pesticide exposure, making them a viable candidate for pollution detection in agricultural water bodies (Benbouzid et al., 2014).

1.1. Background

Traditionally, researchers are known to identify a new EM by conducting a myriad of biochemical tests, and checking references or asking experts. This is very time consuming as well as labour-intensive. In recent years, however, deep learning has revolutionized EM image classification by enabling automated identification of microbes. This improved automated classification can accelerate microbial research and environmental monitoring by reducing dependence on expert analysis.

Despite their promising prospects, automated Environmental Microbiological Content-Based Image Analysis (EM-CBIA) models remain scarce. While Content-Based Image Analysis (CBIA) is well-researched in domains like medical imaging and object recognition, its application in environmental microbiology is limited. This creates a clear research gap: the lack of robust, scalable EM-CBIA models specifically tailored for microbial bioindicators, particularly under small dataset constraints. The classification of EMs is still done manually, and no standardized system has become available for use yet. This makes the process time intensive, and the accuracy is not always ensured. Despite advancements in CBIA using deep learning models, existing methods face limitations such as most of the segmentation methods are semi-automated, lack of good open source curations of EMs that are sufficiently large for model training, lack of high accuracy on training over small datasets, and better accuracy overall.

This is critical for three reasons:

1. Urgency of Water Quality Monitoring in Food Safety: Contaminated irrigation water is a direct pathway for pathogen transmission to crops, posing significant public health risks. EM monitoring can help prevent such contamination by detecting early microbial indicators of harmful pollutants.
2. Impact of Microplastics in Aquatic Food Sources: Microplastics serve as surfaces for biofilm formation, which can harbor pathogens and harmful algal species. Monitoring EMs associated with microplastic biofilms is essential in aquaculture to reduce contamination in fish and shellfish.
3. Shellfish Harvesting and Harmful Algal Blooms (HABs): Dinoflagellates like *Gonyaulax* can form harmful blooms (e.g., red tides), releasing neurotoxins that accumulate in shellfish. EM-based detection allows early identification of bloom conditions, protecting both the marine food supply and human health.

To address this gap, we conduct a series of comparison experiments to evaluate and enhance various deep learning models on EMDS-6.

There have been very few Environmental Microbiological Content Based Image Analysis (EM-CBIA) models in the existing literature so far, even though CBIA is a well researched topic with many existing methodologies. However, they all may not be suitable for the study of EM. There are certain relevant feature extraction and classification methods that exist for different types of microorganisms. Thus, the purpose of this work is attempting to find the best approach to building a small-scale identification and classification model which requires no human intervention by doing a comparative study of the different feature extraction and classification models presented in various papers, as well as by designing our own workflow to classify EMs from the EMDS-6 dataset. Once a considerable accuracy is reached, the system can be modified and scaled to be applied for detecting contamination in water or soil samples.

The EMDS-6 dataset contains 21 types of EMs which are a subset of environmental microorganisms, mainly focusing on certain protists and rotifers. Each type or class of EM contains 40 original and 40 GT images, making a total of 1680 EM images. The classes primarily include protozoa (like Paramecium, Arcella) and microalgae/dinoflagellates (like Ceratium, Gymnodinium). However, microbes from bacteria, archaea, fungi, and viruses are not present as these can be tested using simple methods like Gram staining and dyes, and differentiating one protist from another requires more morphological and behavioral analysis.

1.2. Aims and Objectives

Our main focus is on classification and interpretation using various feature extractors and neural network models, to evaluate and enhance deep learning models for EM classification without requiring a huge dataset and exceptionally high computing power.

The specific objectives include:

- Developing a fully automated pipeline to classify microbial images using a two-stage model.
- Creating a segmentation model that automatically extracts the microbe from the background before classification to enhance identification of EMs.
- Improving segmentation and classification accuracy rate by using deep learning models compared to existing methods.
- Evaluating and comparing performance in segmentation, classification and feature fusion using EMDS-6 to find the best working setup for scalable, time and resource efficient automated monitoring applications in food safety and water quality assessments.

Also, there are challenges posed by models trained on small datasets, as is the case with EMDS-6, including overfitting, when a model memorizes the dataset instead of learning patterns from it, poor generalization to new/unseen data and biases due to class imbalance or lack of diversity. These problems are critical and have to be addressed before real-world deployment. They have been addressed and tackled through transfer learning, data augmentation, K-fold Cross Validation and model regularisation. The detailed outline of the workflow has been provided in Section 5.

2. RELATED WORK

(Zou et al., 2015) proposed a novel Content-Based Image Retrieval (CBIR) system for Environmental Microbiology (EM) to improve the efficiency of microorganism information retrieval. They used semi-automated approaches. Their system applied image segmentation to isolate the microorganism's shape, and then extracted features using a contour-based descriptor called Internal Structure Histogram (ISH).

By calculating Euclidean distances between ISHs, the system measures similarities between microorganism images.

(Li et al., 2014) explored the classification of microorganisms using Content-Based Image Analysis (CBIA) by using both texture and shape features for improved accuracy. This paper demonstrated microbe classification by integrating multiple visual descriptors, including shape and texture. They concluded that a fully automated method results in inaccurate segmentation predictions.

Zhao et al. published another paper to evaluate denoising, segmentation, feature extraction, classification, and detection methods on the EMDS-6 dataset. They used mean-variance and similarity indicators to evaluate filter performance for denoising. Their segmentation results show that U-Net provided the best results and demonstrated that EMDS-6 is highly capable of testing and evaluating image segmentation methods as stated in (Zhao et al., 2022a). They experimented with ten feature extractors, and applied the SVM classifier along with deep learning models, to test them. The results showed that EMDS-6 can be used effectively to perform segmentation, feature extraction, and image classification.

Li et al. worked on the fusion of handcrafted feature extractors after performing semi-automated image segmentation. Ground truth images were created by manual delineation of an EM region, and feature extraction was performed on these. For classification, they used a feature fusion SVM classifier on the original images and their ground truth images. They tested 15 EM classes in this paper, for which a mean AP of 83.78% was achieved. In the cases where segmentation results were perfect, it went as high as 95.97%, as highlighted in (Li et al., 2015). This proved the need for image segmentation for microbial classifications.

Expanding on this, (Zou et al., 2017) proposed a double-stage optimization-based fusion method for content-based image analysis (CBIA) of microorganisms. Their system combined multiple image features and optimized fusion strategies to further enhance retrieval accuracy. While effective, this approach mainly focused on improving the feature fusion process, leaving room for exploring new feature extraction techniques.

Most of the research done on EMs has been done by the Li et al. group. In 2022, they published a survey paper on the use of the U-Net model structure for image analysis. They performed training for segmentation on different U-Net structures like Deformable U-Net, 3D-Unet, Attention U-Net, Double U-Net, Dense U-Net, along with many other variants. This paper shows that deep learning models perform extremely well in microbial segmentation, and Attention U-Net being a commonly used method to improve segmentation performance.

This group also explored deep learning-based approaches for classification. In their work titled A Comparative Study of Deep Learning Classification Methods on a Small Environmental Microorganism Image Dataset (EMDS-6) (Zhao et al.,

2022b), they conduct a comprehensive evaluation involving 21 deep learning models. Their experiments include training with imbalanced data, and hyperparameter tuning. The study determines Xception as the best model in terms of classification accuracy, while the Vision Transformer (ViT) had the fastest training time. In addition, ShuffleNet-V2 model was recognized as the most parameter-efficient. The authors further analyze theoretical fusion accuracies by computing performance improvements for all pairwise combinations of the surveyed models. Their results indicate that fusion can enhance accuracy by taking advantage of the complementary strengths of different models.

In 2022, the aforementioned group published another survey paper, giving a detailed analysis of the different existing methodologies from traditional machine learning to cutting-edge deep learning techniques along with visual transformers, for object detection of EMs (Ma et al., 2023). It consists of a survey of more than 1326 papers, discussing techniques throughout the course of development in computer vision. It points out that microbial classification has had a development history exceeding 30 years, and fusion of different feature extractions improve classification results. It highlights that merging current detection architectures with advanced visual transformers holds the greatest potential for progress in this domain.

In another study, the authors of (Kulwaa et al., 2020) introduce a method involving pairwise feature extraction based on interest points and Shi-Tomasi corner detection. These features are subsequently used for classification with algorithms such as Linear-SVM, SVM with an RBF kernel, k-NN, and Random Forest. The approach achieves the highest classification performance across all tested classifiers on EMs images. Specifically, pairwise features yield exceptional results over unpaired deep learning features. For future research, Kulwaa et al. plan to extend their pairing strategy by incorporating deep features from models like ResNet50 and InceptionV3 to enhance classification and segmentation of microorganisms.

Nillmani et al. proposed a Segmentation-Based Classification Deep Learning Model Embedded with Explainable AI for COVID-19 Detection in Chest X-ray Scans in their paper (Nillmani et al., 2022) that used a two-stage structure where they trained the UNet and UNet+ models for segmentation and selected the best model, and applied the results on eight classification models, namely VGG16, VGG19, Xception, InceptionV3, Densenet201, NASNetMobile, Resnet50, and MobileNet to select the best performing combination. The best performing approach was the UNet+Xception model, achieving impressive results. This performance surpassed that of previously established segmentation-driven classification techniques.

A paper by (Huang et al., 2020) provides a comprehensive review on various fusion techniques for different models. The authors highlighted the importance of feature fusion, emphasizing the integration of contextual information from electronic health records (EHR) alongside pixel data, to enhance diagnostic accuracy, support informed clinical decision-making, and ultimately improve patient outcomes. They cover a wide range of multi-modal fusion strategies using deep learning namely, early fusion, joint fusion, and feature fusion. They screen through over 980 studies and review 17 papers (published in between 2012 and 2019) across PubMed, Scopus, and Google Scholar, and categorize these works according to their fusion strategy. They conclude that most works used early fusion. Feature fusion was used in the least number of papers, but showed improved results in comparison with models that used single modalities.

(Nawghare and Prasad, 2025) introduced a hybrid approach that integrates VGG16, a pre-trained deep convolutional neural network, with a Random Forest (RF) classifier. The hybrid model surpassed the performance of individual deep learning models such as VGG16, EfficientNetB0, and AlexNet, as well as traditional machine learning algorithms. This performance boost highlighted the strength of combining the rich feature representations of deep learning with the ensemble-based decision-making of RF classifiers. Their study emphasized the promise of hybrid architectures that blend deep learning and machine learning techniques.

Makmud et al. explored the use of fusion models for fruit classification. They combined multiple pre-trained CNN models—VGG16, MobileNetV2, Xception, ResNet-50, and InceptionV3—using a concatenation layer in the fusion process. This approach significantly improved the model's accuracy. After applying adaptive fine-tuning, the model outperformed individual models both before and after fine-tuning. The study demonstrated that the fusion model, combined with adaptive fine-tuning, was superior in performance compared to the standalone models. This suggests that fusion models, especially with fine-tuning and concatenation layers, can be highly effective in real-world applications (Makmud et al., 2024).

Table 1. Summary of Methods Studied

Author(s)	Year	Title	Method Used	Key Contribution
Li et al.	2022	A state-of-the-art survey of U-Net in microscopic image analysis	Segmentation	Does an extensive study on the U-Net model for segmentation, and surveys related papers on U-Net and its modified versions. It also finds the best improvements on it, highlighting attention gates to be a good and highly used strategy for better results
Nillmani et al.	2022	Segmentation-Based Classification Deep Learning Model Embedded with Explainable AI for COVID-19 Detection in Chest X-ray Scans	Segmentation-based classification, XAI	Designs a deep learning-based diagnostic system with Explainable AI using segmentation-based classification to detect COVID-19 from chest X-ray images. They tested 16 sets with UNet/UNet+ for segmentation and eight classifiers. UNet+Xception achieved the best results with 97.45% accuracy and 5% error rate.
Li et al.	2021	A Comparative Study of Deep Learning Classification Methods on a Small Environmental Microorganism Image Dataset (EMDS-6): From Convolutional Neural Networks to Visual Transformers	Classification with CNNs and ViT	Evaluates 21 deep learning models for classifying the images. Xception outperformed others in accuracy. Fusion of models, especially combining with visual transformers, improved classification accuracy, showcasing complementary strengths across techniques
Li et al.	2021	EMDS-6: Environmental Microorganism Image Dataset Sixth Version for Image Denoising, Segmentation, Feature Extraction, Classification, and Detection Method Evaluation	Denoising, segmentation, feature extraction, classification, object detection, image analysis	Evaluates methods for denoising, segmentation, feature extraction, and classification. U-Net provided the best segmentation results, while SVM was used for classification. Reliable accuracies in object detection using Fast RCNN and Mask RCNN proved EMDS-6's utility for various image analysis tasks.
Li et al.	2014	Application of content-based image analysis to environmental microorganism classification	CBIR + Handcrafted feature fusion	Early use of content-based image retrieval (CBIR) for EM classification, combining handcrafted features with classifiers
Wu et al.	2019	CBIR of Environmental Microorganisms Using Double-stage Optimization-based Fusion	CBIR, fusion	Introduces a combination of double stage fusion with Particle Swarm Optimisation (PSO) and enhances results using a two-stage approach.
Li et al.	2015	Environmental microbiology aided by content-based image analysis	EM classification, CBIR	Improved EM classification using feature-based image retrieval, includes performance evaluation and challenges
Zou et al.	2016	Content-Based Image Retrieval for Environmental Microbiology (EM	CBIR (ISH+Euclidean)	Developed a CBIR system using semi-automated segmentation, used the Internal Structure Histogram (ISH) for feature extraction, measuring similarities based on Euclidean distances, evaluated using mean Average Precision
Kulwaa et al.	2020	A new pairwise deep learning feature for EM image analysis	Pairwise deep learning	Proposes a novel pairwise feature strategy for better EM image classification using deep learning
Huang et al.	2020	Fusion of medical imaging and electronic health records using deep learning	Fusion (medical imaging + EHR)	Framework and review of integrating imaging with EHRs using deep learning; insights applicable to multimodal fusion

continued on following page

Table 1. Continued

Author(s)	Year	Title	Method Used	Key Contribution
Rani et al.	2021	Machine learning and deep learning based computational approaches in automatic microorganism image recognition	Classification	Comprehensive survey of ML/DL methods, challenges, and future trends for microbial image classification
Ma et al.	2023	A state-of-the-art survey of object detection techniques in microorganism image analysis	Segmentation, Object detection	Reviews both traditional and DL-based object detection methods for microbial image analysis
Makmud et al.	2024	Improving rotten fruit classification through fusion of pretrained CNNs	CNN ensemble fusion	Combines multiple pretrained CNNs to improve accuracy; shows effectiveness of fusion in fine-grained classification
Nawghare & Prasad	2025	Hybrid CNN and RF model with late fusion for ASD detection	CNN + RF + fusion	Applies CNN for feature extraction and RF for classification using late fusion, showing improved performance

3. MODELS USED

3.1. Deep Learning Models

Attention U-Net

Attention U-Net is an advanced variant of the U-Net architecture that integrates attention mechanisms to enhance segmentation accuracy by focusing on relevant regions of an image. In this architecture, attention gates are employed to filter the features propagated through the skip connections, enabling the network to concentrate on target structures of interest while suppressing irrelevant regions. This is particularly useful in medical image segmentation, where the object of interest may occupy only a small portion of the image or be surrounded by visually similar structures.

DenseNet121

DenseNet121 connects each layer to all the subsequent layers in order to improve information flow. This dense connectivity ensures that features are effectively reused throughout the network and that gradients are efficiently propagated during training, mitigating the vanishing gradient problem. DenseNet121 strikes a balance between depth and computational efficiency, making it suitable for tasks requiring strong feature propagation and compact models.

DenseNet169

DenseNet169 is deeper than DenseNet121 as it has more layers. It facilitates feature reuse by linking each layer to all others within a dense block, resulting in enhanced gradient flow and more diversified feature representation. The increased depth allows DenseNet169 to capture more complex patterns in the input data, which can be beneficial for high-resolution image analysis and classification tasks requiring a richer hierarchical understanding.

InceptionV3

InceptionV3 decomposes larger convolutions into smaller ones, which is useful for faster learning and smaller datasets. This model reduces the number of parameters and accelerates training without compromising performance. InceptionV3 also incorporates auxiliary classifiers and factorized convolutions, making it efficient in terms of computation and suitable for deployment in scenarios where resource constraints are present.

ResNet50

ResNet50, short for Residual Networks with 50 layers, is a deep learning model used for visual recognition tasks. It introduces skip connections that allow gradients to flow directly through the network by bypassing one or more layers. This residual learning framework enables the training of much deeper networks by addressing the degradation problem, improving convergence and overall accuracy in complex image classification tasks.

MobileNetV2

MobileNetV2 is a compact, lightweight convolutional neural network (CNN) architecture specifically designed for mobile and embedded vision applications. It incorporates depthwise separable convolutions and an inverted residual structure with linear bottlenecks, which reduces model size and computational cost significantly. MobileNetV2 is particularly well-suited for real-time inference on devices with limited hardware capabilities.

Xception

Xception (Extreme Inception) is a deep convolutional neural network that improves upon the Inception architecture by fully replacing standard convolutions with depth wise separable convolutions. This factorization reduces the computational load while preserving model accuracy. By decoupling spatial and channel-wise processing, Xception achieves a high degree of efficiency and performance, making it ideal for large-scale image classification and transfer learning applications.

3.2. Loss Functions

Binary Cross-Entropy

The performance of an image classification model is estimated and measured using the cross entropy loss. The probability output of the cross-entropy loss function is between 0 and 1. The cross-entropy loss decreases when the predicted probability for the correct class increases, and increases when the predicted probability is low for the actual class (Ruby and Yendapalli, 2020). If the predicted probability is close to zero for the correct class, the model is performing poorly, leading to a high loss value.

Sparse Categorical Cross-Entropy

Sparse Categorical Cross-Entropy is used for multi-classification problems. The difference between categorical cross-entropy and sparse categorical cross-entropy is that the latter has integer-encoded class labels while the former is one-hot encoded. The advantage of this particular loss function is that rather than using a whole vector it only uses a single integer for a class, thereby saving time in memory along with computation.

Focal Loss

Focal Loss is a function devised to tackle the class imbalance in classification tasks by downplaying the loss contribution of easy examples and concentrating more on the hard ones. It is an extension of the standard cross-entropy loss.

Dice Loss

Dice Loss is a commonly used function for image segmentation tasks to compute the intersection between the predicted and ground truth masks.

Boundary Loss

Boundary Loss improves segmentation performance by focusing on the mismatch in the boundary regions of predicted and ground truth masks.

3.3. Preprocessing Techniques

Gaussian Blurring

Noise reduction is the key component of image preprocessing. It helps enhance edges and improves object detection. The Gaussian Function is applied to the image to perform Gaussian blurring. This reduces the noise in the image by removing some of the unwanted details, while preserving the low spatial frequency of the image. The two-pass method of Gaussian filterings, requiring lesser calculations is applied, which is generally preferable in implementation.

Histogram Equalization

A histogram maps the intensity distribution of an image according to the pixel values present in it. Higher the contrast, the more spaced out the histogram is. Thus, Histogram equalization stretches out the pixel intensities by redistributing intensity values more uniformly which enhances contrast, making details more visible (Gonalez and Woods, 2020).

Contrast Limited Adaptive Histogram Equalization (CLAHE)

Histogram equalization often may not improve the visibility of the objects. In that case, an adaptive histogram equalization (AHE) approach like CLAHE has to be adopted. It enhances image contrast without overly amplifying noise, but provides sufficient contrast for the visualization of structures within the image, making it easier to compare different regions (Pizer et al., 1987). In this method, the image is divided into small blocks called "tiles", and each tile undergoes histogram equalization independently under a contrast limit. If any histogram bin exceeds this limit, the excess pixels are clipped and evenly redistributed to other bins before equalization which is then followed by bilinear interpolation to smooth out any artifacts between the tiles.

3.4. Evaluation Metrics

Confusion Matrix

Confusion Matrix is an essential guide to visualize class-wise performance of a model. It is structured as a two-dimensional matrix, where one axis represents the actual class labels and the other represents the classes predicted by the model. In the case of binary classification, the matrix includes two classes: positive and negative. The matrix is divided into four sections: true positives (TP), false positives (FP), true negatives (TN), and false negatives (FN), each indicating a different outcome of the model's predictions relative to the actual labels. The diagonal indicates the number of correct predictions for the corresponding classes. This provides information about which class has better accuracy and also pinpoints the classes and features where the model struggles and requires improvement.

Accuracy

Accuracy is defined as the ratio between the correctly classified samples to the total number of samples. Accuracy is a good criterion for well-balanced classes i.e., all types of microbes have the same number of images, but it is not a reliable indicator of the model alone.

$$\text{Accuracy} = TP + TN / TP + TN + FP + FN \quad (1)$$

Precision

Precision represents the proportion of positive samples that were correctly classified to the total number of positively predicted samples. Precision is vital for this model as it directly reflects the number of false positives, Hence, a higher precision indicates fewer misidentifications of microbes, which is important for ensuring accurate bioindication.

$$\text{Precision} = TP / TP + FP \quad (2)$$

Recall

Recall, sensitivity, True Positive Rate (TPR), or hit rate, of a classifier represents the positive correctly classified samples to the total number of positive samples. Recall is critical in microbial classification because it reflects the model's ability to detect all relevant microbes of a certain class. A high recall means fewer false negatives, which ensures that no bioindicator or pathogen which must be detected is not overlooked. It increases the reliability of the system.

Recall = TP / TP + FN (3)

F1-score

F1-score represents the harmonic mean of precision and recall, balancing both metrics. The value of F-measure is ranged from zero to one, and high values of F-measure indicate high classification performance. A high F1-score makes the system more trustworthy overall, as it ensures minimum false positives as well as false negatives.

F1 Score = 2TP / 2TP + FP + FN (4)

Dice Coefficient

Dice coefficient or Dice Similarity Coefficient (DSC) is a statistical metric used to measure the similarity between two sets, usually used in image segmentation to compare a predicted segmentation (A) with the ground truth (B). A high Dice score implies the predicted segmentation mask and the ground truth image properly resemble each other.

Dice = 2TP / 2TP + FP + FN (5)

Jaccard Index

The Jaccard index measures the size of the intersection divided by the size of the union of two sets. Also used to assess segmentation performance, it works by explicitly ignoring true negatives, focusing only on the region of interest. This is important in images where the background pixels are irrelevant and only the microbe shape matters.

Jaccard = TP / TP + FP + FN (6)

4. DATA DESCRIPTION

The dataset that has been used to compare model performance is Environmental Microorganism Image Dataset Sixth Version (EMDS-6). It consists of 840 microscopic images of different sizes, and are distributed among 21 class folders, each containing 40 images of the corresponding class.The 21 classes covered in EMDS-6 are: Actinophrys, Arcella, Aspidisca, Codosiga, Colpoda, Epistylis, Euglypha, Paramecium, Rotifera, Vorticella, Noctiluca, Ceratium, Stentor, Siprostomum, K.Quadrala, Euglena, Gymnodinium, Gonyaulax, Phacus, Stylonychia, Synchaeta. Figure 1 shows one such image and its ground truth mask image from the 21 classes each. The ground truth (GT) image shows the area of the microorganism marked as white and the rest of the background as black. The white region is the Region of Interest (ROI).

To reduce bias in the training process and improve accuracy of the results in the small testing dataset used, the images in EMDS-6 were augmented. Only these three geometric transformations have been applied-rotation by 180o, horizontal flipping and vertical flipping. The resultant dataset size after augmentation is 3360 images. The reason why the dataset is not created dynamically is to save memory costs as dynamic generation takes up more RAM as well as computational time. The augmented dataset is then divided into two partitions- one for model training, and one for model testing. The train-test splitting is done by 85.5% and 12.5% of the total images in a class. This results in 140 images for training and 20 for testing per class, and a total of 2940 images for training, and 420 for testing, as shown in Table 1 below.

Figure 1. Demonstrations of transformation of original image to get rotation by 180 degrees, horizontal flipping and vertical flipping

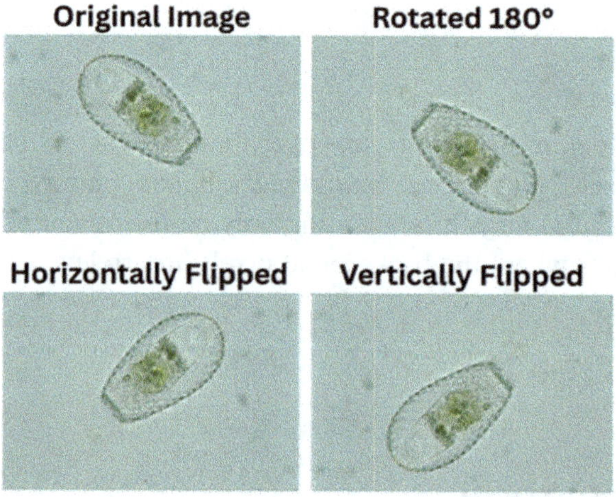

Figure 2. The 21 classes of EMs: one original image and its corresponding ground truth mask

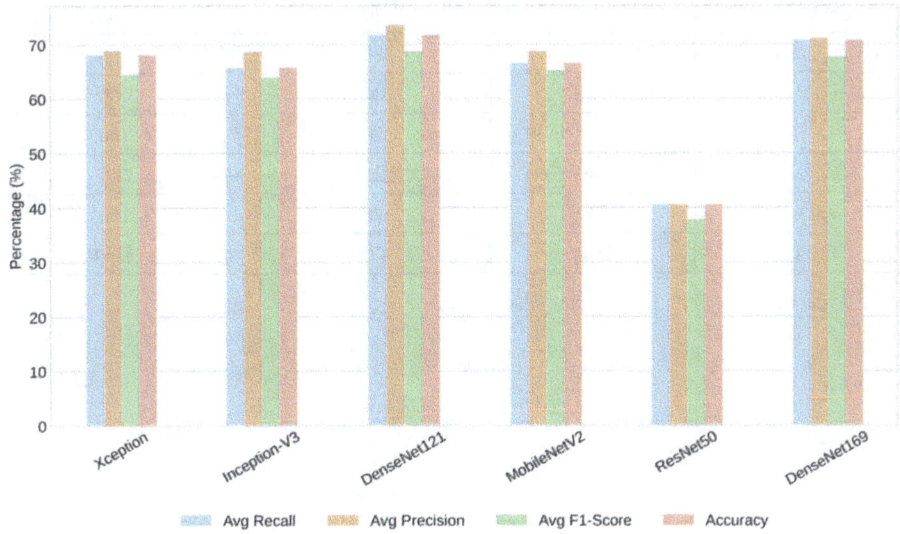

Table 2. The Augmented Dataset Split for Each Class

Class	Train	Test	Total
Actinophrys	140	20	160
Arcella	140	20	160
Aspidisca	140	20	160
Codosiga	140	20	160
Colpoda	140	20	160
Epistylis	140	20	160
Euglypha	140	20	160
Paramecium	140	20	160
Rotifera	140	20	160
Vorticella	140	20	160
Noctiluca	140	20	160
Ceratium	140	20	160
Stentor	140	20	160
Siprostomum	140	20	160
K. Quadrala	140	20	160

continued on following page

Table 2. Continued

Class	Train	Test	Total
Euglena	140	20	160
Gymnodinium	140	20	160
Gonyaulax	140	20	160
Phacus	140	20	160
Stylonychia	140	20	160
Synchaeta	140	20	160
Total	**2940**	**420**	**3360**

5. PROPOSED METHODOLOGY

Figure 3. Workflow Diagram

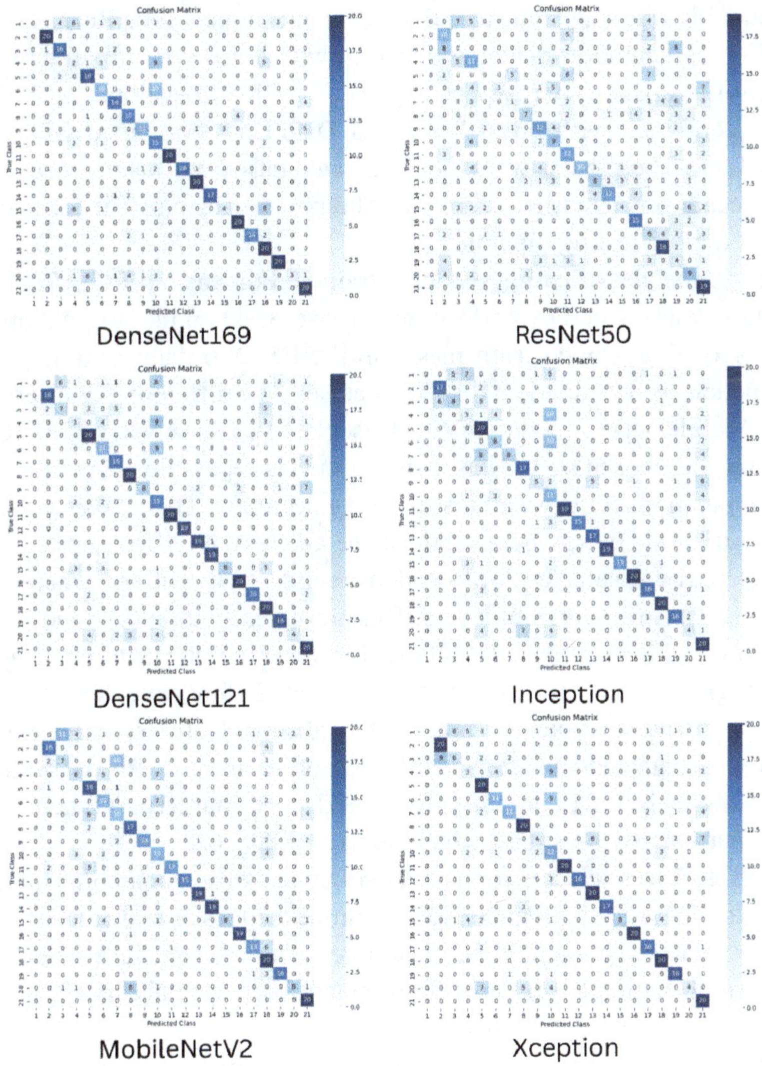

A modular approach has been adopted in order to perform the study with interchangeable units so that it can increase scalability, and enhance parallel training. With this system, multiple segmentation models can be trained in the first stage simultaneously and the best model to predict the segmentation mask can be selected. After obtaining the segmented image (containing the region of interest), it is fed to multiple classifiers parallelly which gives the evaluation results for classification on the test dataset. From this, the best working pipeline for the system can be obtained.

In stage 1, different segmentation models are trained on the training dataset having the original images with their corresponding ground truth masks to automate image segmentation. In this stage, an Attention UNet backbone with a ResNet50 encoder is used. The segmentation model was trained with 6 different loss functions: Dice Loss, Focal Loss, Binary Cross-Entropy (BCE), Dice Loss combined with BCE, Dice Loss combined with Focal Loss, and Dice Loss combined with BCE and Boundary Loss. The models were then evaluated based on various metrics such as accuracy, recall, precision, F1 score etc. The best performing model is chosen to be a part of the full testing pipeline.

In stage 2, the classification model is trained. For this, an ROI (Region of Interest) dataset was created by performing bitwise AND on the original images and their corresponding ground truth masks in the EMDS6 dataset. 6 different deep learning models were trained on this ROI dataset. The different classifiers used are Xception, MobileNetV2, InceptionV3, DenseNet121, DenseNet169 and ResNet50.

Xception was chosen as it is based on depthwise separable convolutions, which reduce computational cost without sacrificing accuracy. This makes it well-suited for the classification task, where subtle shape and texture differences are critical.

DenseNet121 was selected due to its dense connectivity pattern, where each layer receives inputs from all previous layers. This promotes feature reuse and strengthens gradient flow, making it an excellent choice for capturing fine textures and gradients.

DenseNet169 offers more layers than DenseNet121 for deeper feature extraction. While deeper networks typically capture more abstract features, both DenseNet121 and 169 were evaluated to understand the trade-off between depth and performance. In practice, DenseNet121 can outperform its deeper counterpart when overfitting can occur, particularly on smaller datasets like EMDS-6.

MobileNetV2 was selected for its lightweight and efficient architecture, making it highly suitable for deployment on low-resource devices or real-time applications. Despite its efficiency, it maintains competitive accuracy, especially when used with transfer learning on small datasets, making it ideal for microbial classification, where quick and accurate inference is needed.

The ability of InceptionV3 to capture multi-scale spatial features made it a suitable choice. The models are allowed to process both local and global patterns effectively, which is a valuable trait in microbial image classification, where microbes can vary in size, texture, and structure across scales.

ResNet50 serves as a baseline model with a well-established track record in image classification tasks. ResNet50 has been tested across many datasets and tasks, and so it helps in analyzing if a new model is better due to its architectural improvements or just dataset quirks.

In the testing pipeline, the input image is passed on to the best segmentation model which produces its predicted ground truth mask. By performing bitwise AND on the predicted mask and the original input image, its ROI is generated. This ROI is then fed as an input to the 6 classifiers to get the final predicted class in each case. The entire testing dataset undergoes this process and then the performance of the full pipeline is evaluated using standard metrics to then obtain the best model.

5.1. Hyperparameters

Hyperparameters are the values that are set before the training of a model, and are used to control its behavior. These control the learning process and are usually fine-tuned based on trial-and-error to get the best training results. These are usually constant with standard ranges, but adaptive changes can also be applied. The Hyperparameters that have been set in the system are given in the following subsections.

Batch Size

Batch size is the number of training images a model will process before updating its weights based on the loss and gradients computed. The common values for batch sizes are 8, 16, 32, 64, 128 (small batch sizes like 8–32 work better for limited memory, while larger ones improve stability).

Epochs

Epochs denote the number of times the model iterates over the entire training dataset during the training process. One epoch implies all the images of the training dataset will be passed only once, and the model weights will be updated based on the computed loss for each batch.

The typical range for epochs is 10–200 (depends on dataset size and complexity; early stopping is often used to prevent overfitting). The number of epochs was fixed to 100 for segmentation models, and to 40 for the classifiers. Higher epoch values could not be tested owing to computational resource limitations.

Learning Rate

The learning rate of a model is used to control to what extent the model updates its parameters/ weights during training. The typical values are 0.001 (default for Adam), 0.01 (SGD), 0.0001 (fine-tuning deep networks). Lower values ensure stable convergence, while higher ones speed up training but risk divergence. For the models used here, a learning rate of 0.0001 or 1e−4 was maintained.

Optimizer

The optimizer is responsible for adjusting model parameters to minimize the loss function.

Common choices include Adam (adaptive, well-suited for deep networks), SGD (stochastic gradient descent, often used with momentum), and RMSprop (useful for recurrent networks).

The Adam optimizer is used here due to its ability to handle sparse gradients and its efficiency in deep networks.

L2 Regularization

L2 regularization is applied to prevent overfitting by discouraging large weight values. It is done by adding a penalty to the loss function used in the model. This penalty is proportional to the square of the magnitude of the model's weights. Typical values range from 0.0001 to 0.01, where higher values increase regularization and lower values retain model flexibility. In the models used here, an L2 regularization value of 0.0001 has been used to maintain a balance between generalization and performance.

Dropout Rate

Dropout is another regularization technique where a portion of neurons are randomly dropped during training to reduce overfitting. Standard values are typically within 0.1 to 0.5, with CNNs mostly using 0.2–0.3 and RNNs often using 0.5. For the segmentation models, the dropout rate was set to 0.4, and that for the classifiers to 0.6 to improve generalization without significantly affecting learning capacity.

Folds

Folds refer to the number of splits in k-fold cross-validation, which helps evaluate model performance. The common values are 5 or 10, with 5-fold being standard for deep learning and 10-fold often used for smaller datasets. For the experiments performed here, a 5-fold cross-validation strategy was used to ensure robust performance evaluation while keeping computational requirements manageable.

5.2. Preprocessing Steps

In this system, the input dataset undergoes three levels of preprocessing to improve the quality of the image, and enhance meaningful features for the segmentation model to learn. The images are first resized to 224 × 224 pixels.

Gaussian blurring is used to remove high-frequency noise and smooth out the images. It helps the model focus on larger features rather than small pixel clusters whose variations can act as noise. As microbial images contain a lot of fine and detailed texture, this method is highly beneficial.

Histogram Equalization (HE) enhances the global contrast of an image by redistributing its pixel values. It treats an image as a histogram, and stretches it out and renders it back to the resultant image. This technique highlights regions of an image which are not that clear, and have low-contrast. It improves feature extractions by bringing these out, however, sometimes it can over-enhance certain areas and lead to loss of details, so HE alone is not sufficient for preprocessing.

Contrast Limited Adaptive Histogram Equalization or CLAHE is an improved version of HE. It tackles the issue of over-enhancement of an image by applying contrast enhancement locally instead of doing it globally. It preserves fine details while improving contrast, making edges more distinguishable without increasing noise.

Gaussian blurring is followed by CLAHE and then Histogram Equalization for contrast enhancement.

The image is then normalized to adjust the range of the pixel values.

Figure 4. An image undergoing the different stages of preprocessing before entering the segmentation pipeline.

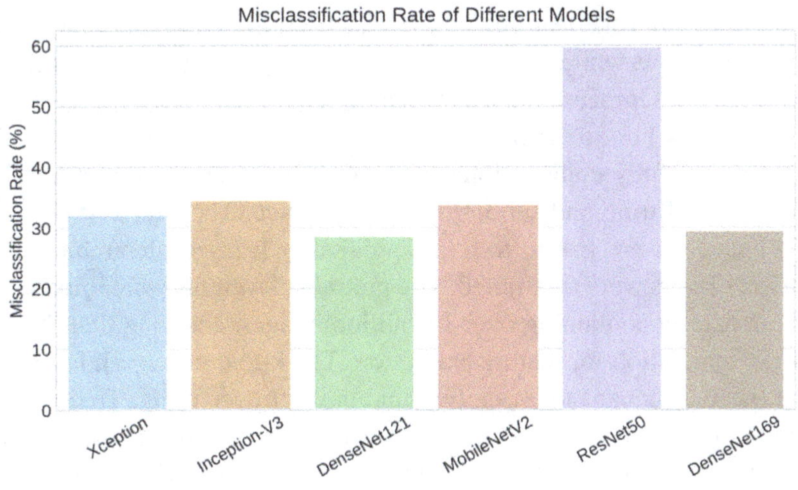

5.3. Segmentation

Figure 5. The segmentation pipeline.

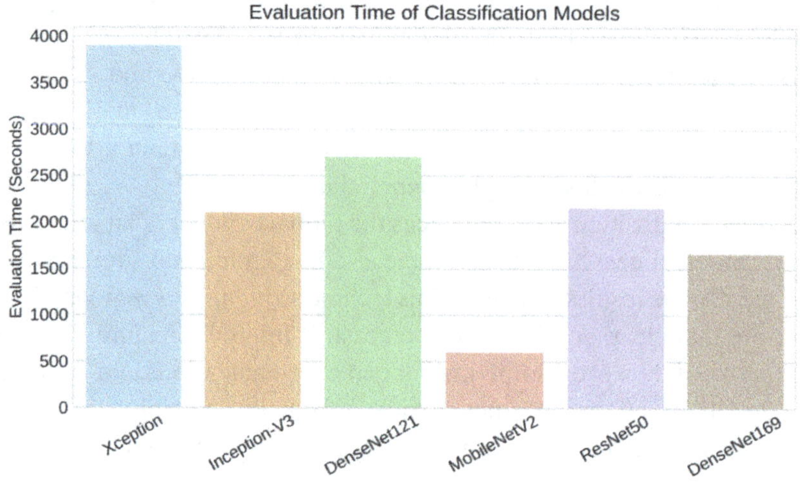

Following preprocessing, the dataset images are used for segmentation model training. The ground truth images along with the original images are given as the training and validation dataset. The model learns from these images and predicts the ground truth mask for an image from the testing dataset. The base segmentation model chosen is Attention U-Net with a ResNet50 encoder. The Attention U-Net architecture incorporates an attention mechanism, which enhances the model's ability to focus on critical features. Several studies have demonstrated that attention mechanisms improve feature representation by directing the network's focus to informative regions while ignoring the unnecessary ones (Nillmani et al., 2022). To further improve segmentation performance, a ResNet50 encoder was integrated into the U-Net back bone to apply transfer learning. Given the dataset size of 3,360 images, training a fully convolutional network from scratch was not feasible due to the risk of overfitting and suboptimal convergence. The first 140 layers of the ResNet50 encoder were frozen to retain previously learned hierarchical features, while the later layers were fine-tuned. The pretrained weights were initialized from ImageNet, a dataset containing over 14 million images, ensuring that the encoder had a strong foundation for feature extraction. The k-fold cross-validation method is applied (keeping k = 5) to check bias and prevent overfitting. To compare segmentation performance, six different loss functions were applied: Dice Loss, Focal Loss, Binary Cross-Entropy (BCE), Dice Loss combined with BCE, Dice Loss combined with Focal Loss, and Dice Loss combined with BCE and Boundary Loss.

The performance was assessed using the following metrics: accuracy, loss, Dice coefficient, and Jaccard index (IoU). After the best model is selected, the predicted mask for a test image can be obtained.

5.4. Classification

Figure 6. The classification pipeline.

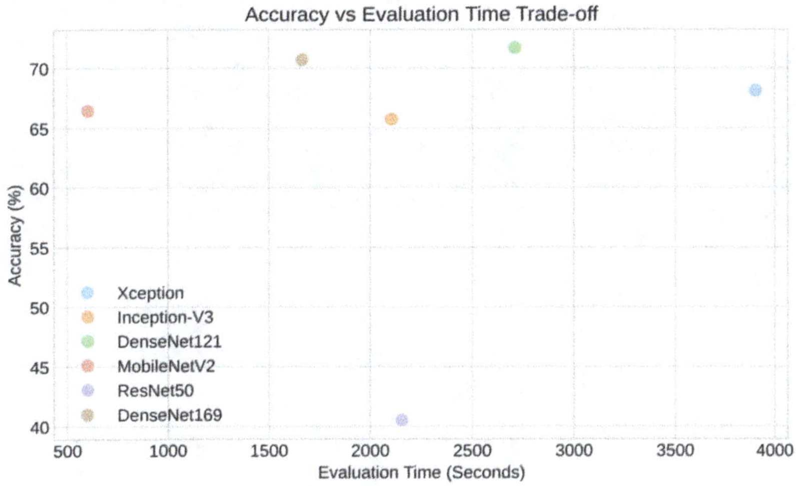

From the Region of Interest (ROI), the classification network extracts different features, which are stored in the form of high dimensional feature maps. The classification dataset is prepared by applying the bitwise AND operation on all pixels of the ground truth image with the original image to obtain the ROI for all images of the training dataset. These images are used for training each classifier model. The labels (the class number) for the dataset are obtainable from the folder structure they are present in. The training starts with downsampling the images so that they can be processed and feature maps can be obtained from them. The global average pooling function is applied to them so that only the most prominent features are learnt. This is repeated for many layers and subsequently, in the final dense layer, the features are combined into a single feature vector from which the probability distribution of the prediction corresponding to each class is obtained. From there, the model makes the final prediction. The class of the microbe is extracted from the file path of the training images. The network minimizes a predefined loss function (categorical cross-entropy) using backpropagation and gradient descent optimization. The performance of the model is evaluated using standard metrics such as accuracy, precision, recall, F1 score and misclassification rate. By combining segmentation and classification, the deep learning pipeline effectively localizes and identifies objects of interest, and predicts the class.

5.5. FEATURE FUSION

Figure 7. The pairwise feature fusion pipeline.

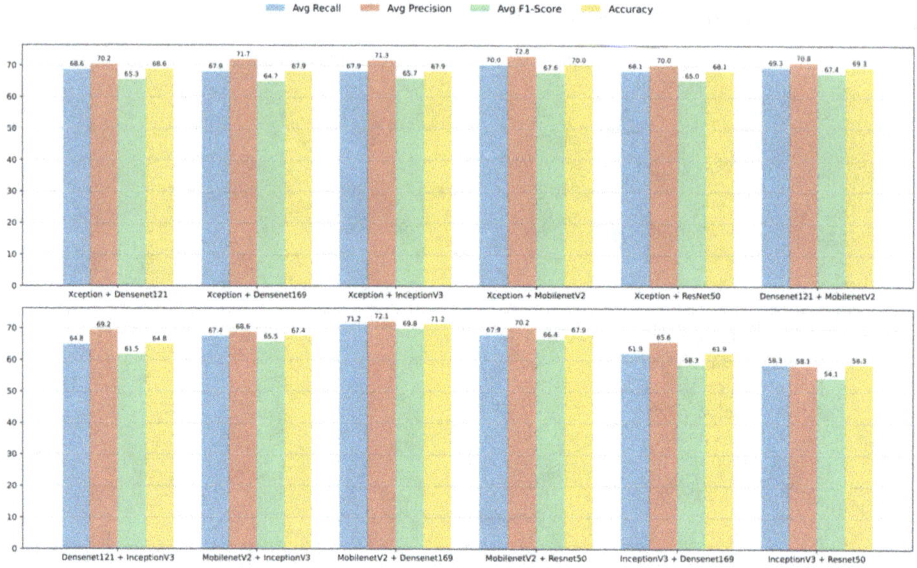

A separate classification approach is to apply fusion. It is an extension of the previous classification network, but the same models are chosen in pairwise combinations where they are independently trained on the same data, and their extracted features are stored parallelly. A global average pooling step is applied to the feature maps. This step condenses each model's output by summarizing spatial information into a compact feature vector, capturing the overall patterns in the image without focusing on specific locations.

These feature vectors, each representing the image as seen through a different model's lens, are then combined into a single, unified representation. This fusion is done by joining the two vectors together side-by-side, allowing the model to analyze from both sets of learned features simultaneously. This is called feature concatenation, and it enables the fusion model to utilize the complementary strengths of both models.

The combined vector is then passed through a fully connected layer that helps the model learn how to best interpret this fused information, which is known as the concatenation layer. A dropout step follows to reduce overfitting, ensuring the model does not become too reliant on any one part of the feature space.

In the final step, the softmax function ensures that the probabilities across all classes add up to one. The class with the highest probability is selected as the predicted category for the input image, enabling multi class classification across 21 microbial classes.

6. PERFORMANCE AND ANALYSIS

6.1. Comparison of Segmentation Experiments

1) Experimental Setup: K-fold cross-validation with k = 5 was used to ensure minimal bias during evaluation, with an 80-20% training-validation split. The segmentation model was trained with the hyperparameters listed in *Table 2.*

Table 3. Hyperparameters for Segmentation Model Training

Parameter	Value
Batch Size	16
Epochs	100
Learning Rate	1e-4
Optimizer	Adam
L2 Regularization	1e-5
Dropout Rate	0.4
Image Size	(224,224)
Folds	5

2) Performance and Results: The segmentation model was trained using multiple loss functions to compare their effectiveness. The training versus validation metrics over the five folds for all six loss functions have been depicted in the graphs below in *Figure 9.* The performance was evaluated based on Dice coefficient, Jaccard index, accuracy and loss. *Table 3* presents the comparative results for different loss functions. The best loss function was found to be Dice +

BinaryCrossEntropy. The prediction masks generated by the segmentation network using the same are given in *Figure 8.*

Table 4. Comparative performance of Attention U-Net with pretrained ResNet50 using different loss functions. Dice + BCE shows the best results.

Loss Function	**Test Accuracy(%)**	**Test Loss(%)**	**Dice(%)**	**Jaccard(%)**
Dice + BinaryCrossEntropy	93.93	54.31	87.50	78.04
Dice	93.98	12.65	87.35	77.76
Dice+ Boundary + BinaryCrossEntropy	93.87	116.00	87.02	77.35
Dice + Focal	93.62	41.45	83.04	74.91
Focal	91.76	31.54	80.44	70.85
BinaryCrossEntropy	93.82	99.30	82.85	74.99

Figure 8. Prediction of segmentation masks by the best segmentation model: Attention U-Net backbone + ResNet50 encoder with Dice + BinaryCrossEntropy as the loss function.

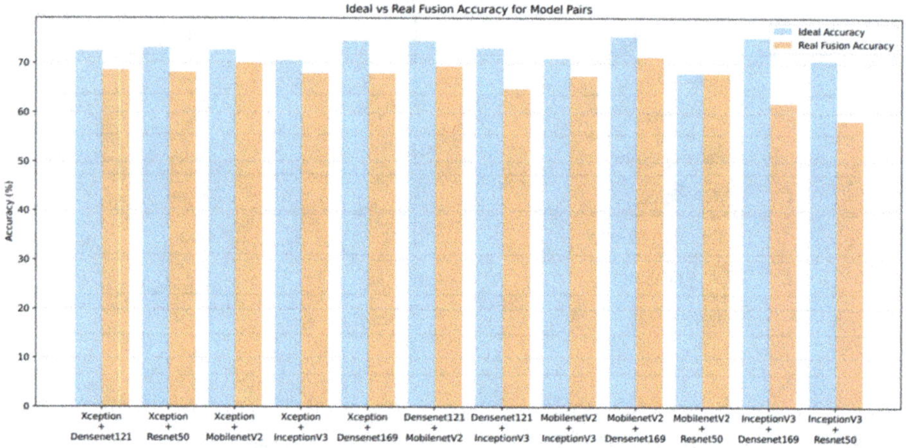

Figure 9. Training versus validation metrics over the five folds for Attention U-Net backbone + ResNet50 encoder with different loss functions.

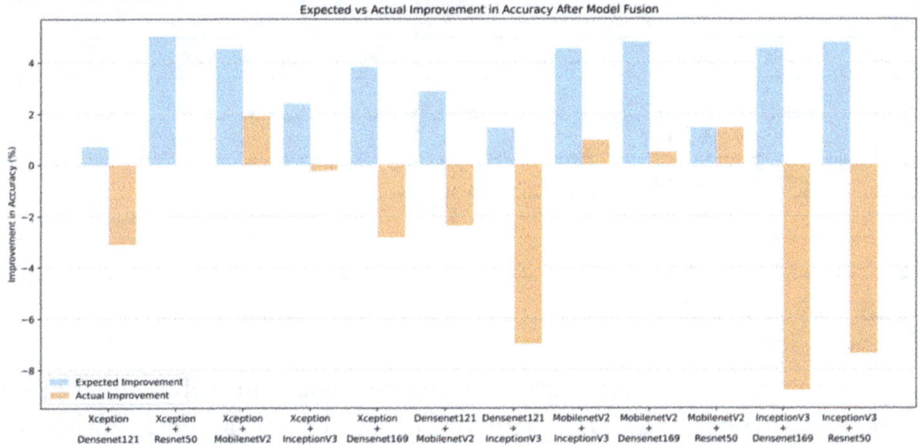

The segmentation results showed that Dice with BCE loss function is better at class boundary precision than Dice alone, even if accuracy is marginally lower. Combining Dice with BCE helps the model optimize both region overlap and pixel-wise precision, which is helpful in cases with class imbalance.

On the other hand, adding Boundary loss to Dice with BCE might have introduced too much complexity or instability, possibly overemphasizing edge pixels and hurting overall performance.

6.2. Comparison of Combined Segmentation and Classification Pipeline Experiments

1) Experimental Setup: The classification network was trained using features extracted from the segmented Regions of Interest (ROIs). Six models were trained—ResNet50, InceptionV3, MobileNetV2, DenseNet121, DenseNet169, and Xception. The hyperparameters used for classification training are listed in *Table 4.* K-Fold cross validation and dataset partition ratio for training and validation were kept the same as done in the segmentation network.

Table 5. Hyperparameters for the Classification Model Training

Parameter	Value
Batch Size	16
Epochs	40
Learning Rate	1e-4
Optimizer	Adam
L2 Regularization	0.001
Dropout Rate	0.6
Image Size	(224,224)
Folds	5

2) Performance and Results: Performance evaluation of the six classifier models were done based on their average recall, precision, F1 score, accuracy, misclassification rates, and the time for evaluation on the testing dataset. *Table 5* presents the comparative results.

The results below indicate that MobileNetV2 offers the greatest speed, a mere 600s, and can be highly beneficial where evaluation time is an important requirement, especially in industrial settings. The best performance was given by DenseNet121 in the current pipeline, giving an accuracy of 71.67%, average recall of 71.67%, average precision of 73.44%, F1-score of 68.66%, as well as maintaining the lowest misclassification rate— 28.33%, which implies better coverage of all classes and fewer wrong predictions. DenseNet121 is the most beneficial among the six classifiers when combined with a segmentation network of Attention-UNet with ResNet50 encoder, to give the best results on small and complex datasets. DenseNet169 has a performance very close to DenseNet169 with a much faster evaluation time (1665s vs. 2700s).

The ResNet50 classifier model showed a very high misclassification rate of 59.52%, which suggests that it either struggles with class separability or has an architecture inappropriate for the complexity of this segmentation task.

The balanced depth and feature reuse of DenseNet121 made it ideal for this task. Its dense connectivity helped it retain finer details and gradients. Also, it is shallower than DenseNet169, making it less prone to overfitting on EMDS-6. The slight performance drop in DenseNet169 proves that deeper isn't always better, as it can overfit or lose fine-grained details if the dataset is small.

MobileNetV2 has been designed for speed and low-resource environments and retains good precision and accuracy due to inverted residual and bottleneck layers. On the other hand, Xception may have taken the longest time for evaluation due to its complex architecture and depth. It is a strong feature extractor but computationally heavy.

InceptionV3 offers a moderate performance, showing that even though it is a good general model, it may not be suited for finer biological structures.

ResNet50 is not explicitly designed to capture subtle texture and shape differences which are critical in microbial images, which is why it failed to live up to the task and may not be suitable for this domain.

Table 6. Performance Comparison of the Different Classification Models

Model	Average Recall(%)	Avg. Precision(%)	Avg. F1-Score(%)	Accuracy(%)	Avg. Misclassification Rate(%)	Evaluation Time (s)
Xception	68.10	68.90	64.49	68.10	31.90	3900
Inception-V3	65.71	68.63	63.84	65.71	34.28	2100
De DenseNet121	**71.67**	**73.44**	**68.66**	**71.67**	**28.33**	**2700**
MobileNetV2	66.43	68.70	65.15	66.43	33.57	600
ResNet50	40.48	40.47	37.73	40.48	59.52	2152
DenseNet169	70.71	71.06	67.57	70.71	29.28	1665

Figure 10. Bar Graph Representing the Performance of the Different Classification Models

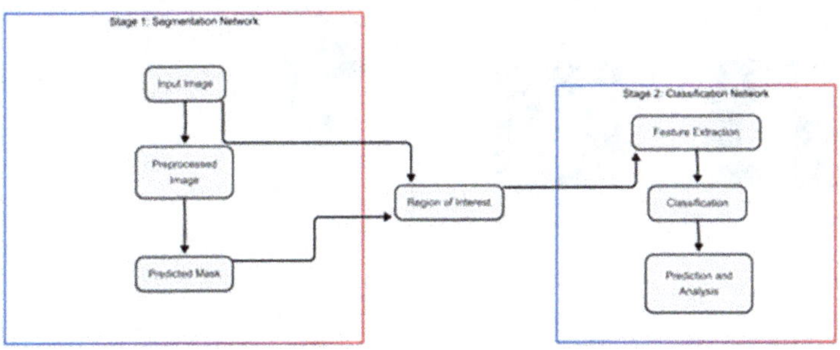

Figure 11. Confusion matrix comparison of different networks on test set, Xception, MobileNetV2, InceptionV3, DenseNet121, ResNet50, and DenseNet169. The labels 1, 2, 3, 4, 5, 6, 7, 8, 9, 10, 11, 12, 13, 14, 15, 16, 17, 18, 19, 20, 21 for true and predicted classes represent Actinophrys, Arcella, Aspidisca, Codosiga, Colpoda, Epistylis, Euglypha, Paramecium, Rotifera, Vorticella, Noctiluca, Ceratium, Stentor, Siprostomum, K. Quadrala, Euglena, Gymnodinium, Gonyaulax, Phacus, Stylonychia, and Synchaeta, respectively.

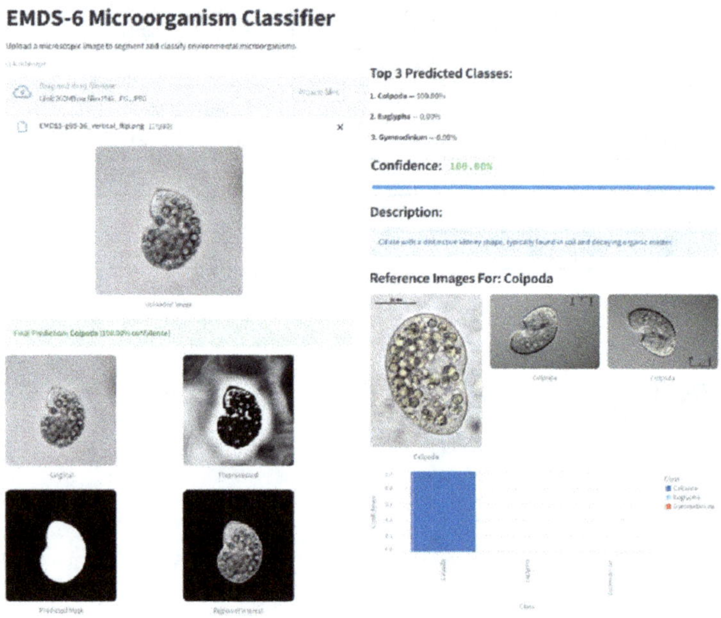

Figure 12. Bar Graph Representing the Misclassification Rate of the Different Classification Models. DenseNet121 shows the lowest, while ResNet50 shows a very high Misclassification Rate of 59.52%.

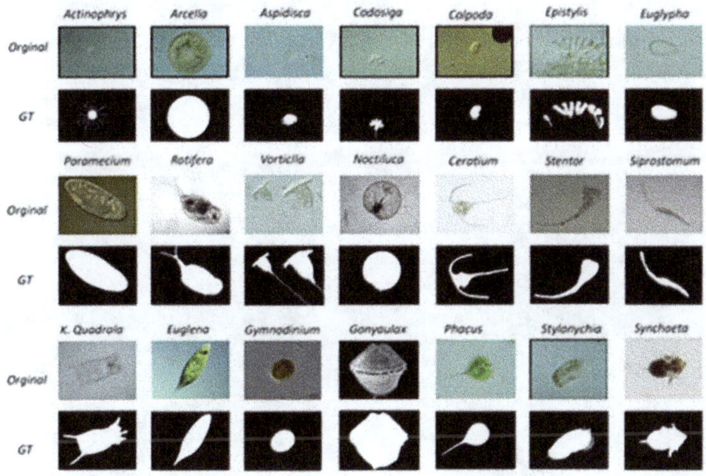

Figure 13. Bar Graph Representing Evaluation Time of the Different Classification Models. Xception takes the longest time to evaluate.

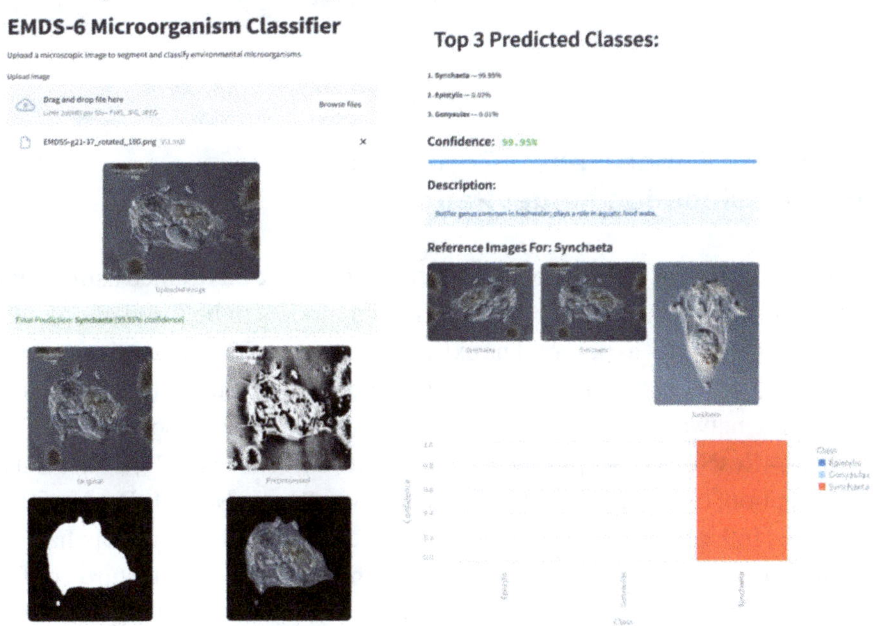

Figure 14. Graph Representing Accuracy vs Evaluation Time of the Different Classification Models. MobileNetV2 and DenseNet169 offer a favourable trade-off between accuracy and evaluation time.

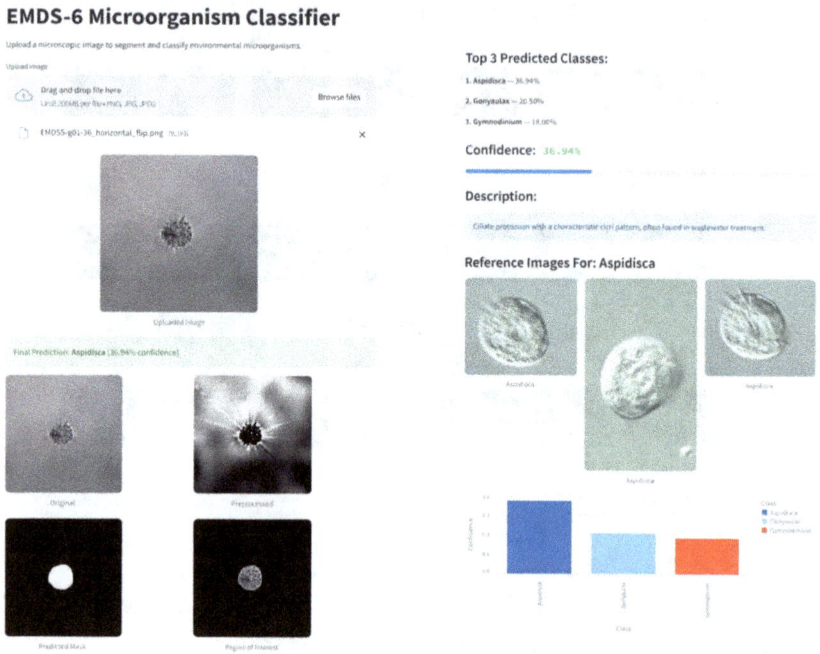

6.3. Extended Experiments: Comparison of Fusion Classification Model Pairs

1) *Experimental Setup:* The fusion classification network was trained using features extracted from the segmented ROIs. The two models to be fused are parallelly trained on the dataset as before. The extracted features of the models are separately stored, and then concatenated, resulting in a new layer called the concatenation layer in the network. This was performed on all pairwise combinations of the 6 models— ResNet50, InceptionV3, MobileNetV2, DenseNet121, DenseNet169, and Xception, except the 3 pairs- ResNet50 with DenseNet121, ResNet50 with DenseNet169, and DenseNet121 with DenseNet169. These models had overlapping layer names and direct feature fusion attempts led to feature conflicts. As attempting to resolve these conflicts by cloning or renaming often led to disruption of pretrained weights, these specific combinations were excluded

from the study. The hyperparameters used for fused classification training were kept the same as the previous settings.

2) *Performance Results:* The performance of the twelve fusion classifiers was evaluated using standard classification metrics: average recall, precision, F1 score, accuracy, and misclassification rate. A comparative summary of these results is provided in *Table 6*. Among the evaluated combinations, the fusion of MobileNetV2 with DenseNet169 achieved the best overall performance. However, it is worth noting that several other model combinations also produced comparable results.

Table 7. Performance Comparison of the Different Feature Fusion Classification Models

Model	**Average Recall(%)**	**Avg. Precision (%)**	**Avg. F1-Score (%)**	**Accuracy (%)**	**Avg. Misclassification Rate(%)**
Xception + Densenet121	68.57	70.17	65.28	68.57	31.42
Xception + Densenet169	67.86	71.68	64.69	67.86	32.14
Xception + InceptionV3	67.86	71.34	65.65	67.86	32.14
Xception + MobilenetV2	70.00	72.78	67.56	70.00	30.00
Xception + ResNet50	68.10	70.04	65.04	68.10	31.90
Densenet121 + MobilenetV2	69.29	70.85	67.36	69.29	30.71
Densenet121 + InceptionV3	64.76	69.18	61.46	64.76	35.24
MobilenetV2 + InceptionV3	67.38	68.62	65.48	67.38	32.62
MobilenetV2 + Densenet169	71.19	72.06	69.75	71.19	28.81
MobilenetV2 + Resnet50	67.86	70.16	66.40	67.86	32.14
InceptionV3 + Densenet169	61.90	65.64	58.28	61.90	38.10
InceptionV3 + Resnet50	58.33	58.08	54.07	58.33	41.67

Figure 15. The Models Combinations and their Evaluation Metrics in a Bar Graph Representation

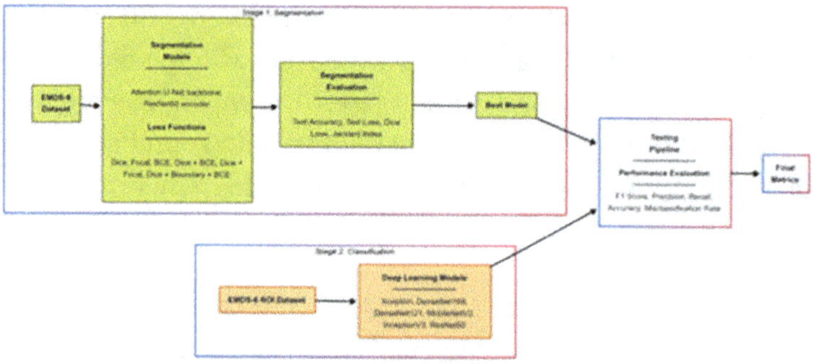

To assess the theoretical performance upper bound of fusion models, the analysis method proposed by (Zhao et al., 2022) was followed. This method estimates the ideal classification performance of a fusion model pair based on the individual performance of each model on the test dataset, as detailed in Table V. Specifically, for each class, the maximum number of correct predictions achieved by either of the two models is identified. These maxima are summed across all classes and then divided by the total number of test images to obtain the ideal accuracy.

The formal mathematical formulation for computing the ideal accuracy is as follows:

$$\text{Ideal accuracy} = \sum_{i=1}^{C} \max(c_1^{(A)}, c_i^{(B)} / N$$

where:

C is the number of classes

$c_1^{(A)}$ and $c_i^{(B)}$ are the number of correct predictions for class i by models A and B respectively,

N is the total number of test images

Using this, the ideal improvement due to fusion can be calculated by subtracting the maximum accuracy between the two models from the ideal accuracy value.

Let:

- IA be the Ideal Accuracy (in),
- Acc_A be the accuracy of model A (in %),
- Acc_B be the accuracy of model B (in %).

Then, the Improvement in Accuracy (&) is given by:

Improvement (%) = IA – max ($Acc_{A'}$, Acc_B)

These results were tabulated along with the difference between ideal and real fusion improvement percentages in *Table 7.*

Table 8. Comparison of Ideal and Real Accuracy Improvements after Model Fusion

Models Fused	Ideal Accuracy (%)	Improvement in Accuracy (%)	Real Fusion Accuracy (%)	Actual improvement in Accuracy after performing fusion (%)
Xception + Densenet121	72.38	0.71	68.57	-3.10
Xception + Resnet50	73.10	5.00	68.10	0.00
Xception + MobilenetV2	72.62	4.52	70.00	+1.90
Xception + InceptionV3	70.48	2.38	67.86	-0.24
Xception + Densenet169	74.52	3.81	67.86	-2.85
Densenet121 + MobilenetV2	74.52	2.85	69.29	-2.38
Densenet121 + InceptionV3	73.10	1.43	64.76	-7.00
MobilenetV2 + InceptionV3	70.95	4.52	67.38	+0.95
MobilenetV2 + Densenet169	75.48	4.77	71.19	+0.48
MobilenetV2 + Resnet50	67.86	1.43	67.86	+1.43
InceptionV3 + Densenet169	75.24	4.53	61.90	-8.81
InceptionV3 + Resnet50	70.48	4.77	58.33	-7.38

The visual representation of the differences in ideal and real accuracies of the different fusion models, and their respective improvement percentages, are given in the diagrams below.

Figure 16. Ideal vs Real Accuracy Values of the Respective Model Combinations

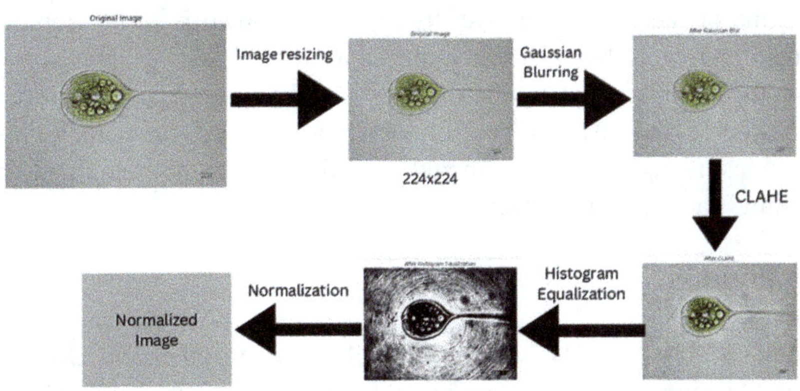

Figure 17. Ideal vs Real Improvement in Accuracy Values of the Respective Model Combinations

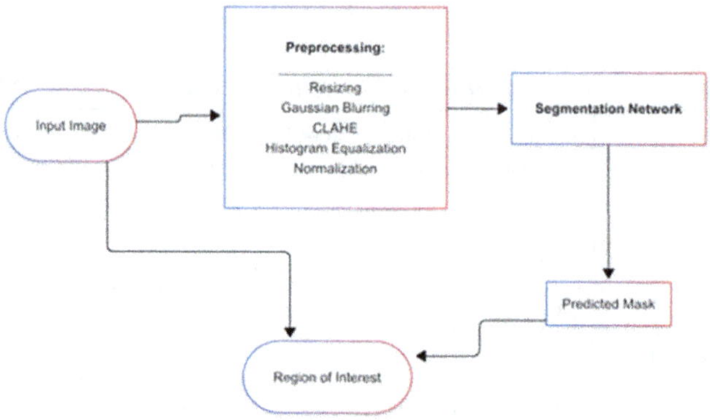

From the above results, it is clear that the fusion of features of two models ideally improves the accuracy in classification. However, there was considerable difference in the expected values compared to the values obtained from the training and testing process. The fusion of DenseNet121 and MobileNetV2 gave the best precision (79. 19%) in the EMDS-6 testing data set, as highlighted in Table VII, making it the most reliable fusion combination in this dataset. However, the highest improvement on fusion was shown by Xception and MobileNetV2 having an improvement score of 1.90%, which means these two models had the maximum complementarity.

Improved accuracy after fusion is a likely indicator of the complementarity between the two models, i.e., which two models work together the best and have the least amount of feature overlap, where one model can compensate for the other model's weakness (Zhao et al., 2022).

From Table 9, although ideal accuracy estimates are high for several fused models, the actual improvement in accuracy after fusion is often very low, or even negative. One of the major negative discrepancies shown is in the combination InceptionV3 and DenseNet169 which shows a drop of -8.81%. This shows that this fusion methodology may not be the most helpful one. The combinations with such high drop of rates may suffer from incompatibility or redundancy in feature representations owing to the base models initialized with the pretrained weights from the same dataset— ImageNet.

6.4. Challenges

1. Limited Dataset Size and Diversity: Only 21 classes are present in EMDS-6, and the system cannot generalize microbes outside this fixed dataset. The dataset includes very similar looking microbes, making classification difficult. The

one-microbe-per-image constraint limits real world usability where multiple microbes may be present in a single water sample.

2. Class Imbalance: Class 1 (Actinophrys) was not correctly classified even once across all models and fusion strategies. This was likely due to lack of distinctive visual features or high similarity with other class microbes. This is a major drawback as it skews overall performance and addressing this issue can improve the performance of the models significantly.
3. Model Accuracy: The accuracy of classifiers is still below 90%, and so it is not yet sufficient for deployment.
4. Computational Constraints: Only six classifier models, and a few loss functions could be tried and tested. Better performing architectures may still exist. Memory and GPU constraints prevented more exhaustive training and testing.
5. Fusion Inefficiency: The results of the proposed fusion techniques did not reach close to the theoretical ideal improvement in fusion accuracies. Models with shared layer names could not be fused together easily either.

7. SAMPLE OUTPUT

Figure 18. Demonstration of the testing pipeline for the two-stage model with one microbial image

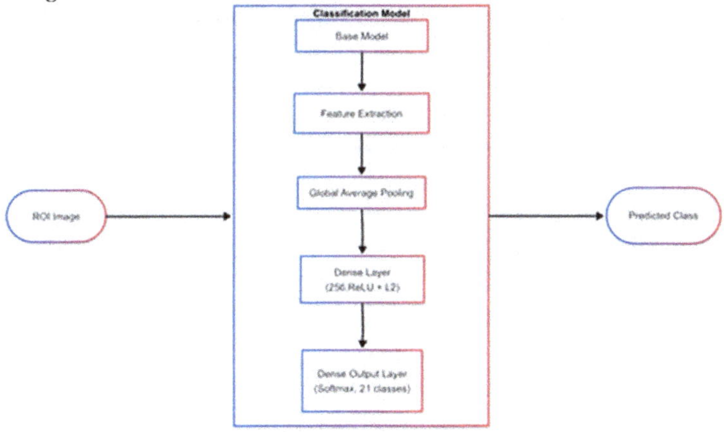

The testing pipeline starts by taking an input image which first goes to the Segmentation Network. In this module, the image is preprocessed using the preprocessing functions: Gaussian Blur, Histogram Equalization and Contrast Limited Adaptive Histogram Equalization. This preprocessed image is sent to the segmentation module as input and the segmentation model predicts the Segmented Mask for the test image. The mask is used to extract the Region of Interest (ROI) for the input image. This gets sent to the Classification Network where the ROI is tested by the classification model, and the probabilities of the class predictions are calculated.

By processing and analyzing these, a detailed analysis report is displayed, which includes the best three class predictions, visualization of the confidence scores for the three classes, some reference images of the final predicted class, and some basic information regarding the microbe type. Figures 19-21 demonstrate the testing pipeline on one image. The output shows the original image, the predicted mask by the segmentation network, the region of interest, and the classification prediction by DenseNet121, as shown below.

Figure 19. The output shown here gives the correct results.

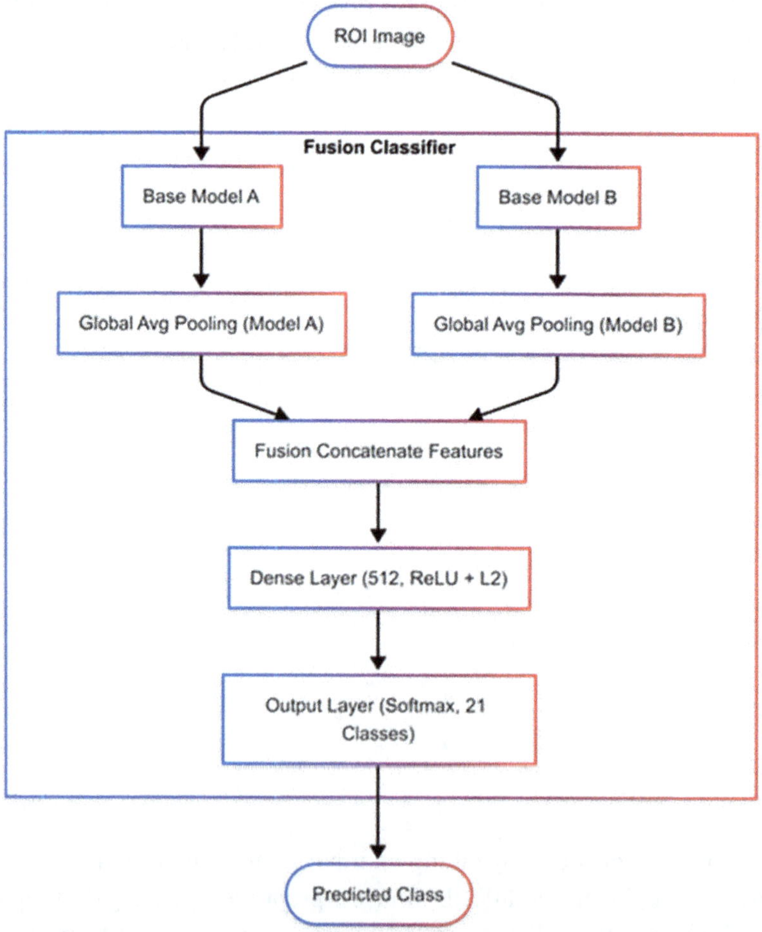

Figure 20. The output shown here gives the correct results.

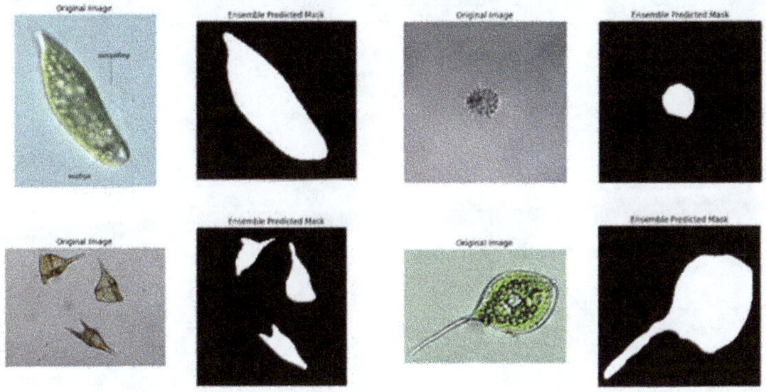

Figure 21. The output shown here gives wrong results.

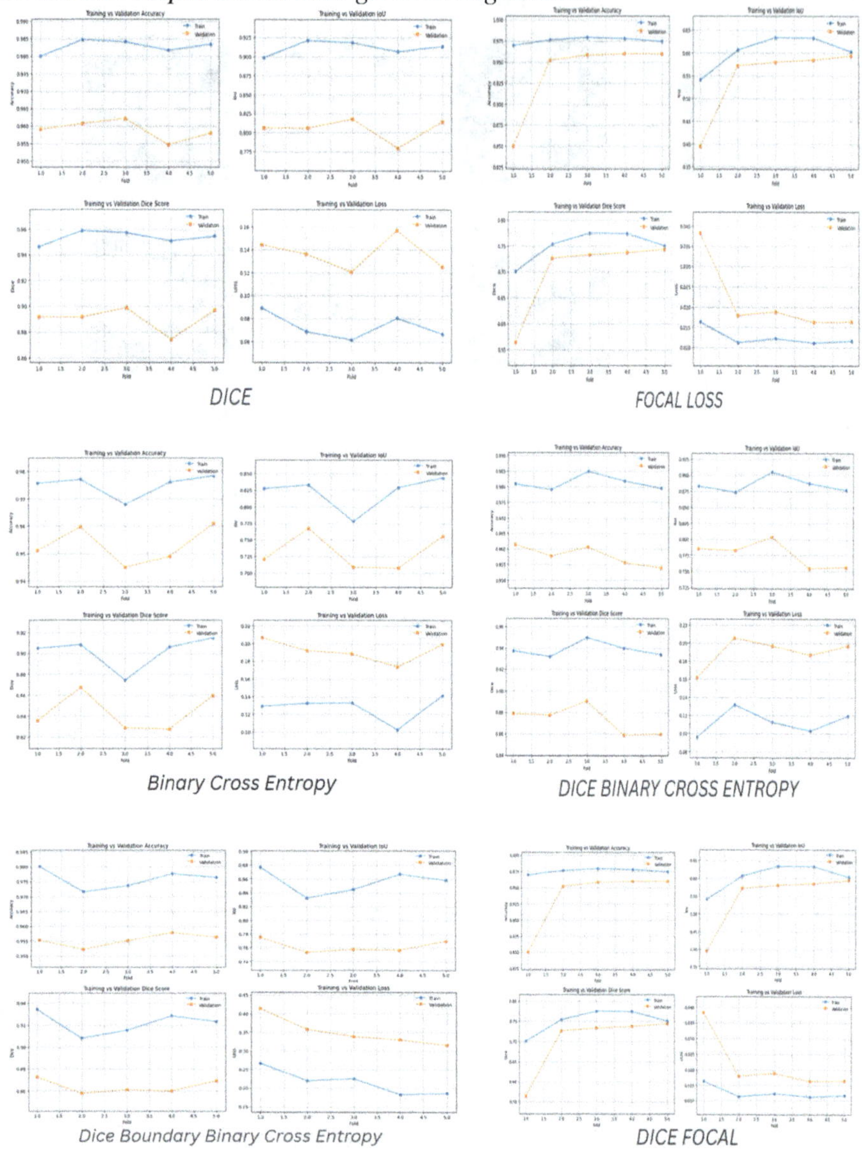

8. CONCLUSION

In a practical application, using deep learning for environmental microbial classification is a scalable approach and provides an efficient way to monitor water quality in food production systems. The process begins by capturing microscopic images of organisms present in the water by taking water samples. These images are then analyzed using a trained deep learning model, which classifies the microbes present based on visual features. Inferences can be drawn based on the microbes detected and their abundance, for example, a high concentration of Euglena in the sample may indicate elevated nitrate or phosphate levels. Species like *Gonyaulax* and *Noctiluca* are responsible for harmful algal blooms, which release toxins into water used in aquaculture. Automated detection can help predict and mitigate bloom events, protecting seafood quality and aquatic ecosystems.

Such insights can be used to prevent using contaminated water in food production early and help signal when water treatment may be required. This can help avoid outbreaks by preventing contaminated food from entering the food chain. Moreover, this method is cost-effective and does not require constant manual intervention, making it suitable for continuous monitoring across multiple sites. On integrating the system with a database, new EMs and their behaviours can be recorded and tracked.

Based on the comparative performance of different classification models, DenseNet121 emerged as the best-performing classifier, achieving the highest average recall (71.67%), precision (73.44%), F1-score (68.66%), and accuracy (71.67%), while also maintaining the lowest misclassification rate (28.33%). For segmentation, the Attention U-Net with a ResNet50 encoder, trained with Dice Loss and Binary Cross-Entropy (BCE) loss functions, produced the most accurate segmented regions of interest (ROI), with a Jaccard Index of 0.78. These accurate segmentations were subsequently used as inputs for the classification network. The optimal end-to-end pipeline for segmentation and classification was achieved by combining the Attention U-Net (ResNet50 encoder) with Dice + BCE loss for segmentation, followed by the DenseNet121 classifier. This combination resulted in the highest classification accuracy for microbial images in this study.

In the cases where evaluation time is a critical factor (e.g., in an industrial setting or real-time applications), MobileNetV2 offers a substantial advantage, as it is the fastest among the rest, with an evaluation time of 600 seconds on the testing dataset. There is still the possibility of improving model performance through hyperparameter tuning. For instance, adjusting parameters like learning rate, batch size, or layer configuration could push models like Xception or DenseNet121 to even higher accuracy levels while managing time better.

This study also explored feature fusion techniques combining pre-trained convolutional neural networks to enhance image classification performance. Fusion models such as Mo- bileNetV2 + DenseNet169 demonstrated significant improvements by utilizing the complementary strengths of lightweight and deep feature extractors. This model outperforms others across the most important performance indicators, making it the most reliable fusion combination in this dataset, with an accuracy of 71.19%.

It highlighted that model fusion, while often effective, does not universally enhance outcomes. Although ideal accuracy estimates are high for several fused models, the actual improvement in accuracy after performing fusion is often negative. Major negative discrepancies were found across several fusion combinations, such as InceptionV3 + Densenet169, which noticed an 8.81% drop in the accuracy. The combinations with such high drop rates most probably suffered from incompatibility or redundancy in feature representations. MobileNetV2 was consistently present in all positively improving combinations, highlighting its strength as a highly efficient fusion partner even though other combinations did not show any improvements. This system has significant potential in tackling real-world challenges such as food contamination and water quality monitoring.

While the models trained on the current dataset do not yet achieve the level of accuracy required for real-world application, their performance suggests strong potential. With access to a significantly larger and more diverse dataset, these models could be further refined to deliver highly accurate results, making them suitable for reliable microbial identification and effective water quality monitoring.

One of the main limitations of this system is the size of the dataset *EMDS-6*. It only contains *21 classes*, and the system predicts a class among these 21 even when a test image beyond these classes is given. The *EMDS-6 dataset* is specially curated for image processing techniques, so it contains very similar looking microbes with complex morphological details which are smaller in size and are harder to detect (e.g., rotifers and protists). This system can only detect one class per image, and cannot detect multiple microbes of different classes yet. It is not that convenient to take microscopic images of different EMs in order to detect and classify them. It is important that in quality assessment of samples, multiple microbes are detected at once. Integrating the system with a database is essential for that.

The classification accuracy is not yet close to 90%. Only 6 variations of classifiers and loss functions have been compared when better combinations may exist. Due to memory and GPU constraints, the system relies on pretrained weights, rather than training all models from scratch. For the same reason, k-fold cross validation could be extended only to 5 folds, and not 10, which is the ideal number for smaller datasets. It might be possible that the model best suited for EM classification are not among these six models.

A major operational limitation of this study was the inability of all evaluated models to correctly classify any instances of Class 1 ("Actinophrys"). Despite employing a range of powerful pretrained architectures and fusion strategies, Class 1 remained entirely unrecognized across the entire evaluation. This suggests a deeper issue beyond model choice, likely stemming from an inherent lack of distinguishable or class-specific features in the available data for this class. It is possible that the features of Class 1 are highly similar to those of other classes, making it difficult for even advanced models to learn meaningful patterns, especially since the dataset used was not large enough. As a result, this class was persistently misclassified, which negatively impacted the overall performance metrics and fairness of the classification system. Future work should explore targeted approaches such as class-specific augmentation or feature engineering to address this limitation. Since circular shapes are the most common feature among the EMs, there is a strong likelihood that the model is unable to distinguish circular microbes. Hence, further feature extraction like color correction, texture features for similar shapes can be performed. This would lead to significant improvement in the overall accuracy of the system.

The feature-level fusion for the twelve model combinations were not that satisfactory, demonstrating the need for a different fusion approach, especially for datasets as small and limited as the EMDS-6. The current approach to feature fusion could not be extended to models with overlapping layer names, such as DenseNet121, DenseNet169, and ResNet50. Moreover, the observed improvement from feature fusion did not come too close to the upper limit of ideal improvement. However, there is a lot of scope for improvement for this study, and the prospects are quite promising.

In the future, the aim is to conduct broader combination studies involving diverse classifier architectures, such as vision transformers. The hyperparameters were fixed based on standard choices for segmentation or classification. However, these hyperparameters can be fine-tuned to suit the current dataset. To reduce the bias of the microbe shapes, especially the circular shapes, class grouping and hierarchical classification can be applied to fine tune sub-classification. The models are quite possibly getting distracted by irrelevant image regions and staining in the microscopic images. Patch-based segmentation and classification is a viable method to counter this. The preprocessing technique used here does not include color normalization, and with better resources, more types of models with different fusion techniques like late fusion, ensemble level detection, and handcrafted feature fusion with the deep features can be applied along with the methods mentioned priorly. Explainable AI is gaining more popularity at current times, and this setup would also benefit from it to perform better error analysis and understand where the models are lacking. By trying out these different strategies and techniques, an even better system can be obtained.

9. REFERENCES

Benbouzid, H., Berrebbah, H., & Berredjem, M. (2014). Induction of oxidative stress in Paramecium aurelia exposed to a novel Phosphoramidate. *Fresenius Environmental Bulletin.*

Huang, S.-C., Pareek, A., Seyyedi, S., Banerjee, I., & Lungren, M. P. (2020). *Fusion of medical imaging and electronic health records using deep learning: a systematic review and implementation guidelines. npj Digital Medicine, 3:136.* URL., DOI: 10.1038/s41746-020-00341-z

Kulwa, F., Li, C., Zhang, J., Shirahama, K., Kosov, S., Zhao, X., & Grzegorzek, M. (2022). A new pairwise deep learning feature for environmental microorganism image analysis. *Environmental Science and Pollution Research International, 29*(34), 51909–51926.

Li, C., Shirahama, K., & Grzegorzek, M. (2014). Application of content-based image analysis to environmental microorganism classification. In *Proceedings of a Conference.*

Li, C., Shirahama, K., & Grzegorzek, M. (2015). Environmental microbiology aided by content-based image analysis. *Pattern Analysis & Applications, 19*(2), 531–547.

Li, X., Huang, X., & Grzegorzek, M. (2022). A state-of-the-art survey of u-net in microscopic image analysis: From simple usage to structure mortification. arXiv, 2202.06465,. URL https://arxiv.org/abs/2202. 06465.

Ma, P., Li, C., Rahaman, M. M., Yao, Y., Zhang, J., Zou, S., & Grzegorzek, M. (2023). A state-of-the-art survey of object detection techniques in microorganism image analysis: From classical methods to deep learning approaches. *Artificial Intelligence Review, 56*(2), 1627–1698.

Makmud, S. T., Haryanto, A. N., & Prasetyo, S. Y. Improving rotten fruit classification accuracy through fusion of multiple pre-trained cnn models. Engineering and Applied Science Research, 51(3):313–320, 2024. doi: . 2024.30. URL https://www.tci-thaijo.org/index.php/easr/indexDOI: 10.14456/easr

Nawghare, P., & Prasad, J. (2025). Hybrid CNN and random forest model with late fusion for detection of autism spectrum disorder in Toddlers. *MethodsX, 14*, 103278.

Nillmani, N. S., Saba, L., Khanna, N. N., Kalra, M. K., Fouda, M. M., & Suri, J. S. (2022). Segmentation-based classification deep learning model embedded with explainable ai for covid-19 detection in chest x-ray scans. *Diagnostics (Basel), 12*(9), 2132. https://www.mdpi.com/journal/diagnostics. DOI: 10.3390/diagnostics12092132 PMID: 36140533

Pizer, S. M., Amburn, E. P., Austin, J. D., Cromartie, R., Geselowitz, A., Greer, T., & Zuiderveld, K. (1987). Adaptive histogram equalization and its variations. *Computer Vision Graphics and Image Processing, 39*(3), 355–368.

Rani, P., Kotwal, S., Manhas, J., Sharma, V., & Sharma, S. (2021). Machine learning and deep learning based computational approaches in automatic microorganisms image recognition: Methodologies, challenges, and developments. *Archives of Computational Methods in Engineering, 29*, 1801–1837. DOI: 10.1007/s11831-021-09639-x PMID: 34483651

Ruby, U., & Yendapalli, V. (2020). Binary cross entropy with deep learning technique for image classification. *International Journal of Advanced Trends in Computer Science and Engineering, 9*(10).

Wu, H., Liu, Q., & Liu, X. (2019). A review on deep learning approaches to image classification and object segmentation. *Computers, Materials & Continua, 60*(2), 575–597.

Zhao, P., Li, C., Rahaman, M. M., Xu, H., Ma, P., Yang, H., Sun, H., Jiang, T., Xu, N., & Grzegorzek, M. (2022a). Emds-6: Environmental microorganism image dataset sixth version for image denoising, segmentation, feature extraction, classification, and detection method evaluation. *Frontiers in Microbiology, 13*. Advance online publication. DOI: 10.3389/fmicb.2022.829027 PMID: 35547119

Zhao, P., Li, C., Rahaman, M. M., Xu, H., Yang, H., Sun, H., Jiang, T., & Grzegorzek, M. (2022b). A comparative study of deep learning classification methods on a small environmental microorganism image dataset (emds-6): From convolutional neural networks to visual transformers. *Frontiers in Microbiology, 13*. Advance online publication. DOI: 10.3389/fmicb.2022.792166 PMID: 35308350

Zou, Y., Li, C., Shirahama, K., Jiang, T., & Grzegorzek, M. (2017). Content-based image retrieval of environmental microorganisms using double-stage optimisation-based fusion. *Information Engineering Express, 3*(4), 43–53.

Chapter 9
Molecular Diagnostics and Genomic Tools for Food Contaminant Detection

Odangowei Inetiminebi Ogidi
https://orcid.org/0000-0002-9961-893X
Bayelsa Medical University, Yenagoa, Nigeria

Ngozi Blessing Okoro
National Biotechnology Research and Development Agency, Nigeria

Afe Victor Dania
https://orcid.org/0000-0002-0055-7234
Bayelsa Medical University, Yenagoa, Nigeria

ABSTRACT

Molecular diagnostics and genomic tools have recently been recognised as game-changing resources for contaminant detection in environmental, agricultural, clinical, and industrial settings. These tools use nucleic acid-based detection methods like PCR, qPCR, digital PCR, and next-generation sequencing (NGS) to detect contaminants at minute concentrations, frequently before traditional phenotypic methods can pick them up. Genomic tools such as metagenomics, transcriptomics, and CRISPR-based assays further enable the identification of unculturable organisms and the surveillance of resistance genes, virulence factors, and microbial community dynamics. In industrial contexts, especially in food production, these diagnostics contribute

DOI: 10.4018/979-8-3373-3982-5.ch009

to quality control, contamination prevention, and supply chain integrity. As these technologies become more affordable and user-friendly, their adoption is expected to grow, offering robust solutions for real-time, field-deployable contaminant detection.

INTRODUCTION

The dependability and recognition of culture-based microbiological methods as "gold standards" for microbial enumeration and identification have been upheld for more than a century in the analysis of foods and other materials (Bonnet et al., 2020). The imperative for rapid detection tools for the food industry and public health agencies is highlighted by the protracted duration (hours to days) required to attain conclusive results, potentially resulting in substantial economic losses for food producers due to increased holding costs. An optimal method would swiftly ascertain the level of contamination and the presence of pathogens within a few hours (Ogidi et al., 2025a).

Nonetheless, the rapid detection of hazardous bacteria in food samples remains challenging due to the complicated physical and chemical nature of food matrices, as pathogenic germs typically present in contaminated foods exist at low concentrations and sometimes possess compromised cells that are difficult to cultivate. A new epoch in food microbiology has commenced, propelled by molecular techniques that depend on sequencing microbial genomes or genes (Stasiewicz et al., 2015; Walsh et al., 2017). These molecular approaches identify microbial taxa and functional features such as antibiotic resistance or pathogenicity by detecting specific gene sequences. For the detection, characterisation, and molecular typing of foodborne microbial isolates, polymerase chain reaction (PCR) technology has been the backbone of most nucleic acid amplification tests (NAATs) and sequencing approaches. In the last ten years, next-generation sequencing (NGS) and whole-genome sequencing (WGS) have revolutionised the methods used to identify microbes in food, water, and environmental samples. Molecular epidemiology research has also relied on sequencing-based typing methods like cgMLST (core-genome) and wgMLST (whole-genome multi-locus sequence typing) (Allard et al., 2018; Stevens et al., 2022).

WGS-based methodologies are progressively replacing different electrophoretic mobility techniques used for DNA fingerprinting and typing in public health contexts. These techniques include ribotyping, restriction fragment length polymorphism (RFLP), and pulsed-field gel electrophoresis (PFGE). The predominant method for identifying bacteria in clinical and public health laboratories is culture-independent diagnostic tests (CIDTs), encompassing several nucleic acid amplification tests (NAATs). These assays can identify both foodborne and gastrointestinal pathogens (Stevens et al., 2022). Moreover, other domains of functional genomics, such as microbiome research and microbial community analysis, have been profoundly transformed by next-generation sequencing methodologies. Therefore, this chapter discusses the recent advancements in computational genomics and bioinformatics techniques that have facilitated the rapid detection and characterisation of foodborne pathogens, and the challenges, limitations, and advantages of these diagnostic methods in food contamination detection.

GENOMICS AND MOLECULAR DIAGNOSTIC TOOLS IN CONTAMINATION DETECTION

Genomics and molecular diagnostic tools are revolutionizing the detection of food contamination by enabling rapid, precise identification of pathogens at the genetic level. These tools, including PCR (Polymerase Chain Reaction), DNA microarrays, and NGS, can detect minute quantities of microbial DNA or RNA, making them highly effective for early and accurate contamination screening (Table 1). By analyzing genetic markers, these technologies not only distinguish between harmful and benign microorganisms but also track the sources of outbreaks with exceptional specificity. Their integration into food safety protocols enhances both public health protection and regulatory compliance, fostering a safer food supply chain. These approaches are extensively discussed below.

Table 1. Genomics and molecular diagnostic tools used in food contamination detection

Tool/Technique	Description	Applications in Food Safety	Advantages	References
PCR (Polymerase Chain Reaction)	Amplifies specific DNA sequences to detect the presence of pathogens.	Identifying bacterial, viral, and fungal contaminants.	High sensitivity and specificity.	(Law et al., 2015)
qPCR (Quantitative PCR)	Provides quantitative data on DNA amplification.	Measures microbial load in food samples.	Real-time monitoring and quantification.	(Postollec et al., 2011)
NGS (Next-Generation Sequencing)	Enables whole-genome sequencing and pathogen tracking.	Detects and characterizes known and emerging pathogens.	Broad scope, detailed microbial profiling.	(Forbes et al., 2017)
DNA Microarrays	Allows simultaneous detection of many DNA sequences via hybridization.	Broad-spectrum pathogen screening.	High-throughput, multiplex detection.	(Kim & Blaschek, 2008)

continued on following page

Table 1. Continued

Tool/Technique	Description	Applications in Food Safety	Advantages	References
LAMP (Loop-Mediated Isothermal Amplification)	Amplifies DNA at a constant temperature using a simple setup.	Rapid, on-site detection of contaminants.	Quick, inexpensive, minimal equipment required.	(Notomi et al., 2000)
CRISPR-based Diagnostics	Uses CRISPR-Cas systems to target and identify genetic sequences.	Innovative tool for specific pathogen detection.	High specificity, adaptable to portable formats.	(Gootenberg et al., 2017)
Metagenomics	Analyzes DNA from entire microbial communities in a sample.	Detects diverse or unculturable microbes in complex food matrices.	Culture-independent, comprehensive analysis.	(Quince et al., 2017)
RFLP (Restriction Fragment Length Polymorphism)	Uses restriction enzymes to cut DNA at specific sequences for analysis.	Differentiates between strains of microbial pathogens.	Cost-effective, suitable for strain typing.	(Oliveira et al., 2009)
Whole Genome Sequencing (WGS)	Maps the complete DNA sequence of pathogens for precise identification.	Traceback investigations and outbreak source tracking.	Highly detailed, enabling genetic comparison across samples.	(Allard et al., 2016)
RT-PCR (Reverse Transcription PCR)	Detects RNA viruses by converting RNA to DNA before amplification.	Identification of RNA-based pathogens like norovirus or hepatitis A.	Effective for RNA viruses; highly specific.	(Rodríguez-Lázaro et al., 2007)
Biosensors (DNA-based)	Use immobilized DNA probes to detect pathogen-specific sequences.	Real-time detection of pathogens in food processing environments.	Rapid, often suitable for point-of-care testing.	(Velusamy et al., 2010)
SNP Genotyping	Detects single-nucleotide polymorphisms for microbial identification.	Differentiates between pathogenic strains with genetic variability.	Precision in strain-level identification.	(Taboada et al., 2013)

Traditional PCR Technology and Its Extensions

Numerous food identification techniques involve polymerase chain reaction (PCR) or its derivatives that depend on nucleic acid detection. Many conventional methods based on polymerase chain reaction (PCR) have emerged as a result of modern gene detection technologies. These include a wide range of PCR biosensors, including ddPCR, fluorescent probe technologies, multiplex PCR, RFLPs, and real-time quantitative PCR. You can use these approaches to find out where different kinds of food came from, identify different varieties, and spot adulteration. By amplifying particular DNA bands with primers developed based on genetic sequence variances among other species, these approaches make it easier to detect, analyse, and identify the origins of dietary components (Ji et al., 2023).

There are many benefits to the current state of conventional PCR technology, including its reliability, speed, accuracy, and short processing time. Unfortunately, accurate quantification is not provided by standard PCR, so the results are not guaranteed. A typical detection threshold at the nanogramme level is observed in research that requires large quantities of template DNA, suggesting somewhat limited sensitivity. Traditional polymerase chain reaction (PCR) cannot distinguish between species having identical sequences because it can only amplify specific DNA regions. Combining polymerase chain reaction amplification with restriction endonuclease digestion, PCR-RFLP makes sequence variation genotyping much easier. This approach is very cost-effective, despite its low throughput (Kim et al., 2024a).

With a moderate sensitivity and detection limits typically ranging from picogrammes to nanogrammes, PCR-RFLP is an excellent tool for identifying particular sequence variants. When it comes to tracking changes in RNA expression levels in real time, qRT-PCR is a very sensitive method. Quantitative reverse transcription polymerase chain reaction (qRT-PCR) is an excellent approach for assessing trace materials and identifying rare transcripts, owing to its exceptional sensitivity and detection limit, which can attain the femtogramme level. It can quickly and accurately tell the difference between dairy products made from various animals. Choose the most suitable PCR algorithm according to the given conditions (Kim at al., 2024a).

Figure 1. Traditional PCR technologies

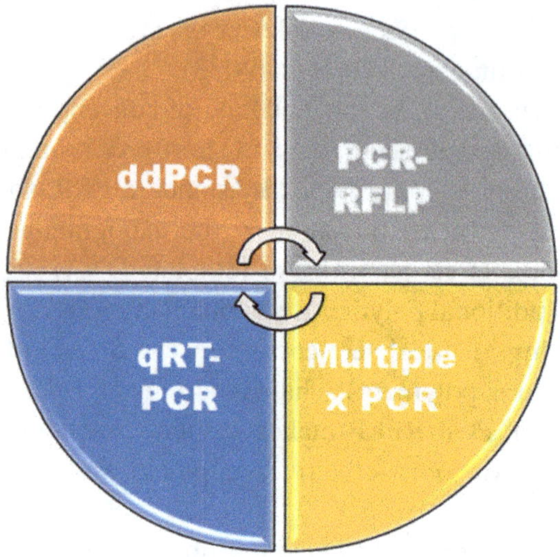

PCR-RFLP

Quick, easy, and inexpensive, PCR-RFLP is an excellent method for differentiating microbial strains and detecting genetic polymorphisms, making it highly suitable for routine screening and epidemiological studies in food safety and microbial diagnostics. Quantitative analysis is infeasible; however, PCR amplification, enzymatic digestion, electrophoresis, and imaging can ascertain the origin of the milk. Multiplex PCR is associated with several advantages, including cost-effectiveness, high throughput, and efficiency. To identify dairy products with multiple sources, it can simultaneously enhance several targets. For instance, it possesses a detection limit of 0.1–0.5% and is capable of identifying animal origin in both commercially produced Italian cheese and laboratory-manufactured cheese (Li at al., 2023).

The possibility of false positive results is a disadvantage. The sensitivity, specificity, and reliability of real-time PCR are exceptionally high. It is proficient in both quantitative and qualitative analyses. The TaqMan-based method offers the capability to set numerous detection targets with high specificity, and the SYBR Green-based approach is quick and cost-effective. TaqMan probes possess a detection limit of 0.001 ng/μL and can be employed to identify DNA from various species in milk. Despite its poor specificity, the SYBR Green approach relies heavily on melting curve analysis. Digital PCR has a detection limit as low as 0.001%, is sensitive at the single-molecule level, and can enable absolute quantification. It is also immune to inhibitors, PCR amplification efficiency, and Ct values, unlike traditional PCR methods (Kim et al., 2024b).

Multiplex PCR

An improvement over traditional PCR, multiplex PCR uses a reaction system with several primer pairs to amplify different DNA fragments at the same time. Detecting food adulteration and identifying harmful microbes are the principal applications of this technology in the food industry. In comparison to other detection methods, this approach has the potential to accurately evaluate the zoonotic risk of materials with a single PCR. Animal and human health can also be assessed in this way (Roig et al., 2022).

The inclusion of the pre-enrichment stage enhances sensitivity to levels comparable to real-time PCR, eliminating the necessity for expensive materials and equipment. Nonetheless, the detection duration is somewhat protracted (6-8 hours for pre-enrichment in addition to PCR time), and the procedure is comparatively complex due to the method's multiple stages, including pre-enrichment and PCR. Potentially limiting its detection efficiency and generalisability are the demanding operational requirements of experiments (Roig et al., 2022). Given that multiplex PCR can identify many kinds of meat, seafood, and processed meals all at once, it is quite useful for detecting meat adulteration. While multiplex PCR is a reliable and effective tool for detecting food fraud, it does have a few drawbacks, such as the need for extremely pure DNA samples and the possibility of interference from different species (Feng et al., 2023).

qRT-PCR

One way to measure DNA levels is with qRT-PCR, which stands for real-time fluorescence quantitative PCR. By adding fluorescent groups to the PCR system, the buildup of fluorescent signals allows for real-time monitoring of the entire PCR process (Xu et al., 2024). When compared to traditional PCR, qRT-PCR offers better sensitivity, specificity, and detection efficiency for quantitative detection. Researchers found that qRT-PCR was quite sensitive when studying gene expression in lemon balm. Subtle alterations in PAL gene expression brought on by jasmonic acid and gibberellin therapy may be detected. The methodology could handle the demands of high-throughput research, and accurate and reliable quantification of gene expression levels was possible. It was based, nonetheless, on reference genes (Ji et al., 2024).

Various algorithms propose optimal pairings of reference genes in diverse manners. Discrepancies in results may arise from inadequate selection. The experimental cost is considerable due to the requirement for specific equipment and chemicals.

Personnel involved in experiments must possess a high level of competence due to the complexity of data analysis. Conventional polymerase chain reaction (PCR) is comprehensible and easy to master. Gene amplification and other prevalent applications are cost-effective and necessitate only fundamental reagents and apparatus. While it offers a satisfactory approximation of gene expression state, it cannot accurately quantify gene expression. Due to its inadequate sensitivity, data from genes exhibiting low expression levels may be overlooked. Moreover, it is susceptible to contamination, which may influence the assessment of subsequent tests (Bharati et al., 2023). This approach has established detection standards for food adulteration and is currently employed in that domain. Real-time PCR is utilised for detecting adulteration in diverse food products, such as seafood, dairy, and meat (Suh et al 2023).

ddPCR

Improved sensitivity and accuracy compared to qRT-PCR have resulted from recent developments in direct quantitative PCR (ddPCR), which allows for direct quantitative measurement independent of reference materials. For studies involving genes with low abundance, allele-specific differential expression, or small relative expression fluctuations, ddPCR is the method of choice. Using ddPCR, allergens derived from soybeans can be detected in meat, fatty creams, milk and wheat (He et al., 2023; Huang et al., 2024). As a result, it can be used to differentiate between different species by highlighting their unique characteristics. The amplification target in an experiment meant to detect pig components was the recombinant pUC57 plasmid, which carried gene fragments peculiar to pigs. Then, qPCR and ddPCR were used for detection after gradient dilution. With an R2 value of 0.9971 for absolute quantification and 0.9998 for linearity assessment, both approaches demonstrated a high level of precision. But their limits of detection were different. While qPCR had a detection limit of five copies per response, ddPCR had a limit of one. qPCR was able to identify pig DNA in many samples, but ddPCR showed improved sensitivity at low levels. The precision and utility of qPCR, a device that sees extensive use, are worthy of praise (Nuraeni et al., 2023).

Next-Generation Sequencing (NGS)

The next-generation sequencing technique, or NGS, has been widely used in genomic research. In order to deduce DNA sequences ranging from hundreds of thousands to millions, next-generation sequencing (NGS) is significantly more efficient than conventional Sanger sequencing. Thanks to several next-generation sequencing (NGS) techniques (Figure 2), microbiologists may now investigate microbial communities in great detail. 16S rRNA gene amplicon sequencing is used to study the conserved areas of bacteria and archaea (Diaz et al., 2019). Gather samples of foods that have undergone fermentation, make note of pertinent information, and then pre-process in order to separate microbial cells for this methodology. Collect genomic DNA from microbes and evaluate its amount and quality. Apply Illumina's high-throughput sequencing technology to PCR-amplify the 16S rRNA gene's V4 hypervariable region, and then sequence the library paired-end. This will produce an extensive amount of sequencing data (Diaz et al., 2019).

Lastly, evaluate the samples using bioinformatics to find out what sorts of bacteria are there and how diverse they are. This approach is great for fast evaluations of microbial diversity and community structure because it is simple to use and doesn't cost a fortune. On the other hand, it can be influenced by priming bias and has low resolution. Whole-genome sequencing (WGS) enables precise identification of bacteria at the strain level by including the entire genome without bias. In the examination of pharmacological resistance, alimentary contamination, and pathogen tracking, it serves as a crucial instrument (Baert et al., 2021).

Nonetheless, its advanced data analysis methodologies and elevated expense render it inappropriate for extensive applications. Gene panel sequencing achieves an optimal balance between cost and sequencing depth by utilising customised probes to enrich specific gene regions. Its versatility is limited and reliant on pre-existing sequence data; however, it is effective for analysing already identified functional genes. Without the need for PCR amplification, shotgun metagenomics may randomly sequence the DNA of all microbes in a sample, providing a thorough picture of the make-up and roles played by microbial communities. Finding new kinds of bacteria is a breeze with this method. However, because of its expensive price and large data volume, it necessitates more processing power. There are clear benefits and drawbacks to each of these technology options (Diaz et al., 2019).

An essential method in next-generation sequencing is pyrosequencing. Watching the release of pyrophosphate determines the base sequence, and the read time is extensive. Using fluorescently labelled nucleotides in sequencing by synthesis allows for low-cost and high-throughput. Using a semiconductor chip to detect changes in hydrogen ion concentration, semiconductor sequencing is characterised by low cost and high speed. Because of DNA nanoballs and combinatorial probe-anchor manufacturing processes, nanoball sequencing provides extended read lengths and high throughput. Sequencing by ligation is a widely used approach in specialised industries. It relies on the ligase reaction to determine a sequence. Genomic sequencing has become the gold standard for oregano identification. Ascertain the species characteristics of Oregano products and possibly polluted plant materials by means of DNA extraction, library construction, and sequencing. Next, the 5,432 entries in the database of plant species can be compared with the sequences (Flügge et al., 2023).

It was discovered in the study that oregano samples contained sweet marjoram, olive leaves, myrtle leaves, and other often polluted plants. Among the Oregano samples evaluated for exotic plant adulteration, 27% had olive leaf, 4% contained sweet marjoram, and 1% contained myrtle leaf. Even though next-generation sequencing (NGS) can detect certain plant components, it can't pick up on non-aromatic endogenous compounds or alien substances lacking DNA. The proportion of DNA readings to the total plant mass has no direct correlation. Nuclear magnetic resonance (NMR) and other complementary analytical techniques are essential. The NGS will be used to select and identify the items to be examined before the NMR database is calibrated. The following stage is to use nuclear magnetic resonance (NMR) analysis on the oregano samples to find out the kind, where it came from, and whether or not there are any other plants in there. By combining the two, you can tell the difference between real and fake oregano with better accuracy. Sequencing is carried out using the Illumina MiSeq platform. When trying to trace the history of honey, this technique usually yields satisfactory results. However, it has issues when working with crystallised honey or honey that is rich in polyphenols, suggesting that there are cases where the precision of species identification is compromised (Prosser et al., 2017).

The trnL-UAA gene sequence was also enhanced in one investigation utilising honey-isolated DNA. Using Ion Torrent technology, the library was produced and sequenced. More than 150,000 reference sequences were used to filter and evaluate the data in a bioinformatics workflow. According to the results, honey has a plant origin and contains several kinds of DNA, although in different proportions. All samples effectively amplified the trnL-UAA area, yielding 52,519 filtered values in total. The identification of 254 distinct plant species was achieved with an average of 1,467 readings per sample (Utzeri et al., 2018). Using NGS technology to detect food fraud is a huge step forward. Linden honey had a maximum of 16,948 reads. The ability to screen DNA from numerous species at once allows for the detection of components in complex processed foods that come from more than one source. Circularisation is a method that can be used to make complex genomic analysis more accurate. The issue of error rates persists despite the fact that technologies developed by Oxford Nanopore, such as the portable MinION, can deliver extremely long read durations. To meet a variety of detection needs, these technologies provide adaptable solutions for food authenticity testing. Access to reliable databases is key for implementing next-generation sequencing (NGS) (Haynes et al., 2019).

Public databases such as GenBank and SRA contain over 8 petabytes of data and are rich resources for species identification. Conversely, more than 240,000 sequences pertaining to genetically modified organisms are housed in specialised databases such as JRC GMO-Amplicons. Creating real-time databases that are specific to commodities or regions can help enhance detection accuracy and reduce the impact of environmental variables. When it comes to checking the legitimacy of food, next-generation sequencing (NGS) technology is a reliable tool. It has surpassed the shortcomings of previous methods in detecting processed or mixed meats, demonstrating exceptional sensitivity and accuracy in distinguishing various varieties of meat (Utzeri et al., 2018).

Figure 2. The main types of Next-Generation Sequencing (NGS) techniques

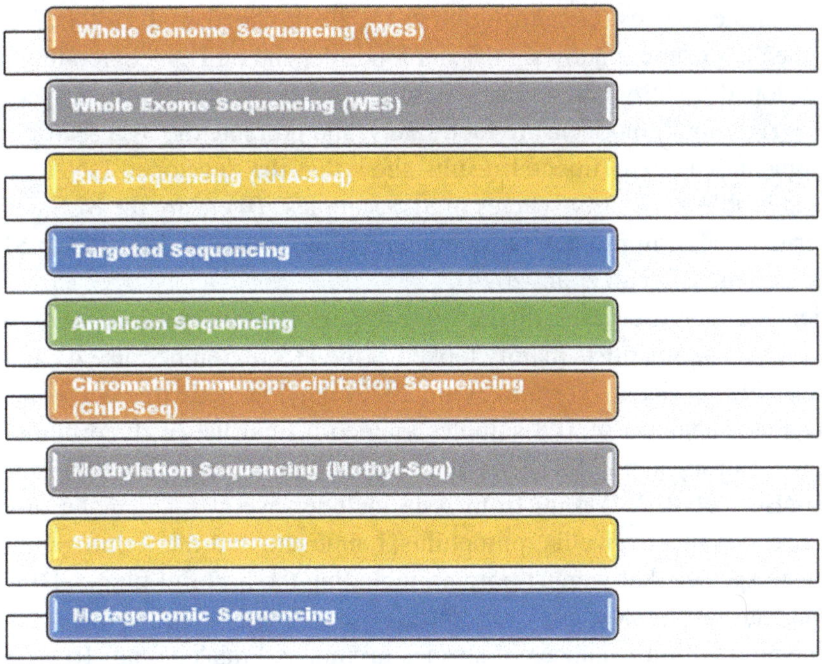

DNA Barcoding

DNA barcoding is a technique for species identification that makes use of bioinformatics and PCR. Using universal primers and newly found mitochondrial DNA sequences, it can detect previously undiscovered species in food samples. The Food and Drug Administration (FDA) authorised the use of U.S. Genomic barcoding in 2011 to prevent seafood label fraud. This system employs standardised gene segments to identify species. Some items marked "sturgeon caviar" turned out to be Mississippi pallid sturgeon, according to the study. The Chinese Academy of Sciences is one of several groups working on cloud computing platforms, testing kits, and meat product DNA barcoding identification. Using conserved sequences in some regions of mitochondrial DNA, universal primers can be designed to produce effective polymerase chain reaction (PCR) amplification (Qayoom et al., 2018).

Certain species can be identified due to the diversity found in certain regions. Mitochondrial DNA is reliable for species identification due to its fixed genetic characteristics caused by its lack of recombination and high maternal inheritance rate. Mitochondrial DNA segments like the COI gene are ideal for PCR amplification and sequencing because they are precisely the right length for primer construction. DNA barcoding has shown to be a highly effective method for distinguishing between various marine species (Giagkazoglou et al., 2024; Currò et al., 2024). Numerous economically or functionally significant foods depend on DNA barcoding. Studies has developed a DNA barcoding system to positively identify SLS species; this will help fight food fraud, ensure food safety, and promote the SLS sector's healthy development. The experimental results show that the developed DNA barcoding approach is able to identify species in SLS samples. To ensure the SLS company's long-term success, improve food safety, and avoid food fraud related to SLS, this method is critical (Zhao et al., 2024).

Scientists identified nine distinct dry porcini (Boletus sp.) items by utilising DNA barcoding. Scientists mainly looked at the ITS region because it can identify all kinds of fungi (Rivas-Ferreiro et al., 2023). Phylogenetic tree construction and ITS sequence analysis of 108 samples yielded a final list of three items, one of which was European porcini (B. edulis). Along with the other conserved samples, B. was also included. Porcini from Asia include several species of Borrelia, including reticulatus, boswellia, pinophilus, bainiugan, and meiweiniuganjun. Then there were species that weren't porcini, including the harmful fungus Hypomyces chlorinigenus and the less common Imleria badia. The research found that ITS barcoding can help find instances of species mixing and mislabeling (Rivas-Ferreiro et al., 2023). It follows that various cuisines probably make use of distinct sets of genes made possible by barcodes. The mitochondrial COI gene fragment is a typical animal food barcode. This 650-base pair gene fragment is essential for the species identification of animal-based foods. Chloroplast gene regions matK and rbcL are common locations for DNA barcodes in plant-based foods. Microbes, including fungi, often utilise the ITS region of ribosomal DNA as a barcode for genetic information (Rivas-Ferreiro et al., 2023).

High-Resolution Melting

A novel approach to genotyping, SNP identification, and gene mutation detection is High-Resolution Melting (HRM) analysis. The overarching concept is to make use of the melting points of DNA sequences, which are affected by base composition and length variations, which in turn affect the thermal stability of double-stranded

DNA and how it melts in response to changes in its sequence (Ogidi & Izah, 2024). Essential to HRM technology is the use of certain fluorescent dyes incorporated into double-stranded DNA. A decrease in fluorescence intensity and a slow unravelling of the DNA occur as a result of increasing temperature. By keeping an eye on changes in fluorescence intensity as a function of temperature, one can construct a melting curve. Unique melting curve geometries for DNA fragments of different sequences are produced by variations in melting temperature (Tm) and changes in fluorescence signals during unwinding. Genotypes can be more easily distinguished in this way. In order to help with species identification, thermal denaturation creates melting curves that are specific to each nucleic acid molecule. These curves are a result of variations in fragment lengths, GC concentration, and GC distribution (Zeng et al., 2024).

By collecting specimens of Mercenaria campechiensis, Mercenaria mercenaria, and their hybrids in both controlled and natural habitats, one study used HRM technology to identify species. Afterwards, HRM analysis and PCR amplification were used to generate derivative melting curves and melting temperature calculations. In order to confirm the results from the wild samples, we used sequencing, in addition to linear discriminant analysis and cross-validation. Ultimately, every sample achieved a predicted classification accuracy of 100% and the overall cross-validation success rate was 100% (Ogidi & Felagha, 2024). It may be more challenging to classify species due to intraspecific genetic differences, since many different species have the same thermal denaturation characteristics in specific gene segments. Pieces were chosen based on their stability, relevance to research goals, sequence differences, and other relevant factors. There had to be a lot of sequence diversity across species for the target segments to work. An example of this is the fact that melting curves and Tm values vary significantly between Mercenaria mercenaria and Mercenaria campechiensis due to their different GC/AT ratios. Since the CO1, 16S rRNA, and ITS1 gene segments contain numerous single-nucleotide polymorphisms, variations in these regions can be utilised to distinguish between different species. To mitigate the effect of intraspecific differences, the target fragments' relative stability is critical. Even though the research items were usually effective, accuracy was nonetheless affected by intraspecific SNPs. As a result, increasing the identification accuracy often involves combining various components. Because of its rapidity, low cost, and high throughput, HRM is a practical method for detecting different forms of food adulteration, regardless of location. The results of this investigation proved that HRM-SNP testing can distinguish between DNA samples collected from leaves and musts when applied to a large number of grape types. Grape varietals including Barbera, Dolcetto, and Arneis have been investigated by researchers in northwest Italy using TaqMan and HRM genotyping technologies (Barrias et al., 2024).

The flexible HRM post-PCR method may identify variants such as SNPs, INDELs, and basic sequence repetitions (SSRs), among others. It successfully identifies many SNPs or INDELs in the same DNA region, in contrast to TaqMan. When trying to identify wines, SSR markers could encounter problems with incorrect PCR amplification due to DNA degradation and inhibitors such polyphenols. The biallelic structure, low mutation rate, and compatibility with low-quality DNA make single nucleotide polymorphisms (SNPs) ideal markers due to their increased stability and precision. Genotyping commercial and experimental wines, detecting single nucleotide polymorphisms (SNPs), and extracting, quantifying, and experimentally vinifying DNA have all been addressed by TaqMan, which has emerged as the gold standard in a number of investigations. The use of HRM analysis has shown to be a useful tool in the fight against fungi and bacteria. Investigating the potential of SSR markers in HRM studies has attracted more attention (Barrias et al., 2024). An investigation of 12 SSR markers was conducted using DNA extracted from various grape varietals. The results showed that some of the new markers were more effective than the older ones in amplifying DNA in must and wine. After that, HRM analysis proved that it was successful in distinguishing between DNA samples. While issues with varietal differences in wine DNA data persist, new methods of grape variety identification have emerged thanks to HRM and other markers that have shown potential (Barrias et al., 2024).

Loop-Mediated Isothermal Amplification (LAMP)

The isothermal amplification method known as LAMP is widely utilised in scientific study; PCR stands in opposition to it. On top of being more sensitive and selective than PCR, LAMP is also more resistant to inhibitors of PCR. Since the LAMP method maintains a constant temperature of 60 to 65 oC, heat cycling equipment is not necessary for nucleic acid amplification. It relies on four or six primers to target six or eight regions of the target gene, making it very specific and sensitive. Enzymes that move strands and a DNA polymerase from Bacillus stearothermophilus (Bst) facilitate the process by creating stem-loop structures and multiple copies of the desired sequence. Short amplification times, adaptability to low-resource settings, and the possibility of point-of-care diagnostics are just a few of the many advantages of LAMP. Due to the increased prevalence of food fraud caused by the globalisation of food supply chains, there is a growing demand for rapid on-site detection technologies. Meat, vegetables, and dairy products can all be quickly and easily identified using LAMP (Holz et al., 2024).

In terms of practicality, price, and operation, the newly developed LAMP-PD device is light years ahead of its predecessor. The device's colorimetric detection limit is 0.1% and its fluorescence detection limit is 0.5%; it's also cheap, costing only USD 54.65 (Wang et al., 2024). Findings were comparable to those from commercial sensors, and the technology provides great potential in resource-constrained regions. This device has built-in LAMP, colorimetric, and fluorescence detection capabilities. Separating the reaction products into their own containers is all that's needed to detect them. Unlike LAMP-PD, which integrates LAMP with detection capabilities, heat cyclers and other specialised equipment are required for traditional PCR amplification. The device lessens operational problems and equipment expenses as a result of its straightforward design. Commercial devices present a challenge in resource-constrained situations due to their dependence on specialists and complex transfer procedures. The colorimetric/fluorescence dual-channel detection method can isolate specific DNA sequences, improve the reliability of results, and detect meat adulteration with significant sensitivity. Such a flexible and reactive sensing device would be handy in a resource-constrained environment (Wang et al., 2024).

Multiple low-cost detections were made possible by substituting modified LAMP loop primers, oligonucleotide sequences, and amplification terminator spacer C3 for traditional, expensive antibodies. A microfluidic device was equipped with the LAMP and LFS detection technologies to lessen the likelihood of aerosol contamination. Test results showed that the method was very selective in distinguishing between goat and cow milk components, with adulteration detected in authentic goat milk samples at levels as low as 5% and 1 pg of cow DNA. Plant, dairy, and meat items can all be identified with the help of LAMP (Li et al., 2024; Moorthy et al., 2024; Schäfer et al., 2024). Downsides include unsuitability for short sequences, non-specific amplification, challenging primer design, and inadequate multiplex detection. Solving these challenges has been accomplished by researchers using OmniAmp polymerases and enhanced versions of Bst DNA polymerases, such as Bst 2.0/3.0. More adaptable are newly created primers such as SLP, loop, and stem (Yang et al., 2024).

Bubble-Mediated SEA (Strand Exchange Amplification)

The isothermal amplification of RNA and DNA is the primary focus of SEA. With SEA, primers may be made more efficiently, and primer dimer formation can be avoided, in comparison to LAMP. By streamlining primer design and doing away with the need for a thermal cycler, this technique can reduce dimer formation without sacrificing temperature stability. Currently, the detection of beef adulteration is being carried out using a visual isothermal approach (SEA). By utilising mitochondrial sequencing of ducks, this method identifies tainted duck meat. Typically, a highly sensitive detection device is needed to identify even trace amounts of contaminants in food. As PCR has the potential to amplify particular DNA sequences, it could be feasible to locate components present in extremely low quantities of adulteration, which bodes well for the detection of food adulteration. The LAMP reaction procedure is relatively complicated due to the employment of four to six primers (Yan et al., 2020).

New methods for detecting DNA and food-borne illnesses have been developed using SEA technology. If the DNA release is inadequate or if the matrix inhibits the direct detection, the time it takes to get a positive result could be longer in complicated meat matrices (Chen et al., 2022). Electrochemical aptamer sensors with nanoparticles added increase sensitivity and SEA specificity. Fast, difficult sample identification is just one of its many exceptional properties; it also boasts a wide linear range (Nodoushan et al., 2022). In addition, it provides strong evidence for food safety regulations (Liu et al., 2019; Zhang et al., 2024) due to its robust performance in sensitivity, specificity, and detection time.

CRISP–CAS System

Modern biotechnology relies on gene editing technology, which has made great strides. Recent breakthroughs in genetic engineering have occurred in tandem with those in biological research, encompassing both novel approaches like the CRISPR-Cas system and more conventional ones like TALENs and ZFNs. Precision genome editing is made possible by the unique action mechanism of CRISPR-Cas9/12 technology, which can revolutionise the food industry and is at the centre of cutting-edge research and applications. The use of CRISPR-Cas12a to verify the origin of meat from various animals has been reported. By combining microdroplet technology with machine learning, it can detect the IL-2 gene in ducks and the ND1 gene in pigs with high sensitivity and accuracy. This method has great promise for food detection because it can detect DNA, which is typically destroyed in processed meats (Zhao et al., 2024).

These methods provide sensitive and precise detection of food adulteration and genetically modified food products through the use of CRISPR-based tests. Methods like this might use or not use nucleic acid amplification. Both bacteria and fungi use the CRISPR-Cas system. For example, in Saccharomyces cerevisiae, a red phenotype is produced by disrupting the ADE2 gene using the CRISPR-Cas9 system. The door is now open for the screening of effectively altered strains. Staphylococcus aureus and Escherichia coli are two examples of bacteria that can have their genes intentionally deleted using the CRISPR-Cas9 system. In Escherichia coli, most especially the lacZ gene. To identify effectively modified strains, the activity of β-galactosidase is measured in the company of S. The agr gene is the target of gene editing in Staphylococcus aureus, and the activity of plasma coagulase is used to detect the strains that have been successfully edited. Differentiation of various microbes is made possible by these methods (Wang et al., 2024).

A variety of food adulteration methods can now be investigated using portable technologies. Using the Cas12a and Cas13a systems, researchers have developed a plethora of detection strategies that make use of portable technology. Fast and accurate visual identification of genome-edited pigs using fluorescence imaging is now possible thanks to the rapid visualisation CRISPR (RAVI-CRISPR) technology. Investigating multiple gene alterations and co-infections in meat products at the same time is the goal of the multi-CRISPR approach (Wang et al., 2023). Thanks to developments in on-site food authenticity detection, the food supply chain can be monitored in real-time, which has led to a decrease in food adulteration issues. Improving portable detection systems that can recognise multiple targets is an encouraging development in food product analysis (Wang et al., 2023).

Other Techniques

Several state-of-the-art methodologies are utilised in genomics research, including biochip and biosensor technologies, fourth-generation nanopore sequencing, third-generation sequencing (TGS), and a plethora of others. Nanopore sequencing technologies are becoming increasingly important in food analysis. Shellfish are among the marine organisms studied using nanopore sequencing (Shum et al., 2024). Sequencing fish and shark samples from all around the world is now possible thanks to this technology, opening up new possibilities for tracking seafood supplies and wildlife trade. Additionally, commercial surimi products have been identified using nanopore sequencing's metabarcoding capabilities (Ogidi et al., 2025b; Detcharoen et al., 2024). This has highlighted cases of species substitution and adulteration and shown that these items include a variety of fish. When it comes to studying food, biosensor technology is an indispensable tool. We can successfully identify carbapenem-resistant Escherichia coli, a danger to both food and water safety, using glycocan-coated magnetic nanoparticles and parallel plasmonic biosensors (Detcharoen et al., 2024). Table 2 shows the comparison of molecular tools in food safety diagnostics.

Table 2. Comparison of molecular tools in food safety diagnostics

Tool/Technique	Sensitivity	Cost	Turnaround Time	Limitations
PCR	High	Moderate	3–5 hours	Contamination risk; needs prior knowledge of target DNA
qPCR	Very High	High	1.5–3 hours	Expensive equipment; inhibitors in complex samples
LAMP	Moderate to High	Low	30–60 minutes	Limited multiplexing; complex primer design
NGS	Very High (broad)	Very High	1–3 days	High cost; bioinformatics expertise required
CRISPR-based	Very High (specific)	Moderate to High	<1–2 hours	Early development; not suited for unknown targets
DNA Microarrays	Moderate	High	4–6 hours	Lower sensitivity; restricted to known sequences
Metagenomics	Very High	Very High	Several days	Complex analysis; may detect non-viable DNA
RFLP	Moderate	Low	1–2 days	Labor-intensive; outdated; low resolution for similar strains
Whole Genome Sequencing (WGS)	Very High	Very High	2–5 days	Time-consuming; resource intensive
RT-PCR	High (for RNA)	Moderate to High	3–5 hours	Only applicable to RNA viruses; RNA is unstable
Biosensors (DNA-based)	Moderate to High	Low to Moderate	Minutes to 1 hour	May lack sensitivity/selectivity; influenced by environmental factors
SNP Genotyping	High (strain-level)	Moderate to High	1–2 days	Limited to known SNPs; less informative than WGS

CASE STUDIES OF GENOMICS AND MOLECULAR DIAGNOSTIC TOOLS IN FOOD CONTAMINATION DETECTION

Genomics and molecular diagnostic tools have played a critical role in unraveling complex food contamination incidents, allowing for the swift identification of pathogens and the implementation of control measures. For instance, during the 2011 *Listeria monocytogenes* outbreak linked to cantaloupes in the United States, Whole Genome Sequencing (WGS) was employed to identify and track the specific epidemic strain. The genomic data enabled public health authorities to link the infections to a single farm in Colorado, prompting an immediate recall and reducing further health risks (CDC, 2011). Similarly, in the 2009 *Salmonella Typhimurium* outbreak associated with peanut butter, WGS and PCR techniques facilitated source tracing to a Georgia-based processing plant. The incident led to widespread recalls and spurred reforms in food safety regulations (FDA, 2009; CDC, 2009).

Emerging technologies like RT-PCR and CRISPR-based diagnostics have also shown promising outcomes in more recent case studies. In 2013, RT-PCR was effectively used across Europe to detect Norovirus contamination in frozen berry products, enabling swift recalls and preventing further illness (EFSA, 2013). Meanwhile, pilot studies using CRISPR-Cas systems, such as CRISPR-Cas12a, demonstrated potential for rapid and portable detection of *Staphylococcus aureus* in milk, offering a novel approach for monitoring dairy contamination on-site (Gootenberg et al., 2018). These case studies exemplify the evolving landscape of food safety diagnostics, showcasing how molecular and genomic tools are being harnessed to enhance public health responses and ensure food supply integrity.

In 2023, the U.S. Centers for Disease Control and Prevention (CDC) utilized Whole Genome Sequencing (WGS) to investigate a multistate outbreak of *Listeria monocytogenes* infections linked to contaminated deli meats and cheeses. Traditional culture and serotyping methods could not effectively trace the contamination source due to widespread product distribution and strain similarity. WGS enabled public health officials to match clinical isolates from infected patients to environmental samples collected in deli processing plants. This genetic precision facilitated targeted recalls, reduced further spread of infection, and highlighted WGS's role in real-time pathogen traceability during foodborne outbreaks (CDC, 2023).

In another example, researchers in China developed a CRISPR-Cas12a-based biosensor for the rapid detection of *Salmonella enterica* in poultry. The platform used the collateral cleavage activity of Cas12a to trigger fluorescence upon detection of *Salmonella* DNA. It achieved comparable sensitivity to PCR but required no thermal cycling, offering results within 30 minutes. When tested on retail chicken samples, the biosensor proved to be highly accurate and adaptable for field use, especially in resource-limited environments. This case illustrates how CRISPR-based diagnostics can revolutionize on-site microbial detection in food safety monitoring (Li et al., 2022). Various case studies of genomics and molecular diagnostic tools in food contamination detection are shown in Table 3.

Table 3. Case studies of genomics and molecular diagnostic tools in food contamination detection

Case Study	Tool/Technique Used	Contaminant Detected	Outcome	Reference
Listeria outbreak in cantaloupes (USA, 2011)	Whole Genome Sequencing (WGS)	*Listeria monocytogenes*	Identified outbreak strain and linked it to a Colorado farm; enabled rapid intervention.	(CDC, 2011)
Salmonella outbreak in peanut butter (USA, 2009)	WGS + PCR	*Salmonella Typhimurium*	Source traced to processing facility; led to major recalls and reforms.	(FDA, 2009; CDC, 2009)
Norovirus in frozen berries (Europe, 2013)	RT-PCR	Norovirus	Allowed detection and recall of contaminated batches across Europe.	(EFSA, 2013)
E. coli outbreak in leafy greens (Canada, 2018)	qPCR + Metagenomics	*E. coli* O157:H7	Helped trace contamination to the irrigation water source.	(PHAC, 2018)
Microbial profiling in ready-to-eat meats	Metagenomics	Mixed bacterial communities	Assessed hygiene and detected spoilage organisms in the production line.	Zwirzitz et al. (2020)
Rapid detection of *Campylobacter* on poultry farms	LAMP	*Campylobacter jejuni*	Enabled quick field testing in low-resource settings.	Yamazaki et al. (2009)
CRISPR detection in dairy milk (pilot study)	CRISPR-Cas12a	*Staphylococcus aureus*	Demonstrated potential for rapid and portable diagnostics.	Gootenberg et al. (2018)
Vibrio detection in seafood	DNA Microarray	*V. parahaemolyticus*, *V. vulnificus*	Multi-pathogen detection in shellfish enhances seafood safety.	Chen et al. (2011)

MOLECULAR DIAGNOSTIC LIMITATIONS

Compared to more conventional approaches like immunoassays and culture techniques, molecular diagnostics are noticeably faster and more accurate. Rapid diagnostics significantly improve food safety, although they have only partially supplanted conventional procedures in a small number of analyses so far. There is still a lot of resistance to diagnostic tests that use the nucleic acids of the pathogen instead of its phenotypic traits. The increased occurrence of both false positives and negative results is a significant cause. It is common for DNA contamination in the lab, the surrounding environment, and the equipment used to prepare the reaction mixture to cause false positives for pathogens. The removal of DNA from surfaces and laboratory equipment is more difficult than that of living cell contaminants. Inhibitors have the risk of producing false negative results. Cheddar and salami are examples of complex matrices that are not suitable for analysis due to the presence of chemical components that may inhibit enzyme activity. Because enzymes are not always active, a false negative result could happen. As a result, there is a need for ongoing research into nucleic acid extraction protocols for inhibitor elimination in specific dietary mixtures (Ogidi et al., 2023).

Consistent use of positive controls, like the IPC used in the TaqMan PCR, increases confidence in the results of the test. The host organism's genomic DNA frequently hinders the detection technique due to competition for probe and primer annealing. This is especially applicable to ground or processed meat, as well as intricate dietary products. Despite a limited quantity of host cells, this phenomenon remains significant due to the considerable size of vertebrate or plant genomes, which are generally 10^3 to 10^4 times larger than those of bacteria (Ogidi et al., 2025c). This issue can often be mitigated by simplifying DNA by the elimination of genomic repetitive sequences (Lin et al., 2006). Bacterial enrichment through culture is typically necessary due to the prevalence of inhibitors or competitors (Melo et al., 2006).

The application of a stomacher device, which mechanically disintegrates the food matrix in a proliferating broth, is the conventional technique for enrichment. After homogenisation, the sample is filtered and allowed to incubate at the designated growth temperature for either one or two days. An assertive method utilising chemical and enzymatic agents is essential for efficient wall lysis before genomic DNA extraction and amplification in instances of bacteria and fungi with robust cell walls. Moreover, prior knowledge of the target sequence is generally necessary for contamination identification via molecular methods, and sporadic mutations may hinder detection in genomic DNA. To circumvent this difficulty, one may utilise primers that render it exceedingly unlikely for all target sequences to simultaneously vary, apply elongated probes, or ease the hybridisation conditions. An alternative is to employ redundant probes (Taylor et al., 2007).

The DNA-based assays can distinguish between viable and non-viable infections, representing a significant advancement. This has advantages and disadvantages, contingent upon the specific pathogen involved. A false positive result may occur due to the discovery of deceased pathogenic organisms, as their genomic DNA remains in the sample. Due to RNA's inferior durability compared to DNA, molecular back-up is facilitated by RNA-based techniques such as NASBA or RT-PCR in specific scenarios. Microfluidics, a domain of interdisciplinary technologies dedicated to the creation of devices at a sub-millimeter scale, may address some limitations of molecular methodologies. Microfluidics-based "lab-on-chips" facilitate the automation of formerly laborious procedures, including microarray hybridisation, DNA extraction, and polymerase chain reaction (PCR). The likelihood of contamination diminishes when all these reactions are combined and executed sequentially (Jain KK 2007).

LDR promptly succeeds PCR of genomic DNA on a lab-on-a-chip. The method can be completed in under one hour, in contrast to the many hours required for a universal microarray (Hashimoto et al., 2006; Hashimoto et al., 2007; Yi et al., 2006). Microfluidics enhances experimental throughput and sensitivity by utilising minimal volumes and sample quantities. Moreover, there is negligible sample carryover, which reduces the likelihood of contamination. Standardising operations is facilitated by lab-on-chip technology, as it integrates all processes into a single device. Minimising laboratory effort, sample handling, and transportation by miniaturisation and standardisation will significantly facilitate the advancement of portable devices for on-site pathogen testing.

Current Limitations in Molecular and Genomic Tools for Food Safety

Despite their transformative potential, molecular and genomic tools for food contaminant detection face several practical limitations. High cost remains a major barrier, particularly for resource-limited settings and small-scale food producers. Technologies like whole genome sequencing (WGS) and next-generation sequencing (NGS) require significant investment in instruments, reagents, and computational infrastructure. Additionally, many of these tools demand specialized technical expertise in molecular biology and bioinformatics for sample processing, data analysis, and interpretation, limiting their accessibility outside well-equipped laboratories. Another limitation is the risk of false positives or negatives, especially in techniques like PCR and LAMP, where contamination or suboptimal primer design can compromise results. Moreover, inhibitors present in complex food matrices (like fat, protein, or preservatives) can interfere with amplification reactions, reducing test accuracy. Current methods are also often target-specific, meaning they require prior knowledge of the contaminant's genetic sequence, which limits their utility for detecting unknown or emerging pathogens. The limitations of various Genomics and Molecular Diagnostic Tools used in food contamination detection are shown in Table 4.

Table 4. The limitations of various genomics and molecular diagnostic tools used in food contamination detection

Tool/Technique	Limitations	References
PCR (Polymerase Chain Reaction)	Prone to contamination leading to false positives; requires prior knowledge of target DNA sequences.	(Law et al., 2015)
qPCR (Quantitative PCR)	Expensive equipment; inhibitors in food matrices can affect accuracy; still needs known targets.	(Postollec et al., 2011)
NGS (Next-Generation Sequencing)	High cost and data complexity; requires bioinformatics expertise; slower turnaround time.	(Forbes et al., 2017)
DNA Microarrays	Limited to known pathogens; lower sensitivity compared to NGS; complex design and interpretation.	(Kim & Blaschek, 2008)
LAMP (Loop-Mediated Isothermal Amplification)	Can generate non-specific amplification; limited multiplexing capabilities; primer design is complex.	(Notomi et al., 2000)
CRISPR-based Diagnostics	Still in early development for food safety; target-specific—cannot detect unknown pathogens.	(Gootenberg et al., 2017)
Metagenomics	Costly and computationally intensive; interpretation can be challenging; may detect non-viable DNA.	(Quince et al., 2017)
RFLP (Restriction Fragment Length Polymorphism)	Labor-intensive; less discriminatory for closely related strains; outdated compared to WGS.	(Oliveira et al., 2009)
Whole Genome Sequencing (WGS)	Expensive; time-consuming; requires substantial computing resources and skilled personnel.	(Allard et al., 2016)
RT-PCR (Reverse Transcription PCR)	Only detects RNA viruses; RNA is less stable than DNA and degrades easily.	(Rodríguez-Lázaro et al., 2007)
Biosensors (DNA-based)	Often have limited sensitivity and selectivity; environmental factors may interfere with results.	(Velusamy et al., 2010)
SNP Genotyping	Not useful for unknown pathogens; limited scope compared to whole genome approaches.	(Taboada et al., 2013)

CHALLENGES IN APPLYING GENOMIC TECHNOLOGIES TO FOOD SAFETY AND TRACEABILITY

Genomic technologies have revolutionized food identification and safety monitoring, yet significant barriers hinder widespread implementation. A core limitation lies in the high cost of equipment, reagents, maintenance, and skilled personnel, especially for applications in low- and middle-income countries. Tools like next-generation sequencing (NGS) and whole-genome sequencing (WGS) remain cost-prohibitive for many institutions, despite their diagnostic accuracy and versatility (Marklewitz et al., 2025). Additionally, the lack of standardized operating procedures and harmonized global regulations limits consistency and scalability. Differences in DNA extraction methods, amplification protocols, and sequencing workflows can lead to variable outcomes, especially in microbial food safety applications (Park et al., 2021; Talukder et al., 2020). Further challenges include technical complexity and the need for specialized bioinformatics expertise. Effective data analysis and interpretation of complex omics datasets remain limited to well-equipped institutions with computational infrastructure. False positives and inaccurate quantification due to inhibitors in complex food matrices or DNA damage during extraction are also common pitfalls. For example, QIAamp and Easy-DNA kits showed conflicting detection rates for *Campylobacter jejuni* and *Salmonella typhimurium* in fecal samples, illustrating how DNA extraction methods can significantly impact results (Bonomi et al., 2020).

Another major hurdle is regulatory inconsistency. In the U.S., oversight is split between the FDA, USDA, and EPA, leading to fragmented governance. While medical genomics is closely monitored, agricultural applications of gene-editing remain loosely regulated. Many policies are outdated and fail to keep pace with rapid innovations in gene sequencing and genome editing (Ramos et al., 2025). Moreover, privacy and ethical concerns surrounding genetic data are intensifying. With the rise of personalized nutrition and traceability systems using genetic information, unauthorized access or misuse of such data could jeopardize both consumer privacy and industry competitiveness (Bombard et al., 2022; Perga et al., 2023). There is also the challenge of quantifying complex microbial populations in food and environmental samples. Although metagenomics offers broad pathogen detection, its utility in quantitative assessments is currently limited by variations in extraction efficiency, amplification bias, and bioinformatics tools. This hinders its application in routine quality control or risk assessments (Bonomi et al., 2020). Additionally, high-throughput phenotyping technologies while promising often struggle with complex traits and require further refinement for consistency and accuracy.

FUTURE PROSPECTS

Despite current limitations, the future of genomics in food safety and traceability looks promising. Advances in portable, cost-effective sequencing technologies such as nanopore sequencing and handheld devices are democratizing access to real-time diagnostics, especially in resource-limited settings. Case studies using mobile genomics labs in Ethiopia and the UK demonstrated the successful use of metagenomics to assess water quality linked to food contamination, with results generated in just one to two days (Acharya et al., 2020). Similarly, Vis/NIRS tools combined with machine learning and GWAS have shown success in predicting carotenoid levels in cassava, aiding both breeding and food quality assessments (Ikeogu et al., 2019). Future developments will increasingly focus on integrating multi-omics approaches, combining genomics with proteomics, metabolomics, and transcriptomics for a more holistic view of food quality and authenticity (Sadurski et al., 2024). Such integration will enhance diagnostic accuracy and provide deeper insights into food composition, adulteration, and microbial contamination at multiple biological levels.

Innovations in AI and machine learning are also expected to simplify data interpretation and enable faster, more precise identification of pathogens and contaminants. These tools can help automate genome annotation, predict food spoilage risks, and guide preventive measures even in real-time scenarios. Additionally, combining genomics with nanotechnology could produce highly sensitive biosensors capable of detecting trace amounts of adulterants or pathogens quickly and accurately (Ogidi & Akpan, 2023). Spectral analysis tools such as near-infrared spectroscopy (NIRS) are also becoming important in food authentication. When paired with genomic data, these tools can help build models that predict the genetic and chemical makeup of food products more reliably (Sohn et al., 2021; Song et al., 2024). Meanwhile, IoT integration is expected to form the backbone of smart food safety networks, enabling real-time genetic surveillance throughout the supply chain.

One of the most promising directions involves integrating genomics with blockchain-based traceability systems. Projects like IBM Food Trust already use blockchain to log food movements from production to consumption. By linking genetic markers to digital ledgers, manufacturers and consumers can authenticate food origin and detect adulteration with unprecedented accuracy. For example, DNA barcoding of grapes could confirm the authenticity of wine origins (Velmovitsky et al., 2021).

Further reductions in sequencing costs will also facilitate broader adoption. For instance, increasing Illumina and ONT throughput can lower per-sample sequencing costs by 66% and 78%, respectively. Additionally, elastic cloud computing, centralized genomic hubs, and open-source software can help distribute analytical workloads and reduce expenses for low-throughput labs (Marklewitz et al., 2025). Finally, the food industry will benefit from genomic insights in personalized nutrition, allowing for customized dietary recommendations based on consumers' genetic profiles (Okoye et al., 2024; Albini et al., 2023). Genomics will not only improve food safety but also enhance quality assurance, authenticity verification, and nutritional transparency paving the way for a smarter, more secure food system (Kepinski-Pacelik et al., 2023).

CONCLUSION

The implementation of diverse ML-based techniques and subsequent bioinformatics analysis may lead to a significant transformation in food safety monitoring, provided that sequencing-based methods can be rendered more predictive. The advancement of sequencing technologies such as ONT has rendered a point-of-care application focused on food safety a tangible reality. This sequencing method has significantly enhanced read accuracy in recent years because of its expedited and simplified sample processing phases. As a result, it's an excellent choice for use in food monitoring and epidemic detection at the point of care. The ability to do analyses on-site also reduces the time samples must be transported to a sequencing laboratory, which is crucial, and it also reduces the risk of microbial compositional bias that could occur during transportation. Ultimately, the advancement of superior genetic methodologies will provide early interventions and enhanced source monitoring, transforming food safety diagnostics into a proactive discipline.

REFERENCES

Acharya, K., Blackburn, A., Mohammed, J., Haile, A. T., Hiruy, A. M., & Werner, D. (2020). Metagenomic water quality monitoring with a portable laboratory. *Water Research, 184*, 116112.

Albini, S., Palmieri, L., Dubois, A., Bourg, N., Lostal, W., & Richard, I. (2023). Assessment of therapeutic potential of a dual AAV approach for duchenne muscular dystrophy. *International Journal of Molecular Sciences, 24*, 11421.

Allard, M. W., Bell, R., Ferreira, C. M., Gonzalez-Escalona, N., Hoffmann, M., Muruvanda, T., Ottesen, A., Ramachandran, P., Reed, E., & Sharma, S.. (2018). Genomics of foodborne pathogens for microbial food safety. *Current Opinion in Biotechnology, 49*, 224–229.

Allard, M. W., Strain, E., Melka, D., Bunning, K., Musser, S. M., Brown, E. W., & Timme, R. (2016). Practical value of food pathogen traceability through whole-genome sequencing. *Trends in Microbiology, 24*(5), 408–416. DOI: 10.1016/j.tim.2016.02.006

Baert, L., McClure, P., Winkler, A., Karn, J., Bouwknegt, M., & Klijn, A. (2021). Guidance document on the use of whole genome sequencing (WGS) for source tracking from a food industry perspective. *Food Control, 130*, 108148.

Barrias, S., Ibáñez, J., & Martins-Lopes, P. (2024). High resolution melting analysis of microsatellite markers applied to grapevine varietal fingerprinting throughout the wine production chain. *Food Control, 160*, 110368.

Bharati, R., Sen, M. K., Kumar, R., Gupta, A., Ziarovská, J., Fernández-Cusimamani, E., & Leuner, O. (2023). Systematic identification of suitable reference genes for quantitative real-time PCR analysis in Melissa officinalis L. *Plants, 2023*(12), 470.

Bombard, Y., Ginsburg, G. S., Sturm, A. C., Zhou, A. Y., & Lemke, A. A. (2022). Digital health-enabled genomics: Opportunities and challenges. *American Journal of Human Genetics, 109*, 1190–1198.

Bonnet, M., Lagier, J. C., Raoult, D., & Khelaifia, S. (2020). Bacterial culture through selective and nonselective conditions: The evolution of culture media in clinical microbiology. *New Microbes and New Infections, 34*, 100622.

Bonomi, L., Huang, Y., & Ohno-Machado, L. (2020). Privacy challenges and research opportunities for genomic data sharing. *Nature Genetics, 52*, 646–654.

Centers for Disease Control and Prevention. (2009). Outbreak of Salmonella Typhimurium infections linked to peanut butter—United States, 2008–2009. *Morbidity and Mortality Weekly Report, 58*(4), 85–90.

Centers for Disease Control and Prevention. (2011). *Multistate outbreak of listeriosis linked to whole cantaloupes from Jensen Farms, Colorado*. https://www.cdc.gov/listeria/outbreaks/cantaloupes-jensen-farms/index.html

Centers for Disease Control and Prevention. (2023). *Outbreak of Listeria infections linked to deli meat and cheese – United States, 2023*. https://www.cdc.gov/listeria/outbreaks/

Chen, P., Zhou, M., Chen, X., Xiong, S., Su, Y., Zhou, H., Peng, J., & Xiong, Y. (2022). Quantum dot bead-based competitive immunochromatographic assay for enterotoxin aureus A detection in pasteurized milk. *Journal of Dairy Science, 105*, 4938–4945.

Chen, Y., Liu, H., Chen, Q., & Zhang, Y. (2011). Development of a DNA microarray for simultaneous detection of Vibrio parahaemolyticus and Vibrio vulnificus in seafood. *International Journal of Food Microbiology, 149*(2), 177–183. DOI: 10.1016/j.ijfoodmicro.2011.06.014

Currò, S., Babbucci, M., Carletti, P., Fasolato, L., Novelli, E., & Balzan, S. (2024). The globalized fish Industry: Employing DNA-barcoding and NIRS technology to combat counterfeiting and safeguard traditional agro-food products. *Food Control, 158*, 110246.

Detcharoen, M., Khrueakaew, P., Sukkapat, P., Benjakul, S., & Saetang, J. (2024). Metabarcoding for authentication of fish species in surimi-based products by Nanopore sequencing. *Food Bioscience, 61*, 104628.

Diaz, M., Kellingray, L., Akinyemi, N., Adefiranye, O. O., Olaonipekun, A. B., Bayili, G. R., Ibezim, J., du Plessis, A. S., Houngbédji, M., & Kamya, D.. (2019). Comparison of the microbial composition of African fermented foods using amplicon sequencing. *Scientific Reports, 9*, 13863.

European Food Safety Authority. (2013). *Outbreak of hepatitis A virus infection in Italy and other EU countries*. https://www.efsa.europa.eu/en/efsajournal/pub/3550

Feng, J. C., Lan, H. Z., & Pan, D. D. (2023). Triplex-colored nucleic acid lateral flow strip and multiplex polymerase chain reaction coupled method for quantitative identification of beef, pork and chicken. *Journal of Food Composition and Analysis, 123*, 105493.

Flügge, F., Kerkow, T., Kowalski, P., Bornhöft, J., Seemann, E., Creydt, M., Schütze, B., & Günther, U. L. (2023). Qualitative and quantitative food authentication of oregano using NGS and NMR with chemometrics. *Food Control, 145*, 109497.

Food and Drug Administration. (2009). *Peanut Corporation of America recall.* https://www.fda.gov/safety/recalls-market-withdrawals-safety-alerts

Forbes, J. D., Knox, N. C., Ronholm, J., Pagotto, F., & Reimer, A. (2017). Metagenomics: The next culture-independent game changer. *Frontiers in Microbiology, 8*, 1069. DOI: 10.3389/fmicb.2017.01069

Giagkazoglou, Z., Loukovitis, D., Gubili, C., Chatziplis, D., Symeonidis, A., & Imsiridou, A. (2024). Untangling the cephalopod market: Authentication of seafood products in Greece with DNA-barcoding. *Food Control, 163*, 110523.

Gootenberg, J. S., Abudayyeh, O. O., Kellner, M. J., Joung, J., Collins, J. J., & Zhang, F. (2017). Nucleic acid detection with CRISPR-Cas13a/C2c2. *Science, 356*(6336), 438–442. DOI: 10.1126/science.aam9321

Gootenberg, J. S., Abudayyeh, O. O., Lee, J. W., Essletzbichler, P., Dy, A. J., Joung, J., & Collins, J. J. (2018). Nucleic acid detection with CRISPR-Cas13a/C2c2. *Science, 360*(6387), 439–444. DOI: 10.1126/science.aas8836

Hashimoto, M., Barany, F., & Soper, S. A. (2006). Polymerase chain reaction/ligase detection reaction/hybridization assays using flow-through microfluidic devices for the detection of lowabundant DNA point mutations. *Biosensors & Bioelectronics, 21*, 1915–1923.

Hashimoto, M., Barany, F., Xu, F., & Soper, S. A. (2007). Serial processing of biological reactions using flow-through microfluidic devices: Coupled PCR/LDR for the detection of low-abundant DNA point mutations. *Analyst, 132*, 913–921.

Haynes, E., Jimenez, E., Pardo, M. A., & Helyar, S. J. (2019). The future of NGS (Next Generation Sequencing) analysis in testing food authenticity. *Food Control, 101*, 134–143.

He, Y. X., Yan, W., Dong, L. M., Ma, Y., Li, C. C., Xie, Y. B., Liu, N., Xing, Z. J., Xia, W., & Long, L. K.. (2023). An effective droplet digital PCR method for identifying and quantifying meat adulteration in raw and processed food of beef (Bos taurus) and lamb (Ovis aries). *Frontiers in Sustainable Food Systems, 7*, 1180301.

Holz, N., Wax, N., Oest, M., & Fischer, M. (2024). REASSURED test system for food control-preparation of LAMP reaction mixtures for in-field identification of plant and animal species. *Applied Sciences (Basel, Switzerland), 14*, 10946.

Huang, J., Zhai, L. G., Wang, J. Y., Sun, X. T., Wang, B. S., & Wei, Z. H. (2024). An evaluation of the sensitivity and applicability of a droplet digital polymerase chain reaction assay to simultaneously detect pseudomonas aeruginosa and pseudomonas fragi in foods. *Foods*, *13*, 1453.

Ikeogu, U. N., Akdemir, D., Wolfe, M. D., Okeke, U. G., Chinedozi, A., Jannink, J. L., & Egesi, C. N. (2019). Genetic correlation, genome-wide association and genomic prediction of portable NIRS predicted carotenoids in cassava roots. *Front. Plant Sci.*, *10*, 1570.

Jain, K. K. (2007). Applications of nanobiotechnology in clinical diagnostics. *Clinical Chemistry*, *53*, 2002–2009.

Ji, C., He, Y. H., Xing, Y. Y., Liu, W., Xie, Y. X., Ba, H. R., Yang, M., He, X. H., Zheng, W. J., & Lu, X. A. (2023). Development of a PCR-based lateral flow immunoassay for the identification of rainbow trout ingredient in foods. *Food Control*, *154*, 110034.

Ji, C., Petchkongkaew, A., van Ruth, S., Wu, D., & Elliott, C. (2024a). The crucial importance of soy sauce authenticity: Global trade, adulteration risks, and analytical challenges. *Trends in Food Science & Technology*, *152*, 104666.

Ji, L., Du, X., Zhang, H., Ji, F., Nie, Z., & Xia, Y. (2024b). Establishment of a multiplex qPCR method for sensitive detection of proteolytic psychrotrophic bacteria in raw milk. *Food Bioscience*, *62*, 105339.

K˛epi ́nska-Pacelik, J., Biel, W., Natonek-Wi'sniewska, M., & Krzy'scin, P. (2023). Assessment of adulteration in the composition of dog food based on DNA identification by real-time PCR. *Animal Feed Science and Technology*, *298*, 115609.

Kim, K. H., & Kang, T. S. (2024a). Development of advanced PCR-based methods for accurate identification and authentication of commercial shrimp products. *Food Control*, *159*, 110288.

Kim, M. J., Park, S. J., & Park, H. (2024b). Trend in serological and molecular diagnostic methods for Toxoplasma gondii infection. *European Journal of Medical Research*, *29*, 520.

Kim, M. S., & Blaschek, H. P. (2008). Microarray analysis of Clostridium beijerinckii NRRL B-598 gene expression during butanol fermentation. *Applied and Environmental Microbiology*, *74*(22), 7405–7413. DOI: 10.1128/AEM.01006-08

Law, J. W. F., Ab Mutalib, N. S., Chan, K. G., & Lee, L. H. (2015). Rapid methods for the detection of foodborne bacterial pathogens: Principles, applications, advantages and limitations. *Frontiers in Microbiology*, *5*, 770. DOI: 10.3389/fmicb.2014.00770

Li, B. Y., Yu, M. X., Xu, W. P., & Chen, L. (2023). Han, J. Comparison of PCR techniques in adulteration identification of dairy products. *Agriculture*, *13*, 1450.

Li, H., Cui, J., Chen, H., Li, H., Xie, Y., Song, W., & Chen, R. (2024). Rapid colorimetric and fluorescence identification of Pinelliae Rhizoma and adulterate Rhizoma Typhonii Flagelliformis using direct-LAMP assay. *Food Chemistry*, *437*, 137840.

Li, S.-Y., Cheng, Q.-X., Wang, J.-M., Li, X.-Y., Zhang, Z.-L., Gao, S., & Wang, J. (2022). CRISPR-Cas12a-based detection of *Salmonella* in food samples. *Biosensors & Bioelectronics*, *206*, 114117. DOI: 10.1016/j.bios.2022.114117

Lin, B., Wang, Z., Vora, G. J., Thornton, J. A., Schnur, J. M., Thach, D. C., Blaney, K. M., Ligler, A. G., Malanoski, A. P., Santiago, J., Walter, E. A., Agan, B. K., Metzgar, D., Seto, D., Daum, L. T., Kruzelock, R., Rowley, R. K., Hanson, E. H., Tibbetts, C., & Stenger, D. A. (2006). Broad-spectrum respiratory tract pathogen identification using resequencing DNA microarrays. *Genome Research*, *16*, 527–535.

Liu, R., Wang, X., Wang, X., Shi, Y., Shi, C., Wang, W., & Ma, C. (2019). A simple isothermal nucleic acid amplification method for the effective on-site identification for adulteration of pork source in mutton. *Food Control*, *98*, 297–302.

Marklewitz, M., Jaguparov, A., Wilhelm, A., Akande, O. W., Musul, B., Poates, A. L., Afrough, B., Norberg, A., Hull, N. C., & Ehsani, S.. (2025). Genomics costing tool: Considerations for improving cost-efficiencies through cross scenario comparison. *Frontiers in Public Health*, *12*, 1498094.

Melo, S. C., Pungartnik, C., Cascardo, J. C., & Brendel, M. (2006). Rapid and efficient protocol for DNA extraction and molecular identification of the basidiomycete Crinipellis perniciosa. *Genetics and Molecular Research*, *5*, 851–855.

Moorthy, G., Benjakul, S., Sukkapat, P., Senathipathi, D. N., & Saetang, J. (2024). Development and validation of colorimetric LAMP assay for species authentication of economically important Mytildae mussels. *Food Control*, *165*, 110697.

Nodoushan, S. M., Nasirizadeh, N., Sedighian, H., Kachuei, R., Azimzadeh-Taft, M., & Fooladi, A. A. I. (2022). Detection of Staphylococcal Enterotoxin A (SEA) using a sensitive nanomaterial-based electrochemical aptasensor. *Diamond and Related Materials*, *127*, 109042.

Notomi, T., Okayama, H., Masubuchi, H., Yonekawa, T., Watanabe, K., Amino, N., & Hase, T. (2000). Loop-mediated isothermal amplification of DNA. *Nucleic Acids Research*, *28*(12), e63. DOI: 10.1093/nar/28.12.e63

Nuraeni, U., Malau, J., Astuti, R. T., Dewantoro, A., Apriori, D., Lusiana, E. D., & Prasetya, B. (2023). Droplet digital PCR versus real-time PCR for in-house validation of porcine detection and quantification protocol: An artificial recombinant plasmid approach. *PLoS One*, *18*, e0287712.

Ogidi, O. I., & Akpan, U. M. (2023). Impacts of Chemical Use in Agricultural Practices: Perspectives of Soil Microorganism and Vegetation. In *One Health implications of Agrochemicals and their sustainable Alternative*. Springer Nature Singapore Pte Ltd.

Ogidi, O. I., Christopher, E., & Orlu, H. A. (2025b). Environmental DNA and RNA Metabarcoding: Advancing Insights in Cancer Research. In Raghavan, R. (Ed.), *Cancer Exposomics and Environmental Influences on Carcinogenesis* (pp. 279–322). IGI Global Scientific Publishing., DOI: 10.4018/979-8-3373-2165-3.ch007

Ogidi, O. I., & Felagha, I. (2024). Gut Microbiome and Diet Interaction. In *Nutrition Controversies and Advances in Autoimmune Disease. Publisher.* IGI Global.

Ogidi, O. I., & Izah, S. C. (2024). Emerging Technologies in the Global South Food Industry: Prospects And Challenges. In: Food Safety and Quality in Global South. Published by *Springer Nature Singapore Pte Ltd*. https://doi.org/DOI: 10.1007/978-981-97-2428-4_7

Ogidi, O. I., Izah, S. C., & Akpan, U. M. (2023). Food Quality and Agrochemical Use: Integrated Monitoring, Assessment, and Management Policies. In: One Health implications of Agrochemicals and their sustainable Alternative. *Springer Nature Singapore Pte Ltd.*

Ogidi, O. I., Okara, E., & Barwant, M. M. (2025c). Nanocarriers in Plant Genetic Engineering. In Chen, J. (Ed.), *Nanocarriers in Plant Science and Agriculture* (pp. 89–124). IGI Global Scientific Publishing., DOI: 10.4018/979-8-3693-9869-2.ch004

Ogidi, O. I., Wenapere, C. M., & Chukwunonso, O. A. (2025a). Enhancing Food Safety and Quality Control With Computer Vision Systems. In Bora, D., & Bania, R. (Eds.), *Computer Vision Techniques for Agricultural Advancements* (pp. 51–88). IGI Global Scientific Publishing., DOI: 10.4018/979-8-3693-8019-2.ch003

Okoye, C. O., Jiang, H., Nazar, M., Tan, X., & Jiang, J. (2024). Redefining modern food analysis: Significance of omics analytical techniques integration, chemometrics and bioinformatics. TrAC. *Trends in Analytical Chemistry*, *175*, 117706.

Oliveira, C. J. B., Soares, R. D. A., & Andrade, M. A. (2009). Molecular typing of Staphylococcus aureus from nasal colonization in healthcare workers using RFLP of coagulase gene. *Brazilian Journal of Microbiology : Publication of the Brazilian Society for Microbiology, 40*(2), 264–268. DOI: 10.1590/S1517-83822009000200014

Park, S., Lee, D., Kim, Y., Lim, S., Chae, H., & Kim, S. (2021). BioVLAB-Cancer-Pharmacogenomics: Tumor heterogeneity and pharmacogenomics analysis of multi-omics data from tumor on the cloud. *Bioinformatics (Oxford, England), 38*, 275–277.

Perga, S., Biolatti, C., Martini, I., Rossi, F., Benso, A., Acutis, P. L., Bagnato, A., Cognata, D., Caroggio, P., & Peletto, S.. (2023). Application of microsatellites to trace the dairy products back to the farm of origin. *Foods, 12*, 4131.

Postollec, F., Falentin, H., Pavan, S., Combrisson, J., & Sohier, D. (2011). Recent advances in quantitative PCR (qPCR) applications in food microbiology. *Food Microbiology, 28*(5), 848–861. DOI: 10.1016/j.fm.2011.02.008

Prosser, S. W. J., & Hebert, P. D. N. (2017). Rapid identification of the botanical and entomological sources of honey using DNA metabarcoding. *Food Chemistry, 214*, 183–191.

Public Health Agency of Canada. (2018). *E. coli O157 outbreak associated with romaine lettuce*. https://www.canada.ca/en/public-health/services/public-health-notices/2018/outbreak-e-coli-infections-linked-romaine-lettuce.html

Qayoom, U., Pawar, R., Mohite, S., Sawant, M., Vivek, N., Pawar, S., Goswami, M., & Lakra, W. (2018). DNA barcoding of some commonly exploited fishes from the northern Western Ghats, India. *The Indian Journal of Animal Sciences, 88*, 245–250.

Quince, C., Walker, A. W., Simpson, J. T., Loman, N. J., & Segata, N. (2017). Shotgun metagenomics, from sampling to analysis. *Nature Biotechnology, 35*(9), 833–844. DOI: 10.1038/nbt.3935

Ramos, P.D. Almeida, M.S. & Olsson, I.A.S. (2025). Safe and purposeful genome editing under harmonized regulation for responsible use: Views of research experts. New Genet. Soc. 2023, 42, e2237177. Foods, 14, 1116 33 of 33

Rivas-Ferreiro, M., Otero, A., & Morán, P. (2023). It's what's inside that counts: DNA-barcoding of porcini (Boletus sp., Basidiomycota) commercial products reveals product mislabelling. *Food Control, 144*, 109346.

Rodríguez-Lázaro, D., Cook, N., & Hernández, M. (2007). Real-time PCR in food science: Introduction and overview. *Current Issues in Molecular Biology, 9*(1), 1–6.

Roig, A. P., Carmona-Salido, H., Sanjuán, E., & Fouz, B. (2022). Amaro, C. A multiplex PCR for the detection of Vibrio vulnificus hazardous to human and/or animal health from seafood. *International Journal of Food Microbiology*, *377*, 109778.

Sadurski, J., Polak-Berecka, M., Staniszewski, A., & Wa'sko, A. (2024). Step-by-Step metagenomics for food microbiome analysis: A detailed review. *Foods*, *13*, 2216.

Schäfer, L., Völker, E., Eule, M., Ahrens, B., Beyer, K., & Holzhauser, T. (2024). Development and validation of a loop-mediated isothermal amplification (LAMP) method for the rapid, sensitive, and specific detection of hazelnut as an allergen in food matrix. *Journal of Agricultural and Food Chemistry*, *72*, 24093–24100.

Shum, P., Cusa, M., Prasetyo, A., & Mariani, S. (2024). Nanopore sequencing facilitates screening of diversity and provenance of seafood and marine wildlife. *Food Control*, *161*, 110382.

Sohn, S.-I., Pandian, S., Oh, Y. J., Zaukuu, J. L. Z., Kang, H.-J., Ryu, T. H., Cho, W. S., Cho, Y. S., Shin, E. K., & Cho, B. K. (2021). An overview of near infrared spectroscopy and its applications in the detection of genetically modified organisms. *International Journal of Molecular Sciences*, *22*, 9940.

Song, Q., Bai, C., Dong, Y., Chen, M., Wang, S., Hu, J., Qiao, X., Chen, J., Li, S., & Liu, X.. (2024). Highly selective Zn2+ near-infrared fluorescent probe and its application in biological imaging. Spectrochim. Acta Part A Mol. *Biomol. Spectrosc.*, *2024*(322), 124828.

Stasiewicz, M. J., den Bakker, H. C., & Wiedmann, M. (2015). (20150. Genomics tools in microbial food safety. *Current Opinion in Food Science*, *4*, 105–110.

Stevens, EL, Carleton, HA, Beal, J, Tillman, GE, Lindsey, RL, Lauer, AC, & Pightling, A, Jarvis 334 KG, Ottesen A, Ramachandran P, et al. (2022). The Use of Whole-Genome Sequencing by the Federal Interagency Collaboration for Genomics for Food and Feed Safety in the United States. *Journal of Food Protection.*

Suh, S.-M., Kim, E., Kim, M.-J., Yang, S.-M., & Kim, H. Y. (2023). Development of real-time PCR method for rapid and accurate detection of Centipedes (*Scolopendra mutilans*) in food. *Food Science and Biotechnology*, *32*, 979–985.

Taboada, E. N., Ross, S. L., Mutschall, S. K., Mackinnon, J. M., Roberts, M. J., & Buchanan, C. J. (2013). Development and validation of a comparative genomic fingerprinting method for high-resolution genotyping of *Campylobacter jejuni. BMC Microbiology*, *13*(1), 171. DOI: 10.1186/1471-2180-13-171

Talukder, A., Barham, C., Li, X., & Hu, H. (2020). Interpretation of deep learning in genomics and epigenomics. *Briefings in Bioinformatics*, *22*, bbaa177.

Taylor, G. M., Worth, D. R., Palmer, S., Jahans, K., & Hewinson, R. G. (2007). Rapid detection of Mycobacterium bovis DNA in cattle lymph nodes with visible lesions using PCR. *BMC Veterinary Research*, *3*, 12.

Utzeri, V. J., Ribani, A., Schiavo, G., Bertolini, F., Bovo, S., & Fontanesi, L. (2018). Application of next generation semiconductor based sequencing to detect the botanical composition of monofloral, polyfloral and honeydew honey. *Food Control*, *86*, 342–349.

Velmovitsky, P. E., Bublitz, F. M., Fadrique, L. X., & Morita, P. P. (2021). Blockchain applications in health care and public health: Increased transparency. *JMIR Medical Informatics*, *9*, e20713.

Velusamy, V., Arshak, K., Korostynska, O., Oliwa, K., & Adley, C. (2010). An overview of foodborne pathogen detection: In the perspective of biosensors. *Biotechnology Advances*, *28*(2), 232–254. DOI: 10.1016/j.biotechadv.2009.12.004

Walsh, A. M., Crispie, F., Claesson, M. J., & Cotter, P. D. (2017). Translating Omics to Food Microbiology. *Annual Review of Food Science and Technology*, *8*, 113–134.

Wang, N., Zhang, J., Xiao, B., Sun, X., Huang, F., & Chen, A. (2024). Disposable and instrument-free nucleic acid lateral flow cassette for rapid and on-site identification of adulterated goat milk. *Talanta*, *267*, 125205.

Wang, Y., Fu, L., Tao, D., Han, X., Xu, B., Deng, M., Li, S., Zhao, C., Li, X., & Zhao, S. (2023). Development of a naked eye CRISPR-Cas12a and -Cas13a multiplex point-of-care detection of genetically modified swine. *ACS Synthetic Biology*, *12*, 2051–2060.

Xu, Q., Li, J., Zhang, Z., Yang, Q., Zhang, W., Yao, J., Zhang, Y., Zhang, Y., Guo, Z., & Li, C. (2024). Precise determination of reaction conditions for accurate quantification in digital PCR by real-time fluorescence monitoring within microwells. *Biosensors & Bioelectronics*, *244*, 115798.

Yamazaki, W., Taguchi, M., Ishibashi, M., & Inoue, K. (2009). Loop-mediated isothermal amplification for detection of Campylobacter jejuni and Campylobacter coli in food samples. *Journal of Food Protection*, *72*(3), 566–570. DOI: 10.4315/0362-028X-72.3.566

Yan, C., Wang, X., Zhao, X., Wei, M., Shi, C., & Ma, C. (2020). Development of a direct and visual isothermal method for meat adulteration detection in low resource settings. *Food Chemistry*, *319*, 126542.

Yang, N., Zhang, H., Han, X., Liu, Z., & Lu, Y. (2024). Advancements and applications of loop-mediated isothermal amplification technology: A comprehensive overview. *Frontiers in Microbiology, 15*, 1406632.

Yi, C., Zhang, Q., Li, C. W., Yang, J., Zhao, J., & Yang, M. (2006). Optical and electrochemical detection techniques for cell-based microfluidic systems. *Analytical and Bioanalytical Chemistry, 384*, 1259–1268.

Zeng, X. Y., Yee, J. C., & Yang, H. P. (2024). Rapid and accurate identification of Mercenaria mercenaria and Mercenaria campechiensis using high-resolution melting analysis. *Journal of Shellfish Research, 43*, 111–118.

Zhang, X. J., Wang, X. J., Shi, B. H., Li, J. F., Fan, Y. F., Li, Y., Shi, C., & Ma, C. P. (2024). A biphasic accelerated strand exchange amplification strategy for culture-independent and rapid detection of Salmonella enterica in food samples. *Analytical Methods, 16*, 4083–4092.

Zhao, G., Li, L. Y., Shen, X., Zhong, R. M., Zhong, Q. P., & Lei, H. T. (2024). DNA barcoding unveils novel discoveries in authenticating high-value snow lotus seed food products. *Foods, 13*, 2580.

Zhao, Z., Wang, R., Yang, X., Jia, J., Zhang, Q., Ye, S., Man, S., & Ma, L. (2024). Machine learning-assisted, dual-channel CRISPR/Cas12a biosensor-in-microdroplet for amplification-free nucleic acid detection for food authenticity testing. *ACS Nano, 18*, 33505–33519.

Zwirzitz, B., Wetzels, S. U., Dixon, E. D., Stessl, B., Zaiser, A., Rabanser, I., & Selberherr, E. (2020). Culture-independent evaluation of bacterial contamination patterns on pig carcasses at a slaughterhouse. *Scientific Reports, 10*, 1–12. DOI: 10.1038/s41598-020-74842-y

Chapter 10
Biosensors for Rapid and On-Site Detection of Food Contaminants

Vikash Kumar
https://orcid.org/0009-0003-8781-2676
Central Drug Research Institute, Lucknow, India

Ram Prakash Pandey
Chandigarh University, India

Sadaf Rahman
Chandigarh University, India

Gurpreet Singh Panesar
Chandigarh University, India

Noble K Kurian
https://orcid.org/0000-0001-6792-6906
TIBR Biotech Private Limited, India

Sylvester Chibueze Izah
https://orcid.org/0000-0001-5526-006X
Bayelsa Medical University, Nigeria

ABSTRACT

Food safety has become a worldwide concern because of the rising frequency of contamination by pathogens, toxins, pesticides, heavy metals, and other harmful substances. Traditional detection techniques, though precise, tend to involve time-consuming protocols, advanced equipment, and skilled operators, and hence are not appropriate for on-site and rapid applications. In this regard, biosensors have become potent analytical tools with high sensitivity, specificity, and portability for real-time food contaminant detection. Recent progress in nanotechnology, micro-fluidics, and signal amplification methods has improved the performance and utility of biosensors in many food matrices significantly. The present chapter discusses the most recent trends in biosensor technologies regarding their design, sensing mechanism, and real-world implementation for rapid and on-site monitoring of food

DOI: 10.4018/979-8-3373-3982-5.ch010

safety. In addition, their challenges and future directions for commercial deployment are also addressed, focusing on the development of cost-efficient, user-friendly, and multiplexed detection platforms.

1. INTRODUCTION

1.1 Overview of Food Safety and Contamination Issues

Judicial aspects of food safety serve as a basic requirement for public health together with economic security and development sustainability(Fung et al., 2018). Global population increases along with expanding industrial food production have unleashed enormous dangers of food contamination. Food contamination describes the existence of dangerous biological and chemical and physical elements within food which contributes to safety and quality deterioration leading to health problems from illnesses through severe outcomes such as deaths. The World Health Organization reports that foodborne diseases infect around 600 million individuals each year and cause approximately 420,000 related casualties globally(Almaary, 2023). These contaminants arise from multiple origins which include raw materials for production and agricultural practices as well as processing environments and packaging materials and storage and transportation systems(Lebelo et al., 2021). Undercooked meat and dairy products and fresh produce serve as the main sources of outbreaks leading to foodborne illnesses caused by microbial pathogens that consist of *Salmonella* spp., *Listeria monocytogenes, Campylobacter jejuni*, and pathogenic strains of *Escherichia coli.*

Chemical contaminants made up of pesticide residues, heavy metals including lead and cadmium, and industrial pollutants like dioxins together with PCBs put people at risk for long-term health problems that involve cancer development and endocrine disruption and neurotoxicity. Developing countries that have weaker regulatory systems become at higher risk for the fungal toxins known as mycotoxins along with allergens since proper monitoring is limited(Chilaka et al., 2022). Environmentally damaging events have made it imperative to develop accurate and speedy methods for food detection which protects the entire food supply chain.

1.2 Importance of Rapid and On-Site Detection

Food testing utilizes traditional analytical methods based on microbial culture techniques together with Gas Chromatography (GC) and High-Performance Liquid Chromatography (HPLC) and Enzyme-Linked Immunosorbent Assay (ELISA) and Mass Spectrometry (MS) methods which authorities consider as the definitive standards due to their high accuracy and specificity(Sogore et al., 2024). The methods are extended in time duration and need expensive devices together with specialized laboratories equipped with expert staff and demanding sample preparation protocols. The period from sample collection to result availability reaches several hours to days thus making it ill-suited for perishable goods and fast-moving production lines. Rapid on-site detection systems present both scientific responsibilities and operational requirements. Real-time site testing allows for immediate operational choices which decrease health perils and decreases product recall incidents for upholding proper food safety standards (HACCP, ISO 22000, Codex Alimentarius) (Wallace et al., 2018). Resource-limited communities along with rural areas need portable detection solutions as their only option to manage food quality standards effectively. Border and checkpoint screening tools with rapid capabilities prevent the movement of contaminated items during food trade activities thus enhancing biosecurity while maintaining trade efficiency.

1.3 Role of Biosensors in Modern Food Safety Management

For the past few year biosensors have undergone rapid evolution which makes them excel in food analysis because they unite selective enhancement with sensitive and speedy operation and portable capabilities (Zhang et al., 2022). A biosensor functions as an analytical device which pairs biological recognition components that include enzymes and their corresponding antibodies and various other whole cells with physical-transducting elements. The sensor signal emerges in various forms including visual, electrical chemical, piezoelectric and thermal signals based on the sensor type. Biosensors provide food safety management with an economical approach to in-field testing which replaces traditional methods. Due to their capability of testing minimal amounts of contamination without requiring laboratory equipment or processing protocols these sensors prove indispensable for immediate quality assessments. Biosensors based on electrochemistry can identify pesticide contaminants in produce inside short periods then aptamer-based systems achieve bacteriological identification at precise levels (McConnell et al., 2020). Biosensors enable femtomolar and picomolar detection levels because they incorporate nanomaterials including gold nanoparticles with carbon nanotubes and graphene. Through their connection to digital platforms that include smartphones and wireless data systems and cloud platforms scientists can conduct smart diagnostics along with remote monitoring that uses data-driven decisions. The precise food safety mission has become possible due to technologies which offer real-time tracking and management of contaminations from end to end across supply network systems.

2. CLASSIFICATION OF FOOD CONTAMINANTS

Unintentional substances and microorganisms which harm food products fall under the category of food contaminants when introduced through any phase from production to storage and transportation (Lebelo et al., 2021). Food contaminants expand across three main groups with biological contact points beside chemicals and physical and emerging contaminants (Table 1). The detection and risk assessment and control requirements for these contaminants produce specific difficulties. Biosensors have emerged as vital examination devices for ensuring food safety because of their exceptional sensitivity together with their specificity and their portability features.

2.1 Biological Contaminants (e.g., Bacteria, Viruses, Fungi)

Foodborne illnesses develop from biological contaminants which include living organisms together with their generated by-products. Food chain activities beginning at farms through dining tables can introduce these harmful substances to food. Biological impurities present great difficulties because they multiply at undetectable rates and need specialized examinations for identification.

- **Bacteria:** Single-celled living organisms spread contamination to food due to incorrect handling practices and poor storage methods as well as inadequate food cooking temperatures. Stomach infections caused by bacteria can be traced back to *Salmonella and Escherichia coli (E. coli) while Listeria monocytogenes and Campylobacter jejuni and Clostridium botulinum* are also involved in foodborne pathogen transmission (Gourama, 2020). The pathogenic bacteria which thrive in food surfaces and water can trigger intestinal diseases that vary from common stomach discomfort to dangerous infections like septicemia and botulism.
- **Viruses:** The infection of norovirus and hepatitis A viruses happens through contaminated food or water before they trigger gastroenteritis and liver disease after ingestion(O'Shea et al., 2019). The reproduction of viruses depends on cellular host environments because they do not reproduce independently which makes them thrive in improperly handled or unprocessed foods.
- **Fungi:** The classic taxa of fungi encompass yeasts and molds that establish growth on different foods which contain significant amounts of sugar or carbohydrates such as fruits and grains (Pouris et al., 2024). Molds generate mycotoxins as products which endanger human health conditions. Aspergillus flavus produces aflatoxins that function as strong carcinogenic substances.

2.2 Chemical Contaminants (e.g., Pesticides, Heavy Metals, Toxins)

Chemicals that contaminate soil derive from agricultural chemicals as well as industrial processes and environmental pollution. Reduced concentrations of chemicals in food create severe health risks since these substances are toxic to both humans and animals:

- **Pesticides:** Plant protection chemicals provide defense against three categories of agricultural pests including fungi and weeds. The presence of pesticide residues in food can continue beyond application because they remain in what we eat even though pesticides aid yield improvement. Human exposure to particular pesticides over long periods triggers neurological problems while also disrupting hormone systems and producing potential cancerous effects and Organophosphates and carbamates and pyrethroids represent three examples of these chemicals (Upadhayay et al., 2020).
- **Heavy Metals:** Environmental contamination from industrial operations leads to food contamination with lead and cadmium together with arsenic and mercury substances (Rahman & Singh, 2019). The long-lasting environmental pollutants among metals result in systemic retention within living bodies that produces neurological effects along with kidney disease and delayed child development.
- **Mycotoxins and Toxins:** The growth of fungus on food leads to mycotoxin production with aflatoxins alongside ochratoxin A and fumonisins among them. Mycotoxins function as strong cancer-causing chemicals and brain-damaging substances that produce liver problems and weaken immune functions while affecting development when consumed(Mohammad et al., 2018). Other foodborne toxins such as botulinum toxin produced by *Clostridium botulinum* also exist which may result in the fatal disease botulism.

2.3 Physical Contaminants (e.g., Microplastics, Particulate Matter)

Physical contaminants exist as foreign materials that enter food during the harvesting phase as well as processing and packaging and transportation phases. Food contaminants cause neither chemical toxicity to consumers nor do they result in chemical threats. However, they create problems which lead to consumer food rejection and cause discomfort or physical injuries. Imperfections from physical contaminants normally show up during visual examination yet complete removal requires effective contamination prevention techniques.

- **Microplastics:** The environment now harbors numerous small plastic fragments which measure below 5 mm because of their growing presence along with migration into food items (Miraj et al., 2021). The environmental occurrence of microplastics results from degradation of plastic waste as well as from plastic packaging materials and plastic utensils. Plastic consumption leads to two main risks which include tissue damage and harmful substances entering human bodies.
- **Particulate Matter:** Physical contaminants found in food operations consist of metal fragments together with glass shards and wood as well as stones which can enter food during production or final packaging steps. The presence of physical contaminants stands as a main reason behind food safety recalls because they create immediate dangers when taken into the human body. Food industries employ metal detectors and sieves with filtration systems as preventive measures to stop physical contaminations.

2.4 Emerging Contaminants (e.g., Nanomaterials, Pharmaceutical Residues)

New and developing risks to food safety emerged since traditional contaminants were not understood in prior times. Contaminants develop from technological progress and industrial developments as well as environmental discoveries of untracked materials.

- **Nanomaterials:** The food industry utilizes nanotechnology through different methods including formula ingredients and product packaging and additive preservatives. The tiny size of nanomaterials offers better packaging solutions and extended preservation but scientists remain uncertain about their accumulation patterns in body tissues which could present health dangers to people (Wyser et al., 2016). Two major nanomaterials utilized in food products include both silver nanoparticles and titanium dioxide.
- **Pharmaceutical Residues:** The presence of pharmaceutical products including antibiotics and hormones and painkillers in food products occurs because of veterinary drug usage in livestock and wastewater contamination (Beyene, 2016). Antibiotic resistance and endocrine disruption and additional health problems result from these remaining substances.

Table 1. Classes of Food Contaminants

Contaminant Class	Source (Origin)	Examples	Health Effects	Detection Challenges	Source
Biological	Improper hygiene, contaminated water/soil, cross-contamination	Salmonella, Listeria, Norovirus, Aspergillus	Food poisoning, diarrhea, liver failure, immune compromise	Require rapid, selective, and sensitive detection; short incubation time critical	(Kawakami et al., 2018), (Law et al., 2015)
Chemical	Agricultural chemicals, industrial waste, natural fungal growth	Pesticides, lead, cadmium, aflatoxins, dioxins	Carcinogenicity, neurotoxicity, endocrine disruption	Often low concentrations; require sample cleanup; toxic even at trace levels	(Toragall et al., 2020)
Physical	Packaging, machinery, handling	Microplastics, glass shards, textile fibers, metal filings	Physical injury, gastrointestinal irritation	Not chemically reactive; traditional screening insufficient	(Sodersten et al., 2018), ((EFSA), 2016)
Emerging	Nanotech in packaging, treated animals, industrial additives	Silver/TiO_2 nanoparticles, antibiotics, BPA, hormones	Unknown toxicity, hormonal imbalance, antimicrobial resistance	Require advanced sensors; evolving regulatory frameworks	(Singh & Kumar, 2023)

3. FUNDAMENTALS OF BIOSENSOR TECHNOLOGY

Biosensors function as analytical devices which unite biologically recognized elements with transducing components to execute quantitative detection of particular substances or contaminants contained in samples. Biological detection systems used for food security analyze food ingredients by employing both biological recognition agents and signal amplifiers to provide economical results more easily than traditional research laboratory protocols (Zhang et al., 2022). A fundamental breakdown of biosensor technology will be examined through reviewing its essential components and operational principles as well as testing parameters.

3.1 Conceptual Framework and Core Elements of Biosensors

A biosensor functions as a device which converted biological reactions into quantifiable outputs. The functionality of biosensors depends on a combination of two essential components which include the biorecognition element together with the transducer. The combination of detection elements functions together to identify contaminants with high precision and sensitivity levels.

3.1.1 Biorecognition Elements (e.g., Enzymes, Antibodies, DNA, Aptamers)

The biorecognition element acts as the component which directly bonds with target contaminants. This element provides biosensor selectivity so it detects exclusively the relevant contaminant type (chemical, biological or physical) in its target range. Common biorecognition elements include:

- **Enzymes:** Enzymes represent specific biological catalysts which detect particular molecules while they work. Research institutions employ these enzymes to detect chemical pollutants using chemical reactions that produce quantifiable signals. The detection of organophosphate pesticides with cholinesterase relies on the enzyme using the pesticide to produce detectable reaction products through hydrolysis (Kaur & Singh, 2020).
- **Antibodies:** The immune system creates these proteins which possess the ability to recognize distinct antigens and maintain binding interaction with them. Biological contaminants including virus types and bacterial types as well as allergens are commonly detected through antibody testing due to their exact identification capabilities (McGrath et al., 2012). The food industry widely uses monoclonal antibodies to run immunoassays bound to Salmonella and E. coli food contaminants.
- **DNA:** Biosensors that use nucleic acids base their detection mechanism on DNA strand binding of complementary genetic sequences which identify pathogens or genetically modified organisms (GMOs). By combining DNA detecting techniques with biosensors laboratories can perform DNA amplification through methods such as polymerase chain reaction (PCR). These systems allow the detection of scarce DNA targets.
- **Aptamers:** DNA or RNA strands known as aptamers hold the ability to connect highly precisely to particular target molecules despite their brief molecule dimensions. Aptamers represent promising biosensors because they show excellent stability and simple production capabilities alongside their capability to measure different compounds extending from small molecules to proteins and microorganisms (Nguyen et al., 2017). Foodborne pathogen detection constitutes a potential substitute for antibodies through these aptamers.

3.1.2 Transducers (e.g., Electrochemical, Optical, Piezoelectric, Thermal)

A biosensor transducer serves as the signal conversion element for biological recognition events between specific elements and their target molecules. The selection of transducer depends simultaneously on the target contaminant along with the preferred operating features that the biosensor should exhibit. Different types of transducers which appear in food safety biosensors include:

- **Electrochemical Transducers:** Electrochemical biosensors remain the most frequently employed transducers in biosensing applications because they have simple construction and high sensitivity along with affordable price. The interaction of the target analyte with biorecognition elements triggers changes in electrical measurements like current flow or voltage changes and impedance variation(Wu et al., 2023). Among electrochemical sensors amperometric and potentiometric and conductometric devices are the most frequently used setups. Glucose biosensors utilize glucose oxidase enzymes as their biorecognition component which generates an electrochemical change in current that monitors during detection processes.
- **Optical Transducers:** Biosensors that operate by optics monitor changes in light characteristics which result from target-analyte interactions that use biorecognition elements. The sensor detection technology monitors food contaminants consisting of both microbial contaminants along with toxins. The optical biosensors use fluorescence resonance energy transfer together with surface plasmon resonance as their detection methods (Q. Wang et al., 2022).
- **Piezoelectric Transducers:** The resonance frequency of piezoelectric materials varies upon interaction between target analytes and biorecognition elements during biosensing operations. A detectable signal emerges from the frequency shift which is directly related to the amount of analyte found on the sensor surface. Without replacement the application of piezoelectric sensors proves beneficial for biological contaminant monitoring of viruses and bacteria (Kuchmenko & Lvova, 2019). The frequency resonance of piezoelectric materials changes as target analytes and biorecognition elements engage with each other during biosensing operations. The sensor produces a signal that directly reflects the quantity of target substance bonded to its surface. The use of piezoelectric sensors without replacement leads to productive biological contaminant monitoring of viruses and bacteria.

3.2 Working Principle of Biosensors

A biosensor functions through the specific recognition between its biorecognition element and target contaminant. The biochemical or biophysical response occurs when the sample interacts with the biorecognition element which specifically attaches to the contaminant. The transducer uses this change as input for signal conversion. A standard procedure contains several consecutive operations in its methodology (Figure 1):

Figure 1. Schematic representation of the working mechanism of a biosensor, emphasizing the sequential element(Grieshaber et al., 2008).

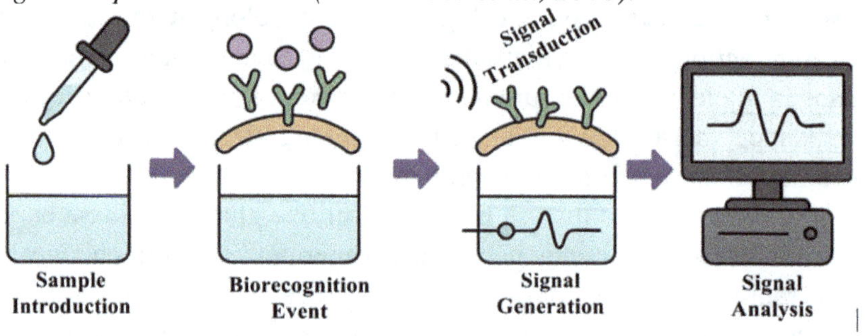

1. **Sample Introduction:** The biosensor system receives a food specimen as either solid or liquid format.
2. **Biorecognition Event:** The target contaminant within the sample finds binding receptors through the biorecognition element such as antibody-antigen or enzyme-substrate pairs.
3. **Signal Generation:** The sensor measurement changes due to its binding interaction with the target element (e.g., current shift, frequency alteration, light variation, etc.).
4. **Signal Transduction:** By using a transducer, the physical or chemical alteration gets transformed into either electronic or optical signals.
5. **Signal Analysis:** Analysis of the output signal transforms the information into quantitative or qualitative measurements about contaminant concentration and detection.

3.3 Critical Performance Metrics (Sensitivity, Specificity, LOD, Response Time)

Biosensors need performance evaluation through several primary criteria when used for food safety applications:

- **Sensitivity:** A biosensor should operate at highly sensitive levels to detect minimal concentrations of target contaminants. Very minute amounts of contaminants can be detected through highly sensibly designed biosensors because they function at very low substance detection levels especially for examining food samples containing small trace contaminants (Eyvazi et al., 2021). The measurement of sensitivity in biosensors usually takes the form of the limit of detection (LOD) which represents the smallest detectable substance level.

- **Specificity:** The biosensor sets itself apart from other substances in the sample to detect only specific target contaminants (X. Wang et al., 2014). Such specific biosensors will detect only their intended target substance while avoiding mistaken identifications of alternative food sample contents. A biosensor must exhibit high specificity for it to detect targets correctly among food samples containing multiple compounds.
- **Limit of Detection (LOD):** A biotechnology sensor uses the LOD to determine the minimum detectable level of contaminants with reliable results. The LOD measurement depends on the sensor's reaction capability together with both the type of analyte and the testing solution makeup (Gauglitz, 2018). Detecting hazardous contaminants at trace levels requires lower detection limits particularly since this leads to better performance.
- **Response Time:** After a sample introduction the biosensor requires response time to generate detectable signals. The requirement for rapid response times remains essential for food testing procedures performed at the testing location to deliver prompt food safety decisions. Biosensors should react within a period ranging between several seconds and a few minutes to meet the requirements of different applications.

4. VARIANTS OF BIOSENSORS UTILIZED FOR DETECTING FOODBORNE CONTAMINANT

Food safety management relies on biosensors to perform quick examinations of foodborne contaminants directly in their positions. Multiple biosensor systems follow separate detection methods for monitoring biological and chemical along with physical contaminants in foods (Eyvazi et al., 2021). The various types of biosensors bring specific advantages which enable their use in different food safety scenarios. The following segment discusses multiple biosensor varieties intended for food contaminant analysis through evaluation of their operational methods in addition to their favorable attributes and corresponding drawbacks.

4.1 Electrochemical Biosensors

Food security applications use electrochemical biosensors as their preferred technology because these devices provide high sensitivity and cost-effective ease of operation. Electricity-based changes in the system properties (including current, voltage and impedance) are detected by biosensors when biorecognition elements link with target contaminants. The basic functional principle of electrochemical biosensors allows us to separate them into three main classifications through their electrical measurement type (Table 2):

- **Amperometric Biosensors:** The sensors track current variations that happen when the biorecognition element connects with the target analyte. The enzymatic reaction between glucose and oxygen inside glucose biosensors leads to current changes which these tools detect through glucose oxidase enzymes(Mandpe et al., 2020). Foodborne pathogens along with toxins and chemical substances such as heavy metals and pesticides can be effectively detected by this particular biosensor type.
- **Potentiometric Biosensors:** The sensors detect voltage changes between two electrodes due to the target binding action with biorecognition elements. Food professionals use Potentiometric sensors because they effectively monitor pH alterations that indicate microbial contamination differences in food items (Ghrissi et al., 2021).
- **Conductometric Biosensors:** The sensors detect electrical conductivity changes within their surfaces which occur as target contaminants bind to the sensor's surfaces. These sensors react strongly to shifts in ionic content which enables them to identify microbial and chemical food pollutants during sample analysis.

Table 2. Comparison of Electrochemical Biosensors

Biosensors	Measurement	Features	Applications	Source
Amperometric Biosensors	Measurement of current	Sensitive, fast response, real-time analysis	Detection of pathogens, pesticide residues, toxin	(Velusamy et al., 2010)
Potentiometric Biosensors	Measurement of potential (voltage)	Simple, ideal for pH changes	Detection of microbial contamination, spoilage indicators	(Ghrissi et al., 2021)
Conductometric Biosensors	Measurement of conductivity	Sensitive to ionic changes, portable	Detection of ions, toxins, pathogens	(Velusamy et al., 2010)

4.2 Optical Biosensors

These sensors operate through monitoring light alterations which occur when target analytes bind to biorecognition elements. The sensors depend on different optical characteristics including absorption fluorescence together with reflectance in order to create detectable signals. There exist three main groups of optical biosensors:

- **Surface Plasmon Resonance (SPR):** The optical method SPR measures the refractive index changes near metal film surfaces following the binding of target elements to biorecognition elements. The surface reflectance angle shifts due to refractive index modification and this change serves as a measurement basis for the target substance concentration level. Due to its extreme sensitivity SPR can analyze pathogens and toxins and allergens that are found in food samples (Ravindran et al., 2023).
- **Fluorescence-Based Biosensors:** The sensors activate by measuring fluorescent emissions that originate from molecules which were attached to the biorecognition element. A change in the fluorescence signal from the biorecognition element occurs after analyte binding triggers a detectable output. Foodborne pathogens detection along with chemical contaminants monitoring benefits highly from fluorescence-based biosensors because of their superior sensitivity (Bhardwaj et al., 2017).
- **Colorimetric Biosensors:** The sensor system uses visible color transformation as the target analyte and biorecognition element interact with each other. Colorimetric assays remain simple to carry out since they do not need complex instruments which makes field tests feasible (Krishnan & others, 2022). These sensors help identify food contaminants including pesticides together with microbial contaminant types.

4.3 Piezoelectric and Acoustic Biosensors

The binding of target contaminants to biorecognition elements on piezoelectric sensors results in alterations of sensor mass and mechanical properties which these devices can detect. The sensors work through the piezoelectric principle which shows that analytes can shift the resonance frequency of quartz crystal or surface acoustic wave device materials. The frequency change directly relates to the amount of analyte mass that has bound to the system. The piezoelectric biosensor functions as an extremely sensitive device which detects biological contaminants that include bacteria, viruses together with foodborne pathogens. These devices present beneficial features because they provide precise detection methods as well as quick response capabilities and can be miniaturized for on-site analysis (Rasheed et al., 2024). Acoustic biosensors represent a version of piezoelectric biosensors and detect acoustic wave shifts resulting from sensor surface analyte bindings. Food products undergo effective detection for small contaminants and pathogens and toxic substances through highly mass-sensitive sensors.

4.4 Thermal Biosensors

The sensor surface temperature measurement underlies thermal biosensors since the biorecognition element produces temperature variations following interaction with the target analyte. The target-biorecognition element interaction may produce either endothermic heat absorption or exothermic heat release depending on their specific chemical compatibility. The detection of food contaminants makes use of thermal biosensors that deliver fast results while maintaining high detection sensitivity. Thermal biosensors serve well in microbial contamination detection because microbial interactions with substrates lead to temperature-based changes (Liu et al., 2024). Thermal sensors have the capability to measure chemical pollutant and toxin reactions.

4.5 Nanomaterial-Based Biosensors (e.g., Graphene, Quantum Dots, Gold Nanoparticles)

Graphene, gold nanoparticles, and quantum dots are examples of nanomaterials that have recently been integrated into biosensing systems. Their unique character includes large surface area alongside conductive nature along with optical properties that supports their excellence in biosensing work.

- **Graphene-Based Biosensors:** Graphene consists of a monolayer sheet of carbon atoms in a two-dimensional hexagonal lattice. It possesses high electric conductivity and enormous surface area that increases the biosensor sensitivity. Graphene-based biosensors for the detection of foodborne diseases, toxins, and chemical pollutants are employed on account of a high surface area-to-volume and ease of affixing the biomolecules onto the graphene surface (L. Fu et al., 2023).
- **Quantum Dot-Based Biosensors:** Quantum dots refer to nanometer-scale semiconductor particles with peculiar optical properties, including fluorescence and emission spectrum tunability (Jelinek, 2016). Such characteristics make them well-suited for application in optical biosensors, where they can be utilized to amplify signal detection. Biosensors based on quantum dots are being utilized for the detection of food contaminants such as pesticides and pathogens.

- **Gold Nanoparticle-Based Biosensors:** Gold nanoparticles are frequently employed in biosensors because they are easy to functionalize, stable, and possess intense optical properties. The sensors usually work by detecting changes in light scattering or color upon conjugation of gold nanoparticles with the biorecognition element and subsequent exposure to the target contaminant. Gold nanoparticle-based biosensors are extremely sensitive and are under investigation for the detection of numerous foodborne contaminants, such as pathogens, toxins, and allergens (Anh et al., 2022).

5. IMPLEMENTATION STRATEGIES FOR DETECTING FOOD CONTAMINANT

5.1 Detection of Pathogenic Microorganisms

Disease-causing bacteria like *Escherichia coli, Salmonella species, and Listeria monocytogenes* are primary agents of foodborne diseases across the globe (Lee & Yoon, 2021). Rapid detection is important to avert outbreaks and maintain consumer protection. Biosensors have emerged as valuable tools to detect microbial spoilage directly from food samples, including meat, dairy, produce, and ready-to-eat foods. Electrochemical biosensors use immobilized antibodies, DNA probes, or aptamers on electrodes to detect specific bacterial antigens or genetic material. These sensors are capable of providing detection within minutes and usually avoid the need for sample enrichment. Optical biosensors, especially those involving surface plasmon resonance (SPR), fluorescence, or colorimetry, provide real-time, label-free detection of bacterial binding reactions. Piezoelectric biosensors respond to mass or viscoelastic changes due to microbial attachment on quartz crystal microbalance surfaces. Moreover, CRISPR-based biosensors are now being developed as ultra-sensitive platforms for identifying bacterial DNA sequences with unprecedented specificity (Kumaran et al., 2023). These biosensing platforms have displayed outstanding potential to be applied to laboratory and in-field use, aiding food manufacturers and regulatory officials in the earlier detection of microbial risk.

5.2 Detection of Chemical Residues

The buildup of chemical residues, such as antibiotics, pesticides, and mycotoxins, is of major health concern due to endocrine disruption, carcinogenicity, and antimicrobial resistance. Food commodities—milk, vegetables, cereals, and fruits—are regularly screened for such residues. Biosensors can be applied as a fast and sensitive tool for residue detection compared to traditional chromatographic methods. Electrochemical biosensors have been extensively used to screen pesticide residues such as organophosphates and carbamates with enzyme inhibition approaches, especially with Acetylcholinesterase (AChE) based formats (Umar & Aisami, 2020). These biosensors also possess superior detection qualities even at trace levels. Likewise, immunosensors have been created to identify antibiotic residues like tetracyclines and sulfonamides in milk products using antibody-antigen interactions. For mycotoxins like aflatoxins and ochratoxins, aptamer-based and antibody-based optical biosensors have been used for real-time analysis (Szelenberger Rafałand Cichoń et al., 2024). Owing to their high throughput, portability, and rapid response, biosensors are being employed more and more by food producers and inspectors for improved compliance with regulatory thresholds and protection of public health.

5.3 Detection of Allergens and Toxins

Food allergens and toxins may be extremely dangerous to those with sensitivities, and in some cases, may cause severe allergic reactions or food poisoning. Peanuts, tree nuts, milk, shellfish, and wheat are typical food allergens. Botulinum toxin or staphylococcal enterotoxins, which are usually the result of microbial contamination, must be monitored very carefully (Rao, 2021). Biosensors allow for rapid and accurate detection of allergens in processed and unprocessed foods. Optical biosensors based on fluorescence or surface plasmon resonance can identify allergenic proteins with no need for labelling or complicated sample preparation. Immunosensors based on particular monoclonal antibodies were developed for detecting peanut and gluten in bakery and packaged foods. Aptamer-based biosensors offer higher specificity, especially for low-molecular-weight toxins that are hard to identify with regular antibodies. Electrochemical biosensors can detect even picogram concentrations of neurotoxins or bacterial toxins for toxin detection, which makes them critical for monitoring in canned foods, dairy, and seafood (Gupta et al., 2021).

5.4 Multiplexed and Multi-analyte Detection Systems

The heterogeneity of food matrices and the variety of possible contaminants require the creation of multiplexed biosensors capable of detecting more than one analyte at a time. These platforms greatly increase throughput and decrease the time and expense of individual assays. Microarray-based biosensors, frequently employing DNA or protein probes, enable parallel detection of multiple pathogens or chemical agents on a single chip. Multi-test line lateral flow assays can detect various toxins or allergens simultaneously and provide a paper-based test that is useful for field applications (Li et al., 2023). Fluorescence biosensors bearing quantum dots or multi-channel electrochemical arrays are able to identify and quantify several targets on the basis of unique signal profiles. Additionally, nanomaterial inclusion such as graphene, carbon nanotubes, and metal nanoparticles improves the sensitivity and signal resolution of these multiplex platforms. Such systems find their greatest worth in food processing facilities and regulatory laboratories where exhaustive, high-throughput screening is needed for quality assurance and monitoring for safety.

6. ADVANCES IN PORTABLE AND ON-SITE BIOSENSING PLATFORMS

6.1 Lab-on-a-Chip and Microfluidic Devices

Lab-on-a-chip (LoC) technology and microfluidic devices are a paradigm shift in the area of biosensing, as the whole laboratory process is being miniaturized onto a tiny chip (Kuru et al., 2023). These platforms combine the sample handling, reagent mixing, reaction, and detection steps within the microchannels etched on glass, silicon, or polymers-based substrates. LoC systems can detect foodborne pathogens, chemical residues, and toxins with utmost accuracy and speed. Microfluidics facilitates the manipulation of very small volumes of samples (usually microliters or smaller), saving reagent use and analysis time. Through the integration of biosensing components like antibodies, aptamers, or enzymes in particular regions of the microchannel, these chips facilitate on-site, real-time analysis. The integration of several detection modes (e.g., optical, electrochemical) into microfluidic devices has facilitated multiplexed analysis and made them best suited for food screening in a high-throughput manner (X. Fu et al., 2021). Also, recent breakthroughs in device fabrication processes such as soft lithography and 3D printing have facilitated their cost-effectiveness and field-scale deployment.

6.2 Smartphone-Integrated Biosensors

Smartphone-integrated biosensors have become an effective means for food safety testing democratization by taking advantage of the ubiquitous availability, connectivity, and processing capability of mobile phones. Smartphone-integrated biosensors are generally made up of a sensing element (e.g., paper strip, chip, or electrode) that is connected to the smartphone through optical readouts, USB, or Bluetooth. Smartphones are used as detectors, analyzers, and data transmitters (Bui et al., 2023). In colorimetric biosensors, for instance, the camera registers the change in intensity of the color in response to the analyte concentration, which is thereafter handled by specialist mobile apps. For chemiluminescent or fluorescence readouts, external lenses or optical filters can be attached to smartphones. These platforms have been used effectively for the detection of pesticide residues on fruits and vegetables, aflatoxins in cereals, and E. coli in water or milk. In addition, the ability to geotag and wirelessly transfer results to regulatory databases provides value for traceability, epidemiological surveillance, and real-time public health response.

6.3 Wearable and Wireless Biosensors

Wearable and wireless biosensors are a future but fast-emerging method of real-time and remote monitoring of food safety and quality. Such devices typically consist of a flexible, thin, and surface- or skin-adaptable substrate implanted with the biosensing elements. While now more prevalent in biomedical use, wearable biosensors are being developed for food safety applications—e.g., wear patches for tracking contamination within food-handling environments or wireless stickers on packaging to track spoilage or gas release. Radio-frequency identification (RFID), near-field communication (NFC), or Bluetooth-enabled wireless biosensors enable data transmission in real-time to surveillance systems (Aledhari et al., 2022). Integration with IoT frameworks allows remote alerts, predictive analysis, and autonomous decision-making for food logistics and cold chain management.

6.4 Paper-Based and Disposable Biosensors

Paper-based biosensors provide a disposable, low-cost, and very portable platform for in-field detection of food contaminants. Cellulose-based materials including filter paper, nitrocellulose, or chromatography paper are used as the substrate upon which biorecognition elements are immobilized. The most common instance is a lateral flow assay (LFA), like a home pregnancy kit, which has been modified for the detection of microbial pathogens, allergens, mycotoxins, and pesticides in food samples (Sohrabi et al., 2022). LFAs give quick qualitative or semi-quantitative answers in minutes and no specialized equipment is needed. Improvements have enabled more advanced formats like microfluidic paper-based analytical devices (μPADs), which use patterned paper channels to enable multiplexed and quantitative analysis by colorimetric or electrochemical responses (Morbioli et al., 2017). Their biodegradability and disposability position them as particularly well suited for single use in resource-limited or remote settings.

7. REGULATORY, COMMERCIAL, AND PRACTICAL CONSIDERATIONS

7.1 Regulatory Guidelines and Acceptance

Regulatory supervision is imperative to guaranteeing the security, consistency, and standardization of biosensor technology applied in food contaminant analysis (Nath, 2024). These include regulatory bodies like the U.S. Food and Drug Administration (FDA), European Food Safety Authority (EFSA), Food Safety and Standards Authority of India (FSSAI), and Codex Alimentarius Commission that stipulate definite procedures for food analysis, such as allowable contaminant levels and detection technology validation specifications. To receive regulatory approval, biosensors should prove stable performance on a number of parameters such as specificity, sensitivity, reproducibility, limit of detection (LOD), and robustness (Chen et al., 2020). Validation procedures usually need head-to-head comparison with the traditional gold standard techniques like HPLC, ELISA, or PCR. In addition, biosensors used for commercial or field applications also need to be compliant with ISO/IEC standards for analytical procedures and device performance (Giorgi & Tonello, 2022). Converging international standards for biosensor validation is still ongoing. Nevertheless, mutual cooperation among regulatory agencies, industries, and research institutions is more and more allowing the use of biosensors in official food safety monitoring programs.

7.2 Industrial Implementation and Challenges

While biosensors hold significant promise, their incorporation into industrial food safety protocols is subject to several practical and technical issues. Food matrices are extremely complex and heterogeneous, and the presence of fats, proteins, or particulates can interfere with the transduction mechanisms or biorecognition elements of the biosensor (Jayan et al., 2020). In addition, large-scale industrial processes require biosensors that not only need to be fast but also scaleable, automatable, and compatible with already established production line protocols. Some main concerns are:

- **Sample preparation complexity –** For most biosensor platforms, little or no sample preparation is ideal, but actual samples in the field may need preprocessing (e.g., filtration, dilution).
- **Durability and reusability –** Sensors deployed in process environments need to be able to withstand variable temperature, humidity, and mechanical stress.

- **Cross-contamination and biofouling** – Reusable biosensors suffer from frequent use by being subject to surface fouling or signal drift, which taints data integrity.

7.3 Cost, Shelf-Life, and User-Friendliness

For biosensors to be feasible substitutes for conventional laboratory methods, they should be economically viable, shelf-stable in varying storage conditions, and usable by non-specialists. These three parameters—cost, stability, and usability—have a crucial influence on their practical uptake.

- **Cost:** The total price of a biosensor involves not only production but also packaging, delivery, data consolidation tools (such as mobile applications), and upkeep (if they are reusable). Although nanomaterial-amplified and smartphone-connected biosensors tend to be more sensitive, their price tag can be prohibitively expensive unless mass-produced.
- **Shelf-life:** Biosensors, especially those with biological recognition units (e.g., enzymes, antibodies), are temperature- and humidity-sensitive, which restricts shelf-life. Stability measures like lyophilization, encapsulation, or synthetic substitutes (e.g., aptamers) are being used to enhance longevity (Odeh et al., 2019).
- **Ease of Use:** Easy-to-use operation protocols, limited sample handling, and easy-to-read data (e.g., color change or smartphone display) are essential for non-laboratory environments. Biosensors to be used on-site in small food enterprises or by consumers need to be made for error-free operation with unambiguous visual or digital signals.

7.4 Case Studies of Commercial Biosensors in the Food Industry

Various biosensor technologies have crossed the boundaries from research demonstrations to market-adopted tools for food safety. These examples give important feedback on the process of adoption as well as experiences during implementation.

- **BioSensor Laboratories' "BioDetect" for Salmonella and E. coli:** This colorimetric lateral flow test has been used by North American poultry processors (Pohanka, 2023). It provides results in 15 minutes and enjoys FDA approval for on-site screening.
- **Romer Labs' AgraStrip®:** A renowned range of lateral flow biosensors for quick detection of mycotoxins (such as aflatoxin and ochratoxin) in grains

(Suo et al., 2023). They are popularly used across grain supply chains and offer high sensitivity with minimal training.

- **QScout MLD from Advanced Animal Diagnostics**: While originally for veterinary diagnostics, this biosensor platform has been repurposed for the detection of antibiotic residues in raw milk (Gaudin, 2017). It integrates microfluidic biosensing with smartphone-based readout.
- **Tellspec Scanner:** A portable spectroscopic device coupled with a smartphone to detect chemical contaminants in food using machine learning (Kapse et al., 2022). Although not a traditional biosensor, it demonstrates the intersection of sensor technology and data analytics for real-time food analysis.

8. CHALLENGES AND FUTURE PERSPECTIVES

8.1 Current Limitations in Sensitivity and Specificity

Although biosensors are intended to be sensitive and specific, in reality, they usually fail to achieve the performance of gold standard laboratory methods like HPLC, LC-MS/MS, or ELISA. One of the main limitations is the complexity of real food matrices, which can interfere with the sensor's ability to detect target analytes accurately (Rodriguez et al., 2020). Such components like fats, proteins, carbohydrates, and pigments can lead to non-specific binding or suppression of the signal and lower the accuracy of the results. Sensitivity, particularly trace sensitivity, remains another longstanding challenge. Most of the food contaminants, for instance, pesticide residues or mycotoxins, are poisonous even at the nanogram or picogram concentrations. Identification of such trace levels necessitates very purified biorecognition units and highly sensitive transducers. False negatives and false positives, as a result of cross-reactivity or deterioration of sensing components, further discount the usefulness of biosensors in high-stakes applications. Time stability, sensor fouling, short shelf-life, and sensitivity to environmental factors (e.g., pH, temperature, humidity) are further practical concerns limiting the general applicability of biosensors (Demir et al., 2024). Ongoing research aims, therefore, to improve robustness, optimize bioelement immobilization methods, and create more selective recognition elements like synthetic aptamers and molecularly imprinted polymers (MIPs).

8.2 Integration with IoT and Artificial Intelligence

The future of monitoring food safety is smart connected systems that facilitate real-time decision-making. The fusion of biosensors with Internet of Things (IoT) platforms and Artificial Intelligence (AI) is very much capable of transforming the detection of food contaminants. With biosensors integrated into wireless networks and cloud-based systems, the real-time information from distant locations—farms, factories, and retail stores—can be monitored and analyzed in real time. IoT-integrated biosensors can notify stakeholders in real time in the event of contamination incidents, enabling prompt interventions to avoid public health threats. Machine learning and artificial intelligence programs can greatly enhance biosensor operation by enhancing noise reduction, result interpretation, and pattern recognition (Cui et al., 2020). Such tools have the ability to correct for sensor drift, learn from past databases, and identify real signals and false alarms accurately. Smart biosensing platforms also have the ability to be incorporated with smartphones, QR-code-based systems, and blockchain for tracing food, enabling consumers and regulatory authorities to gain access to open, tamper-proof safety information (Fernando et al., 2024). This "smart food safety" strategy fits with the principles of Industry 4.0 and is especially apt for globalized supply chains.

8.3 Sustainability and Green Biosensor Design

In the age of climate change and resource depletion, the environmental sustainability of biosensor production and waste disposal is a critical consideration. Traditional biosensors typically incorporate non-biodegradable polymers, toxic solvents, and scarce or poisonous materials, which pose significant concerns over their ecological sustainability. Green biosensor development seeks to mitigate these problems through the use of sustainable materials, environmentally friendly fabrication methods, and biodegradable substrates (Kalambate et al., 2020). For example, low-cost, disposable, biodegradable paper-based biosensors are being looked at as a green alternative. Similarly, the integration of biopolymers, plant-based recognition elements, and natural inks into sensor building is also pursued actively. Designs that are low in energy requirements, such as passive sensors or those that exploit environmental energy sources (e.g., light, heat, or mechanical motion), also contribute to the sustainability strategy. Additionally, miniaturization using microfluidics and lab-on-a-chip technologies keeps material consumption to a minimum while minimizing waste production, and all this helps develop a circular biosensing economy. Integrating sustainability into biosensor innovation not only aligns with international environmental objectives but also renders these devices more appealing to mass production and deployment, particularly in low-resource and rural areas(Obeagu et al., 2024).

8.4 Roadmap for Next-Generation Biosensors

To address the changing demands of the food industry globally and regulatory agencies, next-generation biosensors need to transcend existing limitations and provide a set of superior capabilities. The future of biosensor development has the following in the roadmap:

- **Ultra-High Sensitivity and Specificity:** With enhanced nanostructured materials, engineered enzymes, aptamers, and Molecularity Imprinted Polymer (MIPs), biosensors will be capable of detecting one molecule or ultra-trace levels of pollutants (Ramajayam et al., 2023).
- **Multiplexed and Multi-Analyte Detection:** Future biosensors will detect several contaminants (biological, chemical, and physical) from a single sample simultaneously, saving cost and time with enhanced diagnostic accuracy.
- **Self-Calibrating and Reusable Devices:** With onboard microprocessors and learning algorithms, biosensors will correct for variability and environmental interference automatically, improving reliability on repeated use.
- **Point-of-Need Automation:** Automated sample handling, signal processing, and data analysis will allow non-specialists to use them in decentralized settings such as farms, open markets, or small food producers (Beck-O'Brien & Bringezu, 2021).
- **Cloud-Based Ecosystems and Global Surveillance:** Information from biosensors will feed into cloud databases, allowing global monitoring of foodborne risk, tracing contamination patterns, and informing predictive analytics for food safety policy.
- **Personalized and Consumer-Accessible Sensors:** Next-generation biosensors can be integrated into kitchen appliances, packages, or even handheld devices to equip consumers with the ability to verify food safety in real-time at the point of consumption.

9. CONCLUSION

This chapter offers an extensive review of biosensors for the on-site and rapid identification of food contaminants, highlighting their important role in protecting public health and maintaining food safety. The chapter begins with a consideration of the worldwide threat of different categories of food contaminants—biological, chemical, physical, and emerging—and their implications for human health and regulatory requirements. The principles of biosensor technology are discussed, including the important elements like biorecognition elements and transducers, and performance characteristics such as sensitivity, specificity, limit of detection, and response time. Various biosensor technologies—electrochemical, optical, piezoelectric, thermal, and nanomaterial-based—are discussed in terms of their strengths, weaknesses, and real-world applications. These sensors have been effective in pathogen detection (e.g., E. coli, Salmonella), chemical residues (e.g., antibiotics, pesticides), allergens, and toxins. The chapter also identifies future technologies such as lab-on-a-chip platforms, smartphone-included sensors, wearable biosensors, and paper-based platforms, all driving toward smaller, cheaper, and user-friendly instruments for field deployment. Compared to traditional laboratory-based techniques, biosensors facilitate on-site detection in real time, with a substantially decreased response time during contamination occurrences. Such fast potential aids in swifter recalls and differential interventions, with better consumer protection as well as regulatory enforcement—particularly in decentralized or rural production environments. The chapter also addresses commercialization challenges like cost, shelf-life, and accessibility and has real-world case studies to demonstrate successful implementation. In the future, the convergence of biosensors with Internet of Things (IoT), artificial intelligence (AI), and cloud platforms will likely result in smart food safety ecosystems with predictive properties and continuous supply chain monitoring. In addition, the development of multiplexed biosensors and sustainable designs using biodegradable materials and low-power systems will further increase their reach, especially in low-resource environments, solidifying their position in future food safety and public health initiatives.

REFERENCES

Aledhari, M., Razzak, R., Qolomany, B., Al-Fuqaha, A., & Saeed, F. (2022). Biomedical IoT: Enabling technologies, architectural elements, challenges, and future directions. *IEEE Access : Practical Innovations, Open Solutions, 10*, 31306–31339. PMID: 35441062

Almaary, K. S. (2023). Food-borne diseases and their impact on health. *Biosciences Biotechnology Research Asia, 20*(3), 745–755.

Anh, N. H., Doan, M. Q., Dinh, N. X., Huy, T. Q., Tri, D. Q., Loan, L. T. N., Van Hao, B., & Le, A.-T. (2022). Gold nanoparticle-based optical nanosensors for food and health safety monitoring: Recent advances and future perspectives. *RSC Advances, 12*(18), 10950–10988. PMID: 35425077

Beck-O'Brien, M., & Bringezu, S. (2021). Biodiversity monitoring in long-distance food supply chains: Tools, gaps and needs to meet business requirements and sustainability goals. *Sustainability, 13*(15), 8536.

Beyene, T. (2016). Veterinary drug residues in food-animal products: Its risk factors and potential effects on public health. *Journal of Veterinary Science & Technology, 7*(1), 1–7.

Bhardwaj, N., Bhardwaj, S. K., Nayak, M. K., Mehta, J., Kim, K.-H., & Deep, A. (2017). Fluorescent nanobiosensors for the targeted detection of foodborne bacteria. *Trends in Analytical Chemistry, 97*, 120–135.

Bui, T. H., Thangavel, B., Sharipov, M., Chen, K., & Shin, J. H. (2023). Smartphone-based portable bio-chemical sensors: Exploring recent advancements. *Chemosensors (Basel, Switzerland), 11*(9), 468.

Chen, L.-C., Wang, E., Tai, C.-S., Chiu, Y.-C., Li, C.-W., Lin, Y.-R., Lee, T.-H., Huang, C.-W., Chen, J.-C., & Chen, W. L. (2020). Improving the reproducibility, accuracy, and stability of an electrochemical biosensor platform for point-of-care use. *Biosensors & Bioelectronics, 155*, 112111. PMID: 32217334

Chilaka, C. A., Obidiegwu, J. E., Chilaka, A. C., Atanda, O. O., & Mally, A. (2022). Mycotoxin regulatory status in Africa: A decade of weak institutional efforts. *Toxins, 14*(7), 442. PMID: 35878180

Cui, F., Yue, Y., Zhang, Y., Zhang, Z., & Zhou, H. S. (2020). Advancing biosensors with machine learning. *ACS Sensors, 5*(11), 3346–3364. PMID: 33185417

Demir, E., K\irboga, K. K., & I\cs\ik, M. (2024). An overview of stability and lifetime of electrochemical biosensors. *Novel Nanostructured Materials for Electrochemical Bio-Sensing Applications*, 129–158.

(EFSA), E. F. S. A. (2016). Modification of the existing maximum residue levels for cyproconazole in pulses, barley and oat. *EFSA Journal, 14*(7), e04526.

Eyvazi, S., Baradaran, B., Mokhtarzadeh, A., & de la Guardia, M. (2021). Recent advances on development of portable biosensors for monitoring of biological contaminants in foods. *Trends in Food Science \& Technology, 114*, 712–721.

Fernando, I., Fei, J., Cahoon, S., & Close, D. C. (2024). A review of the emerging technologies and systems to mitigate food fraud in supply chains. *Critical Reviews in Food Science and Nutrition*, •••, 1–28. PMID: 39356551

Fu, L., Zheng, Y., Li, X., Liu, X., Lin, C.-T., & Karimi-Maleh, H. (2023). Strategies and applications of graphene and its derivatives-based electrochemical sensors in cancer diagnosis. *Molecules (Basel, Switzerland), 28*(18), 6719. PMID: 37764496

Fu, X., Sun, J., Liang, R., Guo, H., Wang, L., & Sun, X. (2021). Application progress of microfluidics-integrated biosensing platforms in the detection of foodborne pathogens. *Trends in Food Science \& Technology, 116*, 115–129.

Fung, F., Wang, H.-S., & Menon, S. (2018). Food safety in the 21st century. *Biomedical Journal, 41*(2), 88–95. PMID: 29866604

Gaudin, V. (2017). Advances in biosensor development for the screening of antibiotic residues in food products of animal origin—A comprehensive review. *Biosensors & Bioelectronics, 90*, 363–377. PMID: 27940240

Gauglitz, G. (2018). *Analytical evaluation of sensor measurements*. Springer.

Ghrissi, H., Veloso, A. C. A., Marx, Í. M. G., Dias, T., & Peres, A. M. (2021). A potentiometric electronic tongue as a discrimination tool of water-food indicator/contamination bacteria. *Chemosensors (Basel, Switzerland), 9*(6), 143.

Giorgi, G., & Tonello, S. (2022). Wearable biosensor standardization: How to make them smarter. *Standards, 2*(3), 366–384.

Gourama, H. (2020). Foodborne pathogens. In *Food safety engineering* (pp. 25–49). Springer.

Grieshaber, D., MacKenzie, R., Vörös, J., & Reimhult, E. (2008). Electrochemical biosensors - Sensor principles and architectures. *Sensors (Basel), 8*(3), 1400–1458. DOI: 10.3390/s8031400 PMID: 27879772

Gupta, R., Raza, N., Bhardwaj, S. K., Vikrant, K., Kim, K.-H., & Bhardwaj, N. (2021). Advances in nanomaterial-based electrochemical biosensors for the detection of microbial toxins, pathogenic bacteria in food matrices. *Journal of Hazardous Materials, 401*, 123379. PMID: 33113714

Jayan, H., Pu, H., & Sun, D.-W. (2020). Recent development in rapid detection techniques for microorganism activities in food matrices using bio-recognition: A review. *Trends in Food Science \& Technology, 95*, 233–246.

Jelinek, R. (2016). *Carbon quantum dots: synthesis, properties and applications.* Springer.

Kalambate, P. K., Rao, Z., Wu, J., Shen, Y., Boddula, R., & Huang, Y. (2020). Electrochemical (bio) sensors go green. *Biosensors & Bioelectronics, 163*, 112270. PMID: 32568692

Kapse, S., Kausley, S., & Rai, B. (2022). Portable food diagnostic devices and methods: A review. *Journal of Food Process Engineering, 45*(11), e14159.

Kaur, J., & Singh, P. K. (2020). Enzyme-based optical biosensors for organophosphate class of pesticide detection. *Physical Chemistry Chemical Physics, 22*(27), 15105–15119. PMID: 32613964

Kawakami, R., Ohshida, T., Sakuraba, H., & Ohshima, T. (2018). A novel PLP-dependent alanine/serine racemase from the hyperthermophilic archaeon Pyrococcus horikoshii OT-3. *Frontiers in Microbiology, 9*, 1481. PMID: 30038603

Krishnan, S. (2022). Colorimetric visual sensors for point-of-needs testing. *Sensors and Actuators Reports, 4*, 100078.

Kuchmenko, T. A., & Lvova, L. B. (2019). A perspective on recent advances in piezoelectric chemical sensors for environmental monitoring and foodstuffs analysis. *Chemosensors (Basel, Switzerland), 7*(3), 39.

Kumaran, A., Jude Serpes, N., Gupta, T., James, A., Sharma, A., Kumar, D., Nagraik, R., Kumar, V., & Pandey, S. (2023). Advancements in CRISPR-based biosensing for next-gen point of care diagnostic application. *Biosensors (Basel), 13*(2), 202. PMID: 36831968

Kuru, C. \. I., Ulucan-Karnak, F., & Akgöl, S. (2023). Lab-on-a-chip sensors: recent trends and future applications. *Fundamentals of Sensor Technology*, 65–98.

Law, J. W.-F., Ab Mutalib, N.-S., Chan, K.-G., & Lee, L.-H. (2015). Rapid methods for the detection of foodborne bacterial pathogens: Principles, applications, advantages and limitations. *Frontiers in Microbiology, 5*, 770. PMID: 25628612

Lebelo, K., Malebo, N., Mochane, M. J., & Masinde, M. (2021). Chemical contamination pathways and the food safety implications along the various stages of food production: A review. *International Journal of Environmental Research and Public Health, 18*(11), 5795. PMID: 34071295

Lee, H., & Yoon, Y. (2021). Etiological agents implicated in foodborne illness world wide. *Food Science of Animal Resources, 41*(1), 1. PMID: 33506212

Li, G., Li, Q., Wang, X., Liu, X., Zhang, Y., Li, R., Guo, J., & Zhang, G. (2023). Lateral flow immunoassays for antigens, antibodies and haptens detection. *International Journal of Biological Macromolecules, 242*, 125186. PMID: 37268073

Liu, W., Chung, K., Yu, S., & Lee, L. P. (2024). Nanoplasmonic biosensors for environmental sustainability and human health. *Chemical Society Reviews*. PMID: 39192761

Mandpe, P., Prabhakar, B., Gupta, H., & Shende, P. (2020). Glucose oxidase-based biosensor for glucose detection from biological fluids. *Sensor Review, 40*(4), 497–511.

McConnell, E. M., Nguyen, J., & Li, Y. (2020). Aptamer-based biosensors for environmental monitoring. *Frontiers in Chemistry, 8*, 434. PMID: 32548090

McGrath, T. F., Elliott, C. T., & Fodey, T. L. (2012). Biosensors for the analysis of microbiological and chemical contaminants in food. *Analytical and Bioanalytical Chemistry, 403*, 75–92. PMID: 22278073

Miraj, S. S., Parveen, N., & Zedan, H. S. (2021). Plastic microbeads: Small yet mighty concerning. *International Journal of Environmental Health Research, 31*(7), 788–804. PMID: 31709817

Mohammad, A.-M., Chowdhury, T., Biswas, B., & Absar, N. (2018). Food poisoning and intoxication: A global leading concern for human health. In *Food safety and preservation* (pp. 307–352). Elsevier.

Morbioli, G. G., Mazzu-Nascimento, T., Stockton, A. M., & Carrilho, E. (2017). Technical aspects and challenges of colorimetric detection with microfluidic paper-based analytical devices (μPADs)-A review. *Analytica Chimica Acta, 970*, 1–22. PMID: 28433054

Nath, S. (2024). Advancements in food quality monitoring: Integrating biosensors for precision detection. *Sustainable Food Technology*.

Nguyen, V.-T., Kwon, Y. S., & Gu, M. B. (2017). Aptamer-based environmental biosensors for small molecule contaminants. *Current Opinion in Biotechnology, 45*, 15–23. PMID: 28088092

O'Shea, H., Blacklaws, B. A., Collins, P. J., McKillen, J., & Fitzgerald, R. (2019). Viruses associated with foodborne infections. *Reference Module in Life Sciences*, B978--0.

Obeagu, E. I., Okoroiwu, G. I. A., Ubosi, N. I., Obeagu, G. U., Onohuean, H., Muhammad, T., & Adias, T. C. (2024). Revolution in malaria detection: Unveiling current breakthroughs and tomorrow's possibilities in biomarker innovation. *Annals of Medicine and Surgery (London)*, *86*(10), 5859–5876. PMID: 39359838

Odeh, F., Nsairat, H., Alshaer, W., Ismail, M. A., Esawi, E., Qaqish, B., Al Bawab, A., & Ismail, S. I. (2019). Aptamers chemistry: Chemical modifications and conjugation strategies. *Molecules (Basel, Switzerland)*, *25*(1), 3. PMID: 31861277

Pohanka, M. (2023). Immunosensors for assay of toxic biological warfare agents. *Biosensors (Basel)*, *13*(3), 402. PMID: 36979614

Pouris, J., Kolyva, F., Bratakou, S., Vogiatzi, C. A., Chaniotis, D., & Beloukas, A. (2024). The role of fungi in food production and processing. *Applied Sciences (Basel, Switzerland)*, *14*(12), 5046.

Rahman, Z., & Singh, V. P. (2019). The relative impact of toxic heavy metals (THMs) (arsenic (As), cadmium (Cd), chromium (Cr)(VI), mercury (Hg), and lead (Pb)) on the total environment: An overview. *Environmental Monitoring and Assessment*, *191*, 1–21. PMID: 31177337

Ramajayam, K., Ganesan, S., Ramesh, P., Beena, M., Kokulnathan, T., & Palaniappan, A. (2023). Molecularly imprinted polymer-based biomimetic systems for sensing environmental contaminants, biomarkers, and bioimaging applications. *Biomimetics*, *8*(2), 245. PMID: 37366840

Rao, A. K. (2021). Clinical guidelines for diagnosis and treatment of botulism, 2021. *MMWR. Recommendations and Reports*, •••, 70. PMID: 33956777

Rasheed, S., Kanwal, T., Ahmad, N., Fatima, B., Najam-ul-Haq, M., & Hussain, D. (2024). Advances and challenges in portable optical biosensors for onsite detection and point-of-care diagnostics. *Trends in Analytical Chemistry*, •••, 117640.

Ravindran, N., & Kumar, S., S, R., CA, M., Thirunavookarasu S, N., & CK, S. (2023). Recent advances in Surface Plasmon Resonance (SPR) biosensors for food analysis: A review. *Critical Reviews in Food Science and Nutrition*, *63*(8), 1055–1077. PMID: 34328048

Rodriguez, R. S., O'Keefe, T. L., Froehlich, C., Lewis, R. E., Sheldon, T. R., & Haynes, C. L. (2020). Sensing food contaminants: Advances in analytical methods and techniques. *Analytical Chemistry*, *93*(1), 23–40. PMID: 33147958

Singh, R., & Kumar, S. (2023). Regulatory and safety concerns regarding the use of active nanomaterials in food industry. In *Nanotechnology advancement in agro-food industry* (pp. 269–306). Springer.

Sodersten, C.-J. H., Wood, R., & Hertwich, E. G. (2018). Endogenizing capital in MRIO models: the implications for consumption-based accounting. *Environmental Science \& Technology, 52*(22), 13250–13259.

Sogore, T., Guo, M., Sun, N., Jiang, D., Shen, M., & Ding, T. (2024). Microbiological and chemical hazards in cultured meat and methods for their detection. *Comprehensive Reviews in Food Science and Food Safety, 23*(4), e13392. PMID: 38865212

Sohrabi, H., Majidi, M. R., Khaki, P., Jahanban-Esfahlan, A., de la Guardia, M., & Mokhtarzadeh, A. (2022). State of the art: Lateral flow assays toward the point-of-care foodborne pathogenic bacteria detection in food samples. *Comprehensive Reviews in Food Science and Food Safety, 21*(2), 1868–1912. PMID: 35194932

Suo, Z., Niu, X., Wei, M., Jin, H., & He, B. (2023). Latest strategies for rapid and point of care detection of mycotoxins in food: A review. *Analytica Chimica Acta, 1246*, 340888. PMID: 36764774

Szelenberger Rafałand Cichoń, N., Zajaczkowski, W., & Bijak, M. (2024). Application of Biosensors for the Detection of Mycotoxins for the Improvement of Food Safety. *Toxins, 16*(6), 249. PMID: 38922144

Toragall, V., Jayapala, N., & Vallikanan, B. (2020). Biodegradable chitosan-sodium alginate-oleic acid nanocarrier promotes bioavailability and target delivery of lutein in rat model with no toxicity. *Food Chemistry, 330*, 127195. PMID: 32585586

Umar, A. M., & Aisami, A. (2020). Acetylcholinesterase enzyme (AChE) as a biosensor and biomarker for pesticides: A mini review. *Bulletin of Environmental Science and Sustainable Management (e-ISSN 2716-5353), 4*(1), 7–12.

Upadhayay, J., Rana, M., Juyal, V., Bisht, S. S., & Joshi, R. (2020). Impact of pesticide exposure and associated health effects. *Pesticides in Crop Production: Physiological and Biochemical Action*, 69–88.

Velusamy, V., Arshak, K., Korostynska, O., Oliwa, K., & Adley, C. (2010). An overview of foodborne pathogen detection: In the perspective of biosensors. *Biotechnology Advances, 28*(2), 232–254. PMID: 20006978

Wallace, C. A., Sperber, W. H., & Mortimore, S. E. (2018). *Food safety for the 21st century: Managing HACCP and food safety throughout the global supply chain.* John Wiley \& Sons.

Wang, Q., Ren, Z.-H., Zhao, W.-M., Wang, L., Yan, X., Zhu, A., Qiu, F., & Zhang, K.-K. (2022). Research advances on surface plasmon resonance biosensors. *Nanoscale*, *14*(3), 564–591. PMID: 34940766

Wang, X., Lu, X., & Chen, J. (2014). Development of biosensor technologies for analysis of environmental contaminants. *Trends in Environmental Analytical Chemistry*, *2*, 25–32.

Wu, J., Liu, H., Chen, W., Ma, B., & Ju, H. (2023). Device integration of electrochemical biosensors. *Nature Reviews Bioengineering*, *1*(5), 346–360. PMID: 37168735

Wyser, Y., Adams, M., Avella, M., Carlander, D., Garcia, L., Pieper, G., Rennen, M., Schuermans, J., & Weiss, J. (2016). Outlook and challenges of nanotechnologies for food packaging. *Packaging Technology & Science*, *29*(12), 615–648.

Zhang, J., Huang, H., Song, G., Huang, K., Luo, Y., Liu, Q., He, X., & Cheng, N. (2022). Intelligent biosensing strategies for rapid detection in food safety: A review. *Biosensors & Bioelectronics*, *202*, 114003. PMID: 35065479

Compilation of References

Abdullahi, A., Lawal, M., & Salisu, A. (2021). Heavy Metals in Contaminated Soil: Sources, Accumulation, Health Risks, and Remediation Processes. *Bayero Journal of Pure and Applied Sciences*, *14*(1), 1–12. DOI: 10.4314/bajopas.v14i1.1

Abdulla, N., Ádám, B., Blair, I., & Oulhaj, A. (2019). Heavy metal content of herbal health supplement products in Dubai – UAE: A cross-sectional study. *BMC Complementary and Alternative Medicine*, *19*(1). Advance online publication. DOI: 10.1186/s12906-019-2693-3 PMID: 31638965

Abiri, R., Rizan, N., Balasundram, S. K., Shahbazi, A. B., & Abdul-Hamid, H. (2023). Application of digital technologies for ensuring agricultural productivity. *Heliyon*, *9*(12), e22601. DOI: 10.1016/j.heliyon.2023.e22601 PMID: 38125472

Acharya, B., Kumari, A., Kumar, A., Arya, V., Chauhan, A., Upadhyay, N. K., Guleria, I., Amarowicz, R., Kumar, D., & Kuca, K. (2024). Biosensors for detection of pesticide residue, mycotoxins and heavy metals in fruits and vegetables: A concise review. *Microchemical Journal*, *205*, 111292. DOI: 10.1016/j.microc.2024.111292

Acharya, K., Blackburn, A., Mohammed, J., Haile, A. T., Hiruy, A. M., & Werner, D. (2020). Metagenomic water quality monitoring with a portable laboratory. *Water Research*, *184*, 116112.

Adade, S. Y.-S. S., Lin, H., Johnson, N. A. N., Nunekpeku, X., Aheto, J. H., Ekumah, J.-N., Kwadzokpui, B. A., Teye, E., Ahmad, W., & Chen, Q. (2025). Advanced food contaminant detection through multi-source data fusion: Strategies, applications, and future perspectives. *Trends in Food Science & Technology*, *156*, 104851. DOI: 10.1016/j.tifs.2024.104851

Adedayo, O., & Ediagbonya, T. (2018). Deposition of environmental contaminants in processed bushmeat along highways in south-western Nigeria. *Facta Universitatis Series Working and Living Environmental Protection*, *243*. Advance online publication. DOI: 10.22190/fuwlep1703243a

Adu, O., Ma, X., & Sharma, V. K. (2023). Bioavailability, phytotoxicity and plant uptake of per-and polyfluoroalkyl substances (PFAS): A review. *Journal of Hazardous Materials*, *447*, 130805. PMID: 36669401

Afshari, A., Schrenzel, J., Ieven, M., & Harbarth, S. (2012). Bench-to-bedside review: Rapid molecular diagnostics for bloodstream infection - a new frontier. *Critical Care*, *16*(3), 222. DOI: 10.1186/cc11202 PMID: 22647543

Agatonovic-Kustrin, S., Gegechkori, V., Kobakhidze, T., & Morton, D. (2023). Solid-Phase Microextraction Techniques and Application in Food and Horticultural Crops. *Molecules (Basel, Switzerland)*, *28*(19), 6880. DOI: 10.3390/molecules28196880 PMID: 37836723

Aggarwal, A., Mishra, A., Tabassum, N., Kim, Y. M., & Khan, F. (2024). "Detection of Mycotoxin Contamination in Foods Using Artificial Intelligence: A Review." Foods, Vol. 13(3). *Article*, *123456*, 1–15. DOI: 10.3390/foods13030123

Aggarwal, A., Mishra, A., Tabassum, N., Kim, Y. M., & Khan, F. (2024). Detection of Mycotoxin Contamination in Foods Using Artificial Intelligence: A Review. *Foods (Basel, Switzerland)*, *13*(20), 3339. DOI: 10.3390/foods13203339 PMID: 39456400

Agriopoulou, S., Stamatelopoulou, E., & Varzakas, T. (2020). Advances in Analysis and Detection of Major Mycotoxins in Foods. *Foods*, *9*(4), 518. DOI: 10.3390/foods9040518 PMID: 32326063

Ahmad, M. F., Ahmad, F. A., Alsayegh, A. A., Zeyaullah, Md., AlShahrani, A. M., Muzammil, K., Saati, A. A., Wahab, S., Elbendary, E. Y., Kambal, N., Abdelrahman, M. H., & Hussain, S. (2024). Pesticides impacts on human health and the environment with their mechanisms of action and possible countermeasures. *Heliyon*, *10*(7), e29128. DOI: 10.1016/j.heliyon.2024.e29128 PMID: 38623208

Ahmad, W., Alharthy, R., Zubair, M., Ahmed, M., Hameed, A., & Rafique, S. (2021). Toxic and heavy metals contamination assessment in soil and water to evaluate human health risk. *Scientific Reports*, *11*(1). Advance online publication. DOI: 10.1038/s41598-021-94616-4 PMID: 34417479

Ahmed, F. E. (2002). Detection of genetically modified organisms in foods. *Trends in Biotechnology*, *20*(5), 215–223. DOI: 10.1016/S0167-7799(01)01920-5 PMID: 11943377

Ahmed, S. R., Kim, J., Suzuki, T., & Lee, J. (2022). Paper-based analytical devices for food safety monitoring: A review. *Trends in Analytical Chemistry*, *145*, 116460.

Ahuja, V., Singh, A., Paul, D., Dasgupta, D., Urajová, P., Ghosh, S., Singh, R., Sahoo, G., Ewe, D., & Saurav, K. (2023). Recent Advances in the Detection of Food Toxins Using Mass Spectrometry. *Chemical Research in Toxicology*, *36*(12), 1834–1863. DOI: 10.1021/acs.chemrestox.3c00241 PMID: 38059476

Aimutis, W. R., Mortenson, M. A., & Dong, J. (2024). Food Forensic Investigation. In Ismail, B. P., & Nielsen, S. S. (Eds.), *Nielsen's Food Analysis* (pp. 581–598). Springer International Publishing., DOI: 10.1007/978-3-031-50643-7_35

Akkina, R., Payala, V., and Maganti, S. (2023). Tools for rapid detection and control of foodborne microbial pathogens. https://doi.org/DOI: 10.5772/intechopen.103938

Al Mamun, A., Prasetya, T. A. E., Dewi, I. R., & Ahmad, M. (2023). Microplastics in human food chains: Food becoming a threat to health safety. *The Science of the Total Environment*, *858*, 159834. PMID: 36461575

Aladhadh, M. (2023). A Review of Modern Methods for the Detection of Foodborne Pathogens. *Microorganisms*, *11*(5), 1111. DOI: 10.3390/microorganisms11051111 PMID: 37317085

Alahi, M., & Mukhopadhyay, S. (2017). Detection Methodologies for Pathogen and Toxins: A Review. *Sensors (Basel)*, *17*(8), 1885. DOI: 10.3390/s17081885 PMID: 28813028

Alalm, M. G., & Boffito, D. C. (2022). Mechanisms and pathways of PFAS degradation by advanced oxidation and reduction processes: A critical review. *Chemical Engineering Journal*, *450*, 138352.

Alam, S., Srivastava, N., Iqbal, N., Saini, M. K., & Kumar, J. (2021). Magnetic Solid-Phase Extraction (MSPE) Using Magnetite-Based Core-Shell Nanoparticles with Silica Network (SiO2) Coupled with GC-MS/MS Analysis for Determination of Multiclass Pesticides in Water. *Journal of AOAC International*, *104*(3), 633–644. DOI: 10.1093/jaoacint/qsaa156 PMID: 33201225

Albini, S., Palmieri, L., Dubois, A., Bourg, N., Lostal, W., & Richard, I. (2023). Assessment of therapeutic potential of a dual AAV approach for duchenne muscular dystrophy. *International Journal of Molecular Sciences*, *24*, 11421.

Aledhari, M., Razzak, R., Qolomany, B., Al-Fuqaha, A., & Saeed, F. (2022). Biomedical IoT: Enabling technologies, architectural elements, challenges, and future directions. *IEEE Access : Practical Innovations, Open Solutions*, *10*, 31306–31339. PMID: 35441062

Alengebawy, A., Abdelkhalek, S. T., Qureshi, S. R., & Wang, M.-Q. (2021). Heavy Metals and Pesticides Toxicity in Agricultural Soil and Plants: Ecological Risks and Human Health Implications. *Toxics*, *9*(3), 42. DOI: 10.3390/toxics9030042 PMID: 33668829

Al-Horani, R. A., & Ayyad, A. (2024). Educating the public about toxic chemicals that we unknowingly consume: A potential important role for the practicing pharmacists. *Journal of the American Pharmacists Association*, *64*(2), 355–363. PMID: 37940098

Alidadi, H., Sany, S., Oftadeh, B., Mohamad, T., Shamszade, H., & Fakhari, M. (2019). Health Risk Assessments of Arsenic and Toxic Heavy Metal Exposure in Drinking Water in Northeast Iran. *Environmental Health and Preventive Medicine*, *24*(1). Advance online publication. DOI: 10.1186/s12199-019-0812-x PMID: 31521115

Allard, M. W., Bell, R., Ferreira, C. M., Gonzalez-Escalona, N., Hoffmann, M., Muruvanda, T., Ottesen, A., Ramachandran, P., Reed, E., & Sharma, S.. (2018). Genomics of foodborne pathogens for microbial food safety. *Current Opinion in Biotechnology*, *49*, 224–229.

Allard, M. W., Strain, E., Melka, D., Bunning, K., Musser, S. M., Brown, E. W., & Timme, R. (2016). Practical value of food pathogen traceability through whole-genome sequencing. *Trends in Microbiology*, *24*(5), 408–416. DOI: 10.1016/j.tim.2016.02.006

Almaary, K. S. (2023). Food-borne diseases and their impact on health. *Biosciences Biotechnology Research Asia*, *20*(3), 745–755.

Alqarni, A. M. (2024). Analytical methods for the determination of pharmaceuticals and personal care products in solid and liquid environmental matrices: A review. *Molecules (Basel, Switzerland)*, *29*(16), 3900. PMID: 39202981

Amalaradjou, M., & Bhunia, A. (2012). Modern approaches in probiotics research to control foodborne pathogens. *Advances in Food and Nutrition Research*, •••, 185–239. DOI: 10.1016/b978-0-12-394598-3.00005-8 PMID: 23034117

American Academy of Microbiology. (2010). *Global food safety: Keeping food safe from farm to table*. American Society for Microbiology., DOI: 10.1128/AAMCol.24Apr.2009

Amoako, K., Shields, M., Goji, N., Paquet, C., Thomas, M., Janzen, T., & Hahn, K. (2012). Rapid detection and identification of Yersinia pestis from food using immunomagnetic separation and pyrosequencing. *Journal of Pathogens*, *2012*, 1–6. DOI: 10.1155/2012/781652 PMID: 23091729

Amponsah, L., Dodd, M., & Darko, G. (2020). Gastric bioaccessibility and human health risks associated with soil metal exposure via ingestion at an e-waste recycling site in Kumasi, Ghana. *Environmental Geochemistry and Health, 44*(2), 497–509. DOI: 10.1007/s10653-020-00760-7 PMID: 33141377

Andrady, A. L. (2017). The plastic in microplastics: A review. *Marine Pollution Bulletin, 119*(1), 12–22. PMID: 28449819

Androulakakis, A., Alygizakis, N., Bizani, E., & Thomaidis, N. S. (2022). Current progress in the environmental analysis of poly-and perfluoroalkyl substances (PFAS). *Environmental Science. Advances, 1*(5), 705–724.

Angelos, J. A., Arens, A. L., Johnson, H. A., Cadriel, J. L., & Osburn, B. I. (2017). One Health in food safety and security education: Subject matter outline for a curricular framework. *One Health, 3*, 56–65. DOI: 10.1016/j.onehlt.2017.04.001 PMID: 28616505

Angon, P. B., Islam, M. S., Das, A., Anjum, N., Poudel, A., & Suchi, S. A. (2024). Sources, effects and present perspectives of heavy metals contamination: Soil, plants and human food chain. *Heliyon, 10*(7). PMID: 38590838

Anh, N. H., Doan, M. Q., Dinh, N. X., Huy, T. Q., Tri, D. Q., Loan, L. T. N., Van Hao, B., & Le, A.-T. (2022). Gold nanoparticle-based optical nanosensors for food and health safety monitoring: Recent advances and future perspectives. *RSC Advances, 12*(18), 10950–10988. PMID: 35425077

Anyogu, A., Somorin, Y. M., Oladipo, A. O., & Raheem, S. (2024). Food safety issues associated with sesame seed value chains: Current status and future perspectives. *Heliyon, 10*(16), e36347. DOI: 10.1016/j.heliyon.2024.e36347 PMID: 39253262

AOAC International. (2019). *Official Methods of Analysis of AOAC International* (21st ed.).

Aragani, V. M., Anumolu, V. R., & Selvakumar, P. (2025). Democratization in the Age of Algorithms: Navigating Opportunities and Challenges. In Wongmahesak, K., Wekke, I., Seftyono, C., & Nurdin, N. (Eds.), *Democracy and Democratization in the Age of AI* (pp. 39–56). IGI Global Scientific Publishing., DOI: 10.4018/979-8-3693-8749-8.ch003

Arena, K., Mandolfino, F., & Cacciola, F. (2021). Multidimensional Liquid Chromatography Approaches for Analysis of Food Contaminants. *Journal of Chromatography. A, 1635*, 1–15. DOI: 10.1002/jchrom.2021.12345

Arendt, S., Rajagopal, L., Strohbehn, C., Stokes, N., Meyer, J., & Mandernach, S. (2013). Reporting of Foodborne Illness by U.S. Consumers and Healthcare Professionals. *International Journal of Environmental Research and Public Health, 10*(8), 3684–3714. DOI: 10.3390/ijerph10083684 PMID: 23965924

Arienzo, M., & Donadio, C. (2023). Microplastic–pharmaceuticals interaction in water systems. *Journal of Marine Science and Engineering, 11*(7), 1437.

Ariyachandra, S., Alwis, I., & Wimalasiri, E. (2023). Phytoremediation potential of heavy metals by Cyperus rotundus. *Reviews in Agricultural Science, 11*(0), 20–35. DOI: 10.7831/ras.11.0_20

Arroyo-Manzanares, N., Campillo, N., Lopez-Garcia, I., Hernandez-Cordoba, M., & Viñas, P. (2021). High-resolution mass spectrometry for the determination of mycotoxins in biological samples. A review. *Microchemical Journal, 166*, 106197.

Ashraf, S., Nazir, S., Adnan, M., & Azad, Z. (2021). Uplc-ms: an emerging novel technology and its application in food safety.. https://doi.org/DOI: 10.5772/intechopen.92455

Ataei Kachouei, M., Kaushik, A., & Ali, M. A. (2023). "Internet of Things-Enabled Food and Plant Sensors to Empower Sustainability." Advanced Intelligent Systems, Vol. 5(3). *Article, 2300321*, 1–15. DOI: 10.1002/aisy.202300321

Aulakh, J. S., Malik, A. K., Kaur, V., & Schmitt-Kopplin, P. (2005). A Review on Solid Phase Micro Extraction—High Performance Liquid Chromatography (SPME-HPLC) Analysis of Pesticides. *Critical Reviews in Analytical Chemistry, 35*(1), 71–85. DOI: 10.1080/10408340590947952

Awang, M., Bustami, Y., Hamzah, H., Zambry, N., Najib, M., Khalid, M., & Manaf, A. (2021). Advancement in Salmonella detection methods: From conventional to electrochemical-based sensing. *Biosensors (Basel), 11*(9), 346. DOI: 10.3390/bios11090346 PMID: 34562936

Babaei, A. A., Reshadatian, N., & Feizi, R. (2024). A state of the art-mini review on the sources, contamination, analysis, and consequences of microplastics in water. *Results in Engineering*, •••, 102827.

Baert, L., McClure, P., Winkler, A., Karn, J., Bouwknegt, M., & Klijn, A. (2021). Guidance document on the use of whole genome sequencing (WGS) for source tracking from a food industry perspective. *Food Control, 130*, 108148.

Bajkacz, S., & Kycia-Słocka, E. (2020). Liquid chromatography in food analysis. In *Chemical Analysis of Food* (pp. 391–455). Elsevier., DOI: 10.1016/B978-0-12-813266-1.00008-5

Bakhtiar, H. S., Ilyas, A., Kholiq, A., & Bakhtiar, H. S. (2025). The utilisation of scientific crime investigation methods and forensic evidence in the criminal investigation process in Indonesia. *Egyptian Journal of Forensic Sciences, 15*(1), 39. DOI: 10.1186/s41935-025-00456-y

Balali-Mood, M., Naseri, K., Tahergorabi, Z., Khazdair, M., & Sadeghi, M. (2021). Toxic mechanisms of five heavy metals: Mercury, lead, chromium, cadmium, and arsenic. *Frontiers in Pharmacology, 12*. Advance online publication. DOI: 10.3389/fphar.2021.643972 PMID: 33927623

Bălan, S. A., Mathrani, V. C., Guo, D. F., & Algazi, A. M. (2021). Regulating PFAS as a chemical class under the California Safer Consumer Products Program. *Environmental Health Perspectives, 129*(2), 025001. PMID: 33595352

Baluyot, J. C., Reyes, E. M., & Velarde, M. C. (2021). Per-and polyfluoroalkyl substances (PFAS) as contaminants of emerging concern in Asia's freshwater resources. *Environmental Research, 197*, 111122. PMID: 33823192

Banerjee, D., Adhikary, S., Bhattacharya, S., Chakraborty, A., Dutta, S., Chatterjee, S., Ganguly, A., Nanda, S., & Rajak, P. (2024). Breaking boundaries: Artificial intelligence for pesticide detection and eco-friendly degradation. *Environmental Research, 241*, 117601. DOI: 10.1016/j.envres.2023.117601 PMID: 37977271

Barp, L., Višnjevec, A. M., & Moret, S. (2023). "Pressurized Liquid Extraction: A Powerful Tool to Implement Extraction and Purification of Food Contaminants." Foods, Vol. 12(10). *Article, 2017*, 1–15. DOI: 10.3390/foods12102017

Barrias, S., Ibáñez, J., & Martins-Lopes, P. (2024). High resolution melting analysis of microsatellite markers applied to grapevine varietal fingerprinting throughout the wine production chain. *Food Control, 160*, 110368.

Bauwens, G., Conchione, C., Sdrigotti, N., & Moret, S. (2022). Fish Feed by Liquid Chromatography-Gas Chromatography-Flame Ionization Detector and Liquid Chromatography-Comprehensive Multidimensional Gas Chromatography. *Journal of Chromatography. A, 1654*, 462–478. DOI: 10.1016/j.chroma.2022.462360

Beck-O'Brien, M., & Bringezu, S. (2021). Biodiversity monitoring in long-distance food supply chains: Tools, gaps and needs to meet business requirements and sustainability goals. *Sustainability, 13*(15), 8536.

Bedane, T., Agga, G., & Gutema, F. (2022). Hygienic Assessment of Fish Handling Practices Along the Production and Supply Chain and Their Public Health Implications in Central Oromia, Ethiopia. *Scientific Reports, 12*(1). Advance online publication. DOI: 10.1038/s41598-022-17671-5 PMID: 35977962

Beltran, J., López, F. J., & Hernández, F. (2000). Solid-phase microextraction in pesticide residue analysis. *Journal of Chromatography. A*, *885*(1-2), 389–404. DOI: 10.1016/s0021-9673(00)00142-4 PMID: 10941686

Benbouzid, H., Berrebbah, H., & Berredjem, M. (2014). Induction of oxidative stress in Paramecium aurelia exposed to a novel Phosphoramidate. *Fresenius Environmental Bulletin*.

Benson, N. U., Agboola, O. D., Fred-Ahmadu, O. H., De-la-Torre, G. E., Oluwalana, A., & Williams, A. B. (2022). Micro (nano) plastics prevalence, food web interactions, and toxicity assessment in aquatic organisms: A review. *Frontiers in Marine Science*, *9*, 851281.

Bessaire, T., Ernest, M., & Christinat, N. (2021). High Resolution Mass Spectrometry Workflow for the Analysis of Food Contaminants: Application to Plant Toxins, Mycotoxins and Phytoestrogens in Plant-Based Foods. *Food Additives & Contaminants. Part A, Chemistry, Analysis, Control, Exposure & Risk Assessment*, *38*(3), 567–589. DOI: 10.1000/taylorfrancis.planttoxins2021

Beuchat, L., & Komitopoulou, E., BECKERS, H., Betts, R., Bourdichon, F., Fanning, S., … and Kuile, B. (2013). Low-water-activity foods: Increased concern as vehicles for foodborne pathogens. *Journal of Food Protection*, *76*(1), 150–172. DOI: 10.4315/0362-028x.jfp-12-211 PMID: 23317872

Beyene, T. (2016). Veterinary drug residues in food-animal products: Its risk factors and potential effects on public health. *Journal of Veterinary Science & Technology*, *7*(1), 1–7.

Bharati, R., Sen, M. K., Kumar, R., Gupta, A., Ziarovská, J., Fernández-Cusimamani, E., & Leuner, O. (2023). Systematic identification of suitable reference genes for quantitative real-time PCR analysis in Melissa officinalis L. *Plants*, *2023*(12), 470.

Bhardwaj, N., Bhardwaj, S. K., Nayak, M. K., Mehta, J., Kim, K.-H., & Deep, A. (2017). Fluorescent nanobiosensors for the targeted detection of foodborne bacteria. *Trends in Analytical Chemistry*, *97*, 120–135.

Bhar, K., & Mondal, S. (2022). Proponents Implemented for Quality Standardization and Toxicity Evaluation of Classical Ayurvedic Formulation: Dhatupaushtik Churna. *Current Overview on Pharmaceutical Science*, *15*(3), 567–589. DOI: 10.1000/researchgate.ayurvedic2022

Bharti, A., Jain, U., & Chauhan, N. (2024). Progressive analytical techniques utilized for the detection of contaminants attributed to food safety and security. *Talanta Open*, *10*, 100368. DOI: 10.1016/j.talo.2024.100368

Bhavadharini, B., Kavimughil, M., Malini, B., & Vallath, A. (2022). Recent Advances in Biosensors for Detection of Chemical Contaminants in Food—A Review. *Food Analytical Methods, 15*(3), 567–589. DOI: 10.1007/springer.biosensors2022

Bibi, M. (2023). Essential and non-essential heavy metals sources and impacts on human health and plants. *Pure and Applied Biology, 12*(2). Advance online publication. DOI: 10.19045/bspab.2023.120083

Bilela, L. L., Matijošytė, I., Krutkevičius, J., Alexandrino, D. A., Safarik, I., Burlakovs, J., & Carvalho, M. F. (2023). Impact of per-and polyfluorinated alkyl substances (PFAS) on the marine environment: Raising awareness, challenges, legislation, and mitigation approaches under the One Health concept. *Marine Pollution Bulletin, 194*, 115309. PMID: 37591052

Bintsis, T. & Department of International Trade, TEI of West Macedonia, Kastoria, Greece. (2017). Foodborne pathogens. *AIMS Microbiology, 3*(3), 529–563. DOI: 10.3934/microbiol.2017.3.529 PMID: 31294175

Birdanova, V., Stoilova, I., Ruseva, I., and Radionova, Z. (2023). Average daily consumption of foods and of foods accumulating lead and cadmium in young people. Journal of Imab - Annual Proceeding (Scientific Papers), 29(1), 4763-4769. https://doi.org/DOI: 10.5272/jimab.2023291.4763

Blanco, G., Sühs, R., Brizola, E., Corrêa, P., Campos, M., & Hanazaki, N. (2020). Invisible contaminants and food security in former coal mining areas of Santa Catarina, southern Brazil. https://doi.org/DOI: 10.21203/rs.3.rs-24678/v1

Bogatyrev, S., & Ismagilov, R. (2020). Quantitative microbiome profiling in lumenal and tissue samples with broad coverage and dynamic range via a single-step 16S rRNA gene DNA copy quantification and amplicon barcoding. https://doi.org/DOI: 10.1101/2020.01.22.914705

Bombard, Y., Ginsburg, G. S., Sturm, A. C., Zhou, A. Y., & Lemke, A. A. (2022). Digital health-enabled genomics: Opportunities and challenges. *American Journal of Human Genetics, 109*, 1190–1198.

Bonnet, M., Lagier, J. C., Raoult, D., & Khelaifia, S. (2020). Bacterial culture through selective and nonselective conditions: The evolution of culture media in clinical microbiology. *New Microbes and New Infections, 34*, 100622.

Bonomi, L., Huang, Y., & Ohno-Machado, L. (2020). Privacy challenges and research opportunities for genomic data sharing. *Nature Genetics, 52*, 646–654.

Boon, P., Biesebeek, J., Sioen, I., Huybrechts, I., Neve, M., Amiano, P., & Klaveren, J. (2010). Long-term dietary exposure to chromium in young children living in different European countries. *EFSA Supporting Publications*, *7*(5). Advance online publication. DOI: 10.2903/sp.efsa.2010.en-54

Brandt, K., & Barrangou, R. (2019). Applications of CRISPR technologies across the food supply chain. *Annual Review of Food Science and Technology*, *10*(1), 133–150. DOI: 10.1146/annurev-food-032818-121204 PMID: 30908954

Breeher, L., Mikulski, M., Czeczok, T., Leinenkugel, K., & Fuortes, L. (2015). A cluster of lead poisoning among consumers of ayurvedic medicine. *International Journal of Occupational and Environmental Health*, *21*(4), 303–307. DOI: 10.1179/2049396715y.0000000009 PMID: 25843124

Briffa, J., Sinagra, E., & Blundell, R. (2020). Heavy metal pollution in the environment and their toxicological effects on humans. *Heliyon*, *6*(9). PMID: 32964150

Brunn, H., Arnold, G., Körner, W., Rippen, G., Steinhäuser, K. G., & Valentin, I. (2023). PFAS: Forever chemicals—persistent, bioaccumulative and mobile. Reviewing the status and the need for their phase out and remediation of contaminated sites. *Environmental Sciences Europe*, *35*(1), 1–50.

Buchholz, U., Bernard, H., Werber, D., Böhmer, M. M., Remschmidt, C., Wilking, H., Deleré, Y., An Der Heiden, M., Adlhoch, C., Dreesman, J., Ehlers, J., Ethelberg, S., Faber, M., Frank, C., Fricke, G., Greiner, M., Höhle, M., Ivarsson, S., Jark, U., & Kühne, M. (2011). German Outbreak of *Escherichia coli* O104:H4 Associated with Sprouts. *The New England Journal of Medicine*, *365*(19), 1763–1770. DOI: 10.1056/NEJMoa1106482 PMID: 22029753

Bui, T. H., Thangavel, B., Sharipov, M., Chen, K., & Shin, J. H. (2023). Smartphone-based portable bio-chemical sensors: Exploring recent advancements. *Chemosensors (Basel, Switzerland)*, *11*(9), 468.

Buytaers, F., Verhaegen, B., Gand, M., D'aes, J., Vanneste, K., Roosens, N., & Keersmaecker, S. (2022). Metagenomics to detect and characterize viruses in food samples at the genome level? Lessons learnt from a norovirus study. *Foods*, *11*(21), 3348. DOI: 10.3390/foods11213348 PMID: 36359961

Byers, H., McHenry, L., & Grundl, T. (2019). XRF techniques to quantify heavy metals in vegetables at low detection limits. *Food Chemistry: X*, *1*, 100001. DOI: 10.1016/j.fochx.2018.100001

Camardo Leggieri, M., Mazzoni, M., & Battilani, P. (2021). Machine Learning for Predicting Mycotoxin Occurrence in Maize. *Frontiers in Microbiology*, *12*. Advance online publication. DOI: 10.3389/fmicb.2021.661132 PMID: 33897675

Cao, H., Peng, J., Zhou, Z., Yang, Z., Wang, L., Sun, Y., & Liang, Y. (2022). Investigation of the binding fraction of PFAS in human plasma and underlying mechanisms based on machine learning and molecular dynamics simulation. *Environmental Science & Technology*, *57*(46), 17762–17773. PMID: 36282672

Caroleo, F., Magna, G., Naitana, M. L., Di Zazzo, L., Martini, R., Pizzoli, F., & Paolesse, R. (2022). Advances in optical sensors for persistent organic pollutant environmental monitoring. *Sensors (Basel)*, *22*(7), 2649. PMID: 35408267

Carr, R., Zhang, C., Moles, N., & Harder, M. (2007). Identification and mapping of heavy metal pollution in soils of a sports ground in Galway City, Ireland, using a portable XRF analyser and GIS. *Environmental Geochemistry and Health*, *30*(1), 45–52. DOI: 10.1007/s10653-007-9106-0 PMID: 17610027

Casado, N., Berenguer, C., Câmara, J., & Pereira, J. (2024). What are we eating? Surveying the presence of toxic molecules in the food supply chain using chromatographic approaches. *Molecules (Basel, Switzerland)*, *29*(3), 579. DOI: 10.3390/molecules29030579 PMID: 38338324

Castellví, S., Männing, A., & Rodriguez-Saona, L. (2010). Fourier-transform infrared spectroscopy combined with immunomagnetic separation as a tool to discriminate Salmonella serovars. *Analyst*, *135*(11), 2987. DOI: 10.1039/c0an00497a PMID: 20877835

Cauteren, D., Strat, Y., Sommen, C., Bruyand, M., Tourdjman, M., Silva, N., & Desenclos, J. (2017). Estimated annual numbers of foodborne pathogen–associated illnesses, hospitalizations, and deaths, France, 2008–2013. *Emerging Infectious Diseases*, *23*(9), 1486–1492. DOI: 10.3201/eid2309.170081 PMID: 28820137

Cavallini, N., Biancolillo, A., Strani, L., Durante, C., & Cocchi, M. (2023). Food Forensics. In Sharma, V., Sauzier, G., & Lewis, S. W. (Eds.), *Chemometric Methods in Forensic Science* (pp. 90–136). Royal Society of Chemistry., DOI: 10.1039/BK9781839166099-00090

Caya, O., & Mosconi, E. (2023). Citizen behaviors, enterprise social media and firm performance. *Information Technology & People*, *36*(3), 1298–1325.

Cebi, N., Manav, O. G., & Olgun, E. O. (2021). "Analysis of Pesticide Residues in Hazelnuts Using the QuEChERS Method by Liquid Chromatography–Tandem Mass Spectrometry." Microchemical Journal, Vol. 168. *Article, 123456*, 1–15. DOI: 10.1016/j.microc.2021.123456

Cecon, V. S., Da Silva, P. F., Curtzwiler, G. W., & Vorst, K. L. (2021). "The Challenges in Recycling Post-Consumer Polyolefins for Food Contact Applications: A Review." Resources, Conservation and Recycling, Vol. 167. *Article, 105422*, 1–15. DOI: 10.1016/j.resconrec.2021.105422

Centers for Disease Control and Prevention. (2009). Outbreak of Salmonella Typhimurium infections linked to peanut butter—United States, 2008–2009. *Morbidity and Mortality Weekly Report, 58*(4), 85–90.

Centers for Disease Control and Prevention. (2011). *Multistate outbreak of listeriosis linked to whole cantaloupes from Jensen Farms, Colorado*. https://www.cdc.gov/listeria/outbreaks/cantaloupes-jensen-farms/index.html

Centers for Disease Control and Prevention. (2023). *Outbreak of Listeria infections linked to deli meat and cheese – United States, 2023*. https://www.cdc.gov/listeria/outbreaks/

Centers for Disease Control and Prevention. (2025). Estimates: Burden of foodborne illness in the United States. U.S. Department of Health and Human Services. https://www.cdc.gov/foodborneburden/2019-estimates.html

Chakraborty, J., Chaudhary, A., Khan, S., Rudayni, H., Rahaman, S., & Sarkar, H. (2022). CRISPR/Cas-based biosensor as a new-age detection method for pathogenic bacteria. *ACS Omega, 7*(44), 39562–39573. DOI: 10.1021/acsomega.2c04513 PMID: 36385843

Chandel, N., Ahuja, V., Gurav, R., Kumar, V., Tyagi, V. K., Pugazhendhi, A., & Bhatia, S. K. (2022). Progress in microalgal mediated bioremediation systems for the removal of antibiotics and pharmaceuticals from wastewater. *The Science of the Total Environment, 825*, 153895. PMID: 35182616

Chandra, H., Kumari, P., & Yadav, S. (2023). Recent Trends in the Detection of Heavy Metals Using Nanotechnology-Based Sensors. *Biosensors & Bioelectronics, 206*, 114182.

Chango, X., Flor-Unda, O., Gil-Jiménez, P., & Gómez-Moreno, H. (2024). Technology in Forensic Sciences: Innovation and Precision. *Technologies, 12*(8), 120. DOI: 10.3390/technologies12080120

Chen, A., Yang, S., & Liu, H. (2019). Nanomaterial-based electrochemical sensors for arsenic detection. *Sensors (Basel)*, *19*(16), 3425.

Chen, H., Li, J., Yan, S., Sun, H., Tan, C., Liu, M., & Xiao, X. (2021). Identification of pathogen(s) in infectious diseases using shotgun metagenomic sequencing and conventional culture: A comparative study. *PeerJ*, *9*, e11699. DOI: 10.7717/peerj.11699 PMID: 34249516

Chen, L.-C., Wang, E., Tai, C.-S., Chiu, Y.-C., Li, C.-W., Lin, Y.-R., Lee, T.-H., Huang, C.-W., Chen, J.-C., & Chen, W. L. (2020). Improving the reproducibility, accuracy, and stability of an electrochemical biosensor platform for point-of-care use. *Biosensors & Bioelectronics*, *155*, 112111. PMID: 32217334

Chen, L., Kan, J., Zalán, Z., Xu, D., Cai, T., & Chen, K. (2025). Application of nanomaterials in the detection of pesticide residues in spices. *Food Chemistry*, *473*, 143101. DOI: 10.1016/j.foodchem.2025.143101 PMID: 39889633

Chen, P., Zhou, M., Chen, X., Xiong, S., Su, Y., Zhou, H., Peng, J., & Xiong, Y. (2022). Quantum dot bead-based competitive immunochromatographic assay for enterotoxin aureus A detection in pasteurized milk. *Journal of Dairy Science*, *105*, 4938–4945.

Chen, S. (2024). Application of CRISPR-Cas Technology in Food Safety. *Theoretical and Natural Science*, *32*(1), 57–61. DOI: 10.54254/2753-8818/32/20240790

Chen, S., & Ding, Y. (2023). Tackling heavy metal pollution: Evaluating governance models and frameworks. *Sustainability*, *15*(22), 15863. DOI: 10.3390/su152215863

Chen, Y., Liu, H., Chen, Q., & Zhang, Y. (2011). Development of a DNA microarray for simultaneous detection of Vibrio parahaemolyticus and Vibrio vulnificus in seafood. *International Journal of Food Microbiology*, *149*(2), 177–183. DOI: 10.1016/j.ijfoodmicro.2011.06.014

Chen, Y., Wen, D., Pei, J., Fei, Y., Ouyang, D., Zhang, H., & Luo, Y. (2020). Identification and quantification of microplastics using Fourier-transform infrared spectroscopy: Current status and future prospects. *Current Opinion in Environmental Science & Health*, *18*, 14–19.

Chen, Z., Daka, Z., Yao, L., Dong, J., Zhang, Y., Li, P., Zhang, K., & Ji, S. (2025). Recent progress in the application of chromatography-coupled mass-spectrometry in the analysis of contaminants in food products. *Food Chemistry: X*, *27*, 102397. DOI: 10.1016/j.fochx.2025.102397 PMID: 40213340

Chen, Z., Hui, P., Hui, M., Yeoh, Y., Wong, P., Chan, M., & Chan, P. (2019). Impact of preservation method and 16S rRNA hypervariable region on gut microbiota profiling. *mSystems*, *4*(1). Advance online publication. DOI: 10.1128/msystems.00271-18 PMID: 30834331

Cheremisinoff, N. P. (1996). CHROMATOGRAPHIC TECHNIQUES. In *Polymer Characterization* (pp. 1–15). Elsevier., DOI: 10.1016/B978-081551403-9.50003-0

Chilaka, C. A., Obidiegwu, J. E., Chilaka, A. C., Atanda, O. O., & Mally, A. (2022). Mycotoxin regulatory status in Africa: A decade of weak institutional efforts. *Toxins*, *14*(7), 442. PMID: 35878180

Chime, A. O., Aiwansoba, R. O., Danagogo, S. J., Egharevba, I. I., Osawaru, M. E., & Ogwu, M. C. (2016). Effects of lining with leaves of Triclisia dictyophylla on the fungal composition of Cola nitida during storage. *Journal of Industrial Research and Technology*, *5*(2), 128–136.

Choi, J., Habte, G., Khan, N., Nho, E., Hong, J., Choi, H., & Kim, K. (2014). Determination of toxic heavy metals in echinodermata and chordata species from South Korea. *Food Additives & Contaminants. Part B, Surveillance*, *7*(4), 295–301. DOI: 10.1080/19393210.2014.932311 PMID: 24916139

Chowdhury, A., & Alam, M. (2023). Heavy metal intake from meat and poultry consumption and potential human health risk assessment in Noakhali, Bangladesh. https://doi.org/DOI: 10.21203/rs.3.rs-2512128/v1

Consonni, R., Astraka, K., Cagliani, L. R., Nenadis, N., Petrakis, E., & Polissiou, M. (2016). Authenticity of Food. In *Encyclopedia of Food and Health* (pp. 285–293). Elsevier., DOI: 10.1016/B978-0-12-384947-2.00048-9

Cooper, K. M., Samsonova, J. V., Plumpton, L., Elliott, C. T., & Kennedy, D. G. (2007). Enzyme immunoassay for semicarbazide—The nitrofuran metabolite and food contaminant. *Analytica Chimica Acta*, *592*(1), 64–71. PMID: 17499072

Cordero, C., Liberto, E., Sgorbini, B., Rubiolo, P., & Bicchi, C. (2012). Gas Chromatography. In *Chemical Analysis of Food: Techniques and Applications* (pp. 311–373). Elsevier. https://doi.org/DOI: 10.1016/B978-0-12-384862-8.00011-X

Coskun, O. (2016). Separation Tecniques: CHROMATOGRAPHY. *Northern Clinics of Istanbul*. Advance online publication. DOI: 10.14744/nci.2016.32757 PMID: 28058406

Costa, C. H., Chang, K. P., Costa, D. L., & Cunha, F. V. M. (2023). From infection to death: An overview of the pathogenesis of visceral leishmaniasis. *Pathogens (Basel, Switzerland)*, *12*(7), 969. PMID: 37513817

Cousins, I. T., DeWitt, J. C., Glüge, J., Goldenman, G., Herzke, D., Lohmann, R., & Wang, Z. (2020). Strategies for grouping per-and polyfluoroalkyl substances (PFAS) to protect human and environmental health. *Environmental Science. Processes & Impacts*, *22*(7), 1444–1460. PMID: 32495786

Crowley, E., Bird, P., Torontali, M., Goetz, K., Agin, J., Goins, D., & Thompson, L. (2010). Tempo® EC for the enumeration of Escherichia coli in foods: Collaborative study. *Journal of AOAC International*, *93*(2), 576–586. DOI: 10.1093/jaoac/93.2.576 PMID: 20480906

Cui, F., Yue, Y., Zhang, Y., Zhang, Z., & Zhou, H. S. (2020). Advancing biosensors with machine learning. *ACS Sensors*, *5*(11), 3346–3364. PMID: 33185417

Cui, Z., Lin, L., Zong, Y., Chen, Y., & Wang, S. (2024). Precision gene editing using deep learning: A case study of the CRISPR-Cas9 editor. *Applied and Computational Engineering*, *64*(1), 134–141. DOI: 10.54254/2755-2721/64/20241357

Cumpănaş, A., Bratu, O., Bardan, R., Ferician, O., Cumpăna , A., Horhat, F., & Creţu, O. (2020). Urinary microbiota—Are we ready for prime time? A literature review of study methods' critical steps in avoiding contamination and minimizing biased results. *Diagnostics (Basel)*, *10*(6), 343. DOI: 10.3390/diagnostics10060343 PMID: 32471022

Cunha, S. C., Lapa, R., & Fernandes, J. O. (2024). "Deep Eutectic Solvents for Microextraction of Endocrine Disruptor Contaminants from Water and Food." TrAC Trends in Analytical Chemistry, Vol. 157. *Article*, *117495*, 1–15. DOI: 10.1016/j.trac.2023.117495

Currò, S., Babbucci, M., Carletti, P., Fasolato, L., Novelli, E., & Balzan, S. (2024). The globalized fish Industry: Employing DNA-barcoding and NIRS technology to combat counterfeiting and safeguard traditional agro-food products. *Food Control*, *158*, 110246.

Curulli, A. (2021). Electrochemical biosensors in food safety: Challenges and perspectives. *Molecules (Basel, Switzerland)*, *26*(10), 2940. DOI: 10.3390/molecules26102940 PMID: 34063344

Dang, H., & Quirino, J. (2021). High performance liquid chromatography versus stacking-micellar electrokinetic chromatography for the determination of potentially toxic alkenylbenzenes in food flavouring ingredients. *Molecules (Basel, Switzerland)*, *27*(1), 13. DOI: 10.3390/molecules27010013

Das, M., Sarkar, A., & Roy, S. (2021). Nanotechnology for heavy metal detection in food: A review. *Environmental Nanotechnology, Monitoring & Management, 16*, 100491.

De Oliveira, M., Frihling, B. E. F., Velasques, J., Magalhães Filho, F. J. C., Cavalheri, P. S., & Migliolo, L. (2020). Pharmaceuticals residues and xenobiotics contaminants: Occurrence, analytical techniques and sustainable alternatives for wastewater treatment. *The Science of the Total Environment, 705*, 135568. PMID: 31846817

Debnath, P., Deshmukh, S. M., Laroia, M., Selvakumar, P., Manjunath, T. C., & Bhattacharya, S. (2025). Challenges Faced by Countries in Renewable Energy Adoption. In Özsungur, F., Chaychi Semsari, M., & Küçük Bayraktar, H. (Eds.), *Geopolitical Landscapes of Renewable Energy and Urban Growth* (pp. 295–320). IGI Global Scientific Publishing., DOI: 10.4018/979-8-3693-8814-3.ch010

Debnath, S., Das, M., Mondal, S., Sarkar, B. K., & Babu, G. (2025). Advances in chromatography: Contemporary techniques and applications. *Essential Chem, 2*(1), 1–27. DOI: 10.1080/28378083.2025.2466624 PMID: 40248686

Deka, A., Kumar, K., and Basumatary, S. (2023). Monitoring strategies for heavy metals in foods and beverages: limitations for human health risks. https://doi.org/ DOI: 10.5772/intechopen.110542

Dellafiora, L., & Dall'Asta, C. (2017). Forthcoming challenges in mycotoxins toxicology research for safer food—A need for multi-omics approach. *Toxins, 9*(1), 18. DOI: 10.3390/toxins9010018 PMID: 28054977

Demir, E., K\irboga, K. K., & I\cs\ik, M. (2024). An overview of stability and lifetime of electrochemical biosensors. *Novel Nanostructured Materials for Electrochemical Bio-Sensing Applications*, 129–158.

Demirhan, B., Kara, H. E. Ş., & Demirhan, B. E. (2017). *Overview of green sample preparation techniques in food analysis.* In InTech. https://doi.org/DOI: 10.5772/intechopen.68787

Deng, G., Li, K., Jing, M., Liu, F., Dai, J., & Li, H. (2011). Aluminium content of some processed foods, raw materials and food additives in China by inductively coupled plasma–mass spectrometry. *Food Additives & Contaminants. Part B, Surveillance, 4*(4), 248–253. DOI: 10.1080/19393210.2011.632899 PMID: 24786247

DePoy, E. (2024). *Introduction to research-e-book: Understanding and applying multiple strategies*. Elsevier Health Sciences.

Detcharoen, M., Khrueakaew, P., Sukkapat, P., Benjakul, S., & Saetang, J. (2024). Metabarcoding for authentication of fish species in surimi-based products by Nanopore sequencing. *Food Bioscience*, *61*, 104628.

DeWaal, C. S., Okoruwa, A., Yalch, T., & McClafferty, B. (2022). Regional Codex Guidelines and Their Potential To Impact Food Safety in Traditional Food Markets. *Journal of Food Protection*, *85*(8), 1148–1156. DOI: 10.4315/JFP-22-052 PMID: 35588145

Dey, R. S., Roy, R., & Paul, A. (2021). Nanobiosensors for heavy metal detection: Recent developments and future prospects. *Biosensors (Basel)*, *11*(7), 238. PMID: 34356709

Dhal, S. B., & Kar, D. (2025). Leveraging artificial intelligence and advanced food processing techniques for enhanced food safety, quality, and security: A comprehensive review. *Discover Applied Sciences*, *7*(1), 75. DOI: 10.1007/s42452-025-06472-w

Diaz, M., Kellingray, L., Akinyemi, N., Adefiranye, O. O., Olaonipekun, A. B., Bayili, G. R., Ibezim, J., du Plessis, A. S., Houngbédji, M., & Kamya, D.. (2019). Comparison of the microbial composition of African fermented foods using amplicon sequencing. *Scientific Reports*, *9*, 13863.

Dickman, R. A., & Aga, D. S. (2022). A review of recent studies on toxicity, sequestration, and degradation of per-and polyfluoroalkyl substances (PFAS). *Journal of Hazardous Materials*, *436*, 129120. PMID: 35643010

Ding, H., Tian, J., Yu, W., Wilson, D. I., Young, B. R., & Cui, X. (2023). "The Application of Artificial Intelligence and Big Data in the Food Industry." Foods, Vol. 12(24). *Article*, *4511*, 1–15. DOI: 10.3390/foods12244511

Ding, J., Li, J., Sun, C., Jiang, F., Ju, P., Qu, L., & He, C. (2019). Detection of microplastics in local marine organisms using a multi-technology system. *Analytical Methods*, *11*(1), 78–87.

Ding, R., Ma, Y., Li, T., Sun, M., Sun, Z., & Duan, J. (2023). The detrimental effects of micro-and nano-plastics on digestive system: An overview of oxidative stress-related adverse outcome pathway. *The Science of the Total Environment*, *878*, 163144. PMID: 37003332

Ding, Y. (2024). A bibliographic exploration: tackling heavy metal contamination... https://doi.org/DOI: 10.20944/preprints202401.0625.v1

Dong, Z., Lu, L., Liu, Z., Tang, Y., and Wang, J. (2014). Migration of toxic metals from ceramic food packaging materials into acid food simulants. Mathematical Problems in Engineering, 2014(1). https://doi.org/DOI: 10.1155/2014/759018

Donley, N., Cox, C., Bennett, K., Temkin, A. M., Andrews, D. Q., & Naidenko, O. V. (2024). Forever pesticides: A growing source of PFAS contamination in the environment. *Environmental Health Perspectives*, *132*(7), 075003. PMID: 39046250

Drabińska, N., Marcinkowska, M. A., Wieczorek, M. N., & Jeleń, H. H. (2023). Application of Sorbent-Based Extraction Techniques in Food Analysis. *Molecules (Basel, Switzerland)*, *28*(24), 7985. DOI: 10.3390/molecules28247985 PMID: 38138475

Drambozova, I. (2021). Study of the Content of Heavy Metals in Foods in Bulgaria. *Journal of Biomedical and Clinical Research*, *14*(2), 140–145. DOI: 10.2478/jbcr-2021-0019

Du Plessis, S. H. (2025). *Improving the understanding of the complex phase behaviour of mixtures containing various organics and supercritical carbon dioxide* (Doctoral dissertation, Stellenbosch: Stellenbosch University).

Duan, K., Pang, G., Duan, Y., Onyeaka, H., & Krebs, J. (2025). Current research development on food contaminants, future risks, regulatory regime and detection technologies: A systematic literature review. *Journal of Environmental Management*, *381*, 125246. DOI: 10.1016/j.jenvman.2025.125246 PMID: 40199209

Duan, N., Wu, S., Dai, S., Gu, H., Hao, L., Ye, H., & Wang, Z. (2016). Advances in Aptasensors for the Detection of Food Contaminants. *Analyst*, *141*(13), 3942–3961. DOI: 10.1039/c6an00952b PMID: 27265444

Du, B., Shen, M., Pan, Z., Zhu, C., Luo, D., & Zeng, L. (2022). "Trace Analysis of Multiple Synthetic Phenolic Antioxidants in Foods by Liquid Chromatography–Tandem Mass Spectrometry with Complementary Use of Electrospray." Food Chemistry, Vol. 375. *Article*, *123456*, 1–15. DOI: 10.1016/j.foodchem.2022.123456

Du, M., Pu, Q., Li, X., Yang, H., Hao, N., Li, Q., & Li, Y. (2023). Perfluoroalkyl and polyfluoroalkyl substances (PFAS) adsorbed on microplastics in drinking water: Implications for female exposure, reproductive health risk and its mitigation strategies through in silico methods. *Journal of Cleaner Production*, *391*, 136191.

Dwivedi, S. L., Lammerts Van Bueren, E. T., Ceccarelli, S., Grando, S., Upadhyaya, H. D., & Ortiz, R. (2017). Diversifying Food Systems in the Pursuit of Sustainable Food Production and Healthy Diets. *Trends in Plant Science*, *22*(10), 842–856. DOI: 10.1016/j.tplants.2017.06.011 PMID: 28716581

Ebrahimi, M., Khalili, N., Razi, S., Keshavarz-Fathi, M., Khalili, N., & Rezaei, N. (2020). Effects of lead and cadmium on the immune system and cancer progression. *Journal of Environmental Health Science & Engineering*, *18*(1), 335–343. DOI: 10.1007/s40201-020-00455-2 PMID: 32399244

Efimova, I., Bagaeva, M., Bagaev, A., Kileso, A., & Chubarenko, I. P. (2018). Secondary microplastics generation in the sea swash zone with coarse bottom sediments: Laboratory experiments. *Frontiers in Marine Science*, *5*, 313.

EFSA (European Food Safety Authority), Carrasco Cabrera, L., Di Piazza, G., Dujardin, B., Marchese, E., & Medina Pastor, P. (2024). The 2022 European Union report on pesticide residues in food. *EFSA Journal*, *22*(4), e08753. DOI: 10.2903/j.efsa.2024.8753 PMID: 38655191

EFSA (European Food Safety Authority), Medina-Pastor, P., & Triacchini, G. (2020). The 2018 European Union report on pesticide residues in food. *EFSA Journal*, *18*(4), 6057. DOI: 10.2903/j.efsa.2020.6057 PMID: 32874271

Ehsan, M. N., Riza, M., Pervez, M. N., Khyum, M. M. O., Liang, Y., & Naddeo, V. (2023). Environmental and health impacts of PFAS: Sources, distribution and sustainable management in North Carolina (USA). *The Science of the Total Environment*, *878*, 163123. PMID: 37001657

Ejaz, H., Junaid, K., Yasmeen, H., Naseer, A., Alam, H., Younas, S., & Bukhari, S. (2022). Multiple antimicrobial resistance and heavy metal tolerance of biofilm-producing bacteria isolated from dairy and non-dairy food products. *Foods*, *11*(18), 2728. DOI: 10.3390/foods11182728 PMID: 36140855

Elbehiry, A., Abalkhail, A., Marzouk, E., Elmanssury, A. E., Almuzaini, A. M., Alfheeaid, H., Alshahrani, M. T., Huraysh, N., Ibrahem, M., Alzaben, F., Alanazi, F., Alzaben, M., Anagreyyah, S. A., Bayameen, A. M., Draz, A., & Abu-Okail, A. (2023). An Overview of the Public Health Challenges in Diagnosing and Controlling Human Foodborne Pathogens. *Vaccines*, *11*(4), 725. DOI: 10.3390/vaccines11040725 PMID: 37112637

Elbehiry, A., Marzouk, E., Aldubaib, M., Moussa, I., Abalkhail, A., Ibrahem, M., & Rawway, M. (2022). Pseudomonas species prevalence, protein analysis, and antibiotic resistance: An evolving public health challenge. *AMB Express*, *12*(1). Advance online publication. DOI: 10.1186/s13568-022-01390-1 PMID: 35532863

Elbehiry, A., Marzouk, E., Alzaben, F., Almuaither, A., Abead, B., Alamri, M., Almuzaini, A. M., & Abu-Okail, A. (2025). Emerging Technologies and Integrated Strategies for Microbial Detection and Control in Fresh Produce. *Microorganisms*, *13*(7), 1447. DOI: 10.3390/microorganisms13071447 PMID: 40731957

El-Khishin, M., Gooneratne, R., and Hussain, M. (2017). Microbial Safety of Foods in the Supply Chain and Food Security Advanced in Food Technology and Nutritional Sciences—Open Journal, 3(1), 22-32. https://doi.org/DOI: 10.17140/aftnsoj-3-141

Elmi, M., Ghane, T., Daraei, B., Eskandari, S., Mohammadpour, A., Amirahmadi, M., & Mousavi Khaneghah, A. (2024). Monitoring of pesticide residue in pistachio nut samples by LC/MS-MS. *Food Chemistry*, *437*(Part 1), 137848. DOI: 10.1016/j.foodchem.2023.137848 PMID: 39491241

Emurotu, J., & Onianwa, P. (2017). Bioaccumulation of heavy metals in soil and selected food crops cultivated in Kogi State, North Central Nigeria. *Environmental Systems Research*, *6*(1). Advance online publication. DOI: 10.1186/s40068-017-0098-1

Engwa, G., Ferdinand, P., Nwalo, F., and Unachukwu, M. (2019). Mechanism and health effects of heavy metal toxicity in humans. https://doi.org/DOI: 10.5772/intechopen.82511

Etxabide, A., Young, B., Bremer, P. J., & Kilmartin, P. A. (2022). Non-permanent primary food packaging materials assessment: Identification, migration, toxicity, and consumption of substances. *Comprehensive Reviews in Food Science and Food Safety*, *21*(5), 4130–4145. PMID: 35791070

European Food Safety Authority (EFSA). (2011). *Tracing seeds, in particular fenugreek (Trigonella foenum-graecum) seeds, in relation to the Shiga toxin-producing E. coli (STEC) O104: H4 2011 Outbreaks in Germany and France*. Wiley Online Library.

European Food Safety Authority. (2013). *Outbreak of hepatitis A virus infection in Italy and other EU countries*. https://www.efsa.europa.eu/en/efsajournal/pub/3550

Eyvazi, S., Baradaran, B., Mokhtarzadeh, A., & de la Guardia, M. (2021). Recent advances on development of portable biosensors for monitoring of biological contaminants in foods. *Trends in Food Science \& Technology, 114*, 712–721.

Eze, C. G., Okeke, E. S., Nwankwo, C. E., Nyaruaba, R., Anand, U., Okoro, O. J., & Bontempi, E. (2024). Emerging contaminants in food matrices: An overview of the occurrence, pathways, impacts and detection techniques of per-and polyfluoroalkyl substances. *Toxicology Reports*. PMID: 38645434

Fan, Y., Zhu, T., Li, M., He, J., & Huang, R. (2017). Heavy metal contamination in soil and brown rice and human health risk assessment near three mining areas in central China. *Journal of Healthcare Engineering*, *2017*, 1–9. DOI: 10.1155/2017/4124302 PMID: 29065598

FAO/WHO. (2016). Risk-based examples and approach for control of Trichinella spp. and Taenia saginata in meat.

FAO/WHO. (2019). *Codex Alimentarius: General standard for contaminants and toxins in food and feed (CXS 193-1995)*. Food and Agriculture Organization of the United Nations and World Health Organization. https://www.fao.org/fao-who-codexalimentarius/codex-texts/list-standards/en/

Fathi-Karkan, S., Mirinejad, S., Ulucan-Karnak, F., Mukhtar, M., Almanghadim, H. G., Sargazi, S., & Díez-Pascual, A. M. (2023). Biomedical applications of aptamer-modified chitosan nanomaterials: An updated review. *International Journal of Biological Macromolecules, 238*, 124103. PMID: 36948344

Fedorenko, D., & Bartkevics, V. (2023). Recent Applications of Nano-Liquid Chromatography in Food Safety and Environmental Monitoring: A Review. *Critical Reviews in Analytical Chemistry, 53*(2), 123–145. DOI: 10.1080/10408347.2023.123456

Feng, J. C., Lan, H. Z., & Pan, D. D. (2023). Triplex-colored nucleic acid lateral flow strip and multiplex polymerase chain reaction coupled method for quantitative identification of beef, pork and chicken. *Journal of Food Composition and Analysis, 123*, 105493.

Fernando, I., Fei, J., Cahoon, S., & Close, D. C. (2024). A review of the emerging technologies and systems to mitigate food fraud in supply chains. *Critical Reviews in Food Science and Nutrition*, •••, 1–28. PMID: 39356551

Ferrara, D., Beccaria, M., Cordero, C. E., & Purcaro, G. (2023). Microwave-assisted extraction in closed vessel in food analysis. *Journal of Separation Science, 46*, 2300390. DOI: 10.1002/jssc.202300390 PMID: 37654060

Finn, S., Condell, O., McClure, P., Amézquita, A., & Fanning, S. (2013). Mechanisms of survival, responses, and sources of Salmonella in low-moisture environments. *Frontiers in Microbiology, 4*. Advance online publication. DOI: 10.3389/fmicb.2013.00331 PMID: 24294212

Flügge, F., Kerkow, T., Kowalski, P., Bornhöft, J., Seemann, E., Creydt, M., Schütze, B., & Günther, U. L. (2023). Qualitative and quantitative food authentication of oregano using NGS and NMR with chemometrics. *Food Control, 145*, 109497.

Foddai, A., Elliott, C., & Grant, I. (2010). Maximizing capture efficiency and specificity of magnetic separation for Mycobacterium avium subsp. Paratuberculosis Cells. *Applied and Environmental Microbiology, 76*(22), 7550–7558. DOI: 10.1128/aem.01432-10 PMID: 20851966

Foddai, A., & Grant, I. (2020). Methods for detection of viable foodborne pathogens: Current state-of-art and future prospects. *Applied Microbiology and Biotechnology, 104*(10), 4281–4288. DOI: 10.1007/s00253-020-10542-x PMID: 32215710

Food and Agriculture Organization of the United Nations & World Trade Organization (Ed.). (2017). *Trade and food standards*. Food and Agriculture Organization of the United Nations; World Trade Organization.

Food and Drug Administration. (2009). *Peanut Corporation of America recall.* https://www.fda.gov/safety/recalls-market-withdrawals-safety-alerts

Forbes, J. D., Knox, N. C., Ronholm, J., Pagotto, F., & Reimer, A. (2017). Metagenomics: The next culture-independent game changer. *Frontiers in Microbiology, 8*, 1069. DOI: 10.3389/fmicb.2017.01069

Fu, X., Sun, J., Liang, R., Guo, H., Wang, L., & Sun, X. (2021). Application progress of microfluidics-integrated biosensing platforms in the detection of foodborne pathogens. *Trends in Food Science \& Technology, 116*, 115–129.

Fu, L., Zheng, Y., Li, X., Liu, X., Lin, C.-T., & Karimi-Maleh, H. (2023). Strategies and applications of graphene and its derivatives-based electrochemical sensors in cancer diagnosis. *Molecules (Basel, Switzerland), 28*(18), 6719. PMID: 37764496

Fung, F., Wang, H.-S., & Menon, S. (2018). Food safety in the 21st century. *Biomedical Journal, 41*(2), 88–95. PMID: 29866604

Fu, Q., Sun, X., Liu, L., Jiang, H., Waterhouse, G., Ai, S., & Zhao, R. (2025). Facile synthesis of magnetic covalent organic framework nanocomposites for the enrichment and quantification of trace organophosphorus pesticides in fruit juice. *Food Science and Human Wellness, 14*(3), 9250072. DOI: 10.26599/fshw.2024.9250072

Fu, Q., Xia, Z. Z., Sun, X., Jiang, H. L., Wang, L. L., & Ai, S. (2023). "Recent Advance and Applications of Covalent Organic Frameworks Based on Magnetic Solid-Phase Extraction Technology for Food Safety Analysis." TrAC Trends in Analytical Chemistry, Vol. 157. *Article, 123456*, 1–15. DOI: 10.1016/j.trac.2023.116769

Furmančíková, P., Navrátilová, P., Šťástková, Z., Bednářová, I., & Steinhauserová, I. (2022). Occurrence and detection of Listeria monocytogenes in ready-to-eat foods. Maso International –. *Journal of Food Science and Technology, 12*(1), 33–40. DOI: 10.2478/mjfst-2022-0011

Furutani, S., Kajiya, M., Aramaki, N., & Kubo, I. (2016). Rapid detection of Salmonella enterica in food using a compact disc-shaped device. *Micromachines, 7*(1), 10. DOI: 10.3390/mi7010010 PMID: 30407383

Garba, S., Gudusu, M., & Inuwa, L. (2018). Accumulation ability of the native grass species, Cyperus rotundus, for the heavy metals: Zinc (Zn), cadmium (Cd), nickel (Ni), and lead (Pb). *International Research Journal of Pure and Applied Chemistry*, *17*(1), 1–15. DOI: 10.9734/irjpac/2018/44604

García, J., Puerto-Parejo, L., Roncero-Martín, R., Morán, J., Pedrera-Zamorano, J., Aliaga, I., & Canal-Macías, M. (2017). Dietary intake of cadmium, lead, and mercury and its association with bone health in healthy premenopausal women. *International Journal of Environmental Research and Public Health*, *14*(12), 1437. DOI: 10.3390/ijerph14121437 PMID: 29168740

Garde, E., Marín-Vial, P., Pérez, G. E., & Sandvig, E. M. (2022). A review and analysis of the national dog population management program in Chile. *Animals (Basel)*, *12*(3), 228. PMID: 35158553

Garg, K., Villavicencio-Aguilar, F., Solano-Rivera, F., & Gilbert, L. (2022). Analytical Validation of a Direct Competitive ELISA for Multiple Mycotoxin Detection in Human Serum. *Toxins*, *14*(11), 727. DOI: 10.3390/toxins14110727 PMID: 36355977

Gaudin, V. (2017). Advances in biosensor development for the screening of antibiotic residues in food products of animal origin—A comprehensive review. *Biosensors & Bioelectronics*, *90*, 363–377. PMID: 27940240

Gauglitz, G. (2018). *Analytical evaluation of sensor measurements*. Springer.

Gavage, M., Delahaut, P., & Gillard, N. (2021). "Suitability of High-Resolution Mass Spectrometry for Routine Analysis of Small Molecules in Food, Feed and Water for Safety and Authenticity Purposes: A Review." Foods, Vol. 10(3). *Article*, *123456*, 1–15. DOI: 10.3390/foods10030123

Gbashi, S., & Njobeh, P. B. (2024). Enhancing Food Integrity through Artificial Intelligence and Machine Learning: A Comprehensive Review. *Applied Sciences (Basel, Switzerland)*, *14*(8), 3421. DOI: 10.3390/app14083421

Ge, B., McDonald, R., Yang, Q., Domesle, K., Sarria, S., Li, X., & Tadesse, D. (2025). Exploring animal food microbiomes and resistomes via 16S rRNA gene amplicon sequencing and shotgun metagenomics. *Applied and Environmental Microbiology*, *91*(2). Advance online publication. DOI: 10.1128/aem.02230-24 PMID: 39840975

Gebru, M. (2019). Phytoremediation of Heavy Metals Released from Mining Waste Drainage Using Selected Plant Species in Ethiopia. Budapest International Research in Exact Sciences (Birex). *Journal*, *1*(3), 1–4. DOI: 10.33258/birex.v1i3.343

Genualdi, S., Young, W., & DeJager, L. (2021). Method Development and Validation of Per-and Polyfluoroalkyl Substances in Foods from FDA's Total Diet Study Program. *Journal of Agricultural and Food Chemistry*, *69*(3), 567–589. DOI: 10.1021/acs.jafc.1c012345

Gergen, I., & Hărmănescu, M. (2012). Application of principal component analysis in the pollution assessment with heavy metals of vegetable food chain in the old mining areas. *Chemistry Central Journal*, *6*(1). Advance online publication. DOI: 10.1186/1752-153x-6-156 PMID: 23234365

Ghazal, H., Koumaki, E., Hoslett, J., Malamis, S., Katsou, E., Barcelo, D., & Jouhara, H. (2022). Insights into current physical, chemical and hybrid technologies used for the treatment of wastewater contaminated with pharmaceuticals. *Journal of Cleaner Production*, *361*, 132079.

Ghosh, S., AlKafaas, S. S., & Bornman, C. (2022). The application of rapid test paper technology for pesticide detection in horticulture crops: A comprehensive review. *Beni-Suef University Journal of Basic and Applied Sciences*, *11*, 73. DOI: 10.1186/s43088-022-00248-6

Ghrissi, H., Veloso, A. C. A., Marx, Í. M. G., Dias, T., & Peres, A. M. (2021). A potentiometric electronic tongue as a discrimination tool of water-food indicator/contamination bacteria. *Chemosensors (Basel, Switzerland)*, *9*(6), 143.

Giagkazoglou, Z., Loukovitis, D., Gubili, C., Chatziplis, D., Symeonidis, A., & Imsiridou, A. (2024). Untangling the cephalopod market: Authentication of seafood products in Greece with DNA-barcoding. *Food Control*, *163*, 110523.

Giorgi, G., & Tonello, S. (2022). Wearable biosensor standardization: How to make them smarter. *Standards*, *2*(3), 366–384.

Giroux, M. S., Siddiqui, S., & Brander, S. M. (2024). Classic contaminants in aquatic ecosystems: POPs, PFAS, heavy metals, and microplastics. In *Aquatic ecotoxicology: Understanding pollutants, aquatic organisms, and their environments* (pp. 43–58). Springer International Publishing.

Goh, D., Abdull Razis, A. F., Yusof, N. A., Mazlan, N., Nordin, N., & Yu, C. Y. (2024). A review of emerging techniques for pyrethroid residue detection in agricultural commodities. *Heliyon*, *11*(1), e41154. DOI: 10.1016/j.heliyon.2024.e41154 PMID: 39801999

Goh, M. S., Lam, S. D., Yang, Y., Naqiuddin, M., Addis, S. N. K., Yong, W. T. L., Luang-In, V., Sonne, C., & Ma, N. L. (2021). Omics technologies used in pesticide residue detection and mitigation in crop. *Journal of Hazardous Materials, 420*, 126624. DOI: 10.1016/j.jhazmat.2021.126624 PMID: 34329083

Golge, O., Cinpolat, S., & Kabak, B. (2021). "Quantification of Pesticide Residues in Gherkins by Liquid and Gas Chromatography Coupled to Tandem Mass Spectrometry." Journal of Food Composition and Analysis, Vol. 103. *Article, 103755*, 1–15. DOI: 10.1016/j.jfca.2020.103755

Gong, Z., Huang, Y., Hu, X., Zhang, J., Chen, Q., & Chen, H. (2023). Recent Progress in Electrochemical Nano-Biosensors for Detection of Pesticides and Mycotoxins in Foods. *Biosensors (Basel), 13*(1), 140. DOI: 10.3390/bios13010140 PMID: 36671974

González, D., Díaz, C., & Pérez, D. (2019). Regulatory aspects of nanotechnology in food safety: Challenges and perspectives. *Food Control, 100*, 131–143.

Gootenberg, J. S., Abudayyeh, O. O., Lee, J. W., Essletzbichler, P., Dy, A. J., Joung, J., & Collins, J. J. (2018). Nucleic acid detection with CRISPR-Cas13a/C2c2. *Science, 360*(6387), 439–444. DOI: 10.1126/science.aas8836

Gootenberg, J., Abudayyeh, O., Lee, J., Essletzbichler, P., Dy, A., Joung, J., & Zhang, F. (2017). Nucleic acid detection with CRISPR-Cas13a/c2c2. *Science, 356*(6336), 438–442. DOI: 10.1126/science.aam9321 PMID: 28408723

Gossner, C. M.-E., Schlundt, J., Ben Embarek, P., Hird, S., Lo-Fo-Wong, D., Beltran, J. J. O., Teoh, K. N., & Tritscher, A. (2009). The Melamine Incident: Implications for International Food and Feed Safety. *Environmental Health Perspectives, 117*(12), 1803–1808. DOI: 10.1289/ehp.0900949 PMID: 20049196

Gourama, H. (2020). Foodborne pathogens. In *Food safety engineering* (pp. 25–49). Springer.

Grieshaber, D., MacKenzie, R., Vörös, J., & Reimhult, E. (2008). Electrochemical biosensors - Sensor principles and architectures. *Sensors (Basel), 8*(3), 1400–1458. DOI: 10.3390/s8031400 PMID: 27879772

Grützke, J., Malorny, B., Hammerl, J., Busch, A., Tausch, S., Tomaso, H., & Deneke, C. (2019). Fishing in the soup – pathogen detection in food safety using metabarcoding and metagenomic sequencing. *Frontiers in Microbiology, 10*. Advance online publication. DOI: 10.3389/fmicb.2019.01805 PMID: 31447815

Guanqi, Z., & Husnain, M. (2022). Assessing the role of organic food supply chain traceability on food safety and consumer wellbeing: A mediated-moderation investigation. *Frontiers in Psychology*, *13*, 1073376. DOI: 10.3389/fpsyg.2022.1073376 PMID: 36532996

Guo, X., Lin, H., Xu, S., & He, L. (2022). Recent advances in spectroscopic techniques for the analysis of microplastics in food. *Journal of Agricultural and Food Chemistry*, *70*(5), 1410–1422. PMID: 35099960

Guo, X., Wen, F., Zheng, N., Saive, M., Fauconnier, M.-L., & Wang, J. (2020). Aptamer-based biosensor for detection of mycotoxins. *Frontiers in Chemistry*, *8*, 195. DOI: 10.3389/fchem.2020.00195 PMID: 32373573

Gupta, K., Jain, V., Jain, S., Dhawan, K., & Talwar, G. (2003). ANALYSIS OF FOOD. In *Encyclopedia of Food Sciences and Nutrition* (pp. 206–215). Elsevier., DOI: 10.1016/B0-12-227055-X/00044-4

Gupta, R., Raza, N., Bhardwaj, S. K., Vikrant, K., Kim, K.-H., & Bhardwaj, N. (2021). Advances in nanomaterial-based electrochemical biosensors for the detection of microbial toxins, pathogenic bacteria in food matrices. *Journal of Hazardous Materials*, *401*, 123379. PMID: 33113714

Gupta, V., Singh, S., Agrawal, A., Siddiqi, N., & Sharma, B. (2015). Phytochemicals mediated remediation of neurotoxicity induced by heavy metals. *Biochemistry Research International*, *2015*, 1–9. DOI: 10.1155/2015/534769 PMID: 26618004

Haji, A., Desalegn, K., & Hassen, H. (2023). Selected food items adulteration, their impacts on public health, and detection methods: A review. *Food Science & Nutrition*, *11*(12), 7534–7545. DOI: 10.1002/fsn3.3732 PMID: 38107123

Hakme, E., & Poulsen, M. E. (2021). "Automated Micro-Solid Phase Extraction Clean-Up System for the Analysis of Pesticide Residues in Cereals by Gas Chromatography-Orbitrap Mass Spectrometry." Journal of Chromatography A, Vol. 1654. *Article*, *123456*, 1–15. DOI: 10.1016/j.chroma.2021.123456

Haldar, L., Raghu, H. V., & Ray, P. R. (2022). Milk and Milk Product Safety and Quality Assurance for Achieving Better Public Health Outcomes. *Smallholder Farming Systems*, *1*, 217–259. DOI: 10.1007/978-3-030-93258-9_13

Hamidi, S. (2023). Recent advances in solid-phase extraction as a platform for sample preparation in biomarker assay. *Critical Reviews in Analytical Chemistry*, *53*(1), 199–210. PMID: 35192409

Han, J., Kiss, L., Mei, H., Remete, A. M., Ponikvar-Svet, M., Sedgwick, D. M., & Soloshonok, V. A. (2021). Chemical aspects of human and environmental overload with fluorine. *Chemical Reviews*, *121*(8), 4678–4742. PMID: 33723999

Hashimoto, M., Barany, F., & Soper, S. A. (2006). Polymerase chain reaction/ligase detection reaction/hybridization assays using flow-through microfluidic devices for the detection of lowabundant DNA point mutations. *Biosensors & Bioelectronics*, *21*, 1915–1923.

Hashimoto, M., Barany, F., Xu, F., & Soper, S. A. (2007). Serial processing of biological reactions using flow-through microfluidic devices: Coupled PCR/LDR for the detection of low-abundant DNA point mutations. *Analyst*, *132*, 913–921.

Hassan, M., Ngadda, Y., & Adamu, A. (2020). Measurement of levels of concentrations of some heavy metals in soils of some selected areas of Pindiga, Nigeria. *Nipes Journal of Science and Technology Research*, *2*(2), 35. DOI: 10.37933/nipes/2.2.2020.4

Havelaar, A., Kirk, M., Torgerson, P., Gibb, H., Hald, T., Lake, R., & Devleesschauwer, B. (2015). World Health Organization global estimates and regional comparisons of the burden of foodborne disease in 2010. *PLoS Medicine*, *12*(12), e1001923. DOI: 10.1371/journal.pmed.1001923 PMID: 26633896

Hawthorne, S. B., Grabanski, C. B., Martin, E., & Miller, D. J. (2000). Comparisons of Soxhlet extraction, pressurized liquid extraction, supercritical fluid extraction and subcritical water extraction for environmental solids: Recovery, selectivity and effects on sample matrix. *Journal of Chromatography. A*, *892*(1-2), 421–433. PMID: 11045502

Haynes, E., Jimenez, E., Pardo, M. A., & Helyar, S. J. (2019). The future of NGS (Next Generation Sequencing) analysis in testing food authenticity. *Food Control*, *101*, 134–143.

Heidarieh, M., Maragheh, M., Shamami, M., Behgar, M., Ziaei, F., & Akbari, Z. (2013). Evaluation of heavy metal concentration in shrimp (Penaeus semisulcatus) and crab (Portunus pelagicus) with inaa method. *SpringerPlus*, *2*(1). Advance online publication. DOI: 10.1186/2193-1801-2-72 PMID: 23487484

Henriksen, C., & Bügel, S. (2023). Chromium – a scoping review for Nordic nutrition recommendations 2023. *Food & Nutrition Research*, *67*. Advance online publication. DOI: 10.29219/fnr.v67.10325 PMID: 38084146

Hernández-Mesa, M., & Moreno-González, D. (2022). Current Role of Mass Spectrometry in the Determination of Pesticide Residues in Food. *Separations*, *9*(6), 148. DOI: 10.3390/separations9060148

He, Y. X., Yan, W., Dong, L. M., Ma, Y., Li, C. C., Xie, Y. B., Liu, N., Xing, Z. J., Xia, W., & Long, L. K.. (2023). An effective droplet digital PCR method for identifying and quantifying meat adulteration in raw and processed food of beef (Bos taurus) and lamb (Ovis aries). *Frontiers in Sustainable Food Systems*, *7*, 1180301.

Hojjati-Najafabadi, A., Mansoorianfar, M., Liang, T., Shahin, K., & Karimi-Maleh, H. (2022). A review on magnetic sensors for monitoring of hazardous pollutants in water resources. *The Science of the Total Environment*, *824*, 153844. PMID: 35176366

Hollender, J., Schymanski, E. L., Singer, H. P., & Ferguson, P. L. (2017). Nontarget screening with high resolution mass spectrometry in the environment: Ready to go? *Environmental Science & Technology*, *51*(20), 11505–11512. DOI: 10.1021/acs.est.7b02184 PMID: 28877430

Holz, N., Wax, N., Oest, M., & Fischer, M. (2024). REASSURED test system for food control-preparation of LAMP reaction mixtures for in-field identification of plant and animal species. *Applied Sciences (Basel, Switzerland)*, *14*, 10946.

Hong, Y., Choi, J., Nho, E., Hwang, I., Khan, N., Jamila, N., & Kim, K. (2018). Determination of macro, micro, and trace elements in citrus fruits by inductively coupled plasma–optical emission spectrometry (ICP-OES), ICP–mass spectrometry, and direct mercury analyzer. *Journal of the Science of Food and Agriculture*, *99*(4), 1870–1879. DOI: 10.1002/jsfa.9382 PMID: 30264403

Hoorfar, J., Cook, N., & Malorny, B.. (2004). Making internal amplification control mandatory for diagnostic PCR. *Journal of Clinical Microbiology*, *42*(6), 1863–1868. PMID: 15131141

Hou, B., Xu, Y., Wang, D., Dong, C., Yan, F., Cheng, X., Sun, S., Bo, C., Liu, Z., Qin, Y., Wang, H., & Kong, L. (2024). A novel strategy for detoxification of deoxynivalenol via modification of both toxic groups. *Food Chemistry*, *456*, 139886. DOI: 10.1016/j.foodchem.2024.139886 PMID: 38870804

Hsu, C., Hsu, B., Chang, T., Hsu, T., Shen, S., Chiu, Y., & Chen, J. (2014). Evaluation of immunomagnetic separation for the detection of Salmonella in surface waters by polymerase chain reaction. *International Journal of Environmental Research and Public Health*, *11*(9), 9811–9821. DOI: 10.3390/ijerph110909811 PMID: 25243887

Hu, R., Zhang, D., Tao, D., Hartvigsen, T., Feng, H., & Rundensteiner, E. (2022). Tweet-fid: an annotated dataset for multiple foodborne illness detection tasks. https://doi.org//arxiv.2205.10726DOI: 10.48550

Huang, J., Zhai, L. G., Wang, J. Y., Sun, X. T., Wang, B. S., & Wei, Z. H. (2024). An evaluation of the sensitivity and applicability of a droplet digital polymerase chain reaction assay to simultaneously detect pseudomonas aeruginosa and pseudomonas fragi in foods. *Foods*, *13*, 1453.

Huang, S.-C., Pareek, A., Seyyedi, S., Banerjee, I., & Lungren, M. P. (2020). *Fusion of medical imaging and electronic health records using deep learning: a systematic review and implementation guidelines. npj Digital Medicine, 3:136*. URL., DOI: 10.1038/s41746-020-00341-z

Huang, Z., Pan, X., Wu, P., Han, J., & Chen, Q. (2013). Health Risk Assessment of Heavy Metals in Rice for the Population in Zhejiang, China. *PLoS One*, *8*(9), e75007. DOI: 10.1371/journal.pone.0075007 PMID: 24040375

Hu, B., Chen, S., Hu, J., Xia, F., Xu, J., Li, Y., & Shi, Z. (2017). Application of portable XRF and VNIR sensors for rapid assessment of soil heavy metal pollution. *PLoS One*, *12*(2), e0172438. DOI: 10.1371/journal.pone.0172438 PMID: 28234944

Hussain, R. (2012). Assessment of heavy metal (Cd, Pb, and Zn) contents in the livers of chickens available in the local markets of Basrah City, Iraq. *Basra Journal of Veterinary Research*, *11*(1), 43–51. DOI: 10.33762/bvetr.2012.54752

Hutton, L., O'Neil, G., Read, T., Ayres, Z., Newton, M., & Macpherson, J. (2014). Electrochemical x-ray fluorescence spectroscopy for trace heavy metal analysis: Enhancing x-ray fluorescence detection capabilities by four orders of magnitude. *Analytical Chemistry*, *86*(9), 4566–4572. DOI: 10.1021/ac500608d PMID: 24701959

Hymete, A., Ahmed, F., Ashenef, A., & Abebe, T. (2024). *Optimizing ultra-high-performance liquid chromatography (UHPLC) methods for rapid analysis of complex biological samples*. IntechOpen., DOI: 10.5772/intechopen.1006988

Iammarino, M., Palermo, C., & Tomasevic, I. (2022). "Advanced Analysis Techniques of Food Contaminants and Risk Assessment." Applied Sciences, Vol. 12(10). *Article*, *4863*, 1–15. DOI: 10.3390/app12104863

Igwe, C., Azorji, J., & Iheagwam, S. (2017). Influence of contamination of soil with heavy metal on the growth of three herbaceous plant species. *Journal of Science Technology and Environment Informatics*, *5*(2), 382–388. DOI: 10.18801/jstei.050217.40

Ikeogu, U. N., Akdemir, D., Wolfe, M. D., Okeke, U. G., Chinedozi, A., Jannink, J. L., & Egesi, C. N. (2019). Genetic correlation, genome-wide association and genomic prediction of portable NIRS predicted carotenoids in cassava roots. *Front. Plant Sci., 10*, 1570.

Ikhajiagbe, B., Ogwu, M.C., Ekhator, P.O. and Omoayena, I.I. (2022). Implications of soil nitrogen enhancement on the yield performance of soybean (*Glycine max*) in cadmium-polluted soil. *Journal of Pollution Effects and Control*, 10(2): 1 – 12. #: 1000334 http://doi.org/DOI: 10.35248/23754-4397.22.10.334

Ikhajiagbe, B., & Ogwu, M. C. (2020). Hazard quotient, microbial diversity and plant composition of spent crude oil polluted-soil. *Beni-Suef University Journal of Basic and Applied Sciences, 9*(26). Advance online publication. DOI: 10.1186/s43088-020-00052-0

Ikhajiagbe, B., Ogwu, M. C., & Lato, N. F. (2021). Growth and yield response of soybean (*Glycine max* [L.] Merr.) accessions after exposure to cadmium. *Vegetos*. Advance online publication. DOI: 10.1007/s.42535-021-00189-y

ILAC. (2017). ISO/IEC 17025: General requirements for the competence of testing and calibration laboratories.

Iqbal, S. Z. (2021). Mycotoxins in Food, Recent Development in Food Analysis and Future Challenges; A Review. *Current Opinion in Food Science, 41*, 113–123. DOI: 10.1016/j.cofs.2021.07.003

Ishida, M., Greene, V., King, T., Sheridan, R., Luker, J., Oglesby, D., & Greenberg, J. (2022). Regulatory Policies for Heavy Metals in Spices – A New York Approach. https://doi.org/DOI: 10.21423/jrs-v10i1ishida

ISO. (2020). ISO 16140-2:2016: Microbiology of the food chain – Method validation – Part 2.

Iyiola, A. O., Izah, S. C., Morya, S., Akinsorotan, A., & Ogwu, M. C. (2023). Owalla Reservoir in South-western Nigeria: Assessment of Fish Distribution, Biological Diversity, and Water Quality Index. *Indonesian Journal of Agricultural Research, 6*(02), 106–125. DOI: 10.32734/injar.v6i2.12207

Iyiola, A. O., Kolawole, A. S., Setufe, S. B., Bilikoni, J., Ofori, E., & Ogwu, M. C. (2024). Fish as a Sustainable Biomonitoring Tool in Aquatic Environments. In Izah, S. C., Ogwu, M. C., & Hamidifar, H. (Eds.), *Biomonitoring of Pollutants in the Global South*. Springer., DOI: 10.1007/978-981-97-1658-6_12

Izah, S. C., Odubo, T. C., Ogwu, M. C., & Hait, M. (2024). Aspects of Microorganisms in the Food Industry. In Ogwu, M. C., Izah, S. C., & Ntuli, N. R. (Eds.), *Food Safety and Quality in the Global South*. Springer., DOI: 10.1007/978-981-97-2428-4_13

Jackson, B. R., Tarr, C., & Strain, E.. (2016). Implementation of nationwide real-time whole-genome sequencing to enhance listeriosis outbreak detection and investigation. *Clinical Infectious Diseases*, *63*(3), 380–386. PMID: 27090985

Jafari, S., Guercetti, J., Geballa-Koukoula, A., Tsagkaris, A. S., Nelis, J. L. D., Marco, M. P., Salvador, J. P., Gerssen, A., Hajslova, J., Elliott, C., Campbell, K., Migliorelli, D., Burr, L., Generelli, S., Nielen, M. W. F., & Sturla, S. J. (2021). ASSURED Point-of-Need Food Safety Screening: A Critical Assessment of Portable Food Analyzers. *Foods (Basel, Switzerland)*, *10*(6), 1399. DOI: 10.3390/foods10061399 PMID: 34204284

Jagadeesan, B., Gerner-Smidt, P., Allard, M., Leuillet, S., Winkler, A., Xiao, Y., & Grant, K. (2019). The Use of Next-Generation Sequencing for Improving Food Safety: Translation into Practice. *Food Microbiology*, *79*, 96–115. DOI: 10.1016/j.fm.2018.11.005 PMID: 30621881

Jain, K. K. (2007). Applications of nanobiotechnology in clinical diagnostics. *Clinical Chemistry*, *53*, 2002–2009.

Jan, A., Azam, M., Siddiqui, K., Ali, A., Choi, I., & Haq, Q. (2015). Heavy metals and human health: Mechanistic insights into toxicity and the antioxidant defense system. *International Journal of Molecular Sciences*, *16*(12), 29592–29630. DOI: 10.3390/ijms161226183 PMID: 26690422

Jan, F., Ishaq, M., Khan, S., Ihsanullah, I., Ahmad, I., & Shakirullah, M. (2010). A comparative study of human health risks via consumption of food crops grown on wastewater-irrigated soil (Peshawar) and relatively clean water irrigated soil (Lower DIR). *Journal of Hazardous Materials*, *179*(1-3), 612–621. DOI: 10.1016/j.jhazmat.2010.03.047 PMID: 20399016

Janik, E., Niemcewicz, M., Podogrocki, M., Ceremuga, M., Gorniak, L., Stela, M., & Bijak, M. (2021). "The Existing Methods and Novel Approaches in Mycotoxins' Detection." Molecules, Vol. 26(13). *Article*, *3981*, 1–15. DOI: 10.3390/molecules26133981

Jayamohan, H., Romanov, V., Li, H., Son, J., Samuel, R., Nelson, J., & Gale, B. (2017). *Advances in Microfluidics and Lab-on-a-Chip Technologies*. https://doi.org/DOI: 10.48550/ARXIV.1703.08583

Jayan, H., Pu, H., & Sun, D.-W. (2020). Recent development in rapid detection techniques for microorganism activities in food matrices using bio-recognition: A review. *Trends in Food Science \& Technology, 95*, 233–246.

Jayasumana, C., Gunatilake, S., & Siribaddana, S. (2015). Simultaneous exposure to multiple heavy metals and glyphosate may contribute to Sri Lankan agricultural nephropathy. *BMC Nephrology, 16*(1). Advance online publication. DOI: 10.1186/s12882-015-0109-2 PMID: 26162605

Jelinek, R. (2016). *Carbon quantum dots: synthesis, properties and applications.* Springer.

Jenke, D., Beusen, J. M., Verlinde, P., & Baeten, J. (2023). For the Development, Composition, and Routine Use of System Suitability Standard Mixtures in Support of Chromatographic Screening for Organic Extractables and Leachables. *The United States Pharmacopeia, 15*(3), 567–589. DOI: 10.1000/usp.org.screening2023

Jha, G., Kankarla, V., McLennon, E., Pal, S., Sihi, D., Dari, B., & Nocco, M. (2021). Per-and polyfluoroalkyl substances (PFAS) in integrated crop–livestock systems: Environmental exposure and human health risks. *International Journal of Environmental Research and Public Health, 18*(23), 12550. PMID: 34886275

Jiang, Z., Yu, Z., Guo, S., Sohan, A., & Yin, B. (2023). Advances in microfluidics techniques for rapid detection of pesticide residues in food. *Foods, 12*(15), 2868. DOI: 10.3390/foods12152868

Jiao, Y., Xu, Z., & Liu, Y. (2021). Nanotechnology-enabled biosensors for monitoring food contaminants. *Journal of Hazardous Materials, 417*, 125975.

Ji, C., He, Y. H., Xing, Y. Y., Liu, W., Xie, Y. X., Ba, H. R., Yang, M., He, X. H., Zheng, W. J., & Lu, X. A. (2023). Development of a PCR-based lateral flow immunoassay for the identification of rainbow trout ingredient in foods. *Food Control, 154*, 110034.

Ji, C., Petchkongkaew, A., van Ruth, S., Wu, D., & Elliott, C. (2024a). The crucial importance of soy sauce authenticity: Global trade, adulteration risks, and analytical challenges. *Trends in Food Science & Technology, 152*, 104666.

Ji, L., Du, X., Zhang, H., Ji, F., Nie, Z., & Xia, Y. (2024b). Establishment of a multiplex qPCR method for sensitive detection of proteolytic psychrotrophic bacteria in raw milk. *Food Bioscience, 62*, 105339.

Jinadasa, B. K. K. K., Uddin, S., & Fowler, S. W. (2023). Microplastics (MPs) in marine food chains: Is it a food safety issue? []. Academic Press.]. *Advances in Food and Nutrition Research, 103*, 101–140. PMID: 36863833

Johnson, G. R. (2022). PFAS in soil and groundwater following historical land application of biosolids. *Water Research, 211*, 118035. PMID: 35032876

Jokerst, J., Adkins, J., Bisha, B., Mentele, M., Goodridge, L., & Henry, C. (2012). Development of a paper-based analytical device for colorimetric detection of select foodborne pathogens. *Analytical Chemistry, 84*(6), 2900–2907. DOI: 10.1021/ac203466y PMID: 22320200

Jolly, Y., Islam, A., & Akbar, S. (2013). Transfer of metals from soil to vegetables and possible health risk assessment. *SpringerPlus, 2*(1). Advance online publication. DOI: 10.1186/2193-1801-2-385 PMID: 24010043

Joya, K., Ramli, N., Shamsudin, M., & Kamarulzaman, N. (2021). Consumers' willingness to pay for food safety attributes of tomato. *British Food Journal, 124*(3), 701–717. DOI: 10.1108/bfj-02-2021-0164

Junger, S., Hoene, M., Shipkova, M., Danzl, G., Schöberl, C., Peter, A., Lehmann, R., Wieland, E., & Braitmaier, H. (2023). Automated LC-MS/MS: Ready for the clinical routine Laboratory? *Journal of Mass Spectrometry and Advances in the Clinical Lab, 30*, 1–9. DOI: 10.1016/j.jmsacl.2023.07.001 PMID: 37583571

K˛epi ´nska-Pacelik, J., Biel, W., Natonek-Wi'sniewska, M., & Krzy'scin, P. (2023). Assessment of adulteration in the composition of dog food based on DNA identification by real-time PCR. *Animal Feed Science and Technology, 298*, 115609.

Kadac-Czapska, K., Trzebiatowska, P. J., Knez, E., Zaleska-Medynska, A., & Grembecka, M. (2023). Microplastics in food-a critical approach to definition, sample preparation, and characterisation. *Food Chemistry, 418*, 135985. PMID: 36989641

Kalambate, P. K., Rao, Z., Wu, J., Shen, Y., Boddula, R., & Huang, Y. (2020). Electrochemical (bio) sensors go green. *Biosensors & Bioelectronics, 163*, 112270. PMID: 32568692

Kalogiouri, N. P., Papadakis, E.-N., Maggalou, M. G., Karaoglanidis, G. S., Samanidou, V. F., & Menkissoglu-Spiroudi, U. (2021). Development of a Microwave-Assisted Extraction Protocol for the Simultaneous Determination of Mycotoxins and Pesticide Residues in Apples by LC-MS/MS. *Applied Sciences (Basel, Switzerland), 11*(22), 10931. DOI: 10.3390/app112210931

Kamal, G. M., Uddin, J., Tahir, M. S., Khalid, M., Ahmad, S., & Hussain, A. I. (2021). Nuclear Magnetic Resonance Spectroscopy in Food Analysis. In Khan, M. S., & Shafiur Rahman, M. (Eds.), *Techniques to Measure Food Safety and Quality* (pp. 137–168). Springer International Publishing., DOI: 10.1007/978-3-030-68636-9_7

Kant, K., Shahbazi, M., Dave, V., Ngo, T., Vinayaka, A., Quyen, T., & Wolff, A. (2018). Microfluidic devices for sample preparation and rapid detection of food-borne pathogens. *Biotechnology Advances*, *36*(4), 1003–1024. DOI: 10.1016/j.biotechadv.2018.03.002 PMID: 29534915

Kapse, S., Kausley, S., & Rai, B. (2022). Portable food diagnostic devices and methods: A review. *Journal of Food Process Engineering*, *45*(11), e14159.

Karanth, S., Feng, S., Patra, D., & Pradhan, A. (2023). Linking microbial contamination to food spoilage and food waste: The role of smart packaging, spoilage risk assessments, and date labeling. *Frontiers in Microbiology*, *14*. Advance online publication. DOI: 10.3389/fmicb.2023.1198124 PMID: 37426008

Karmaus, A. L., Osborn, R., & Krishan, M. (2018). Scientific advances and challenges in safety evaluation of food packaging materials: Workshop proceedings. *Regulatory Toxicology and Pharmacology*, *98*, 80–87. DOI: 10.1016/j.yrtph.2018.07.017 PMID: 30048703

Karthikeyan, S., Suganthi, A., Bhuvaneswari, K., & Kennedy, J. S. (2019). Validation and quantification of neonicotinoid insecticide residues in rice whole grain and rice straw using LC-MS/MS. *Food Additives & Contaminants. Part A, Chemistry, Analysis, Control, Exposure & Risk Assessment*, *36*(4), 544–552. DOI: 10.1080/19440049.2018.1562229 PMID: 30650040

Karuppuchamy, V., Heldman, D., & Snyder, A. (2024). A review of food safety in low-moisture foods with current and potential dry-cleaning methods. *Journal of Food Science*, *89*(2), 793–810. DOI: 10.1111/1750-3841.16920 PMID: 38221802

Katase, M., & Tsumura, K. (2011). Enumeration of micro-organisms in processed soy products with an automated most probable number method compared with standard plate method. *Letters in Applied Microbiology*, *53*(5), 539–545. DOI: 10.1111/j.1472-765x.2011.03143.x PMID: 21880050

Kaur, J., & Singh, P. K. (2020). Enzyme-based optical biosensors for organophosphate class of pesticide detection. *Physical Chemistry Chemical Physics*, *22*(27), 15105–15119. PMID: 32613964

Kaur, N., & Jain, A. (2020). Nanobiosensors for heavy metal detection in food and water: A review. *Critical Reviews in Food Science and Nutrition*, *60*(21), 3655–3671.

Kawakami, R., Ohshida, T., Sakuraba, H., & Ohshima, T. (2018). A novel PLP-dependent alanine/serine racemase from the hyperthermophilic archaeon Pyrococcus horikoshii OT-3. *Frontiers in Microbiology*, *9*, 1481. PMID: 30038603

Kawase, J., Etoh, Y., Ikeda, T., Yamaguchi, K., Watahiki, M., Shima, T., & Shirabe, K. (2016). An improved multiplex real-time SYBR Green PCR assay for analysis of 24 target genes from 16 bacterial species in fecal DNA samples from patients with foodborne illnesses. *Japanese Journal of Infectious Diseases*, *69*(3), 191–201. DOI: 10.7883/yoken.jjid.2015.027 PMID: 26166499

Kellner, M. J., Koob, J. G., Gootenberg, J. S., Abudayych, O. O., & Zhang, F. (2019). SHERLOCK: Nucleic acid detection with CRISPR nucleases. *Nature Protocols*, *14*(10), 2986–3012. DOI: 10.1038/s41596-019-0210-2 PMID: 31548639

Kempe, M., Cederfur, J., Zihui, M., Pei, Y., Wistrand, U. J., Fiaccabrino, G. C., & van Rhijn, H. (1999). *CREAM—Cartridges with Molecularly Imprinted Recognition Elements for Antibiotic residues Monitoring in Milk. Lunds Universitet.* Centre for Chemistry and Chemical Engineering.

Khair Biek, S., Khudur, L. S., & Ball, A. S. (2024). Challenges and Remediation Strategies for Per-and Polyfluoroalkyl Substances (PFAS) Contamination in Composting. *Sustainability*, *16*(11), 4745.

Khan, A. (2021). Potential of coupling heavy metal (HM) phytoremediation by bioenergy plants and their associated HM-adapted rhizosphere microbiota (arbuscular mycorrhizal fungi and plant growth-promoting microbes) for bioenergy production. *Journal of Energy and Power Technology*, *3*(4), 1–1. DOI: 10.21926/jept.2104049

Khan, A., Khan, S., Khan, M., Qamar, Z., & Waqas, M. (2015). The uptake and bioaccumulation of heavy metals by food plants, their effects on plants nutrients, and associated health risk: A review. *Environmental Science and Pollution Research International*, *22*(18), 13772–13799. DOI: 10.1007/s11356-015-4881-0 PMID: 26194234

Khan, A., Sahir, K., Ahmed, M., & Khan, S. (2014). Rapid detection of Salmonella in food samples by polymerase chain reaction after a 10 h pre-enrichment. *Journal of Food Safety*, *34*(1), 79–86. DOI: 10.1111/jfs.12099

Khan, N. A., López-Maldonado, E. A., Majumder, A., & Singh, S. (2023). "A State-of-Art-Review on Emerging Contaminants: Environmental Chemistry, Health Effect, and Modern Treatment Methods." Chemosphere, Vol. 321. *Article*, *138123*, 1–15. DOI: 10.1016/j.chemosphere.2023.138123

Kim, H., Cho, J., Song, M., Cho, J., Kim, S., Kim, E., & Lee, J. (2021). Evaluating the prevalence of foodborne pathogens in livestock using metagenomics approach. *Journal of Microbiology and Biotechnology*, *31*(12), 1701–1708. DOI: 10.4014/jmb.2109.09038 PMID: 34675137

Kim, K. H., Kabir, E., & Jahan, S. A. (2017). Exposure to pesticides and the associated human health effects. *The Science of the Total Environment*, *575*, 525–535. DOI: 10.1016/j.scitotenv.2016.09.009 PMID: 27614863

Kim, K. H., & Kang, T. S. (2024a). Development of advanced PCR-based methods for accurate identification and authentication of commercial shrimp products. *Food Control*, *159*, 110288.

Kim, M. J., Park, S. J., & Park, H. (2024b). Trend in serological and molecular diagnostic methods for Toxoplasma gondii infection. *European Journal of Medical Research*, *29*, 520.

Kim, M. S., & Blaschek, H. P. (2008). Microarray analysis of Clostridium beijerinckii NRRL B-598 gene expression during butanol fermentation. *Applied and Environmental Microbiology*, *74*(22), 7405–7413. DOI: 10.1128/AEM.01006-08

Kim, S., & Kim, S. (2020). Bacterial pathogen detection by conventional culture-based and recent alternative (polymerase chain reaction, isothermal amplification, enzyme-linked immunosorbent assay, bacteriophage amplification, and gold nanoparticle aggregation) methods in food samples: A review. *Journal of Food Safety*, *41*(1). Advance online publication. DOI: 10.1111/jfs.12870

King, T., Cole, M., Farber, J. M., Eisenbrand, G., Zabaras, D., Fox, E. M., & Hill, J. P. (2017). Food safety for food security: Relationship between global megatrends and developments in food safety. *Trends in Food Science & Technology*, *68*, 160–175. DOI: 10.1016/j.tifs.2017.08.014

Kittlaus, S., Kardos, M. K., Dudás, K. M., Weber, N., Clement, A., Petkova, S., & Zoboli, O. (2024). A harmonized Danube basin-wide multi-compartment concentration database to support inventories of micropollutant emissions to surface waters. *Environmental Sciences Europe*, *36*(1), 52.

Koca, N., Erbay, Z., & Urgu Öztürk, M. (2022). Regulations and legislations on processed cheese. In *Processed Cheese Science and Technology* (pp. 449–490). Elsevier., DOI: 10.1016/B978-0-12-821445-9.00004-2

Kocoń, A., & Jurga, B. (2016). The evaluation of growth and phytoextraction potential of Miscanthus x giganteus and Sida hermaphrodita on soil contaminated simultaneously with Cd, cu, Ni, Pb, and Zn. *Environmental Science and Pollution Research International*, *24*(5), 4990–5000. DOI: 10.1007/s11356-016-8241-5 PMID: 27995509

Koo, O., Mertz, A., Akins, E., Sirsat, S., Neal, J., Morawicki, R., & Ricke, S. (2012). Analysis of microbial diversity on deli slicers using polymerase chain reaction and denaturing gradient gel electrophoresis technologies. *Letters in Applied Microbiology, 56*(2), 111–119. DOI: 10.1111/lam.12021 PMID: 23121623

Koutsoumanis, K., Allende, A., Álvarez-Ordóñez, A., Bolton, D., Bover-Cid, S., Chemaly, M., & Herman, L. (2019). Whole genome sequencing and metagenomics for outbreak investigation, source attribution, and risk assessment of food-borne microorganisms. *EFSA Journal, 17*(12). Advance online publication. DOI: 10.2903/j.efsa.2019.5898 PMID: 32626197

Krause, R. M. (2021). Why are we doing this? Issue framing, problem proximity, and cities' rationale for regulating single-use plastics. *Journal of Environmental Policy and Planning, 23*(4), 482–495.

Krishnan, S. (2022). Colorimetric visual sensors for point-of-needs testing. *Sensors and Actuators Reports, 4*, 100078.

Krishnasree, V., Nethra, P., Dheeksha, J., Madumitha, M., Vidyaeswari, R., and Lakshya, P. (2018). A pilot study on assessing the sustainability of food safety and hygienic practices in street food handling systems. Asian Journal of Dairy and Food Research, (of). https://doi.org/DOI: 10.18805/ajdfr.dr-1381

Kuchmenko, T. A., & Lvova, L. B. (2019). A perspective on recent advances in piezoelectric chemical sensors for environmental monitoring and foodstuffs analysis. *Chemosensors (Basel, Switzerland), 7*(3), 39.

Kulwa, F., Li, C., Zhang, J., Shirahama, K., Kosov, S., Zhao, X., & Grzegorzek, M. (2022). A new pairwise deep learning feature for environmental microorganism image analysis. *Environmental Science and Pollution Research International, 29*(34), 51909–51926.

Kumaran, A., Jude Serpes, N., Gupta, T., James, A., Sharma, A., Kumar, D., Nagraik, R., Kumar, V., & Pandey, S. (2023). Advancements in CRISPR-based biosensing for next-gen point of care diagnostic application. *Biosensors (Basel), 13*(2), 202. PMID: 36831968

Kuppusamy, S., Meivelu, M., Praburaman, L., Mujahid Alam, M., & Al-Sehemi, A. G., & K, A. (2024). Integrating AI in food contaminant analysis: Enhancing quality and environmental protection. *Journal of Hazardous Materials Advances, 16*, 100509. DOI: 10.1016/j.hazadv.2024.100509

Kuru, C. \. I., Ulucan-Karnak, F., & Akgöl, S. (2023). Lab-on-a-chip sensors: recent trends and future applications. *Fundamentals of Sensor Technology*, 65–98.

Kwon, J. H., Kim, J. W., Pham, T. D., Tarafdar, A., Hong, S., Chun, S. H., & Jung, J. (2020). Microplastics in food: A review on analytical methods and challenges. *International Journal of Environmental Research and Public Health, 17*(18), 6710. PMID: 32942613

Lanzas, C., Lu, Z., & Gröhn, Y. (2011). Mathematical modeling of the transmission and control of foodborne pathogens and antimicrobial resistance at preharvest. *Foodborne Pathogens and Disease, 8*(1), 1–10. DOI: 10.1089/fpd.2010.0643 PMID: 21043837

Lapris, M., Errico, M., Rocchetti, G., & Gallo, A. (2024). The Potential of Multi-Screening Methods and Omics Technologies to Detect Both Regulated and Emerging Mycotoxins in Different Matrices. *Foods, 13*(11), 1746. DOI: 10.3390/foods13111746 PMID: 38890974

Law, J., Mutalib, N., Chan, K., & Lee, L. (2015). Rapid methods for the detection of foodborne bacterial pathogens: Principles, applications, advantages and limitations. *Frontiers in Microbiology, 5*. Advance online publication. DOI: 10.3389/fmicb.2014.00770 PMID: 25628612

Lebelo, K., Malebo, N., Mochane, M. J., & Masinde, M. (2021). Chemical contamination pathways and the food safety implications along the various stages of food production: A review. *International Journal of Environmental Research and Public Health, 18*(11), 5795. PMID: 34071295

Lee SueJan, L. S., Lin ChinShiung, L. C., & Chen MaoChiang, C. M. (1996). Antibiotic residues in milk following treatment with antibiotics for lactating dairy cows.

Lee, H., and Yoon, Y. (2021). Etiological agents implicated in foodborne illness world

Lee, A., Ng, I., Wong, E., Wong, I., Sze, R., Chan, K., & Siu, G. (2023). Comprehensive identification of pathogenic microbes and antimicrobial resistance genes in food products using nanopore sequencing-based metagenomics. https://doi.org/ DOI: 10.1101/2023.10.15.562131

Lee, H., Cho, Y., Jung, G., Kim, H., & Jeong, W. (2023). Comparison of recovery efficiency and matrix effect reduction in pesticide residue analysis: QuEChERS with d-SPE, SPE, and FaPEx in apples and Korean cabbage. *Analytical Methods, 15*(30), 3709–3716. DOI: 10.1039/d3ay00612c PMID: 37475594

Lee, H., & Yoon, Y. (2021). Etiological agents implicated in foodborne illness world wide. *Food Science of Animal Resources, 41*(1), 1. PMID: 33506212

Lee, H., & Yoon, Y. (2021). Etiological Agents Implicated in Foodborne Illness World Wide. *Food Science of Animal Resources*, *41*(1), 1–7. DOI: 10.5851/kosfa.2020.e75 PMID: 33506212

Lee, J. C., Smaoui, S., Duffill, J., Marandi, B., & Varzakas, T. (2025). Research progress in current and emerging issues of PFASs' global impact: Long-term health effects and governance of food systems. *Foods*, *14*(6), 958. PMID: 40231978

Lee, J. C., Smaoui, S., Duffill, J., Marandi, B., & Varzakas, T. (2025). Research Progress in Current and Emerging Issues of PFASs' Global Impact: Long-Term Health Effects and Governance of Food Systems. *Foods*, *14*(6), 958. PMID: 40231978

Lee, J., Park, S., Khan, F., Jo, D., Lee, D., Kang, M., & Kim, Y. (2021). Simultaneous isolation and enumeration of virulent Vibrio cholerae and Vibrio vulnificus using an advanced MPN-PCR method. https://doi.org/DOI: 10.21203/rs.3.rs-528186/v1

Lee, K., Son, M., Kang, J., Kim, D., Hong, S., Park, T., & Choi, S. (2018). A triangle study of human, instrument and bioelectronic nose for non-destructive sensing of seafood freshness. *Scientific Reports*, *8*(1). Advance online publication. DOI: 10.1038/s41598-017-19033-y

Lehotay, S. J., & Hajšlová, J. (2002). Application of gas chromatography in food analysis. *Trends in Analytical Chemistry*, *21*(9–10), 686–697. DOI: 10.1016/S0165-9936(02)00805-1

Lehotay, S. J., Son, K. A., Kwon, H., Koesukwiwat, U., Fu, W., Mastovska, K., Hoh, E., & Leepipatpiboon, N. (2010). Comparison of QuEChERS sample preparation methods for the analysis of pesticide residues in fruits and vegetables. *Journal of Chromatography. A*, *1217*(16), 2548–2560. DOI: 10.1016/j.chroma.2010.01.044 PMID: 20144460

Lemus, J. A., Blanco, G., Grande, J., Arroyo, B., & Garcia-Montijano, M. (2008). Antibiotics Threaten Wildlife: Circulating Quinolone Residues and Disease in Avian Scavengers. *PLoS One*, *3*(1), e1444. PMID: 18197254

Li, S., Chai, Y., and Chin, B. (2015). High-throughput pathogen screening for food safety using magnetoelastic biosensors. https://doi.org/DOI: 10.1117/12.2176777

Li, X., Huang, X., & Grzegorzek, M. (2022). A state-of-the-art survey of u-net in microscopic image analysis: From simple usage to structure mortification. arXiv, 2202.06465,. URL https://arxiv.org/abs/2202. 06465.

Li, A., Zhang, H., Zhang, X., Wang, Q., Tian, J., Li, Y., & Li, J. (2010). Rapid separation and immunoassay for low levels of Salmonella in foods using magnetosome–antibody complex and real-time fluorescence quantitative PCR. *Journal of Separation Science*, *33*(21), 3437–3443. DOI: 10.1002/jssc.201000441 PMID: 20886524

Liang, Y., Yi, X., Dang, Z., Wang, Q., Luo, H., & Tang, J. (2017). Heavy metal contamination and health risk assessment in the vicinity of a tailing pond in Guangdong, China. *International Journal of Environmental Research and Public Health*, *14*(12), 1557. DOI: 10.3390/ijerph14121557 PMID: 29231884

Li, B. Y., Yu, M. X., Xu, W. P., & Chen, L. (2023). Han, J. Comparison of PCR techniques in adulteration identification of dairy products. *Agriculture*, *13*, 1450.

Li, B., Patel, I., Tall, B., & Elkins, C. (2017). Advancements in microarray utility for detection and tracking of foodborne microbes in the genomic era. *Advanced Techniques in Biology and Medicine*, *05*(03). Advance online publication. DOI: 10.4172/2379-1764.1000239

Liberty, J. T., & Bhuiyan, Md. H. R., & Ngadi, M. (2025). Advancements in rapid and non-destructive approaches for quality assessment of fried foods and frying oil. *Journal of Food Composition and Analysis*, *145*, 107839. DOI: 10.1016/j.jfca.2025.107839

Li, C., Shirahama, K., & Grzegorzek, M. (2014). Application of content-based image analysis to environmental microorganism classification. In *Proceedings of a Conference*.

Li, C., Shirahama, K., & Grzegorzek, M. (2015). Environmental microbiology aided by content-based image analysis. *Pattern Analysis & Applications*, *19*(2), 531–547.

Li, G., Chen, T., & Dahlman, J. (2023). "Current Challenges and Future Directions for Engineering Extracellular Vesicles for Heart, Lung, Blood and Sleep Diseases." Journal of Extracellular Vesicles, Vol. 12(3). *Article*, *12305*, 1–15. DOI: 10.1002/jev2.12305

Li, G., Li, Q., Wang, X., Liu, X., Zhang, Y., Li, R., Guo, J., & Zhang, G. (2023). Lateral flow immunoassays for antigens, antibodies and haptens detection. *International Journal of Biological Macromolecules*, *242*, 125186. PMID: 37268073

Liguori, J., Trübswasser, U., Pradeilles, R., Le Port, A., Landais, E., Talsma, E. F., Lundy, M., Béné, C., Bricas, N., Laar, A., Amiot, M. J., Brouwer, I. D., & Holdsworth, M. (2022). How do food safety concerns affect consumer behaviors and diets in low- and middle-income countries? A systematic review. *Global Food Security*, *32*, 100606. DOI: 10.1016/j.gfs.2021.100606

Li, H., Cui, J., Chen, H., Li, H., Xie, Y., Song, W., & Chen, R. (2024). Rapid colorimetric and fluorescence identification of Pinelliae Rhizoma and adulterate Rhizoma Typhonii Flagelliformis using direct-LAMP assay. *Food Chemistry*, *437*, 137840.

Li, H., Meng, M., Cao, J., Hu, X., Wang, H., Pan, M., & Wang, S. (2025). Review: Nanozyme-based point-of-care testing for portable and on-site monitoring of food contaminants. *Trends in Food Science & Technology*, *161*, 105062. DOI: 10.1016/j.tifs.2025.105062

Li, J., Liu, H., & Chen, J. P. (2018). Microplastics in freshwater systems: A review on occurrence, environmental effects, and methods for microplastics detection. *Water Research*, *137*, 362–374. PMID: 29580559

Lin, B., Wang, Z., Vora, G. J., Thornton, J. A., Schnur, J. M., Thach, D. C., Blaney, K. M., Ligler, A. G., Malanoski, A. P., Santiago, J., Walter, E. A., Agan, B. K., Metzgar, D., Seto, D., Daum, L. T., Kruzelock, R., Rowley, R. K., Hanson, E. H., Tibbetts, C., & Stenger, D. A. (2006). Broad-spectrum respiratory tract pathogen identification using resequencing DNA microarrays. *Genome Research*, *16*, 527–535.

Lin, D. Y., Yu, C. Y., Ku, C. A., & Chung, C. K. (2023). "Design, Fabrication, and Applications of SERS Substrates for Food Safety Detection." Micromachines, Vol. 14(3). *Article*, *123456*, 1–15. DOI: 10.3390/micromachines14030123

Line, J., Stern, N., Oakley, B., & Seal, B. (2011). Comparison of an automated most-probable-number technique with traditional plating methods for estimating populations of total aerobes, coliforms, and Escherichia coli associated with freshly processed broiler chickens. *Journal of Food Protection*, *74*(9), 1558–1563. DOI: 10.4315/0362-028x.jfp-11-024 PMID: 21902928

Ling, Z., Yang, L., Zhang, W., Yao, T., & Xu, H. (2024). Detection of food contaminants: A review of established rapid analytical techniques and their applications and limitations. *Food Safety and Health*, *2*(1), 72–95. DOI: 10.1002/fsh3.12032

Lin, Q., Li, H., Zhong, H., Zhao, Q., Da-hui, X., & Wang, Z. (2014). Determination of titanium in nano-titanium(IV) oxide composite food packaging by microwave digestion and inductively coupled plasma atomic emission spectrometry and inductively coupled plasma mass spectrometry. *Analytical Letters*, *47*(12), 2095–2103. DOI: 10.1080/00032719.2014.895907

Li, S.-Y., Cheng, Q.-X., Wang, J.-M., Li, X.-Y., Zhang, Z.-L., Gao, S., & Wang, J. (2022). CRISPR-Cas12a-based detection of *Salmonella* in food samples. *Biosensors & Bioelectronics*, *206*, 114117. DOI: 10.1016/j.bios.2022.114117

Liu, D. M., & Dong, C. (2023). "Gold Nanoparticles as Colorimetric Probes in Food Analysis: Progress and Challenges." Food Chemistry, Vol. 375. *Article, 123456*, 1–15. DOI: 10.1016/j.foodchem.2023.123456

Liu, J., Zeng, S., Zhu, H., Wan, X., Sohan, A. S. M. M. F., & Yin, B. (2024). A Portable Automated Microfluidic Platform for Point-of-Care Testing for Multiple Mycotoxins in Wine. *Foods*, *13*(13), 2066. DOI: 10.3390/foods13132066 PMID: 38998571

Liu, R., Wang, X., Wang, X., Shi, Y., Shi, C., Wang, W., & Ma, C. (2019). A simple isothermal nucleic acid amplification method for the effective on-site identification for adulteration of pork source in mutton. *Food Control*, *98*, 297–302.

Liu, S., Yang, Y., Zong, Y., Chen, W., Geng, J., Zhao, Y., Du, R., & He, Z. (2025). Ginsenoside Rg3 alleviates brain damage caused by chlorpyrifos exposure by targeting and regulating the microbial-gut-brain axis. *Phytomedicine*, *143*, 156838. Advance online publication. DOI: 10.1016/j.phymed.2025.156838 PMID: 40381500

Liu, W., Chung, K., Yu, S., & Lee, L. P. (2024). Nanoplasmonic biosensors for environmental sustainability and human health. *Chemical Society Reviews*. PMID: 39192761

Liu, Y., Wu, P., & He, C. (2020). Fluorescence-based quantum dot sensors for heavy metal detection in food matrices. *Food Chemistry*, *329*, 127190.

Liu, Z., Wang, J., Wang, Z., Xu, H., Di, S., Zhao, H., Qi, P., & Wang, X. (2022). Development of magnetic solid phase extraction using magnetic amphiphilic polymer for sensitive analysis of multi-pesticides residue in honey. *Journal of Chromatography. A*, *1664*, 462789. DOI: 10.1016/j.chroma.2021.462789 PMID: 35026602

Li, X., Shen, X., Jiang, W., Xi, Y., & Li, S. (2024). Comprehensive review of emerging contaminants: Detection technologies, environmental impact, and management strategies. *Ecotoxicology and Environmental Safety*, *278*, 116420. PMID: 38701654

Li, X., Zhen, Y., Wang, R., Li, T., Dong, S., Zhang, W., & Cheng, J. (2021). "Application of Gas Chromatography Coupled to Triple Quadrupole Mass Spectrometry (GC-(APCI) MS/MS) in Determination of PCBs (Mono-to Deca-) and PCDD/Fs in Food." Chemosphere, Vol. 263. *Article*, *128517*, 1–15. DOI: 10.1016/j.chemosphere.2020.128517

Li, Y., Man, S., Ye, S., Liu, G., & Ma, L. (2022). CRISPR-Cas-based detection for food safety problems: Current status, challenges, and opportunities. *Comprehensive Reviews in Food Science and Food Safety*, *21*(4), 3770–3798. DOI: 10.1111/1541-4337.13000 PMID: 35796408

Li, Y., Qin, G., He, F., Zou, K., Zuo, B., Liu, R., & Zhang, W. (2022). "Investigation and Analysis of Pesticide Residues in Edible Fungi Produced in the Mid-Western Region of China." Food Control, Vol. 134. *Article, 123456*, 1–15. DOI: 10.1016/j.foodcont.2022.123456

Li, Y., Zhang, D., Zeng, X., Liu, C., Wu, Y., & Fu, C. (2024). Advances in Aptamer-Based Biosensors for the Detection of Foodborne Mycotoxins. *Molecules (Basel, Switzerland), 29*(16), 3974. DOI: 10.3390/molecules29163974 PMID: 39203052

Li, Y., Zhao, Z., Liu, Y., Wang, N., Man, S., Ma, L., & Wang, S. (2023). CRISPR/Cas System: The accelerator for the development of non-nucleic acid target detection in food safety. *Journal of Agricultural and Food Chemistry, 71*(37), 13577–13594. DOI: 10.1021/acs.jafc.3c03619 PMID: 37656446

Loaiza, I., Boeck, G., Alcazar, J., Campos, D., Cárdenas-Alayza, S., Ganoza, M., & Troch, M. (2021). Trophic interactions and metal transfer in marine ecosystems driven by the Peruvian scallop Argopecten purpuratus aquaculture. *Journal of the World Aquaculture Society, 53*(2), 452–474. DOI: 10.1111/jwas.12822

López-Campos, G., & Martínez-Suárez, J. V.. (2020). Detection, identification, and analysis of foodborne pathogens. *Frontiers in Microbiology, 11*, 2869.

Lv, L., He, L., Jiang, S., Chen, J., Zhou, C., Qu, J., & Li, C. (2020). In situ surface-enhanced Raman spectroscopy for detecting microplastics and nanoplastics in aquatic environments. *The Science of the Total Environment, 728*, 138449. PMID: 32353796

Maculewicz, J., Kowalska, D., Świacka, K., Toński, M., Stepnowski, P., Białk-Bielińska, A., & Dołżonek, J. (2022). Transformation products of pharmaceuticals in the environment: Their fate,(eco) toxicity and bioaccumulation potential. *The Science of the Total Environment, 802*, 149916. PMID: 34525754

Mahadev Madgule, N. (2023). Vinayaka, Yeshwant M. Sonkhaskar, Dhiren Ramanbhai Patel, R. Karthikeyan, P. Selvakumar, (2023). Mechanical properties and microstructure of activated TIG welded similar joints of Inconel alloys by desirability approaches. *Materials Today: Proceedings, 77*, 528–533.

Mahdavi, V., Behbahan, A. K., Moradi, F., & Aboul-Enein, H. Y. (2021). Analysis of Alternative New Pesticide (Fluopyram, Flupyradifurone, and Indaziflam) Residues in Pistachio, Date, and Soil by Liquid Chromatography Triple Quadrupole Tandem Mass Spectrometry. *Soil & Sediment Contamination, 30*(4), 373–383. DOI: 10.1080/15320383.2020.1854173

Majer-Baranyi, K., Székács, A., & Adányi, N. (2023). Application of electrochemical biosensors for the determination of food spoilage. *Biosensors (Basel), 13*(4), 456. DOI: 10.3390/bios13040456 PMID: 37185531

Makmud, S. T., Haryanto, A. N., & Prasetyo, S. Y. Improving rotten fruit classification accuracy through fusion of multiple pre-trained cnn models. Engineering and Applied Science Research, 51(3):313–320, 2024. doi: . 2024.30. URL https://www.tci-thaijo.org/index.php/easr/indexDOI: 10.14456/easr

Ma, L., Yang, L., Tang, L., Wang, Y., Luo, H., Zhong, Z., Zhang, W., Chen, D., Wei, J., Li, P., & Wang, Y. (2025). Sample preparation techniques for quality evaluation and safety control of medicinal and edible plants: Overview, advances, applications, and future perspectives. *Journal of Pharmaceutical Analysis, 101296*. Advance online publication. DOI: 10.1016/j.jpha.2025.101296

Malavi, D., Muzhingi, T., & Abong, G. (2018). Good Manufacturing Practices and Microbial Contamination Sources in an Orange-Fleshed Sweet Potato Puree Processing Plant in Kenya. *International Journal of Food Sciences, 2018*, 1–11. DOI: 10.1155/2018/4093161 PMID: 29808161

Malhotra, S. (2017). Food safety issues related to street vendors. In *Food Safety in the 21st Century* (pp. 395–402). Elsevier., DOI: 10.1016/B978-0-12-801773-9.00031-5

Mandal, P., Lanaridi, O., Warth, B., & Ansari, K. M. (2024). Metabolomics as an emerging approach for deciphering the biological impact and toxicity of food contaminants: The case of mycotoxins. *Critical Reviews in Food Science and Nutrition, 64*(27), 9859–9883. DOI: 10.1080/10408398.2023.2217451 PMID: 37283072

Mandal, S., Poi, R., Hazra, D. K., Ansary, I., Bhattacharyya, S., & Karmakar, R. (2023). Review of extraction and detection techniques for the analysis of pesticide residues in fruits to evaluate food safety and make legislative decisions: Challenges and anticipations. *Journal of Chromatography. B, Analytical Technologies in the Biomedical and Life Sciences, 1215*, 123587. DOI: 10.1016/j.jchromb.2022.123587 PMID: 36628882

Mandpe, P., Prabhakar, B., Gupta, H., & Shende, P. (2020). Glucose oxidase-based biosensor for glucose detection from biological fluids. *Sensor Review, 40*(4), 497–511.

Maneeboon, T., Chuaysrinule, C., & Mahakarnchanakul, W. (2023). Optimization and Validation of Dispersive Liquid–Liquid Microextraction for Simultaneous Determination of Aflatoxins B1, B2, G1, and G2 in Senna Leaves and Pods Using HPLC-FLD with Pre-Column Derivatization. *Toxins, 15*(4), 277. DOI: 10.3390/toxins15040277 PMID: 37104215

Mangal, M., Bansal, S., Sharma, S., & Gupta, R. (2015). Molecular detection of foodborne pathogens: A rapid and accurate answer to food safety. *Critical Reviews in Food Science and Nutrition, 56*(9), 1568–1584. DOI: 10.1080/10408398.2013.782483 PMID: 25830555

Manning, L., & Soon, J. M. (2014). Developing systems to control food adulteration. *Food Policy, 49*, 23–32. DOI: 10.1016/j.foodpol.2014.06.005

Manojkumar, Y., Pilli, S., Rao, P. V., & Tyagi, R. D. (2023). Sources, occurrence and toxic effects of emerging per-and polyfluoroalkyl substances (PFAS). *Neurotoxicology and Teratology, 97*, 107174. PMID: 36907230

Man, Y., Yang, Z., Sun, Y., Zhao, W., Xiang, G., & He, L. (2024). Simple and rapid ionic liquid-based one-, two-, three-phase transition microextraction for efficient extraction of trace organic pollutants and elimination of lipid co-extractives from fatty food matrices. *Food Chemistry, 439*, 138165. DOI: 10.1016/j.foodchem.2023.138165 PMID: 38091782

Ma, P., Li, C., Rahaman, M. M., Yao, Y., Zhang, J., Zou, S., & Grzegorzek, M. (2023). A state-of-the-art survey of object detection techniques in microorganism image analysis: From classical methods to deep learning approaches. *Artificial Intelligence Review, 56*(2), 1627–1698.

Maphaisa, T. C., Akinmoladun, O. F., Adelusi, O. A., Mwanza, M., Fon, F., Tangni, E., & Njobeh, P. B. (2025). Advances in mycotoxin detection techniques and the crucial role of reference material in ensuring food safety: A review. *Food and Chemical Toxicology, 200*, 115387. DOI: 10.1016/j.fct.2025.115387 PMID: 40081789

Maragos, C. M. (2022). Application of ambient ionization mass spectrometry to detect the mycotoxin Roquefortine C in blue cheese. *Food Analytical Methods, 15*, 751–760. DOI: 10.1007/s12161-021-02165-3

Margot, H., Stephan, R., O'Mahony, E., & Iversen, C. (2013). Comparison of rapid cultural methods for the detection of Salmonella species. *International Journal of Food Microbiology, 163*(2-3), 47–50. DOI: 10.1016/j.ijfoodmicro.2013.02.013 PMID: 23558186

Marin, S., Ramos, A. J., Cano-Sancho, G., & Sanchis, V. (2013). Mycotoxins: Occurrence, toxicology, and exposure assessment. *Food and Chemical Toxicology, 60*, 218–237. DOI: 10.1016/j.fct.2013.07.047 PMID: 23907020

Marklewitz, M., Jaguparov, A., Wilhelm, A., Akande, O. W., Musul, B., Poates, A. L., Afrough, B., Norberg, A., Hull, N. C., & Ehsani, S.. (2025). Genomics costing tool: Considerations for improving cost-efficiencies through cross scenario comparison. *Frontiers in Public Health*, *12*, 1498094.

Masquelier, J., Tangni, E. K., Becker, P., Sanders, J., Laporte, J., & Mertens, B. (2025). Validation of an LC-MS Method for Quantification of Mycotoxins and Characterization of Fungal Strains Occurring in Food and Feed. *Chemosensors (Basel, Switzerland)*, *13*(3), 106. DOI: 10.3390/chemosensors13030106

Matchado, M., Rühlemann, M., Reimeiter, S., Kacprowski, T., Frost, F., Haller, D., & List, M. (2023). On the Limits of 16S rRNA Gene-Based Metagenome Prediction and Functional Profiling. https://doi.org/DOI: 10.1101/2023.11.07.564315

Mateo, F., Mateo, E. M., Tarazona, A., García-Esparza, M. Á., Soria, J. M., & Jiménez, M. (2025). New Strategies and Artificial Intelligence Methods for the Mitigation of Toxigenic Fungi and Mycotoxins in Foods. *Toxins*, *17*(5), 231. DOI: 10.3390/toxins17050231 PMID: 40423314

Mayer, M., Vega, J., DeMarco, I., Eicher, R., Molina, D., Lee, C., … and Naufel, J. (2023). Message convergence as reassurance about the safety of the food supply concerning African swine fever virus. AHTR, 2(1), 02-11. https://doi.org/DOI: 10.31491/ahtr.2023.02.002

Mazarakioti, E., Zotos, A., Thomatou, A., Kontogeorgos, A., Patakas, A., & Ladavos, A. (2022). Inductively coupled plasma-mass spectrometry (ICP-MS) is a useful tool in the authenticity of agricultural products' and foods' origin. *Foods*, *11*(22), 3705. DOI: 10.3390/foods11223705 PMID: 36429296

McConnell, E. M., Nguyen, J., & Li, Y. (2020). Aptamer-based biosensors for environmental monitoring. *Frontiers in Chemistry*, *8*, 434. DOI: 10.3389/fchem.2020.00434 PMID: 32548090

McGrath, T. F., Elliott, C. T., & Fodey, T. L. (2012). Biosensors for the analysis of microbiological and chemical contaminants in food. *Analytical and Bioanalytical Chemistry*, *403*, 75–92. PMID: 22278073

McLain, J., Cytryn, E., Durso, L., & Young, S. (2016). Culture-based methods for detection of antibiotic resistance in agroecosystems: Advantages, challenges, and gaps in knowledge. *Journal of Environmental Quality*, *45*(2), 432–440. DOI: 10.2134/jeq2015.06.0317 PMID: 27065389

Meera, A., Kc, M., John, J., Sudha, B., and Sajeena, A. (2022). Aquatic plants as bioremediators in pollution abatement of heavy metals. https://doi.org/DOI: 10.5772/intechopen.99627

Meira, D. I., Barbosa, A. I., Borges, J., & Reis, R. L. (2024). Recent Advances in Nanomaterial-Based Optical Biosensors for Food Safety Applications: Ochratoxin-A Detection, as Case Study. *Critical Reviews in Food Science and Nutrition*, *64*(3), 567–589. DOI: 10.1000/taylorfrancis.ochratoxin2024

Mellmann, A., Harmsen, D., Cummings, C. A., Zentz, E. B., Leopold, S. R., Rico, A., Prior, K., Szczepanowski, R., Ji, Y., Zhang, W., McLaughlin, S. F., Henkhaus, J. K., Leopold, B., Bielaszewska, M., Prager, R., Brzoska, P. M., Moore, R. L., Guenther, S., Rothberg, J. M., & Karch, H. (2011). Prospective Genomic Characterization of the German Enterohemorrhagic Escherichia coli O104:H4 Outbreak by Rapid Next Generation Sequencing Technology. *PLoS One*, *6*(7), e22751. DOI: 10.1371/journal.pone.0022751 PMID: 21799941

Melo, S. C., Pungartnik, C., Cascardo, J. C., & Brendel, M. (2006). Rapid and efficient protocol for DNA extraction and molecular identification of the basidiomycete Crinipellis perniciosa. *Genetics and Molecular Research*, *5*, 851–855.

Merkle, S., Kleeberg, K. K., & Fritsche, J. (2015). Recent Developments and Applications of Solid Phase Microextraction (SPME) in Food and Environmental Analysis-A Review. *Chromatography (Basel)*, *2*(3), 293–381. DOI: 10.3390/chromatography2030293

Mermiga, E., Pagkali, V., Kokkinos, C., & Economou, A. (2023). An Aptamer-Based Lateral Flow Biosensor for Low-Cost, Rapid and Instrument-Free Detection of Ochratoxin A in Food Samples. *Molecules (Basel, Switzerland)*, *28*(24), 8135. DOI: 10.3390/molecules28248135 PMID: 38138623

Meyers, N., Catarino, A. I., Declercq, A. M., Brenan, A., Devriese, L., Vandegehuchte, M., & Everaert, G. (2022). Microplastic detection and identification by Nile red staining: Towards a semi-automated, cost-and time-effective technique. *The Science of the Total Environment*, *823*, 153441. PMID: 35124051

Mikac, L., Rigó, I., Himics, L., Tolić, A., Ivanda, M., & Veres, M. (2023). Surface-enhanced Raman spectroscopy for the detection of microplastics. *Applied Surface Science*, *608*, 155239.

Miraj, S. S., Parveen, N., & Zedan, H. S. (2021). Plastic microbeads: Small yet mighty concerning. *International Journal of Environmental Health Research*, *31*(7), 788–804. PMID: 31709817

Mirres, A. C. M., Silva, B. E. P. M. D., Tessaro, L., Galvan, D., Andrade, J. C., Aquino, A., Joshi, N., & Conte-Junior, C. A. (2022). Recent Advances in Nanomaterial-Based Biosensors for Pesticide Detection in Foods. *Biosensors (Basel)*, *12*(8), 572. DOI: 10.3390/bios12080572 PMID: 36004968

Mishra, G., Barfidokht, A., Tehrani, F., & Mishra, R. (2018). Food safety analysis using electrochemical biosensors. *Foods*, *7*(9), 141. DOI: 10.3390/foods7090141 PMID: 30200448

Modi, B., Timilsina, H., & Bhandari, S. (2021). Current Trends of Food Analysis, Safety, and Packaging. *International Journal of Food Science & Technology*, *56*(3), 567–589. DOI: 10.1000/wiley.foodsafety2021

Modupalli, N., Naik, M., Sunil, C. K., & Natarajan, V. (2021). Emerging non-destructive methods for quality and safety monitoring of spices. *Trends in Food Science & Technology*, *108*, 133–147. DOI: 10.1016/j.tifs.2020.12.021

Mohammad, A.-M., Chowdhury, T., Biswas, B., & Absar, N. (2018). Food poisoning and intoxication: A global leading concern for human health. In *Food safety and preservation* (pp. 307–352). Elsevier.

Moller, A., & Liang, C. (2017). Metacrast: Reference-guided extraction of CRISPR spacers from unassembled metagenomes. *PeerJ*, *5*, e3788. DOI: 10.7717/peerj.3788 PMID: 28894651

Moorthy, G., Benjakul, S., Sukkapat, P., Senathipathi, D. N., & Saetang, J. (2024). Development and validation of colorimetric LAMP assay for species authentication of economically important Mytildae mussels. *Food Control*, *165*, 110697.

Morbioli, G. G., Mazzu-Nascimento, T., Stockton, A. M., & Carrilho, E. (2017). Technical aspects and challenges of colorimetric detection with microfluidic paper-based analytical devices ($µ$PADs)-A review. *Analytica Chimica Acta*, *970*, 1–22. PMID: 28433054

Mukhi, S., Srikantiah, R., Manjrekar, P., Iyyaswami, R., & Sindhu, H. (2022). Assessment of heavy metals in food and drug packaging materials. *F1000 Research*, *11*, 648. DOI: 10.12688/f1000research.121473.1 PMID: 38779461

Munjanja, B. K., Nomngongo, P. N., & Mketo, N. (2025). In Vegetable Oils: A Review of Recent Developments, Current Challenges and Future Perspectives in Sample Preparation, Chromatographic Determination, and Analysis of Real Samples. *Critical Reviews in Analytical Chemistry*, *55*(3), 567–589. DOI: 10.1080/10408347.2023.2286642

Muratova, A., Lyubun, E., Голубев, С., & Turkovskaya, O. (2022). Effect of copper ions on the associations of Azospirillum bacteria with wheat seedlings (Triticum aestivum L.). *Vavilov Journal of Genetics and Breeding*, *26*(5), 477–485. DOI: 10.18699/vjgb-22-58 PMID: 36128570

Murphy, C., Carson, C., Smith, B., Chapman, B., Marrotte, J., McCann, M., & Parmley, E. (2018). Factors potentially linked with the occurrence of antimicrobial resistance in selected bacteria from cattle, chickens, and pigs: A scoping review of publications for use in modelling of antimicrobial resistance (IAM.amr project). *Zoonoses and Public Health*, *65*(8), 957–971. DOI: 10.1111/zph.12515 PMID: 30187682

Mu, W., Kleter, G. A., Bouzembrak, Y., Dupouy, E., Frewer, L. J., Al Natour, F. N. R., & Marvin, H. J. P. (2024). Making food systems more resilient to food safety risks by including artificial intelligence, big data, and internet of things into food safety early warning and emerging risk identification tools. *Comprehensive Reviews in Food Science and Food Safety*, *23*, e13296. DOI: 10.1111/1541-4337.13296 PMID: 38284601

Nafea, A. M., Wang, Y., Wang, D., Salama, A. M., Aziz, M. A., Xu, S., & Tong, Y. (2024). Application of next-generation sequencing to identify different pathogens. *Frontiers in Microbiology*, *14*, 1329330. DOI: 10.3389/fmicb.2023.1329330 PMID: 38348304

Nahar, K., Zulkarnain, N. A., & Niven, R. K. (2023). A review of analytical methods and Technologies for Monitoring per-and Polyfluoroalkyl Substances (PFAS) in water. *Water (Basel)*, *15*(20), 3577.

Narváez, A., Izzo, L., Castaldo, L., Lombardi, S., Rodríguez-Carrasco, Y., & Ritieni, A. (2023). Multi-Mycotoxin Method Development Using Ultra-High Liquid Chromatography with Orbitrap High-Resolution Mass Spectrometry Detection in Breakfast Cereals from the Campania Region, Italy. *Toxins*, *15*(2), 148. DOI: 10.3390/toxins15020148 PMID: 36828462

Naseem, S., & Rizwan, M. (2025a). The role of artificial intelligence in advancing food safety: A strategic path to zero contamination. *Food Control*, *175*, 111292. DOI: 10.1016/j.foodcont.2025.111292

Nasseri, B., Soleimani, N., Rabiee, N., Kalbasi, A., Karimi, M., & Hamblin, M. (2018). Point-of-care microfluidic devices for pathogen detection. *Biosensors & Bioelectronics*, *117*, 112–128. DOI: 10.1016/j.bios.2018.05.050 PMID: 29890393

Nath, S. (2024). Advancements in food quality monitoring: Integrating biosensors for precision detection. *Sustainable Food Technology*.

Nawghare, P., & Prasad, J. (2025). Hybrid CNN and random forest model with late fusion for detection of autism spectrum disorder in Toddlers. *MethodsX*, *14*, 103278.

Neff, J., Xie, Y., Gill, J., & Liu, M. (2019). Complete genome sequence of Salmonella enterica serovar Heidelberg myophage meda. *Microbiology Resource Announcements*, *8*(17). Advance online publication. DOI: 10.1128/mra.00253-19 PMID: 31023796

Nelis, J. L. D., Tsagkaris, A. S., Dillon, M. J., Hajslova, J., & Elliott, C. T. (2020). Smartphone-based optical assays in the food safety field. *Trends in analytical chemistry. TRAC*, *129*, 115934. DOI: 10.1016/j.trac.2020.115934 PMID: 32904649

Nguyen, H., Oh, H., & Kim, M. (2021). Impacts of antioxidant vitamins, curry consumption and heavy metal levels on the metabolic syndrome with comorbidities: a national cross-sectional study. https://doi.org/DOI: 10.21203/rs.3.rs-225694/v1

Nguyen, V., Linh, T., Vo, T., Nguyen, Q., & Van, T. (2023). Analytical techniques for determination of heavy metal migration from different types of locally made plastic food packaging materials using ICP-MS. *Food Science & Nutrition*, *11*(7), 4030–4037. DOI: 10.1002/fsn3.3391 PMID: 37457188

Nguyen, V.-T., Kwon, Y. S., & Gu, M. B. (2017). Aptamer-based environmental biosensors for small molecule contaminants. *Current Opinion in Biotechnology*, *45*, 15–23. PMID: 28088092

Nillmani, N. S., Saba, L., Khanna, N. N., Kalra, M. K., Fouda, M. M., & Suri, J. S. (2022). Segmentation-based classification deep learning model embedded with explainable ai for covid-19 detection in chest x-ray scans. *Diagnostics (Basel)*, *12*(9), 2132. https://www.mdpi.com/journal/diagnostics. DOI: 10.3390/diagnostics12092132 PMID: 36140533

Nilsson, R. H., Larsson, K. H., Taylor, A. F. S., Bengtsson-Palme, J., Jeppesen, T. S., Schigel, D., & Abarenkov, K. (2019). The UNITE database for molecular identification of fungi: Handling dark taxa and parallel taxonomic classifications. *Nucleic Acids Research*, *47*(D1), D259–D264. DOI: 10.1093/nar/gky1022 PMID: 30371820

Nkwunonwo, U., Odika, P., & Onyia, N. (2020). A review of the health implications of heavy metals in food chain in Nigeria. *TheScientificWorldJournal*, *2020*, 1–11. DOI: 10.1155/2020/6594109 PMID: 32351345

Nodoushan, S. M., Nasirizadeh, N., Sedighian, H., Kachuei, R., Azimzadeh-Taft, M., & Fooladi, A. A. I. (2022). Detection of Staphylococcal Enterotoxin A (SEA) using a sensitive nanomaterial-based electrochemical aptasensor. *Diamond and Related Materials*, *127*, 109042.

Nolvachai, Y., Amaral, M. S. S., & Marriott, P. J. (2023). Foods and Contaminants Analysis Using Multidimensional Gas Chromatography: An Update of Recent Studies, Technology, and Applications. *Analytical Chemistry*, *95*(1), 238–263. DOI: 10.1021/acs.analchem.2c04680

Notomi, T., Okayama, H., Masubuchi, H., Yonekawa, T., Watanabe, K., Amino, N., & Hase, T. (2000). Loop-mediated isothermal amplification of DNA. *Nucleic Acids Research*, *28*(12), e63. DOI: 10.1093/nar/28.12.e63

Nuraeni, U., Malau, J., Astuti, R. T., Dewantoro, A., Apriori, D., Lusiana, E. D., & Prasetya, B. (2023). Droplet digital PCR versus real-time PCR for in-house validation of porcine detection and quantification protocol: An artificial recombinant plasmid approach. *PLoS One*, *18*, e0287712.

Nwizugbo, K. C., Ogwu, M. C., Eriyamremu, G. E., & Ahana, C. M. (2023). Alterations in energy metabolism, total protein, uric and nucleic acids in African sharptooth catfish (*Clarias gariepinus* Burchell.) exposed to crude oil and fractions. *Chemosphere*, *316*, 137778. DOI: 10.1016/j.chemosphere.2023.137778 PMID: 36640975

O'Connor, J., Bolan, N. S., Kumar, M., Nitai, A. S., Ahmed, M. B., Bolan, S. S., & Kirkham, M. B. (2022). Distribution, transformation and remediation of poly- and per-fluoroalkyl substances (PFAS) in wastewater sources. *Process Safety and Environmental Protection*, *164*, 91–108.

O'Shea, H., Blacklaws, B. A., Collins, P. J., McKillen, J., & Fitzgerald, R. (2019). Viruses associated with foodborne infections. *Reference Module in Life Sciences*, B978--0.

O157. (n.d.). H7 using immunomagnetic separation and mPCR in Turkish foods of animal origin. *Letters in Applied Microbiology*, *57*(4), 373–379. DOI: 10.1111/lam.12124 PMID: 23809061

Obeagu, E. I., Okoroiwu, G. I. A., Ubosi, N. I., Obeagu, G. U., Onohuean, H., Muhammad, T., & Adias, T. C. (2024). Revolution in malaria detection: Unveiling current breakthroughs and tomorrow's possibilities in biomarker innovation. *Annals of Medicine and Surgery (London)*, *86*(10), 5859–5876. PMID: 39359838

Obongodot, N. U., & Ogwu, M. C. (2023). Plant Food for Human Health: Case Study of Indigenous Vegetables in Akwa Ibom State, Nigeria. In: Izah, S.C., Ogwu, M.C., Akram, M. (eds) *Herbal Medicine Phytochemistry*. Reference Series in Phytochemistry. Springer, Cham. https://doi.org/DOI: 10.1007/978-3-031-21973-3_2-1

Odeh, F., Nsairat, H., Alshaer, W., Ismail, M. A., Esawi, E., Qaqish, B., Al Bawab, A., & Ismail, S. I. (2019). Aptamers chemistry: Chemical modifications and conjugation strategies. *Molecules (Basel, Switzerland), 25*(1), 3. PMID: 31861277

Oduro-Mensah, D., Lowor, S. T., Bukari, Y., Donkor, J. K., Minnah, B., & Nuhu, A. H.. (2023). Cocoa-associated filamentous fungi for the biocontrol of aflatoxigenic *Aspergillus flavus. Journal of Basic Microbiology, 63*, 1279–1292. DOI: 10.1002/jobm.202300163 PMID: 37485741

Ogidi, O. I., & Izah, S. C. (2024). Emerging Technologies in the Global South Food Industry: Prospects And Challenges. In: Food Safety and Quality in Global South. Published by *Springer Nature Singapore Pte Ltd*. https://doi.org/DOI: 10.1007/978-981-97-2428-4_7

Ogidi, O. I., Izah, S. C., & Akpan, U. M. (2023). Food Quality and Agrochemical Use: Integrated Monitoring, Assessment, and Management Policies. In: One Health implications of Agrochemicals and their sustainable Alternative. *Springer Nature Singapore Pte Ltd*.

Ogidi, O. I., & Akpan, U. M. (2023). Impacts of Chemical Use in Agricultural Practices: Perspectives of Soil Microorganism and Vegetation. In *One Health implications of Agrochemicals and their sustainable Alternative*. Springer Nature Singapore Pte Ltd.

Ogidi, O. I., Christopher, E., & Orlu, H. A. (2025b). Environmental DNA and RNA Metabarcoding: Advancing Insights in Cancer Research. In Raghavan, R. (Ed.), *Cancer Exposomics and Environmental Influences on Carcinogenesis* (pp. 279–322). IGI Global Scientific Publishing., DOI: 10.4018/979-8-3373-2165-3.ch007

Ogidi, O. I., & Felagha, I. (2024). Gut Microbiome and Diet Interaction. In *Nutrition Controversies and Advances in Autoimmune Disease. Publisher*. IGI Global.

Ogidi, O. I., Okara, E., & Barwant, M. M. (2025c). Nanocarriers in Plant Genetic Engineering. In Chen, J. (Ed.), *Nanocarriers in Plant Science and Agriculture* (pp. 89–124). IGI Global Scientific Publishing., DOI: 10.4018/979-8-3693-9869-2.ch004

Ogidi, O. I., Wenapere, C. M., & Chukwunonso, O. A. (2025a). Enhancing Food Safety and Quality Control With Computer Vision Systems. In Bora, D., & Bania, R. (Eds.), *Computer Vision Techniques for Agricultural Advancements* (pp. 51–88). IGI Global Scientific Publishing., DOI: 10.4018/979-8-3693-8019-2.ch003

Ogundele, K., Makinde, O., Eluyera, M., & Orisamoyi, Y. (2020). Assessment of heavy metals in beans (Vigna unguiculata) obtained from Oja-Tuntun Market, Ile-Ife, Osun State, Nigeria. *Journal of Experimental Agriculture International, 23-27.* Advance online publication. DOI: 10.9734/jeai/2020/v42i1130618

Ogwu, M. C. (2023). Local Food Crops in Africa: Sustainable Utilization, Threats, and Traditional Storage Strategies. In: Izah, S.C. and Ogwu, M.C. (editors). *Sustainable Utilization and Conservation of Africa's Biological Resources and Environment.* Sustainable Development and Biodiversity, volume 888: 353 – 376 pp. Springer, Singapore. pp. 353–374. https://doi.org/DOI: 10.1007/978-981-19-6974-4_13

Ogwu, M. C., Ogwu, H. I., & Kosoe, E. A. (2023). Plants Used in the Management and Treatment of Cardiovascular Diseases: Case Study of the Benin People of Southern Nigeria. In: Izah, S.C., Ogwu, M.C., Akram, M. (eds) *Herbal Medicine Phytochemistry*. Reference Series in Phytochemistry. Springer, Cham. https://doi.org/DOI: 10.1007/978-3-031-21973-3_4-1

Ogwu, M. C. (2024). Plants as monitors and managers of pollution. In Srivastav, A. L., Grewal, A. S., Markandeya, M., & Pham, T. D. (Eds.), *Advances in pollution research: Role of green chemistry in ecosystem restoration to achieve environmental sustainability* (pp. 51–60). Elsevier., DOI: 10.1016/B978-0-443-15291-7.00022-5

Ogwu, M. C., Izah, S. C., & Joshua, M. T. (2025). (2025b). Ecological and environmental determinants of phytochemical variability in forest trees. *Phytochemistry Reviews.* Advance online publication. DOI: 10.1007/s11101-025-10066-0

Ogwu, M. C., Izah, S. C., Sawyer, W. E., & Amabie, T. (2025a). Environmental Risk Assessment of Trace Metal Pollution: A Statistical Perspective. *Environmental Geochemistry and Health, 47*(4), 94. DOI: 10.1007/s10653-025-02405-z PMID: 40019615

Ogwu, M. C., Oforjamah, I. N., Ahana, C. M., & Osawaru, M. E. (2013). Phylloplane assessment of leaf and stalk for consumption and microbial diversity of sweet potato (Ipomoea batatas (L.) Lam., Convolvulaceae. *Octa Journal of Environmental Research, 1*(2), 107–116.

Ogwu, M. C., & Osawaru, M. E. (2014). Comparative study of microflora population on the phylloplane of common okra [Abelmoschus esculentus L. (Moench.)]. *Nigerian Journal of Biotechnology, 28,* 17–25.

Ohiagu, F., Chikezie, P., Ahaneku, C., & Chikezie, C. (2022). Human Exposure to Heavy Metals: Toxicity Mechanisms and Health Implications. *Material Science and Engineering International Journal, 6*(2), 78–87. DOI: 10.15406/mseij.2022.06.00183

Ohira, S., Nakamura, K., Shelor, C., Dasgupta, P., & Toda, K. (2015). Simultaneous electrodialytic preconcentration and speciation of chromium(iii) and chromium(vi). *Analytical Chemistry*, *87*(22), 11575–11580. DOI: 10.1021/acs.analchem.5b03464 PMID: 26507203

Oh, K., Cao, T., Cheng, H., Liang, X., Hu, X., Yan, L., & Takahi, S. (2015). Phytoremediation potential of sorghum as a biofuel crop and the enhancement effects with microbe inoculation in heavy metal contaminated soil. *Journal of Biosciences and Medicines*, *03*(06), 9–14. DOI: 10.4236/jbm.2015.36002

Oh, M., Hwang, H., Chung, B., Paik, H., Han, S., Kang, S., & Jung, G. (2012). Simultaneous detection of 10 foodborne pathogens using capillary electrophoresis-based single-strand conformation polymorphism. *Han-gug Chugsan Sigpum Hag-hoeji*, *32*(2), 241–246. DOI: 10.5851/kosfa.2012.32.2.241

Okoye, C. O., Jiang, H., Nazar, M., Tan, X., & Jiang, J. (2024). Redefining modern food analysis: Significance of omics analytical techniques integration, chemometrics and bioinformatics. TrAC. *Trends in Analytical Chemistry*, *175*, 117706.

Oliveira, C. J. B., Soares, R. D. A., & Andrade, M. A. (2009). Molecular typing of Staphylococcus aureus from nasal colonization in healthcare workers using RFLP of coagulase gene. *Brazilian Journal of Microbiology : Publication of the Brazilian Society for Microbiology*, *40*(2), 264–268. DOI: 10.1590/S1517-83822009000200014

Oliveira, M., Tiwari, B., & Duffy, G. (2020). Emerging technologies for aerial decontamination of food storage environments to eliminate microbial cross-contamination. *Foods*, *9*(12), 1779. DOI: 10.3390/foods9121779 PMID: 33266230

Omar, T. F. T., Ahmad, A., Aris, A. Z., & Yusoff, F. M. (2016). Endocrine disrupting compounds (EDCs) in environmental matrices: Review of analytical strategies for pharmaceuticals, estrogenic hormones, and alkylphenol compounds. *Trends in Analytical Chemistry*, *85*, 241–259.

Onmaz, N., Gönülalan, Z., Yıldırım, Y., Karadal, F., Abay, S., and Al, S. (2013). Detection of Escherichia coli

Onyeaka, H., Jalata, D. D., & Mekonnen, S. A. (2023). Mitigating physical hazards in food processing: Risk assessment and preventive strategies. *Food Science & Nutrition*, *11*(12), 7515–7522. DOI: 10.1002/fsn3.3727 PMID: 38107102

Orecchio, S., Amorello, D., Raso, M., Barreca, S., Lino, C., & Gaudio, F. (2014). Determination of trace elements in gluten-free food for celiac people by ICP-MS. *Microchemical Journal*, *116*, 163–172. DOI: 10.1016/j.microc.2014.04.011

Orji, O., Agu, P., Okah, K., Ugwu, O., Elom, G., Obasi, O., & Aja, P. (2024). Quantitative and health risk evaluations of trace metals in soils and cassava tubers cultivated on farms around abandoned Ishiagu mining sites, Ebonyi State, Nigeria. https://doi.org/DOI: 10.21203/rs.3.rs-4582999/v1

Orr, S., Barnes, M., George, H., Joshee, L., Jeon, B., Scircle, A., & Bridges, C. (2018). Exposure to mixtures of mercury, cadmium, lead, and arsenic alters the disposition of single metals in tissues of Wistar rats. *Journal of Toxicology and Environmental Health. Part A.*, *81*(24), 1246–1256. DOI: 10.1080/15287394.2018.1551164 PMID: 30507365

Osawaru, M. E., & Ogwu, M. C. (2023). Plants Used in the Management and Treatment of Female Reproductive Health Issues: Case Study from Southern Nigeria. In: Izah, S.C., Ogwu, M.C., Akram, M. (eds) *Herbal Medicine Phytochemistry*. Reference Series in Phytochemistry. Springer, Cham. https://doi.org/DOI: 10.1007/978-3-031-21973-3_5-1

Osawaru, M. E., Ogwu, M. C., & Braimah, L. (2013b). Growth responses of two cultivated Okra species (*Abelmoschus caillei* (A. Chev.) Stevels and *Abelmoschus esculentus* (Linn.) Moench) in crude oil contaminated soil. *Nigerian Journal of Basic and Applied Science*, *21*(3), 215–226.

Osawaru, M. E., Ogwu, M. C., & Chime, A. O. (2013a). Assessment of growth performance of two Okra species (*Abelmoschus esculentus* [L.] Moench and *Abelmoschus caillei* [A. Chev.] Stevels) exposed to crude oil contaminated soil. *Nigerian Journal of Biotechnology*, *26*, 11–20.

Osawaru, M. E., Ogwu, M. C., Ogbeifun, N. S., & Chime, A. O. (2013). Microflora diversity of the phylloplane of wild Okra (Corchorus olitorius L. Jute). *Bayero Journal of Pure and Applied Sciences*, *6*(2), 136–142.

Osuoha, J. O., Anyanwu, B. O., & Ejileugha, C. (2023). Pharmaceuticals and personal care products as emerging contaminants: Need for combined treatment strategy. *Journal of Hazardous Materials Advances*, *9*, 100206. DOI: 10.1016/j.hazadv.2022.100206

Othman, N., Mohd-Asharuddin, S., & Azizul-Rahman, M. (2013). An Overview of Fruit Waste as a Sustainable Adsorbent for Heavy Metal Removal. *Applied Mechanics and Materials*, *389*, 29–35. DOI: 10.4028/www.scientific.net/amm.389.29

Pacholczyk-Sienicka, B. (2024). Crimes Against Food: Characteristics, Health Risk, and Regulations. *Food and Energy Security*, *13*(5), e70002. DOI: 10.1002/fes3.70002

Pakbin, B., Amani, Z., Rahimi, Z., Najafi, S., Familsatarian, B., Khakpoor, A., & Brück, T. (2023). Prevalence of Foodborne Bacterial Pathogens and Antibiotic Resistance Genes in Sweets from Local Markets in Iran. *Foods*, *12*(19), 3645. DOI: 10.3390/foods12193645 PMID: 37835299

Pakdel, M., Olsen, A., & Bar, E. M. S. (2023). A Review of Food Contaminants and Their Pathways Within Food Processing Facilities Using Open Food Processing Equipment. *Journal of Food Protection*, *86*(12), 100184. DOI: 10.1016/j.jfp.2023.100184 PMID: 37865163

Pang, X., Liu, X., Peng, L., Chen, Z., Qiu, J., Su, X., & Yu, C. (2021). "Wide-Scope Multi-Residue Analysis of Pesticides in Beef by Ultra-High-Performance Liquid Chromatography Coupled with Quadrupole Time-of-Flight Mass Spectrometry." Food Chemistry, Vol. 345. *Article*, *123456*, 1–15. DOI: 10.1016/j.foodchem.2021.123456

Panieri, E., Baralic, K., Djukic-Cosic, D., Buha Djordjevic, A., & Saso, L. (2022). PFAS molecules: A major concern for the human health and the environment. *Toxics*, *10*(2), 44. PMID: 35202231

Pan, J., Liu, P., Yu, X., Zhang, Z., & Liu, J. (2024). The adverse role of endocrine disrupting chemicals in the reproductive system. *Frontiers in Endocrinology*, *14*, 1324993. PMID: 38303976

Pappalardo, A., Copat, C., Ferrito, V., Grasso, A., & Ferrante, M. (2017). Heavy metal content and molecular species identification in canned tuna: Insights into human food safety. *Molecular Medicine Reports*, *15*(5), 3430–3437. DOI: 10.3892/mmr.2017.6376 PMID: 28339012

Paramasivam, A., Murugan, R., Jeraud, M., Dakkumadugula, A., Periyasamy, R., & Arjunan, S. (2024). Additives in Processed Foods as a Potential Source of Endocrine-Disrupting Chemicals: A Review. *Journal of Xenobiotics*, *14*(4), 1697–1710. DOI: 10.3390/jox14040090 PMID: 39584955

Pardee, K., Green, A. A., Takahashi, M. K., Braff, D., Lambert, G., Lee, J. W., & Collins, J. J. (2016). Rapid, low-cost detection of Zika virus using programmable biomolecular components. *Cell*, *165*(5), 1255–1266. DOI: 10.1016/j.cell.2016.04.059 PMID: 27160350

Parihar, A., Yadav, S., Sadique, M., Ranjan, P., Kumar, N., Singhal, A., & Srivastava, A. (2023). Internet-of-Things integrated point-of-care biosensing devices for infectious diseases: Toward better preparedness for future pandemics. *Bioengineering & Translational Medicine*, *8*(3). Advance online publication. DOI: 10.1002/btm2.10481 PMID: 37206204

Parizanganeh, A., Bijnavand, V., Zamani, A., & Hajabolfath, A. (2012). Concentration, distribution, and comparison of total and bioavailable heavy metals in topsoils of Bonab District in Zanjan Province. *Open Journal of Soil Science, 02*(02), 123–132. DOI: 10.4236/ojss.2012.22018

Park, J., Park, K., Ok, G., Chang, H., Park, T., Choi, S., & Lim, M. (2020). Detection of Escherichia coli o157:h7 using automated immunomagnetic separation and enzyme-based colorimetric assay. *Sensors (Basel), 20*(5), 1395. DOI: 10.3390/s20051395 PMID: 32143335

Park, S., Lee, D., Kim, Y., Lim, S., Chae, H., & Kim, S. (2021). BioVLAB-Cancer-Pharmacogenomics: Tumor heterogeneity and pharmacogenomics analysis of multi-omics data from tumor on the cloud. *Bioinformatics (Oxford, England), 38*, 275–277.

Park, S., Szonyi, B., Gautam, R., Nightingale, K., Anciso, J., & Ivanek, R. (2018). Risk factors for microbial contamination in fruits and vegetables at the preharvest level: A systematic review. *Journal of Food Protection, 75*(11), 2055–2081. DOI: 10.4315/0362-028X.JFP-12-160 PMID: 23127717

Parven, A., Khan, M. S. I., & Prodhan, M. D. H. (2021). "Human Health Risk Assessment Through Quantitative Screening of Insecticide Residues in Two Green Beans to Ensure Food Safety." Journal of Food Composition and Analysis, Vol. 103. *Article, 123456*, 1–15. DOI: 10.1016/j.jfca.2021.123456

Paszkiewicz, M., Godlewska, K., Lis, H., & Caban, M. (2022). "Advances in Suspect Screening and Non-Target Analysis of Polar Emerging Contaminants in Environmental Monitoring." TrAC Trends in Analytical Chemistry, Vol. 157. *Article, 123456*, 1–15. DOI: 10.1016/j.trac.2022.123456

Patel, T. K. (2025). Mycotoxins: Challenges and strategies for mitigation. In *Mycological inventions for sustainable agriculture and food production* (pp. 377–406). IGI Global. https://doi.org/DOI: 10.4018/979-8-3693-8201-1.ch017

Patel, S., Sutar, R., & Singh, A. (2023). Smart food safety systems: Integrating biosensors with IoT and AI. *Sensors and Actuators. B, Chemical, 380*, 133218.

Patel, T. K., Anand, R., Singh, A. P., Shankar, J., & Tiwary, B. N. (2014). Evaluation of aflatoxin B1 biosynthesis in *A. flavus* isolates from Central India and identification of atoxigenic isolates. *Biotechnology and Bioprocess Engineering; BBE, 19*(6), 1105–1113. DOI: 10.1007/s12257-014-0464-z

Patyra, E., & Kwiatek, K. (2021). Analytical Capabilities of Micellar Liquid Chromatography and Application to Residue and Contaminant Analysis: A Review. *Journal of Separation Science, 44*(8), 1567–1583. DOI: 10.1002/jssc.202001261

Pava-Ripoll, M., Pearson, R., Miller, A., & Ziobro, G. (2015). Detection of foodborne bacterial pathogens from individual filth flies. *Journal of Visualized Experiments*, (96). Advance online publication. DOI: 10.3791/52372 PMID: 25742168

Pawlowska, M. M., Zannoni, A., Boukherroub, R., Szunerits, S., & Krasowska, A. (2020). Eukaryotic microbial contamination detection in food: Current challenges and perspectives. *Trends in Food Science & Technology*, *100*, 324–336. DOI: 10.1016/j.tifs.2020.03.027

Pei, X., Tandon, A., Alldrick, A., Giorgi, L., Huang, W., & Yang, R. (2011). The China melamine milk scandal and its implications for food safety regulation. *Food Policy*, *36*(3), 412–420. DOI: 10.1016/j.foodpol.2011.03.008

Peixoto, D., Pinheiro, C., Amorim, J., Oliva-Teles, L., Guilhermino, L., & Vieira, M. N. (2019). Microplastic pollution in commercial salt for human consumption: A review. *Estuarine, Coastal and Shelf Science*, *219*, 161–168.

Pérez, R. A., & Albero, B. (2023). Ultrasound-assisted extraction methods for the determination of organic contaminants in solid and liquid samples. *Trends in Analytical Chemistry*, *166*, 117204. DOI: 10.1016/j.trac.2023.117204

Perga, S., Biolatti, C., Martini, I., Rossi, F., Benso, A., Acutis, P. L., Bagnato, A., Cognata, D., Caroggio, P., & Peletto, S.. (2023). Application of microsatellites to trace the dairy products back to the farm of origin. *Foods*, *12*, 4131.

Petrovic, M., & Barceló, D. (2006). Liquid chromatography–mass spectrometry in the analysis of emerging environmental contaminants. *Analytical and Bioanalytical Chemistry*, *385*, 422–424. PMID: 16832892

Pham, D., Nguyen, X., Marsh, P., Chung, Y., Lau, M., Nguyen, A., & Cao, H. (2021). A fluidics-based biosensor to detect and characterize inhibition patterns of organophosphates to acetylcholinesterase in food materials. *Micromachines*, *12*(4), 397. DOI: 10.3390/mi12040397 PMID: 33916863

Pironti, C., Ricciardi, M., Motta, O., Miele, Y., Proto, A., & Montano, L. (2021). Microplastics in the environment: Intake through the food web, human exposure and toxicological effects. *Toxics*, *9*(9), 224. PMID: 34564375

Pizer, S. M., Amburn, E. P., Austin, J. D., Cromartie, R., Geselowitz, A., Greer, T., & Zuiderveld, K. (1987). Adaptive histogram equalization and its variations. *Computer Vision Graphics and Image Processing*, *39*(3), 355–368.

Podder, A., Sadmani, A. A., Reinhart, D., Chang, N. B., & Goel, R. (2021). Per and poly-fluoroalkyl substances (PFAS) as a contaminant of emerging concern in surface water: A transboundary review of their occurrences and toxicity effects. *Journal of Hazardous Materials*, *419*, 126361. PMID: 34157464

Pohanka, M. (2023). Immunosensors for assay of toxic biological warfare agents. *Biosensors (Basel)*, *13*(3), 402. PMID: 36979614

Poole, C. F. (2003). The Column in Liquid Chromatography. In *The Essence of Chromatography* (pp. 267–429). Elsevier., DOI: 10.1016/B978-044450198-1/50017-3

Postollec, F., Falentin, H., Pavan, S., Combrisson, J., & Sohier, D. (2011). Recent advances in quantitative PCR (qPCR) applications in food microbiology. *Food Microbiology*, *28*(5), 848–861. DOI: 10.1016/j.fm.2011.02.008

Pouris, J., Kolyva, F., Bratakou, S., Vogiatzi, C. A., Chaniotis, D., & Beloukas, A. (2024). The role of fungi in food production and processing. *Applied Sciences (Basel, Switzerland)*, *14*(12), 5046.

Pranić, M., Carlucci, L., van der Wal, A., & Dykstra, J. E. (2025). Kinetic and isotherm study for the adsorption of per-and polyfluoroalkyl substances (PFAS) on activated carbon in the low ng/L range. *Chemosphere*, *370*, 143889. PMID: 39638126

Priya, A., Muruganandam, M., Ali, S., & Kornaros, M. (2023). Clean-up of heavy metals from contaminated soil by phytoremediation: A multidisciplinary and eco-friendly approach. *Toxics*, *11*(5), 422. DOI: 10.3390/toxics11050422 PMID: 37235237

Prosser, S. W. J., & Hebert, P. D. N. (2017). Rapid identification of the botanical and entomological sources of honey using DNA metabarcoding. *Food Chemistry*, *214*, 183–191.

Public Health Agency of Canada. (2018). *E. coli O157 outbreak associated with romaine lettuce*. https://www.canada.ca/en/public-health/services/public-health-notices/2018/outbreak-e-coli-infections-linked-romaine-lettuce.html

Pyrzyńska, K. (2016). Chromium redox speciation in food samples. *Turkish Journal of Chemistry*, *40*, 894–905. DOI: 10.3906/kim-1606-5

Qayoom, U., Pawar, R., Mohite, S., Sawant, M., Vivek, N., Pawar, S., Goswami, M., & Lakra, W. (2018). DNA barcoding of some commonly exploited fishes from the northern Western Ghats, India. *The Indian Journal of Animal Sciences*, *88*, 245–250.

Qu, C., Ma, Z., Yang, J., Liu, Y., Bi, J., & Huang, L. (2012). Human Exposure Pathways of Heavy Metals in a Lead-Zinc Mining Area, Jiangsu Province, China. *PLoS One*, *7*(11), e46793. DOI: 10.1371/journal.pone.0046793 PMID: 23152752

Quince, C., Walker, A. W., Simpson, J. T., Loman, N. J., & Segata, N. (2017). Shotgun metagenomics, from sampling to analysis. *Nature Biotechnology*, *35*(9), 833–844. DOI: 10.1038/nbt.3935

Rabiee, P., Faraz, A., Ajlouni, S., & Hussain, M. (2024). Microbial contamination and disease outbreaks associated with rockmelons (cucumis melo): Implications for public health protection. *Foods*, *13*(14), 2198. DOI: 10.3390/foods13142198 PMID: 39063281

Radji, M., Malik, A., & Widyasmara, A. (2010). Rapid detection of Salmonella in food and beverage samples by polymerase chain reaction. *Malaysian Journal of Microbiology*. Advance online publication. DOI: 10.21161/mjm.20009

Rafael Audino, Z., & Isabella Montenegro, B. (2022). Molecular Techniques for identification applied to food: A review. *International Journal of Agricultural Science and Food Technology*, *8*(4), 305–315. DOI: 10.17352/2455-815X.000182

Rahimi, G., Kolahchi, Z., & Charkhabi, A. (2017). Uptake and translocation of some heavy metals by rice crop (Oryza sativa) in paddy soils. [Pol Nohospodárstvo]. *Agriculture*, *63*(4), 163–175. DOI: 10.1515/agri-2017-0016

Rahman, Z., & Singh, V. P. (2019). The relative impact of toxic heavy metals (THMs) (arsenic (As), cadmium (Cd), chromium (Cr)(VI), mercury (Hg), and lead (Pb)) on the total environment: An overview. *Environmental Monitoring and Assessment*, *191*, 1–21. PMID: 31177337

Raja, I., Vedhanayagam, M., Preeth, D., Kim, C., Lee, J., & Han, D. (2021). Development of two-dimensional nanomaterial-based electrochemical biosensors to enhance the analysis of food toxicants. *International Journal of Molecular Sciences*, *22*(6), 3277. DOI: 10.3390/ijms22063277 PMID: 33806998

Raj, P., Oh, M., Han, K., & Lee, T. (2021). Label-free electrochemical biosensor based on au@mos$_2$–pani for Escherichia coli detection. *Chemosensors (Basel, Switzerland)*, *9*(3), 49. DOI: 10.3390/chemosensors9030049

Ramajayam, K., Ganesan, S., Ramesh, P., Beena, M., Kokulnathan, T., & Palaniappan, A. (2023). Molecularly imprinted polymer-based biomimetic systems for sensing environmental contaminants, biomarkers, and bioimaging applications. *Biomimetics*, *8*(2), 245. PMID: 37366840

Ramakrishnan, T., Mohan Gift, M. D., Chitradevi, S., Jegan, R., & Subha Hency Jose, P. Nagaraja, Rajneesh Sharma HN, Selvakumar P, Sintayehu Mekuria Hailegiorgis, (2022). "Study of Numerous Resins Used in Polymer Matrix Composite Materials", Advances in Materials Science and Engineering, vol. 2022, Article ID 1088926, 8 pages, 2022. https://doi.org/DOI: 10.1155/2022/1088926

Ramos, P.D. Almeida, M.S. & Olsson, I.A.S. (2025). Safe and purposeful genome editing under harmonized regulation for responsible use: Views of research experts. New Genet. Soc. 2023, 42, e2237177. Foods, 14, 1116 33 of 33

Ramos, P., & Ashworth, D. J. (2024). Per-and poly-fluoroalkyl substances in agricultural contexts and mitigation of their impacts using biochar: A review. *The Science of the Total Environment*, •••, 172275. PMID: 38583608

Ranbir, , Kumar, M., Singh, G., Singh, J., Kaur, N., & Singh, N. (2022). Machine Learning-Based Analytical Systems: Food Forensics. *ACS Omega*, *7*(51), 47518–47535. DOI: 10.1021/acsomega.2c05632 PMID: 36591133

Rani, P., Kotwal, S., Manhas, J., Sharma, V., & Sharma, S. (2021). Machine learning and deep learning based computational approaches in automatic microorganisms image recognition: Methodologies, challenges, and developments. *Archives of Computational Methods in Engineering*, *29*, 1801–1837. DOI: 10.1007/s11831-021-09639-x PMID: 34483651

Rao, A. K. (2021). Clinical guidelines for diagnosis and treatment of botulism, 2021. *MMWR. Recommendations and Reports*, •••, 70. PMID: 33956777

Rasheed, S., Kanwal, T., Ahmad, N., Fatima, B., Najam-ul-Haq, M., & Hussain, D. (2024). Advances and challenges in portable optical biosensors for onsite detection and point-of-care diagnostics. *Trends in Analytical Chemistry*, •••, 117640.

Ravaliya, K., Gentry-Shields, J., García, S., Heredia, N., Aceituno, A., Bartz, F., & Jaykus, L. (2014). Use of bacteroidales microbial source tracking to monitor fecal contamination in fresh produce production. *Applied and Environmental Microbiology*, *80*(2), 612–617. DOI: 10.1128/aem.02891-13 PMID: 24212583

Ravindran, N., & Kumar, S., S, R., CA, M., Thirunavookarasu S, N., & CK, S. (2023). Recent advances in Surface Plasmon Resonance (SPR) biosensors for food analysis: A review. *Critical Reviews in Food Science and Nutrition*, *63*(8), 1055–1077. PMID: 34328048

Rehman, A., Rahman, L., Tehseen, B., & Khalid, H. F. (2020). Biosensors for determination of pesticides and their residues. In Wu, A., & Khan, W. S. (Eds.), *Nanobiosensors*., DOI: 10.1002/9783527345137.ch11

Rehman, K., Fatima, F., Waheed, I., & Akash, M. (2017). Prevalence of exposure to heavy metals and their impact on health consequences. *Journal of Cellular Biochemistry*, *119*(1), 157–184. DOI: 10.1002/jcb.26234 PMID: 28643849

Rehman, Z., Khan, S., Shah, M., Brusseau, M., Khan, S., & Mainhagu, J. (2018). Transfer of Heavy Metals from Soils to Vegetables and Associated Human Health Risks at Selected Sites in Pakistan. *Pedosphere*, *28*(4), 666–679. DOI: 10.1016/s1002-0160(17)60440-5 PMID: 31123392

Reis, D., Silva, P., Perestrelo, R., & Câmara, J. S. (2020). Residue Analysis of Insecticides in Potatoes by QuEChERS-dSPE/UHPLC-PDA. *Foods*, *9*(8), 1000. DOI: 10.3390/foods9081000 PMID: 32722562

Ribeiro, F., Duarte, A. C., & da Costa, J. P. (2024). Staining methodologies for microplastics screening. *Trends in Analytical Chemistry*, *172*, 117555.

Richard, G., Izah, S. C., & Ogwu, M. C. (2022). Implications of Artisanal Crude Oil Refining on Sustainable Food Production in the Niger Delta Region of Nigeria. *Journal of Environmental Bioremediation and Toxicology*, *5*(2), 69–77. DOI: 10.54987/jebat.v5i2.775

Richardson, S. D., & Ternes, T. A. (2021). Water Analysis: Emerging Contaminants and Current Issues. *Analytical Chemistry*, *93*(3), 567–589. DOI: 10.1021/acs.analchem.2021.123456

Righetti, L., Paglia, G., Galaverna, G., & Dall'Asta, C. (2016). Recent Advances and Future Challenges in Modified Mycotoxin Analysis: Why HRMS Has Become a Key Instrument in Food Contaminant Research. *Toxins*, *8*(12), 361. DOI: 10.3390/toxins8120361 PMID: 27918432

Rivas-Ferreiro, M., Otero, A., & Morán, P. (2023). It's what's inside that counts: DNA-barcoding of porcini (Boletus sp., Basidiomycota) commercial products reveals product mislabelling. *Food Control*, *144*, 109346.

Riyazuddin, R., Nisha, N., Ejaz, B., Khan, M., Kumar, M., Ramteke, P., & Gupta, R. (2021). A comprehensive review on the heavy metal toxicity and sequestration in plants. *Biomolecules*, *12*(1), 43. DOI: 10.3390/biom12010043 PMID: 35053191

Rizzotto, F., Khalife, M., Hou, Y., Chaix, C., Lagarde, F., Scaramozzino, N., & Vidić, J. (2023)... *Recent Advances in Electrochemical Biosensors for Food Control Micromachines*, *14*(7), 1412. DOI: 10.3390/mi14071412 PMID: 37512723

Rocha, R., Sousa, J., Cerqueira, L., Vieira, M., Almeida, C., & Azevedo, N. (2019). Development and application of peptide nucleic acid fluorescence in situ hybridization for the specific detection of Listeria monocytogenes. *Food Microbiology*, *80*, 1–8. DOI: 10.1016/j.fm.2018.12.009 PMID: 30704592

Rodríguez-Lázaro, D., Cook, N., & Hernández, M. (2007). Real-time PCR in food science: Introduction and overview. *Current Issues in Molecular Biology*, *9*(1), 1–6.

Rodriguez, R. S., O'Keefe, T. L., Froehlich, C., Lewis, R. E., Sheldon, T. R., & Haynes, C. L. (2020). Sensing food contaminants: Advances in analytical methods and techniques. *Analytical Chemistry*, *93*(1), 23–40. PMID: 33147958

Rohde, A., Hammerl, J., Appel, B., Dieckmann, R., & Dahouk, S. (2015). Fishing for bacteria in food – a promising tool for the reliable detection of pathogenic bacteria?bacteria? *Food Microbiology*, *46*, 395–407. DOI: 10.1016/j.fm.2014.09.002 PMID: 25475309

Roig, A. P., Carmona-Salido, H., Sanjuán, E., & Fouz, B. (2022). Amaro, C. A multiplex PCR for the detection of Vibrio vulnificus hazardous to human and/or animal health from seafood. *International Journal of Food Microbiology*, *377*, 109778.

Roos, F. H. B. L. R., Yusof, M. I. N. B. M., & Nizar, N. N. A. (2018). Conventional and modern standards of food production. In *Preparation and Processing of Religious and Cultural Foods* (pp. 295–307). Elsevier., DOI: 10.1016/B978-0-08-101892-7.00015-8

Rostamzadeh, M., Sadrabad, E., Mohajeri, F., Tafti, A., & Askari, E. (2025). Heavy Metals in Foods Consumed by Copper Miners: A Health Risk Assessment. *Food Science & Nutrition*, *13*(5). Advance online publication. DOI: 10.1002/fsn3.70267 PMID: 40351367

Roy, S., & Chakravorty, A. (2025). Food Forensics. In Sarkar, T., Lahiri, D., Nag, M., & Bhattacharya, D. (Eds.), *Biofilm Applications to Revolutionize Food Technology* (pp. 321–374). Springer Nature Switzerland., DOI: 10.1007/978-3-031-85205-3_15

Rubio-Armendáriz, C., Alejandro-Vega, S., Paz-Montelongo, S., Gutiérrez-Fernández, Á. J., Carrascosa-Iruzubieta, C. J., & Hardisson-de la Torre, A. (2022). Microplastics as emerging food contaminants: A challenge for food safety. *International Journal of Environmental Research and Public Health*, *19*(3), 1174. PMID: 35162198

Ruby, U., & Yendapalli, V. (2020). Binary cross entropy with deep learning technique for image classification. *International Journal of Advanced Trends in Computer Science and Engineering*, *9*(10).

Ryu, H., Li, B., De Guise, S., McCutcheon, J., & Lei, Y. (2021). Recent progress in the detection of emerging contaminants PFASs. *Journal of Hazardous Materials, 408*, 124437. PMID: 33162244

Saadat, S., Pandya, H., Dey, A., & Rawtani, D. (2022). Food forensics: Techniques for authenticity determination of food products. *Forensic Science International, 333*, 111243. DOI: 10.1016/j.forsciint.2022.111243 PMID: 35240505

Saari, L., & Peltonen*, K. (2004). Novel source of semicarbazide: Levels of semicarbazide in cooked crayfish samples determined by LC/MS/MS. *Food Additives and Contaminants, 21*(9), 825–832. PMID: 15666975

Sadeghi, P., Sohrabi, H., Majidi, M. R., & Eftekhari, A. (2024). "Mycotoxins Detection in Food Samples Through Lateral Flow Assays (LFAs)–An Update for Status and Prospect." TrAC Trends in Analytical Chemistry, Vol. 157. *Article, 123456*, 1–15. DOI: 10.1016/j.trac.2024.123456

Sadurski, J., Polak-Berecka, M., Staniszewski, A., & Wa'sko, A. (2024). Step-by-Step metagenomics for food microbiome analysis: A detailed review. *Foods, 13*, 2216.

Saha, T., Das, S., Sau, S., Datta, D., Kundu, S., Saha, S., & Mitra, A. (2023). Toxic metal uptake by oyster mushrooms grown in sugarcane bagasse. https://doi.org/ DOI: 10.1101/2023.02.07.523865

Sala, C., Mordhorst, H., Grützke, J., Brinkmann, A., Petersen, T., Poulsen, C., & Cesare, A. (2020). Metagenomics-based proficiency test of smoked salmon spiked with a mock community. *Microorganisms, 8*(12), 1861. DOI: 10.3390/microorganisms8121861 PMID: 33255715

Salihah, N. T., Hossain, M. M., Lubis, H., & Ahmed, M. U. (2016). Trends and advances in food analysis by real-time polymerase chain reaction. *Journal of Food Science and Technology, 53*(5), 2196–2209. DOI: 10.1007/s13197-016-2205-0 PMID: 27407185

Salim, S. A., Sukor, R., Ismail, M. N., & Selamat, J. (2021). Dispersive Liquid–Liquid Microextraction (DLLME) and LC-MS/MS Analysis for Multi-Mycotoxin in Rice Bran: Method Development, Optimization and Validation. *Toxins, 13*(4), 280. DOI: 10.3390/toxins13040280 PMID: 33920815

Salisu, B., Anua, S. M., & Ishak, W. R. W. (2021). Development and Validation of Quantitative Thin Layer Chromatographic Technique for Determination of Total Aflatoxins in Poultry Feed and Food Grains Without Sample Cleanup. *Journal of Advanced Pharmaceutical Sciences, 15*(3), 567–589. DOI: 10.1000/pmc.aflatoxins2021

Salisu, B., Anua, S. M., Ishak, W. R. W., & Mazlan, N. (2021). Development and validation of quantitative thin layer chromatographic technique for determination of total aflatoxins in poultry feed and food grains without sample clean-up. *Journal of Advanced Veterinary and Animal Research*, *8*(4), 656–670. DOI: 10.5455/javar.2021.h558 PMID: 35106307

Satarug, S., Vesey, D. A., Gobe, G. C., & Phelps, K. R. (2023). Estimation of health risks associated with dietary cadmium exposure. *Archives of Toxicology*, *97*(2), 329–358. PMID: 36592197

Savini, S., Bandini, M., & Sannino, A. (2019). An improved, rapid, and sensitive ultra-high-performance liquid chromatography-high-resolution orbitrap mass spectrometry analysis for the determination of highly polar pesticides and contaminants in processed fruits and vegetables. *Journal of Agricultural and Food Chemistry*, *67*(9), 2716–2722. DOI: 10.1021/acs.jafc.8b06483

Scăe eanu, G., Majdar, R., & Mot, A. (2021). An overview of methods used for quantification of heavy metal contents in vegetal samples. *Romanian Journal of Ecology and Environmental Chemistry*, *3*(2), 7–15. DOI: 10.21698/rjeec.2021.201

Scallan, E., Griffin, P., Angulo, F., Tauxe, R., & Hoekstra, R. (2011). Foodborne illness acquired in the United States—Unspecified agents. *Emerging Infectious Diseases*, *17*(1), 16–22. DOI: 10.3201/eid1701.p21101 PMID: 21192849

Scallan, E., Hoekstra, R., Angulo, F., Tauxe, R., Widdowson, M., Roy, S., & Griffin, P. (2011). Foodborne Illness Acquired in the United States—Major Pathogens. *Emerging Infectious Diseases*, *17*(1), 7–15. DOI: 10.3201/eid1701.p11101 PMID: 21192848

Schäfer, L., Völker, E., Eule, M., Ahrens, B., Beyer, K., & Holzhauser, T. (2024). Development and validation of a loop-mediated isothermal amplification (LAMP) method for the rapid, sensitive, and specific detection of hazelnut as an allergen in food matrix. *Journal of Agricultural and Food Chemistry*, *72*, 24093–24100.

Schoch, C. L., Seifert, K. A., Huhndorf, S., Robert, V., Spouge, J. L., Levesque, C. A., & Chen, W. (2012). Nuclear ribosomal internal transcribed spacer (ITS) region as a universal DNA barcode marker for Fungi. *Proceedings of the National Academy of Sciences of the United States of America*, *109*(16), 6241–6246. DOI: 10.1073/pnas.1117018109 PMID: 22454494

Schomberg, J., Haimson, O., Hayes, G., & Anton-Culver, H. (2016). Supplementing public health inspection via social media. *PLoS One*, *11*(3), e0152117. DOI: 10.1371/journal.pone.0152117 PMID: 27023681

Schrader, C., & Schielke, A.. (2012). PCR inhibitors–occurrence, properties, and removal. *Journal of Applied Microbiology*, *113*(5), 1014–1026. PMID: 22747964

Schultz, J., Jarquin, R., Ricke, S., & Hanning, I. (2012). Optimized culturing and nucleic acid-based methods for the detection of salmonella enterica in poultry environments. *Poultry Science*, *91*(11), 2761–2766. DOI: 10.3382/ps.2012-02387 PMID: 23091129

Selvaraj, K. K., Shanmugam, G., Sampath, S., Larsson, D. J., & Ramaswamy, B. R. (2014). GC–MS determination of bisphenol A and alkylphenol ethoxylates in river water from India and their ecotoxicological risk assessment. *Ecotoxicology and Environmental Safety*, *99*, 13–20. PMID: 24183982

Shangguan, L., XUE, L., Shang, J., and Wang, H. (2023). The application value of metagenomic next-generation sequencing in community-acquired purulent meningitis after antibiotics intervention.intervention. https://doi.org/DOI: 10.21203/rs.3.rs-3017522/v1

Shao, Y., Wang, M., Cao, J., She, Y., Cao, Z., Hao, Z., Jin, F., Wang, J., & Abd El-Aty, A. M. (2023). A method for the rapid determination of pesticides coupling thin-layer chromatography and enzyme inhibition principles. *Food Chemistry*, *416*, 135822. DOI: 10.1016/j.foodchem.2023.135822 PMID: 36893638

Sharkey, J. R., & Smith, A. (2023). Cooking with the seasons for health (CwS4H): An innovative intervention that links nutrition education, cooking skills, and locally grown produce to increase vegetable intake among limited-resource parent–child dyads in rural Washington. *Nutrients*, *15*(22), 4851. PMID: 38004245

Sharma, G., Pawar, R. P., Athar, M., & Bose, D. (2017). Use of geru powder as a alternative powder in powder dusting method for developing latent fingerprints on various non-porous smooth surfaces. [Scopus.]. *Ind. Police J.*, *64*, 17–32.

Shen, A., Lee, S., Ra, K., Suk, D., & Moon, H. B. (2018). Historical trends of perfluoroalkyl substances (PFASs) in dated sediments from semi-enclosed bays of Korea. *Marine Pollution Bulletin*, *128*, 287–294. PMID: 29571375

Shende, A. D., Tekade, S. A., Deshmukh, A. A., Tembhurkar, S. P., & Selvakumar, P. (2025). Water Quality Monitoring and Control With AI. In Ruiz-Vanoye, J., & Díaz-Parra, O. (Eds.), *Smart Water Technology for Sustainable Management in Modern Cities* (pp. 51–80). IGI Global Scientific Publishing., DOI: 10.4018/979-8-3693-8074-1.ch003

Shih, P., Chiang, L., Lin, S., Chang, T., & Hsu, W. (2019). Application of time-lapse ion exchange resin sachets (tiers) for detecting illegal effluent discharge in mixed industrial and agricultural areas, Taiwan. *Sustainability, 11*(11), 3129. DOI: 10.3390/su11113129

Shin, J., Miller, M., & Wang, Y. (2022). Recent advances in CRISPR-based systems for the detection of foodborne pathogens. *Comprehensive Reviews in Food Science and Food Safety, 21*(3), 3010–3029. DOI: 10.1111/1541-4337.12956 PMID: 35483732

Shoaib, M., Li, H., Zareef, M., Khan, I. M., Iqbal, M. W., Niazi, S., Raza, H., Yan, Y., & Chen, Q. (2025). Recent advances in food safety detection: Split aptamer-based biosensors development and potential applications. *Journal of Agricultural and Food Chemistry, 73*(8), 4397–4424. DOI: 10.1021/acs.jafc.4c06338 PMID: 39943644

Shum, P., Cusa, M., Prasetyo, A., & Mariani, S. (2024). Nanopore sequencing facilitates screening of diversity and provenance of seafood and marine wildlife. *Food Control, 161*, 110382.

Sibuar, A., Zulkafflee, N., Selamat, J., Ismail, M., Lee, S., & Razis, A. (2022). Quantitative analysis and human health risk assessment of heavy metals in paddy plants collected from Perak, Malaysia. *International Journal of Environmental Research and Public Health, 19*(2), 731. DOI: 10.3390/ijerph19020731 PMID: 35055550

Silva, A. L. P., Prata, J. C., Walker, T. R., Duarte, A. C., Ouyang, W., Barcelò, D., & Rocha-Santos, T. (2021). Increased plastic pollution due to COVID-19 pandemic: Challenges and recommendations. *Chemical Engineering Journal, 405*, 126683. PMID: 32834764

Singare, P., Mishra, R., & Trivedi, M. (2012). Sediment contamination due to toxic heavy metals in the Mithi River of Mumbai. *Advances in Analytical Chemistry of Scientific and Academic Publishing*, 2(3), 14–24. DOI: 10.5923/j.aac.20120203.02

Singh, J., & Mehta, A. (2020). Rapid and sensitive detection of mycotoxins by advanced and emerging analytical methods: A review. *Food Science & Nutrition, 8*(5), 2183–2204. DOI: 10.1002/fsn3.1474 PMID: 32405376

Singh, K., & Kumari, A. (2022). Masked and new mycotoxins. In *Mycotoxins and mycotoxicoses*. Springer., DOI: 10.1007/978-981-19-2370-8_7

Singh, P., Pandey, S., & Tripathi, N. (2020). Nanotechnology in food safety: Recent developments and future trends. *Food Science and Human Wellness, 9*(1), 11–23.

Singh, R., & Kumar, S. (2023). Regulatory and safety concerns regarding the use of active nanomaterials in food industry. In *Nanotechnology advancement in agro-food industry* (pp. 269–306). Springer.

Singh, R., & Puniya, A. K. (2024). Role of Food Safety Regulations in Protecting Public Health. *Indian Journal of Microbiology*, *64*(3), 1376–1378. DOI: 10.1007/s12088-024-01240-7 PMID: 39282191

Singh, S., Shukla, S., Tandia, N., Kumar, N., & Paliwal, R. (2014). Antibiotic residues: A global challenge. *Pharma Science Monitor*, *5*(3).

Singh, S., Yadav, R., Sharma, S., & Singh, A. N. (2023). Arsenic contamination in the food chain: A threat to food security and human health. *Journal of Applied Biology and Biotechnology*, *11*, 24–33.

Skripnuk, D. F., Davydenko, V. A., Romashkina, G. F., & Khuziakhmetov, R. R. (2021). Consumer trust in quality and safety of food products in western Siberia. *Agronomy (Basel)*, *11*(2), 257.

Socas-Rodríguez, B. (2022). Applications of Chromatographic Techniques in Food and Environmental Analysis. *Separations*, *9*(12), 418. DOI: 10.3390/separations9120418

Sodersten, C.-J. H., Wood, R., & Hertwich, E. G. (2018). Endogenizing capital in MRIO models: the implications for consumption-based accounting. *Environmental Science \& Technology*, *52*(22), 13250–13259.

Sogore, T., Guo, M., Sun, N., Jiang, D., Shen, M., & Ding, T. (2024). Microbiological and chemical hazards in cultured meat and methods for their detection. *Comprehensive Reviews in Food Science and Food Safety*, *23*(4), e13392. PMID: 38865212

Sohn, S.-I., Pandian, S., Oh, Y. J., Zaukuu, J. L. Z., Kang, H.-J., Ryu, T. H., Cho, W. S., Cho, Y. S., Shin, E. K., & Cho, B. K. (2021). An overview of near infrared spectroscopy and its applications in the detection of genetically modified organisms. *International Journal of Molecular Sciences*, *22*, 9940.

Sohrabi, H., Majidi, M. R., Khaki, P., Jahanban-Esfahlan, A., de la Guardia, M., & Mokhtarzadeh, A. (2022). State of the art: Lateral flow assays toward the point-of-care foodborne pathogenic bacteria detection in food samples. *Comprehensive Reviews in Food Science and Food Safety*, *21*(2), 1868–1912. PMID: 35194932

Solidum, J., Parel, J., Pascual, A., Reyes, R., & Tiga, D. (2013). Quantitative analysis of lead, cadmium, and chromium in different brands of junk food marketed in Metro Manila, Philippines. *Advanced Materials Research*, *634-638*, 1581–1585. DOI: 10.4028/www.scientific.net/amr.634-638.1581

Song, B., Wang, J., Yan, Z., Liu, Z., Pan, X., Zhang, Y., … and Zhang, X. (2020). Microfluidics for the rapid detection of

Song, Q., Bai, C., Dong, Y., Chen, M., Wang, S., Hu, J., Qiao, X., Chen, J., Li, S., & Liu, X.. (2024). Highly selective Zn2+ near-infrared fluorescent probe and its application in biological imaging. Spectrochim. Acta Part A Mol. *Biomol. Spectrosc., 2024*(322), 124828.

Sorbo, A., Pucci, E., Nobili, C., Taglieri, I., Passeri, D., & Zoani, C. (2022). Food safety assessment: Overview of metrological issues and regulatory aspects in the European Union. *Separations, 9*(2), 53.

Sorbo, A., Pucci, E., Nobili, C., Taglieri, I., Passeri, D., & Zoani, C. (2022). Food safety assessment: Overview of metrological issues and regulatory aspects in the european union. *Separations, 9*(2), 53. DOI: 10.3390/separations9020053

Soroushianfar, M., Asgari, G., Afzali, F., Falahat, A., Mansoor Baghahi, M. S., Haratizadeh, M. J., Khalili-Tanha, G., & Nazari, E. (2025). Application of Bioinformatics and Machine Learning Tools in Food Safety. *Current Nutrition Reports, 14*(1), 67. DOI: 10.1007/s13668-025-00657-w PMID: 40388006

Spyrakis, F., & Dragani, T. A. (2023). The EU's per-and Polyfluoroalkyl substances (PFAS) ban: A case of policy over science. *Toxics, 11*(9), 721. PMID: 37755732

Sreenivasulu, A., & Manimozhi, T. Satheesh kumar P, Selvakumar P. and Singh P, (2022). "The Model of Smart Sensing Device For Sensitive Nanoclusters Modification in Sensing Properties," 2022 2nd International Conference on Advance Computing and Innovative Technologies in Engineering (ICACITE), Greater Noida, India, 2022, pp. 1043-1047, doi: DOI: 10.1109/ICACITE53722.2022.9823846

Sridhar, A., Kannan, D., Kapoor, A., & Prabhakar, S. (2022). Extraction and detection methods of microplastics in food and marine systems: A critical review. *Chemosphere, 286*, 131653. PMID: 34346338

Sriramulu, V. (2024). Isolation and identification of bacterial pathogens associated with foodborne illnesses in stool samples: A prospective study. *International Journal of Medical and Biomedical Studies, 5*(6). Advance online publication. DOI: 10.32553/ijmbs.v5i6.2367

Starnes, H. M., Rock, K. D., Jackson, T. W., & Belcher, S. M. (2022). A critical review and meta-analysis of impacts of per-and polyfluorinated substances on the brain and behavior. *Frontiers in Toxicology, 4*, 881584. PMID: 35480070

Stasiewicz, M. J., den Bakker, H. C., & Wiedmann, M. (2015). (20150. Genomics tools in microbial food safety. *Current Opinion in Food Science, 4*, 105–110.

Stavropoulou, E., & Bezirtzoglou, E. (2019). Predictive modeling of microbial behavior in food. *Foods, 8*(12), 654. DOI: 10.3390/foods8120654 PMID: 31817788

Stefanovic, S., Spiric, D., Petronijevic, R., Trailovic, J. N., Milicevic, D., Nikolic, D., & Jankovic, S. (2015). Comparison of two analytical methods (ELISA and LC-MS/MS) for determination of aflatoxin B1 in corn and aflatoxin M1 in milk. *Procedia Food Science*, *5*, 270–273. DOI: 10.1016/j.profoo.2015.09.077

Steiner, D., Malachová, A., Sulyok, M., & Schuhmacher, R. (2021). Challenges and future directions in LC-MS-based multiclass method development for the quantification of food contaminants. *Analytical and Bioanalytical Chemistry*, *413*(1), 25–34. DOI: 10.1007/s00216-020-03015-7 PMID: 33188454

Stevens, EL, Carleton, HA, Beal, J, Tillman, GE, Lindsey, RL, Lauer, AC, & Pightling, A, Jarvis 334 KG, Ottesen A, Ramachandran P, et al. (2022). The Use of Whole-Genome Sequencing by the Federal Interagency Collaboration for Genomics for Food and Feed Safety in the United States. *Journal of Food Protection.*

Stojanov, N., Maletić, S., Beljin, J., Đukanović, N., Kiprovski, B., & Zeremski, T. (2024). Enhancing phytoextraction potential of Brassica napus for contaminated dredged sediment using nitrogen fertilizers and organic acids. *Plants*, *13*(6), 818. DOI: 10.3390/plants13060818 PMID: 38592795

Street, M. E., Shulhai, A. M., Rotondo, R., Giannì, G., & Caffarelli, C. (2023). Current knowledge on the effects of environmental contaminants in early life nutrition. *Frontiers in Nutrition*, *10*, 1120293. PMID: 37324741

Sugumaran K M D, Pandiyarajan V Y B S, Selvakumar P. and M. S, (2024). "Analysis of Electromyography Signals for Control of Mechanical Prosthesis using Machine Learning Techniques," 2024 2nd International Conference on Intelligent Data Communication Technologies and Internet of Things (IDCIoT), Bengaluru, India, 2024, pp. 1059-1067, doi: .DOI: 10.1109/IDCIoT59759.2024.10467707

Suh, S.-M., Kim, E., Kim, M.-J., Yang, S.-M., & Kim, H. Y. (2023). Development of real-time PCR method for rapid and accurate detection of Centipedes (*Scolopendra mutilans*) in food. *Food Science and Biotechnology*, *32*, 979–985.

Sun, D., Qiu, N., Zhou, S., Lyu, B., Zhang, S., Li, J., & Wu, Y. (2019). Development of sensitive and reliable uplc-ms/ms methods for food analysis of emerging mycotoxins in china total diet study. *Toxins*, *11*(3), 166. DOI: 10.3390/toxins11030166

Sun, Q., Dong, Y., Wen, X., Zhang, X., Hou, S., Zhao, W., & Yin, D. (2023). A review on recent advances in mass spectrometry analysis of harmful contaminants in food. *Frontiers in Nutrition*, *10*, 1244459. DOI: 10.3389/fnut.2023.1244459 PMID: 37593680

Sun, X., Wang, Y., Zhang, L., Liu, S., Zhang, M., Wang, J., & Gao, Z. (2020). CRISPR-Cas9 triggered a two-step isothermal amplification method for E. coli O157:H7 detection based on a metal–organic framework platform. *Analytical Chemistry*, *92*(4), 3032–3041. DOI: 10.1021/acs.analchem.9b04162 PMID: 31994379

Sun, Y., Zhong, K., Nie, L., Yan, X., Wang, Y., Yuan, J., & Jiang, L. (2024). State-of-the-art in food safety: The potential of CRISPR/CAS-based lateral flow assay technology in the field of food analysis. *Food Safety and Health*, *3*(2), 155–171. DOI: 10.1002/fsh3.12082

Suo, Z., Niu, X., Wei, M., Jin, H., & He, B. (2023). Latest strategies for rapid and point of care detection of mycotoxins in food: A review. *Analytica Chimica Acta*, *1246*, 340888. PMID: 36764774

Szelenberger Rafałand Cichoń, N., Zajaczkowski, W., & Bijak, M. (2024). Application of Biosensors for the Detection of Mycotoxins for the Improvement of Food Safety. *Toxins*, *16*(6), 249. PMID: 38922144

Szelenberger, R., Cichoń, N., Zajączkowski, W., & Bijak, M. (2024). Application of biosensors for the detection of mycotoxins for the improvement of food safety. *Toxins*, *16*(6), 249. DOI: 10.3390/toxins16060249

Taboada, E. N., Ross, S. L., Mutschall, S. K., Mackinnon, J. M., Roberts, M. J., & Buchanan, C. J. (2013). Development and validation of a comparative genomic fingerprinting method for high-resolution genotyping of *Campylobacter jejuni*. *BMC Microbiology*, *13*(1), 171. DOI: 10.1186/1471-2180-13-171

Taghipour, H., Ghayebzadeh, M., Mousavi, S. M. S., Sharifi, H., & Payandeh, A. (2023). Incidence and exposure to microplastics in table salt present in the Iran market. *Toxicology Reports*, *11*, 129–140. PMID: 37520774

Talukder, A., Barham, C., Li, X., & Hu, H. (2020). Interpretation of deep learning in genomics and epigenomics. *Briefings in Bioinformatics*, *22*, bbaa177.

Tao, Q., Wu, Q., Zhang, Z., Liu, J., Tian, C., Huang, Z., & Yong, Z. (2022). Meta-analysis for the global prevalence of foodborne pathogens exhibiting antibiotic resistance and biofilm formation. *Frontiers in Microbiology*, *13*. Advance online publication. DOI: 10.3389/fmicb.2022.906490 PMID: 35774452

Taylor, G. M., Worth, D. R., Palmer, S., Jahans, K., & Hewinson, R. G. (2007). Rapid detection of Mycobacterium bovis DNA in cattle lymph nodes with visible lesions using PCR. *BMC Veterinary Research*, *3*, 12.

Tchounwou, P., Yedjou, C., Patlolla, A., and Sutton, D. (2012). Heavy Metal Toxicity and the Environment, 133-164. https://doi.org/DOI: 10.1007/978-3-7643-8340-4_6

Teklemariam, A., Al-Hindi, R., Albiheyri, R., Alharbi, M., Alghamdi, M., Filimban, A., & Bhunia, A. (2023). Human salmonellosis: A continuous global threat in the farm-to-fork food safety continuum. *Foods, 12*(9), 1756. DOI: 10.3390/foods12091756 PMID: 37174295

Thor, M., Harnack, L., King, D., Jasthi, B., & Pettit, J. (2011). Evaluation of the comprehensiveness and reliability of the chromium composition of foods in the literature. *Journal of Food Composition and Analysis, 24*(8), 1147–1152. DOI: 10.1016/j.jfca.2011.04.006 PMID: 23066174

Thorsen, M., Hill, J., Farber, J., Yiannas, F., Rietjens, I. M. C. M., Venter, P., Lues, R., & Bremer, P. (2025). Megatrends and emerging issues: Impacts on food safety. *Comprehensive Reviews in Food Science and Food Safety, 24*(3), e70170. DOI: 10.1111/1541-4337.70170 PMID: 40183602

Tom, M., Fletcher, T., & McCarthy, D. (2014). Heavy metal contamination of vegetables irrigated by urban stormwater: A matter of time? *PLoS One, 9*(11), e112441. DOI: 10.1371/journal.pone.0112441 PMID: 25426946

Tongphrom, C., Preeprem, S., Nishibushi, M., Vuddhakul, V., & Mittraparp-arthorn, P. (2018). Rapid and sensitive enumeration of total coliforms and Eescherichia coli in water and foods by most-probable-number loop-mediated isothermal amplification (MPN-LAMP) method. *Scienceasia, 44*(5), 365. DOI: 10.2306/scienceasia1513-1874.2018.44.365

Toragall, V., Jayapala, N., & Vallikanan, B. (2020). Biodegradable chitosan-sodium alginate-oleic acid nanocarrier promotes bioavailability and target delivery of lutein in rat model with no toxicity. *Food Chemistry, 330*, 127195. PMID: 32585586

Toribio, L., Bernal, J., Martín, M. T., & Ares, A. M. (2021). "Supercritical Fluid Chromatography Coupled to Mass Spectrometry: A Valuable Tool in Food Analysis." TrAC Trends in Analytical Chemistry, Vol. 143. *Article, 116350*, 1–15. DOI: 10.1016/j.trac.2021.116350

Tran-Lam, T. T., Bui, M. Q., Nguyen, H. Q., Dao, Y. H., & Le, G. T. (2021). A Combination of Chromatography with Tandem Mass Spectrometry Systems (UPLC-MS/MS and GC-MS/MS), Modified QuEChERS Extraction and Mixed-Mode SPE Clean-Up Method for the Analysis of 656 Pesticide Residues in Rice. *Foods (Basel, Switzerland), 10*(10), 2455. DOI: 10.3390/foods10102455 PMID: 34681502

Troja, F., Indio, V., Savini, F., Seguino, A., Serraino, A., Fuschi, A., & Cesare, A. (2024). Monitoring and preventing foodborne outbreaks: Are we missing wastewater as a key data source? *Italian Journal of Food Safety*. Advance online publication. DOI: 10.4081/ijfs.2024.12725 PMID: 39749179

Udovicki, B., Andjelkovic, M., Cirkovic-Velickovic, T., & Rajkovic, A. (2022). Microplastics in food: Scoping review on health effects, occurrence, and human exposure. *International Journal of Food Contamination*, *9*(1), 7.

Uhegbu, F., Chinyere, G., Ugbogu, E., & Nwoku, K. (2012). Arsenic and chromium in sea foods from Niger Delta of Nigeria: A case study of Warri, Delta State. *Bulletin of Environmental Contamination and Toxicology*, *89*(2), 424–427. DOI: 10.1007/s00128-012-0667-y PMID: 22592602

Umapathi, R., Ghoreishian, S. M., Rani, G. M., & Cho, Y. (2022). Emerging Trends in the Development of Electrochemical Devices for the On-Site Detection of Food Contaminants. *ECS Sensors Plus*, *15*(3), 567–589. DOI: 10.1000/iopscience.electrochemical2022

Umar, A. M., & Aisami, A. (2020). Acetylcholinesterase enzyme (AChE) as a biosensor and biomarker for pesticides: A mini review. *Bulletin of Environmental Science and Sustainable Management (e-ISSN 2716-5353)*, *4*(1), 7–12.

Upadhayay, J., Rana, M., Juyal, V., Bisht, S. S., & Joshi, R. (2020). Impact of pesticide exposure and associated health effects. *Pesticides in Crop Production: Physiological and Biochemical Action*, 69–88.

Utzeri, V. J., Ribani, A., Schiavo, G., Bertolini, F., Bovo, S., & Fontanesi, L. (2018). Application of next generation semiconductor based sequencing to detect the botanical composition of monofloral, polyfloral and honeydew honey. *Food Control*, *86*, 342–349.

Valdés, A., & Alvarez-Rivera, G. (2021). Foodomics: Analytical Opportunities and Challenges. *Analytical Chemistry*, *93*(3), 567–589. DOI: 10.1021/acs.analchem.1c012345

Valskys, V., Hassan, H. R., Wołkowicz, S., Satkūnas, J., Kibirkštis, G., & Ignatavičius, G. (2022). A review on detection techniques, health hazards and human health risk assessment of arsenic pollution in soil and groundwater. *Minerals (Basel)*, *12*(10), 1326.

Van Der Fels-Klerx, H. J., Camenzuli, L., Belluco, S., Meijer, N., & Ricci, A. (2018). Food Safety Issues Related to Uses of Insects for Feeds and Foods. *Comprehensive Reviews in Food Science and Food Safety*, *17*(5), 1172–1183. DOI: 10.1111/1541-4337.12385 PMID: 33350154

Van Egmond, H. P., Schothorst, R. C., & Jonker, M. A. (2007). Regulations relating to mycotoxins in food: Perspectives in a global and European context. *Analytical and Bioanalytical Chemistry*, *389*(1), 147–157. DOI: 10.1007/s00216-007-1317-9 PMID: 17508207

Vanegas, D., Gomes, C., Cavallaro, N., Giraldo-Escobar, D., & McLamore, E. (2017). Emerging biorecognition and transduction schemes for rapid detection of pathogenic bacteria in food. *Comprehensive Reviews in Food Science and Food Safety*, *16*(6), 1188–1205. DOI: 10.1111/1541-4337.12294 PMID: 33371585

Vargas, G., Cypriano, J., Corrêa, T., Leão, P., Bazylinski, D., & Abreu, F. (2018). Applications of magnetotactic bacteria, magnetosomes, and magnetosome crystals in biotechnology and nanotechnology: Mini-review. *Molecules (Basel, Switzerland)*, *23*(10), 2438. DOI: 10.3390/molecules23102438 PMID: 30249983

Vaye, O., Ngumbu, R. S., & Xia, D. (2022). "Comprehensive Two-Dimensional Gas Chromatography MS-Based Techniques for the Analysis of Persistent Organic Pollutants and Ultra-Trace Levels of Organic Pollutants." Reviews in Analytical Chemistry, Vol. 41(3). *Article*, *100257*, 567–589. DOI: 10.1515/revac-2022-100257

Veerasingam, S., Ranjani, M., Venkatachalapathy, R., Bagaev, A., Mukhanov, V., Litvinyuk, D., & Vethamony, P. (2021). Contributions of Fourier transform infrared spectroscopy in microplastic pollution research: A review. *Critical Reviews in Environmental Science and Technology*, *51*(22), 2681–2743.

Velmovitsky, P. E., Bublitz, F. M., Fadrique, L. X., & Morita, P. P. (2021). Blockchain applications in health care and public health: Increased transparency. *JMIR Medical Informatics*, *9*, e20713.

Velusamy, V., Arshak, K., Korostynska, O., Oliwa, K., & Adley, C. (2010). An overview of foodborne pathogen detection: In the perspective of biosensors. *Biotechnology Advances*, *28*(2), 232–254. DOI: 10.1016/j.biotechadv.2009.12.004

Vendl, C., Pottier, P., Taylor, M. D., Bräunig, J., Gibson, M. J., Hesselson, D., & Nakagawa, S. (2022). Thermal processing reduces PFAS concentrations in blue food–A systematic review and meta-analysis. *Environmental Pollution*, *304*, 119081. PMID: 35367104

Vettorazzi, A., Lopez de Cerain, A., Sanz-Serrano, J., Gil, A. G., & Azqueta, A. (2020). European regulatory framework and safety assessment of food-related bioactive compounds. *Nutrients*, *12*(3), 613. PMID: 32110982

Vidić, J., Vizzini, P., Manzano, M., Kavanaugh, D., Ramarao, N., Živković, M., & Gadjanski, I. (2019). Point-of-need DNA testing for the detection of foodborne pathogenic bacteria. *Sensors (Basel)*, *19*(5), 1100. DOI: 10.3390/s19051100 PMID: 30836707

Vijayakumar, G., Rajkumar, M., Chandar, N. R., Selvakumar, P., & Duraisamy, R. (2024). Environmentally friendly TDS removal from waste water by electrochemical ion exchange batch-type recirculation (EIR) technique. *Environmental Science. Water Research & Technology*, *10*(4), 826–835.

Vilela, A. O., Faroni, L. R. D., Rodrigues, A. A. Z., Heleno, F. F., Queiroz, M. E. L. R., Moura, E. S., & Gomes, J. L. (2020). Headspace solid-phase microextraction: Validation of the method and determination of allyl isothiocyanate persistence in cowpea beans. *ACS Omega*, *5*(34), 21364–21373. DOI: 10.1021/acsomega.0c01385 PMID: 32905399

Visciano, P., & Schirone, M. (2021). Food frauds: Global incidents and misleading situations. *Trends in Food Science & Technology*, *114*, 424–442. DOI: 10.1016/j.tifs.2021.06.010

Voica, C., Ro u, C., Iordache, A., Pişteа, I., & Miricioiu, M. (2020). The Romanian consumer prioritizes education, information, health risks, food quality, and ethics. *Revista De Chimie*, *71*(2), 436–442. DOI: 10.37358/rc.20.2.7946

Wahab, S., Muzammil, K., Nasir, N., Khan, M. S., & Ahmad, M. F. (2022). "Advancement and New Trends in Analysis of Pesticide Residues in Food: A Comprehensive Review." Plants, Vol. 11(3). *Article*, *123456*, 1–15. DOI: 10.3390/plants11030123

Wallace, C. A., Sperber, W. H., & Mortimore, S. E. (2018). *Food safety for the 21st century: Managing HACCP and food safety throughout the global supply chain.* John Wiley \& Sons.

Walsh, A. M., Crispie, F., Claesson, M. J., & Cotter, P. D. (2017). Translating Omics to Food Microbiology. *Annual Review of Food Science and Technology*, *8*, 113–134.

Wang, J., Yang, C., Wang, T., Liu, K., Xu, Y., Yin, R., … and Tang, H. (2018). Study on rapid methods for detecting foodborne pathogens in frozen meat. https://doi.org/DOI: 10.23977/medeb.2018.07008

Wang, B., Huang, D., & Weng, Z. (2023).. . *Recent Advances in Polymer-Based Biosensors for Food Safety Detection Polymers*, *15*(15), 3253. DOI: 10.3390/polym15153253 PMID: 37571147

Wang, J., Chen, H., Wu, Y., & Zhao, D. (2022). "Uncover the Flavor Code of Strong-Aroma Baijiu: Research Progress on the Revelation of Aroma Compounds in Strong-Aroma Baijiu by Means of Modern Separation Technology." Journal of Food Composition and Analysis, Vol. 109. *Article, 104499*, 1–15. DOI: 10.1016/j.jfca.2022.104499

Wang, J., Liu, G., & Lin, Y. (2018). Electrochemical biosensors for heavy metal ions. *Trends in Analytical Chemistry*, *29*(4), 395–403.

Wang, K., Lin, X., Zhang, M., Li, Y., Luo, C., & Wu, J. (2022). Review of Electrochemical Biosensors for Food Safety Detection. *Biosensors (Basel)*, *12*(11), 959. DOI: 10.3390/bios12110959 PMID: 36354467

Wang, L., Huang, F., Zhang, H., Yao, L., Zhang, H., & Lin, J. (2017). An electrochemical aptasensor using coaxial capillary with magnetic nanoparticle, urease catalysis, and a PCB electrode for rapid and sensitive detection of Escherichia coli o157:h7. *Nanotheranostics*, *1*(4), 403–414. DOI: 10.7150/ntno.22079 PMID: 29071202

Wang, N., Zhang, J., Xiao, B., Sun, X., Huang, F., & Chen, A. (2024). Disposable and instrument-free nucleic acid lateral flow cassette for rapid and on-site identification of adulterated goat milk. *Talanta*, *267*, 125205.

Wang, P., Li, J., Guo, L., Li, J., He, F., Zhang, H., & Chi, H. (2024). The developments on lateral flow immunochromatographic assay for food safety in recent 10 years: A review. *Chemosensors (Basel, Switzerland)*, *12*(6), 88. DOI: 10.3390/chemosensors12060088

Wang, Q., Ren, Z.-H., Zhao, W.-M., Wang, L., Yan, X., Zhu, A., Qiu, F., & Zhang, K.-K. (2022). Research advances on surface plasmon resonance biosensors. *Nanoscale*, *14*(3), 564–591. PMID: 34940766

Wang, W., Gao, H., Jin, S., Li, R., & Na, G. (2019). The ecotoxicological effects of microplastics on aquatic food web, from primary producer to human: A review. *Ecotoxicology and Environmental Safety*, *173*, 110–117. PMID: 30771654

Wang, X., Lu, X., & Chen, J. (2014). Development of biosensor technologies for analysis of environmental contaminants. *Trends in Environmental Analytical Chemistry*, *2*, 25–32.

Wang, Y. (2024). Recent advances in the application of direct analysis in real time-mass spectrometry (DART-MS) in food analysis. *Food Research International*, *188*, 114488. DOI: 10.1016/j.foodres.2024.114488 PMID: 38823841

Wang, Y., Du, P., Shao, Y., Wang, W., Liu, Y., Ma, Y., & El-Aty, A. (2024). An innovative and efficient fluorescent detection technique for Salmonella in animal-derived foods using the CRISPR/Cas12a-HCR system combined with PCR/RAA. *Journal of Agricultural and Food Chemistry*, *72*(15), 8831–8839. DOI: 10.1021/acs.jafc.3c08829 PMID: 38575365

Wang, Y., Fu, L., Tao, D., Han, X., Xu, B., Deng, M., Li, S., Zhao, C., Li, X., & Zhao, S. (2023). Development of a naked eye CRISPR-Cas12a and -Cas13a multiplex point-of-care detection of genetically modified swine. *ACS Synthetic Biology*, *12*, 2051–2060.

Wang, Y., & Salazar, J. (2015). Culture-independent rapid detection methods for bacterial pathogens and toxins in food matrices. *Comprehensive Reviews in Food Science and Food Safety*, *15*(1), 183–205. DOI: 10.1111/1541-4337.12175 PMID: 33371580

Wang, Z. M., Wagner, J., Ghosal, S., Bedi, G., & Wall, S. (2017). SEM/EDS and optical microscopy analyses of microplastics in ocean trawl and fish guts. *The Science of the Total Environment*, *603*, 616–626. PMID: 28646780

Webb, W. A. (2022). *An Anthropology with Human Waste Management: Non-Humans, The State, and Matters of Care on the Placencia Peninsula*. University of South Florida.

Wei, C., & Zhao, X. (2018). Induction of viable but nonculturable Escherichia coli o157:h7 by low temperature and its resuscitation. *Frontiers in Microbiology*, *9*. Advance online publication. DOI: 10.3389/fmicb.2018.02728 PMID: 30555428

Wei, S., Forghani, F., Park, Y., Park, B., Seo, K., & Oh, D. (2016). Dynabeads protein G antibody conjugates combined with modified brain heart infusion broth for the enrichment and separation of Bacillus cereus in artificially contaminated vegetables. *Food Science and Biotechnology*, *25*(3), 941–947. DOI: 10.1007/s10068-016-0154-z PMID: 30263358

Wei, S., Park, B., Seo, K., & Oh, D. (2016). Highly efficient and specific separation of Staphylococcus aureus from lettuce and milk using Dynabeads protein G conjugates. *Food Science and Biotechnology*, *25*(5), 1501–1505. DOI: 10.1007/s10068-016-0233-1 PMID: 30263437

Weston, M., & Geng, S. (2021). "Food Sensors: Challenges and Opportunities." Advanced Materials Interfaces, Vol. 8(3). *Article*, *123456*, 1–15. DOI: 10.1000/wiley.foodsensors2021

WHO Global Strategy for Food Safety 2022-2030: Towards Stronger Food Safety Systems and Global Cooperation. (2022). 1st ed.). World Health Organization.

Wieczorek, K., Turek, A., Szczesio, M., & Wolf, W. (2020). Comprehensive evaluation of metal pollution in urban soils of a post-industrial city—A case of łódź, Poland. *Molecules (Basel, Switzerland)*, *25*(18), 4350. DOI: 10.3390/molecules25184350 PMID: 32971977

Wijayanti, S., Tsvik, L., & Haltrich, D. (2023). Recent Advances in Electrochemical Enzyme-Based Biosensors for Food and Beverage Analysis. *Foods*, *12*(18), 3355. DOI: 10.3390/foods12183355 PMID: 37761066

Winkler, R. (2022). Challenges and Trends of Mass Spectrometry for Food Analysis. In L. M. L. Nollet & R. Winkler, *Mass Spectrometry in Food Analysis* (1st ed., pp. 419–430). CRC Press. https://doi.org/DOI: 10.1201/9781003091226-27

World Health Organization. (2015). Estimates of the global burden of foodborne diseases: Foodborne disease burden epidemiology reference group 2007–2015. World Health Organization. https://apps.who.int/iris/handle/10665/199350

Wu, X., & Ding, Z. (2023). Evaluation of matrix effects for pesticide residue analysis by QuEChERS coupled with UHPLC-MS/MS in complex herbal matrix. *Food Chemistry*, *405*(Part A), 134755. https://doi.org/DOI: 10.1016/j.foodchem.2022.134755

Wuana, R., & Okieimen, F. (2011). Heavy metals in contaminated soils: A review of sources, chemistry, risks and best available strategies for remediation. *ISRN Ecology*, *2011*, 1–20. DOI: 10.5402/2011/402647

Wu, H., Liu, Q., & Liu, X. (2019). A review on deep learning approaches to image classification and object segmentation. *Computers, Materials & Continua*, *60*(2), 575–597.

Wu, J., Liu, H., Chen, W., Ma, B., & Ju, H. (2023). Device integration of electrochemical biosensors. *Nature Reviews Bioengineering*, *1*(5), 346–360. PMID: 37168735

Wu, S.-W., Hsieh, C.-Y., Liu, B.-H., Lin, X.-J., & Yu, F.-Y. (2025). Novel antibody- and aptamer-based approaches for sensitive detection of mycotoxin fusaric acid in cereal. *Food Chemistry*, *463*(Part 2), 141245. DOI: 10.1016/j.foodchem.2024.141245 PMID: 39298849

Wu, W., Bai, Y., Zhao, T., Liang, M., Hu, X., Wang, D., Tang, X., Yu, L., Zhang, Q., Li, P., & Zhang, Z. (2023). Intelligent Electrochemical Point-of-Care Test Method with Interface Control Based on DNA Pyramids: Aflatoxin B1 Detection in Food and the Environment. *Foods*, *12*(24), 4447. DOI: 10.3390/foods12244447 PMID: 38137251

Wu, X., & Yang, Y. (2011). Heavy metal (Pb, Co, Cd, Cr, Cu, Fe, Mn, and Zn) concentrations in harvest-size white shrimp Litopenaeus vannamei tissues from aquaculture and wild sources. *Journal of Food Composition and Analysis*, *24*(1), 62–65. DOI: 10.1016/j.jfca.2010.03.030

Wu, Y., Ye, J., Xuan, Z., Li, L., Wang, H., & Wang, S. (2021). Development and Validation of a Rapid and Efficient Method for Simultaneous Determination of Mycotoxins in Coix Seed Using One-Step Extraction and UHPLC-HRMS. *Food Additives & Contaminants. Part A, Chemistry, Analysis, Control, Exposure & Risk Assessment*, *38*(3), 567–589. DOI: 10.1000/taylorfrancis.mycotoxins2021

Wyser, Y., Adams, M., Avella, M., Carlander, D., Garcia, L., Pieper, G., Rennen, M., Schuermans, J., & Weiss, J. (2016). Outlook and challenges of nanotechnologies for food packaging. *Packaging Technology & Science*, *29*(12), 615–648.

Xiao, F., Li, W., & Xu, H. (2022). Advances in Magnetic Nanoparticles for the Separation of Foodborne Pathogens: Recognition, Separation Strategies, and Applications. *Comprehensive Reviews in Food Science and Food Safety*, *21*(5), 4478–4504. DOI: 10.1111/1541-4337.13023 PMID: 36037285

Xie, S., Yue, Y., and Yang, F. (2024). Recent Advances in CRISPR/Cas System-Based Biosensors for the Detection of Foodborne Pathogenic Microorganisms Micromachines, 15(11), 1329. https://doi.org/DOI: 10.3390/mi15111329

Xin, X., Su, J., Hao-ran, C., Wang, L., & Song, S. (2025). Recent Advances in Clustered Regularly Interspaced Short Palindromic Repeats (CRISPR)-. *Associated Protein System-Based Biosensors*, *15*(3), 155. DOI: 10.3390/bios15030155 PMID: 40136952

Xiong, Y., Zhang, J., Yang, Z., Mou, Q., Ma, Y., Xiong, Y., & Lu, Y. (2019). Functional DNA-regulated CRISPR-Cas12a sensors for point-of-care diagnostics of non-nucleic-acid targets. *Journal of the American Chemical Society*, *142*(1), 207–213. DOI: 10.1021/jacs.9b09211 PMID: 31800219

Xu, M.-L., Gao, Y., Wang, X., Han, X. X., & Zhao, B. (2021). Comprehensive Strategy for Sample Preparation for the Analysis of Food Contaminants and Residues by GC–MS/MS: A Review of Recent Research Trends. *Foods*, *10*(10), 2473. DOI: 10.3390/foods10102473 PMID: 34681522

Xu, Q., Li, J., Zhang, Z., Yang, Q., Zhang, W., Yao, J., Zhang, Y., Zhang, Y., Guo, Z., & Li, C. (2024). Precise determination of reaction conditions for accurate quantification in digital PCR by real-time fluorescence monitoring within microwells. *Biosensors & Bioelectronics*, *244*, 115798.

Yamazaki, W., Taguchi, M., Ishibashi, M., & Inoue, K. (2009). Loop-mediated isothermal amplification for detection of Campylobacter jejuni and Campylobacter coli in food samples. *Journal of Food Protection*, *72*(3), 566–570. DOI: 10.4315/0362-028X-72.3.566

Yanagisawa, H., Obayashi, K., Furuta, M., & Fujimaki, S. (2025). Comprehensive screening of per-and polyfluoroalkyl substances in consumer products using pyrolysis GC–MS. *Discover Chemistry*, *2*(1), 1–7.

Yan, C., Wang, X., Zhao, X., Wei, M., Shi, C., & Ma, C. (2020). Development of a direct and visual isothermal method for meat adulteration detection in low resource settings. *Food Chemistry*, *319*, 126542.

Yang, J., Howe, A., Lee, J., Yoo, K., & Park, J. (2020). An improved approach to identifying bacterial pathogens in humans from environmental metagenomes. *Journal of Microbiology and Biotechnology*, *30*(9), 1335–1342. DOI: 10.4014/jmb.2005.05033 PMID: 32627750

Yang, N., Zhang, H., Han, X., Liu, Z., & Lu, Y. (2024). Advancements and applications of loop-mediated isothermal amplification technology: A comprehensive overview. *Frontiers in Microbiology*, *15*, 1406632.

Yang, W., Cao, C., Wang, W., & Wang, Y. (2016). Rapid detection of Streptococcus agalactiae infection using a loop-mediated isothermal amplification method. *Jundishapur Journal of Microbiology*, *9*(12). Advance online publication. DOI: 10.5812/jjm.37171

Yi, C., Zhang, Q., Li, C. W., Yang, J., Zhao, J., & Yang, M. (2006). Optical and electrochemical detection techniques for cell-based microfluidic systems. *Analytical and Bioanalytical Chemistry*, *384*, 1259–1268.

Yordanova, R. and Andonova, A. (2024). Rapid methods for detection of foodborne pathogens and their toxins - a review of current trends and future perspectives. Journal of Imab - Annual Proceeding (Scientific Papers), 30(2), 5587-5592. https://doi.org/DOI: 10.5272/jimab.2024302.5587

Yoshitomi, K., Jinneman, K., Zapata, R., Weagant, S., & Fedio, W. (2012). Detection and isolation of low levels of E. coli o157:h7 in cilantro by real-time PCR, immunomagnetic separation, and cultural methods with and without an acid treatment. *Journal of Food Science*, *77*(8). Advance online publication. DOI: 10.1111/j.1750-3841.2012.02813.x PMID: 22860597

Yulandi, A., Suwanto, A., Waturangi, D., & Wahyudi, A. (2020). Shotgun metagenomic analysis reveals new insights into bacterial community profiles in tempeh. https://doi.org/DOI: 10.21203/rs.3.rs-42618/v2

Zafarzadeh, A., Rahimzadeh, H., & Mahvi, A. (2018). Health risk assessment of heavy metals in vegetables in an endemic esophageal cancer region in Iran. *Health Scope*, *7*(3). Advance online publication. DOI: 10.5812/jhealthscope.12340

Zaheer, R., Noyes, N., Polo, R., Cook, S., Marinier, E., Domselaar, G., & McAllister, T. (2018). The Impact of Sequencing Depth on Characterizing the Microbiome and Resistome. *Scientific Reports*, *8*(1). Advance online publication. DOI: 10.1038/s41598-018-24280-8 PMID: 29651035

Zahir, A., Ge, Z., & Khan, I. A. (2025). Public Health Risks Associated with Food Process Contaminants – A Review. *Journal of Food Protection*, *88*(2), 100426. DOI: 10.1016/j.jfp.2024.100426 PMID: 39643160

Zaib, M., Ali, Z., Akram, H., Hameed, S., Wakeel, A., Qasim, S., & Aslam, S. (2023). Soil contamination and human health: exploring the heavy metal landscape: a comprehensive review. *JHRR*, *3*(2), 351–356. DOI: 10.61919/jhrr.v3i2.123

Zaidi, N., Mir, M. A., Chang, S. K., Abdelli, N., Hasnain, S. M., Ali Khan, M. A., & Andrews, K. (2025). Pharmaceuticals and personal care products as emerging contaminants: Environmental fate, detection, and mitigation strategies. *International Journal of Environmental Analytical Chemistry*, •••, 1–29.

Zakhar, R., Derco, J., & Cacho, F. (2018). An overview of main arsenic removal technologies. *Acta Chimica Slovenica*, *11*(2), 107–113.

Zarazúa, G., Girón-Romero, K., Tejeda, S., León, C., & Ávila-Pérez, P. (2014). Total reflection x-ray fluorescence analysis of toxic metals in fish tissues. *American Journal of Analytical Chemistry*, *05*(12), 805–811. DOI: 10.4236/ajac.2014.512089

Zareiyan, F., Khajehsharifi, H., & Rajabi, H. R. (2024). Dispersive liquid–liquid microextraction based on deep eutectic solvent followed by high-performance liquid chromatography-ultraviolet for imidacloprid, hexythiazox, and fenpyroximate determination: Experimental design approach. *Microchemical Journal*, *201*, 110519. DOI: 10.1016/j.microc.2024.110519

Zeng, D., Chen, Z., Jiang, Y., Xue, F., & Li, B. (2016). Advances and Challenges in Viability Detection of Foodborne Pathogens. *Frontiers in Microbiology*, *7*. Advance online publication. DOI: 10.3389/fmicb.2016.01833 PMID: 27920757

Zeng, X. Y., Yee, J. C., & Yang, H. P. (2024). Rapid and accurate identification of Mercenaria mercenaria and Mercenaria campechiensis using high-resolution melting analysis. *Journal of Shellfish Research, 43*, 111–118.

Zhang, J., Zhang, X., Zhang, Y., Liu, Y., Zhang, L., & Wu, X. (2025). Emerging biosensors integrated with microfluidic devices: A promising analytical tool for on-site detection of mycotoxins. *npj Science of Food, 9*, 84. https://doi.org/DOI: 10.1038/s41538-025-00444-5

Zhang, D., Sun, B., Luan, Y., Zhou, X., Huang, J., Sun, Y., Feng, S., & Yang, H. (2025). Simultaneous broad-spectrum detection of over 50 mycotoxins in maize silage via liquid chromatography-tandem mass spectrometry: Method development and application. *Food Control, 172*, 111143. DOI: 10.1016/j.foodcont.2025.111143

Zhang, J., Huang, H., Song, G., Huang, K., Luo, Y., Liu, Q., He, X., & Cheng, N. (2022). Intelligent biosensing strategies for rapid detection in food safety: A review. *Biosensors & Bioelectronics, 202*, 114003. PMID: 35065479

Zhang, J., Yang, H., Li, J., Hu, J., Lin, G., Tan, B., & Lin, S. (2023). Current perspectives on viable but non-culturable foodborne pathogenic bacteria: A review. *Foods, 12*(6), 1179. DOI: 10.3390/foods12061179 PMID: 36981106

Zhang, L., Yang, X., Liang, M., Tian, Z., & Zhang, Z. (2023). Intelligent point of care test for food safety via a smartphone. *Exploration of Foods and Foodomics, 1*, 143–161. DOI: 10.37349/eff.2023.00012

Zhang, M., Liu, J., Shen, Z., Liu, Y., Song, Y., Liang, Y., & Zhao, Y. (2021). A newly developed paper-embedded microchip based on a lamp for rapid multiple detections of foodborne pathogens. *BMC Microbiology, 21*(1). Advance online publication. DOI: 10.1186/s12866-021-02223-0 PMID: 34182947

Zhang, Q., Xu, E. G., Li, J., Chen, Q., Ma, L., Zeng, E. Y., & Shi, H. (2020). A review of microplastics in table salt, drinking water, and air: Direct human exposure. *Environmental Science & Technology, 54*(7), 3740–3751. PMID: 32119774

Zhang, X. J., Wang, X. J., Shi, B. H., Li, J. F., Fan, Y. F., Li, Y., Shi, C., & Ma, C. P. (2024). A biphasic accelerated strand exchange amplification strategy for culture-independent and rapid detection of Salmonella enterica in food samples. *Analytical Methods, 16*, 4083–4092.

Zhang, X., Huang, Q., & Yu, S. (2022). Role of nanomaterials in biosensors for food safety detection. *Advanced Materials Research, 1171*, 12–22.

Zhang, X., Shi, Y., Chen, G., Wu, D., Wu, Y., & Li, G. (2022). CRISPR/CAS systems-inspired nano/biosensors for detecting infectious viruses and pathogenic bacteria. *Small Methods*, *6*(10). Advance online publication. DOI: 10.1002/smtd.202200794 PMID: 36114150

Zhang, X., Wang, Y., Wu, S., Han, J., Raghavan, V., & Wang, J. (2025). Portable sensor devices based on multifunctional framework materials: Recent advances for food safety assurance by on-site detection. *Trends in Food Science & Technology*, *159*, 104965. DOI: 10.1016/j.tifs.2025.104965

Zhang, Y.-y., Zhao, M.-j., Liu, C.-y., Ma, K., Liu, T.-y., Chen, F., Wu, L.-n., Hu, D.-j., & Lv, G.-p. (2023). Comparison of two commercial methods with a UHPLC–MS/MS method for the determination of multiple mycotoxins in cereals. *Food Chemistry*, *406*, 135056. DOI: 10.1016/j.foodchem.2022.135056 PMID: 36455316

Zhao, G., Li, L. Y., Shen, X., Zhong, R. M., Zhong, Q. P., & Lei, H. T. (2024). DNA barcoding unveils novel discoveries in authenticating high-value snow lotus seed food products. *Foods*, *13*, 2580.

Zhao, P., Li, C., Rahaman, M. M., Xu, H., Ma, P., Yang, H., Sun, H., Jiang, T., Xu, N., & Grzegorzek, M. (2022a). Emds-6: Environmental microorganism image dataset sixth version for image denoising, segmentation, feature extraction, classification, and detection method evaluation. *Frontiers in Microbiology*, *13*. Advance online publication. DOI: 10.3389/fmicb.2022.829027 PMID: 35547119

Zhao, P., Li, C., Rahaman, M. M., Xu, H., Yang, H., Sun, H., Jiang, T., & Grzegorzek, M. (2022b). A comparative study of deep learning classification methods on a small environmental microorganism image dataset (emds-6): From convolutional neural networks to visual transformers. *Frontiers in Microbiology*, *13*. Advance online publication. DOI: 10.3389/fmicb.2022.792166 PMID: 35308350

Zhao, X., Lin, C. W., Wang, J., & Oh, D. H. (2014). Advances in rapid detection methods for foodborne pathogens. *Journal of Microbiology and Biotechnology*, *24*(3), 297–312. DOI: 10.4014/jmb.1310.10013 PMID: 24375418

Zhao, Z., Wang, R., Yang, X., Jia, J., Zhang, Q., Ye, S., Man, S., & Ma, L. (2024). Machine learning-assisted, dual-channel CRISPR/Cas12a biosensor-in-microdroplet for amplification-free nucleic acid detection for food authenticity testing. *ACS Nano*, *18*, 33505–33519.

Zheng, L., Zhang, H., Wang, S., Wang, M., & Lin, J. (2019). Optical biosensor for rapid detection of Salmonella typhimurium based on porous gold@platinum nanocatalysts and a 3d fluidic chip. *ACS Sensors*, *5*(1), 65–72. DOI: 10.1021/acssensors.9b01472 PMID: 31875386

Zhou, J., Brereton, P., & Campbell, K. (2024). "Progress Towards Achieving Intelligent Food Assurance Systems." Food Control, Vol. 145. *Article, 123456*, 1–15. DOI: 10.1016/j.foodcont.2024.123456

Zhou, J., Deng, T., Zeng, Q., Wang, H., & Deng, C. (2024). An electrochemical biosensor for sensitive detection of live Salmonella in food via MXene amplified methylene blue signals and electrostatic immobilization of bacteriophages. https://doi.org/DOI: 10.21203/rs.3.rs-4649888/v1

Zhou, J., Gui, Y., Lv, X., He, J., Xie, F., Li, J., & Cai, J. (2022). Nanomaterial-Based Fluorescent Biosensor for Food Safety Analysis. *Biosensors (Basel)*, *12*(12), 1072. DOI: 10.3390/bios12121072 PMID: 36551039

Zhu, L., Chirase, N., & Ravishankar, S. (2014). Use of fulvic acid or sodium silicate-based sanitizers to inactivate Listeria monocytogenes, Salmonella typhimurium, and Pseudomonas aeruginosa on food contact surfaces. *Journal of Food Safety*, *34*(2), 132–140. DOI: 10.1111/jfs.12106

Ziani, I., Bouakline, H., El Guerraf, A., & El Bachiri, A., Marie-Laure Fauconnier, & Sher, F. (2025). Integrating AI and advanced spectroscopic techniques for precision food safety and quality control. *Trends in Food Science & Technology*, *156*, 104850. DOI: 10.1016/j.tifs.2024.104850

Ziani, K., Ioni ă-Mîndrican, C. B., Mititelu, M., Neac u, S. M., Negrei, C., Moro an, E., & Preda, O. T. (2023). Microplastics: a real global threat for environment and food safety: a state of the art review. *Nutrients*, *15*(3), 617. PMID: 36771324

Zoellner, C., Al-Mamun, M., & Gröhn, Y., JACKSON, P., and Worobo, R. (2018). Postharvest supply chain with microbial travelers: A farm-to-retail microbial simulation and visualization framework. *Applied and Environmental Microbiology*, *84*(17). Advance online publication. DOI: 10.1128/aem.00813-18 PMID: 29959243

Zou, Y., Li, C., Shirahama, K., Jiang, T., & Grzegorzek, M. (2017). Content-based image retrieval of environmental microorganisms using double-stage optimisation-based fusion. *Information Engineering Express*, *3*(4), 43–53.

Zuo, T., Zhang, L., Wang, Y., Nie, L., Shen, M., Liu, L., & Ma, S. (2023). Technical Guidelines for Risk Assessment of Heavy Metals in Traditional Chinese Medicines. *Chinese Medicine*, *18*(1). Advance online publication. DOI: 10.1186/s13020-023-00771-3 PMID: 37287001

Zwirzitz, B., Wetzels, S. U., Dixon, E. D., Stessl, B., Zaiser, A., Rabanser, I., & Selberherr, E. (2020). Culture-independent evaluation of bacterial contamination patterns on pig carcasses at a slaughterhouse. *Scientific Reports*, *10*, 1–12. DOI: 10.1038/s41598-020-74842-y

About the Contributors

Matthew Chidozie Ogwu is an Assistant Professor of Integrated Ecology and Sustainable Development at Appalachian State University, USA. He is an accomplished interdisciplinary scholar with transdisciplinary expertise spanning One Health, EcoHealth, and sustainability science. His research is deeply rooted in the assessment of coupled human and natural systems and socio-ecological dynamics, with a strong focus on resilience, environmental justice, and sustainable development. Dr. Ogwu has earned numerous prestigious awards, competitive research grants, and international scholarships in recognition of his academic excellence and impactful contributions. He is widely published and serves as a board member and peer reviewer for several high-impact journals across environmental science, public health, and ecological disciplines. Passionate about science-policy integration and community-engaged scholarship, Dr. Ogwu consistently lends his expertise to international collaborative projects and volunteers his time to promote education, health equity, and environmental sustainability worldwide. His leadership and mentoring have inspired students, early-career researchers, and community stakeholders in various global regions.

Sylvester Izah is a distinguished faculty member at Bayelsa Medical University in Yenagoa, Nigeria. Dr. Izah holds dual Ph.D. in Public Health, and Applied Microbiology and Environmental Health, an M.Sc. in Applied Microbiology, and a B.Sc. in Biological Sciences from Niger Delta University, Nigeria. In addition to his teaching role, he holds the position of Assistant Director of Academic Planning, Research, and Innovations at the university. Dr. Izah is a licensed Environmental Health Specialist in Nigeria. His expertise spans various disciplines, and he has a particular focus on Sustainable Human-Environmental Health Interactions. This broad area of study involves the assessment and management of air, soil, and water quality, as well as toxicology, hygiene, sanitation, food science, waste management, biodiversity, and sustainability. Dr. Izah's academic prowess is highlighted by an impressive research and publication portfolio, which includes more than 300 peer-reviewed publications. His work extends to the publication of edited books with renowned publishers such as Springer and Elsevier, contributions to book chapters in publications by Elsevier, Springer, and Taylor and Francis, as well as numerous journal articles.

***

Afe Victor Dania, born on the 14th day of April 1980. Currently Lecturer 1/ Head of Department of Optometry. Bayelsa Medical University. I have a Doctor of Optometry degree, Master's in Public Health and Fellowship of the Nigerian Post Graduate College of Optometrists.

Debabrata Datta has many years of teaching experience both at the undergraduate and postgraduate level of Computer Science and Applications. His main research interest is in the domain of data analysis. He takes courses on Machine Learning, Data Mining, Data Security, Computer Networks, Data Structures and Algorithms, Object Oriented Programming etc. He is also a Life Member of IETE.

Anjali Dubey is an Assistant Professor in the Department of Forensic Science at Medicaps University, Indore. With expertise in Crime scene management, Forensic Toxicology, Forensic Biology/Serology, and Forensic Anthropology, she holds three years of teaching and research experience in Jain University, Bangalore and 1.5 years in Medicaps University. Miss Anjali Dubey has made significant contributions to her field, including the publication of research papers and book chapters. She completed her B.Sc. in Forensic Science, Chemistry and Life Science and her M.Sc. in Forensic Science from Sam Higginbottom University of Agriculture, Technology and Sciences, Prayagraj. In addition, she qualified for the FACT Plus in March 2022 and has Qualified UGC NET 2018, 2023.

Soumia El Malahi is a Doctor of Agronomic Sciences and Plant Biotechnologies, as well as a trained horticultural engineer. Her research focuses on the sustainable cultivation and valorization of native Moroccan plant species, including Argania spinosa (argan tree) and Rosa damascena (Damask rose). She specializes in the development of agronomic strategies aimed at improving plant performance, stress resilience, and product quality. Her work places particular emphasis on the use of organic amendments, biostimulants, and microbial inoculants to enhance soil fertility, nutrient use efficiency, and the biological integrity of agricultural systems. This expertise naturally extends to the interactions between plants, soil microbiota, and microbial safety, especially in the context of optimizing input use while safeguarding food and plant-derived product quality. Dr. EL MALAHI is committed to advancing sustainable agriculture by integrating biotechnology and microbiology. Her interdisciplinary approach supports the development of resilient agroecosystems and contributes to food safety and environmental stewardship in arid and semi-arid regions.

Jyothsna Devi Kuchipudi is a dedicated academic and professional in the field of Nutrition and Dietetics. She is affiliated with the Department of Nutrition and Dietetics under the School of Allied Healthcare and Sciences at Jain (Deemed-to-be University), located in Bengaluru, Karnataka. With a commitment to promoting health and wellness through evidence-based nutritional practices, she contributes to both teaching and research in her area of expertise.

Maurya Manjunath is a Reader in the Department of Public Health Dentistry at JSS Dental College & Hospital, under the JSS Academy of Higher Education & Research, Mysuru. With over 16 years of experience, Dr. Manjunath is a seasoned Dental Public Health Consultant and Oral Physician. His professional journey reflects a strong commitment to advancing oral health through academic excellence, community service, and research in Public Health Dentistry.

T.C. Manjunath was born on 6th February 1967. He did his BE from RV College of Engg. in 1989, ME in Robotics & Automation from LD College of Engg. in 1992 with First Rank and PhD in Systems & Control Engg. from IIT Bombay in the year 2007. He is currently working as the Dean Research R D and as Professor in the Dept. of Computer Science & Engineering with specialization in IoT, Cyber Security & Blockchain Technology in Rajarajeswari College of Engineering, Bengaluru, Karnataka. He has completed B.E. Degree from R.V. College of Engg., M.E. Govt.'s LD College of Engg., Gujarat Univ., Ahmedabad in First Rank and Ph.D. in Systems and Control Engineering from the Indian Institute of Technology Bombay IIT Bombay He has got a teaching academic, research and administrative experience of more than 35+ years 10 years as Principal, 8 years as HoD rest as faculty and has published more than 1200+ research papers in various national & international journals and conferences, with 175+ patents, 25 book-chapters, 25 awards and also published 25 text books in various field of science, engineering and technology, completed projects worth Rs. 9 Crores till date apart from guiding 10 PhD scholars. He is a member of 30 Professional Society Chapters and is a fellow of IE, IETE, AMIEE and a chartered engineer. He got the best research scholar award in 2006 in IIT Bombay amongst all the research scholars of the institute. He has got 10 awards in academics and research activities.

Odangowei Inetiminebi Ogidi is a Lecturer, Acting Head, Department of Biochemistry and Sub-Dean, Faculty of Basic Medical Sciences, Bayelsa Medical University, Yenagoa, Bayelsa State, Nigeria. He holds a Bachelor of Science degree in Biochemistry, Master of Science and Doctor of Philosophy degrees in Biotechnology. He has also served as a Lecturer, Head of Departments of Science Laboratory Technology and Biochemistry, and Deputy Director, Centre for Continuing Education in Federal Polytechnic Ekowe, Bayelsa State. He has authored over 80 peer-reviewed academic articles in reputable local and international journals, published over 50 book chapters and presented over 40 conference papers. He is a member of various Journal Editorial Boards. He is indexed in World Scientist and University Ranking amongst other international bodies. He have supervised over 30 ND and HND students and currently supervising over 20 undergraduate students. His research interest has been Medical/Pharmaceutical and Environmental Biotechnology/Biochemistry, Phytomedicine, Phytochemistry.

Fredrick Ojija earned his Ph.D. in Biodiversity and Ecosystem Management from the Nelson Mandela African Institution of Science and Technology (NM-AIST) in Arusha, Tanzania, in 2020. He obtained his M.Sc. in Biology–Ecology from Vrije Universiteit Brussel (VUB) in Brussels, Belgium, in 2014, and his B.Sc. in Wildlife Science and Conservation from the University of Dar es Salaam (UDSM), Tanzania, in 2009. Dr. Ojija is currently a Senior Lecturer in the Department of Earth Sciences at Mbeya University of Science and Technology (MUST). At MUST, he teaches courses in Biodiversity, Ecology, Environment, Zoology, and Research Methods.

Ngozi Blessing Okoro is an upcoming researcher. Ngozi base degree is Biotechnology and she has a Master's degree on Environmental Health Biology from Federal University of Technology Owerri, Imo State. Ngozi is a professional member of Biotechnology Society of Nigeria, she has also attended several workshops and conferences.

P. Selvakumar is currently an Associate Professor in the Department of Science and Humanities at Nehru Institute of Technology, Coimbatore, Tamil Nadu, India. Prior to joining Nehru Institute of Technology, he held positions as an Assistant Professor in the Department of Science and Humanities various engineering colleges in Coimbatore, Tamil Nadu, India, spanning from 2011 to 2023. Dr.P. Selvakumar's academic and research interests lie primarily in the field of phytochemistry. He has made significant contributions to his field, with an impressive record that includes 30 journal publications, 12 patents, and 20 conference papers. Additionally, he has organized 5 seminars and conferences, further demonstrating his commitment to academic and scientific exchange. Throughout his career, Dr. P.Selvakumar has been recognized for his dedication and excellence in teaching, having been honored with several awards, including the Best Teacher Awards. He holds a prominent position in the academic community, contributing actively to research.

Vivek Pazhamalai working as Associate Professor at Vels Institute of Science, Technology and Advanced Studies, and has teaching experience of 16 years. He has published more than 69 research articles in international and national journals to his credit.

Tarun Kumar Patel is a distinguished academician and researcher in the field of biotechnology, currently serving as an Assistant Professor and Head of the Department of Biotechnology at Sant Guru Ghasidas Government P.G. College, Kurud, Dhamtari, Chhattisgarh, India. With a robust academic background and a decade-long teaching career, Dr. Patel has made significant contributions to fungal biology, mass spectrometry, and microbial biotechnology. Educational Background and Qualifications Dr. Patel earned his M.Sc. in Biotechnology and subsequently his Ph.D. in Biotechnology from Guru Ghasidas Vishwavidyalaya, Bilaspur, with research focused on detecting and analyzing aflatoxin production in Aspergillus flavus strains using advanced mass spectrometry techniques. His academic excellence is further demonstrated by qualifying the CSIR-NET (Life Sciences) and the GATE examination. Research Contributions Dr. Patel's research encompasses a wide array of topics within biotechnology. His notable works include studies on the biosynthesis of aflatoxin B1 in Aspergillus flavus, identification of genetic relationships in Chlorophytum species, and the exploration of entomopathogenic fungi as biocontrol agents in agriculture. He has contributed to several peer-reviewed journals and book chapters, including publications in Biologia, Biotechnology and Bioprocess Engineering, and contributions to books published by Elsevier and Springer. Professional Experience and Teaching With over ten years of teaching experience at the undergraduate level and seven years at the postgraduate level, Dr. Patel has been instrumental in shaping the academic journey of numerous students. His commitment to education is evident in his active participation in various workshops and training programs, enhancing both his knowledge and teaching methodologies. Professional Affiliations and Recognitions Dr. Patel is an active member of esteemed scientific societies, including the Asian Federation of Biotechnology (AFOB) and the Association of Microbiologists of India (AMI). His research has garnered citations and recognition within the scientific community, reflecting the impact of his work on the field of biotechnology. Dr. Tarun Kumar Patel's dedication to research and education in biotechnology has significantly contributed to advancements in fungal biology and microbial biotechnology. His work continues to inspire students and researchers alike, fostering innovation and excellence in the scientific community.

Chitra Rengarajan is a distinguished researcher at the Centre for Agricultural Nanotechnology, Tamil Nadu Agricultural University, Coimbatore. Her work focuses on the innovative application of nanotechnology in agriculture, aiming in crop productivity, Stabilization of Plant bioactives, Food systems, sustainability, and environmental safety. With a strong commitment to interdisciplinary research, Dr. Rengarajan contributes significantly to advancing agricultural science and technology. Her expertise and dedication play a vital role in addressing modern challenges in agriculture through cutting-edge nanotechnological solutions.

Meenambiga S S Assistant Professor completed 6 years of teaching experience and 4 years of research experience in the field of Biotechnology. Passionate about teaching and have trained many young students in the field of Mycology, Natural Product Research, Nanoparticle synthesis, Drug Discovery from Plants and fungal sources and in-silico analysis of drug targets using modern Bio-informatics tools. Skilled in scientific writing covering the following aspects: Proposal submission, research articles suitable for national and international publications,drafting, proof reading and editing research thesis and expertise in critically reviewing the research manuscript for indexed journals. Member of Board of Studies for B.Tech Biotechnology curriculum in VISTAS. Guided more than 20 undergraduate students projects and served as Principal Investigator in the student's project funded by Tamilnadu State Council for Science and Technology. Published more than 30 research and review articles in national and international indexed journal and has volunteered as reviewer for many reputed journals. Life member of Mycological Society of India, BRSI, IARA, IAENG and many other professional organizations. Completed

Mahipal Singh Sankhla is an Assistant Professor in the Dept. of Forensic Science, School of Basic and Applied Science, K.R. Mangalam University, Gurugram, Haryana. With a profound expertise in forensic science, Dr. Sankhla has made significant contributions to the field, particularly in heavy metals toxicity, forensic toxicology, environmental toxicology, environmental forensics, nanotechnology, and fingerprint analysis. Dr. Sankhla earned his Bachelor of Science (Hons.) degree in Forensic Science, Master of Science degree in Forensic Science, and Ph.D. in Forensic Science from Galgotias University, Greater Noida, Uttar Pradesh. His academic journey is marked by extensive training at renowned institutions, including the Forensic Science Laboratory (FSL) in Lucknow, Central Bureau of Investigation (CBI) Central Forensic Science Laboratory (CFSL) in New

Delhi, Codon Institute of Biotechnology in Noida, and the Rajasthan State Mines and Minerals Limited (R&D Division) in Udaipur. Before his tenure at Chandigarh University, His exceptional work has been recognized with numerous awards, including a Junior Research Fellowship (JRF) from the Department of Science and Technology (DST)-funded project at Malaviya National Institute of Technology (MNIT), Jaipur, and accolades such as the Young Scientist Award, Young Researcher Award, and Forensic Researcher Award. In addition to his teaching and research responsibilities, Dr. Sankhla has been honored by the Government of India with the nomination to mentor the 'Atal Innovation Mission' of NITI Aayog, aimed at setting up Atal Tinkering Labs (ATL) across Indian schools. His commitment to research and innovation is reflected in his extensive publication record, with over 200+ peer-reviewed articles in national and international journals and book chapters. He has also edited 10 books published by renowned publishers such as Springer Nature and CRC Press. Additionally, he holds two Indian copyrights, published and grant over 70+ Indian patents. Dr. Sankhla actively engages in the scientific community by serving on editorial boards, reviewing for prestigious journals such as Springer, Elsevier, and Bentham, and participating in organizing committees for national and international conferences. His work continues to influence the field of forensic science and environmental toxicology research, making him a prominent figure in both academic and professional circles.

Girraj Sharma is a prominent forensic scientist specializing in forensic chemistry and toxicology. His career began as an assistant professor at the Government Institute of Forensic Science in Nagpur, Maharashtra, India, where he laid the foundation for his expertise in forensic science. In December 2014, Dr. Sharma qualified for the National Eligibility Test-Junior Research Fellowship (NET-JRF), which paved the way for his advanced studies and research. Dr. Sharma earned his Ph.D. from Dr. Harisingh Gour Vishwavidyalaya (A Central University) in Sagar, Madhya Pradesh, India. His doctoral research focused on developing a rapid, simple, and highly sensitive liquid chromatography method to analyze commonly used antibiotics—amikacin, amoxicillin, ciprofloxacin, ceftriaxone, and metronidazole—in hospital wastewater. This work led to the innovative application of micellar liquid chromatography (MLC) for detecting these antibiotics. His MLC method is distinguished by its ability to analyze complex biological samples directly without pre-treatment, providing high sensitivity and selectivity. In 2022, he further demonstrated his expertise by qualifying for the Forensic Aptitude and Caliber Test (FACT Plus) in Forensic Chemical Sciences, organized by the National Forensic Sciences University (NFSU), Ministry of Home Affairs, Government of India. Dr. Sharma's contributions extend beyond research into education. He co-authored e-Contents for undergraduate forensic science courses with the Educational

Multimedia Research Center at Dr. Harisingh Gour Vishwavidyalaya. His role included subject mapping, course preview, and text editing, which supports the academic growth of students in forensic science. His research addresses the growing concern of antibiotic pollution and emerging contaminants, wastewater, soil, food hospital wastewater etc. By developing a method to detect unmetabolized antibiotics in hospital wastewater, Dr. Sharma's work helps tackle the environmental impact of pharmaceutical pollutants and supports efforts to protect ecosystems and public health. Dr. Sharma's innovative approaches and dedication to forensic science and forensic toxicology and chemistry continue to impact research and education, reflecting his commitment to advancing knowledge and addressing real-world challenges in forensic and environmental sciences.

S. Udayakumar M.com., M.L., Ph.D.(law). He is currently a Assistant Professor in The Department of School of Law at the Vel Tech Rangarajan Dr. Sagunthala R&D Institute of Science and Technology, Tamilnadu, Chennai.

Baljeet Yadav is an Forensic expert in the area of Forensic Toxicology and background of basics of Forensic Sciences and Forensic Ballistics. He has worked for more than 11 years at universities such as the Central University, Lucknow, Amity University, Gurugram and GD Goenka University. He has research experience of more than 5 years and published 3 patents & 15 research papers, 04 book Chapters, of high reputed databases such (Web of Science & Scopus Indexed in field of Forensic Toxicology and Forensic Pharmacognosy and Forensic Odontology. He has been awarded "Young Researcher's Award on World Environment Summit 2021 held at IIAC Center New Delhi. He is in Editorial Board Member in Referred journal such as Journal Forensic Chemistry & Toxicology. Life member in professional bodies such ESDA (Environmental Sustainability Development Association, New Delhi. Since jan 2025, Dr. Baljeet Yadav is working as Programme Coordinator of Forensic Science in School of Basic & Applied Sciences (SBAS) at K.R. Mangalam University University teaching undergraduates and postgraduates and Ph.D. Scholars students of Forensic Sciences. Baljeet Yadav completed PhD in Forensic Sciences from University of Bundelkhand, Jhansi, U.P (India); MSc. Forensic Science University of Bundelkhand, Jhansi, U.P (India); BSc. (Hons.) Forensic Science University of Bundelkhand, Jhansi, U.P (India) and qualified UGC NET- JRF Examination in 2014 and started career in teaching and research activities.

Index

A

B

C

D

E

F

G

H

I

L

M

N

P

Q

R

S

T

W

www.ingramcontent.com/pod-product-compliance
Lightning Source LLC
LaVergne TN
LVHW062330300926
849161LV00007B/1636
* 9 7 9 8 3 3 7 3 3 9 8 2 5 *